Genetic Improvement of Solanaceous Crops

Volume I: Potato

Genetic Improvement of Solanaceous Crops

Volume I: Potato

Editors

Maharaj K. Razdan
University of Delhi
Delhi
India

Autar K. Mattoo
Vegetable Laboratory
Building 010 A, US Department of Agriculture
Beltsville Agricultural Research Center
Beltsville, MD
USA

Science Publishers, Inc.
Enfield (NH), USA   Plymouth, UK

SCIENCE PUBLISHERS, INC.
Post Office Box 699
Enfield, New Hampshire 03748
United States of America

Internet site: *http://www.scipub.net*

sales@scipub.net (marketing department)
editor@scipub.net (editorial department)
info@scipub.net (for all other enquiries)

ISBN 1-57808-184-X

Cover Illustrations

Bottom:
Plant of the dihaploid of potato cv. Kuba, resistant to Potato Virus Y (PVY), leafroll virus (PLRV), mid resistant to late blight and nematodes

Top Inset:
Tubers of dihaploid (left) and tetraploid cv. Kuba (right)
(Courtesy: Dr. Ewa Zimnoch-Guzowska)

Published by Science Publishers Inc., Enfield, NH, USA
Printed in India.

Foreword

Genetic improvement of crops has undergone an unparalleled transition over the past two decades. Practically everything we thought we knew about genes and trait expression has been turned upside down and inside out by technological revolutions at the cellular and molecular level. It's not that previous research led us to the wrong conclusions; it's that so little of the actual process by which cells carry out their functions was revealed. And even though we now have such a grander insight into those processes, we still have to use our new knowledge in real-world, field applications, if they are going to result in benefit to the world. This book provides the context by which our knowledge, from one end of the spectrum of crop improvement to the other, can be drawn together and applied to solve some of the most difficult problems facing improvement of a major global food crop-the potato. Because of its global importance at a time when efforts to eliminate hunger and malnutrition are at the forefront, this book comes at an opportune time to take stock of how well we have integrated the new into the old and formed new approaches that will contribute to increased food security in many parts of the world.

The book includes chapters by an impressive list of authors—those most well known in the potato research community. It covers the breadth and depth of potato improvement from the gene to the field and provides a collective knowledge that will be valued by all potato researchers around the world. The topics discussed are relevant to the most pressing problems facing potato growers, processors, and consumers. The volume highlights, from the beginning chapter, the crucial importance of genetic diversity and the dangers of loss or erosion of that diversity both within and outside genebanks. This is possibly the most complete chapter on genetic diversity of potatoes in existence and provides an excellent foundation for the following chapters which cover the eventual utilization of germplasm in solving the myriad of problems presented by such a genetically complex crop.

And finally, the authors and editors have done an excellent job in demonstrating how new technological approaches are becoming integrated throughout the research spectrum and providing new avenues for improving one of the world's most important crop plants. The book should quickly become a necessary addition to the reference libraries of scientists and research organizations worldwide.

Wanda W. Collins, Ph. D.
Director, Plant Sciences Institute
USDA-ARS
Beltsville, Maryland

Preface

The past few decades have witnessed a marked change in attitudes toward technological applications in crop research and development. Conventional methods of genetic improvement are now supplemented with so called "nonconventional" molecular and biotechnological approaches to gene modifications, particularly to induce higher crop yields, resistance to various types of pathogens, or abiotic environmental stresses, etc. Gene mapping and molecular-assisted selection facilitate improving quantitative traits of interest in plant breeding. Potato is the fourth largest crop after rice, wheat and maize. Agronomists, breeders, and farmers have always had to be concerned with quality and productivity of this important noncereal vegetable and food crop in order to meet the demands of consumers and food processors. It is imperative that potato growers be provided with integrated technical advice enabling them to make the best use of available resources. They also need an understanding of the technological principles for supporting their traditional skills and the art of potato husbandry.

Notwithstanding the fact that areas of genetic manipulation in crops are becoming increasingly specialized, major institutions now recognize that research on potato genetics and improvement be organized as a team activity by maintenance of closer links in R&D (Research and Development), extension programs, and farming through constant flow and exchange of information among them. Considering the fact that rapid advances have been made in improving potatoes using traditional breeding methods as well as genetic engineering technology, the aim of the present volume is to provide a critical appraisal of the state-of-the-art findings on this crop. The book starts with an insight into the origin, history, and conservation of potato germplasm (both *in situ* and *ex situ*) along with data on the existing status of its genetic resources and use of online database for exchange of relevant information (Chapter 1). Application of TPS (True Potato Seed) as a technological alternative to conventional propagation of tubers as a planting source, economic viability and on-farm profitability of TPS-related technology are discussed based on farmer experience at various agroecological sites in Egypt, India, Indonesia, Peru, and Vietnam (Chapter 2). A critical analysis of the shortcomings in traditional potato-breeding strategies due to complexity of tetrasomic inheritance, susceptibility to pests and diseases, fertility problems associated with cultivated potato (tetraploid) are assessed with combining ability of desirable attributes in a single clone (Chapter 3). Furthermore, breeding potential of 4x–2x matings in transmission and combining ability of quantitative traits are highlighted (Chapter 4). Of the 200 species in the genus *Solanum*, from Central and South America, an incredible wealth of genetic diversity represented by Ca. 70% diploid species remains vastly underutilized in potato-breeding programs. This wealth of diploid potato germplasm could contribute to allelic diversity, which may be useful for genetic enhancement of important traits in new commercial cultivars. Keeping this objective in mind, breeding strategies have been adopted over the years at North Carolina State University, Raleigh, NC, with a follow-up at USDA/ARS-Beltsville Center, MD, and diploid germplasm base increased for commercial use (Chapter 5), with similar breeding programs executed in Poland, Scotland, and CIP (Peru). Application of molecular markers in gene mapping and fingerprinting of the quantitatively inherited important agronomic traits have contributed to enhancement of germplasm base, recombining ability, and interpretation of taxonomic and phylogenetic interrelationship among various potato taxa (Chapter 6); ploidy manipulation, e.g. application of monoploids, haploids from tetraploid parents, first division restitution (FDR) and second division restitution (SDR) mechanisms of 2n gamete formation, present special unique methods of gene mapping and gene action in potatoes (Chapter 7). Genetic manipulation for economically important traits can be better understood by identification and functional characterization of gene(s) controlling the expression of these traits. Map-based cloning, candidate gene approach, and transposon tagging are some of the novel approaches followed with application to potato (Chapter 8). Cell and tissue culture methods have long been applied in production of monoploid and dihaploid potatoes that are now the subject of intensive genetic analysis in molecular mapping of genes and genetic transformation studies, in particular, introgression of desirable traits through somatic hybridization for release of new improved varieties (Chapter 9). Physiological and biomolecular changes during storage at low temperatures greatly affect the tuber quality of potato. Enzymes regulating starch-sugar interconver-sions in response to changes in temperature play an important role and could be the target of genetic engineering, especially antisensing invertase gene may have long-lasting positive effects in transgenic potato (Chapter 10). Finally, advances made in transgenic technology in production of potato cultivars resistant to insects, nematodes, viruses, bacteria and fungal pathogens are aspects significantly contributing to genetic improvement of potatoes (Chapters 11–15).

This book thus presents the research findings of experts from international institutes and organizations who have coordinated efforts over the past several decades for improvement of the potato crop. It is hoped that this compendium will not only find favor with potato breeders and specialists, but will also have value to teachers and students seeking recent information on potato genetics, physiology and pathology. The Editors sincerely thank the publisher and contributors of the respective chapters for their sincere cooperation and support.

July 2004

Maharaj K. Razdan
Delhi, India

Autar K. Mattoo
Beltsville, USA

List of Contributors

Bamberg, J., USDA, Agricultural Research Service, Vegetable Crops Research Unit, Inter-Regional Potato Introduction Station, 4312 Highway. 42, Sturgeon Bay, WI 54235, USA, *e-mail: nr6jb@ars-grin.gov.*

Brown, Charles R., USDA/ARS, 24106 N. Bunn Rd, Prosser, WA 99350, USA, *e-mail: cbrown@pars.ars.usda.gov.*

Celibi-Toprak, F., University of Tsukuba, Gene Research Center and Institute of Biological Sciences, 1-1-1 Tennoudai, Tsukuba, Ibaraki, 305-8572, Japan; Pamukale University, Department of Biology, Fen Edebiyat Fakultest, Kinikii, Denizii 20010, Turkey

Chilver, A., DFID, c/o FCO, Kampala, Uganda

de Jong, Walter, Cornell University, Dept. Plant Breeding, 309 Bradfield Hall, Ithaca, NY 14853-1901, USA, *e-mail: wsd2@cornell.edu*

del Rio, A., University of Wisconsin-Madison, Dept. Horticulture, 1575 Linden Drive, Madison, WI 53706, USA

Douches, David S., Michigan State University, Depts. Crop & Soil Science and Entomology, East Lansing, MI 48824-1325, USA, *e-mail: douchesd@msu.edw*

Fano, H., Proyceto INCAGRO, Ministry of Agriculture, Lima, Peru

Grafius, Edward J., Michigan State University, Depts. Crop & Soil Science and Entomology, East Lansing, MI 48824-1325, USA

Haynes, Kathleen G., USDA/ARS, Vegetable Laboratory, Plant Sciences Institute, 10300 Baltimore Ave., Building 010A, Beltsville, MD 20705, USA, *e-mail: haynesk@ba.ars.usda.gov*

Jones, Richard W., USDA-ARS, Vegetable Laboratory, Plant Sciences Institute, 10300 Baltimore Ave., Building 010A, Beltsville, MD 20705, USA

Khatana, V., Symbiosis of Technology Environment and Management, New Delhi, India

Lojkowska, Ewa, University of Gdansk & Medical University of Gdansk, Dept. Plant Protection and Biotechnology, Kladki 24, 80-822 Gdansk, Poland

Lu, Wenhe, Northeast Agricultural University, Dept. Agronomy, Harbin, Heilongjiang 150030, People's Republic of China

Mackay, George R., Scottish Crop Research Institute, Invergowrie, Dundee DD2 5DA, Scotland, *e-mail: gmacka@scri.sari.ac.uk*

Mattoo, Autar K., US Dept. Agriculture, Vegetable Laboratory, Building 010A, Beltsville Agricultural Research Center, Beltsville, MD 20705-2350, USA

Mojtahedi, Hassan, USDA/ARS, 24106 N. Bunn Rd, Prosser, WA 99350, USA

Ortiz, Rodomiro, Intl. Inst. Tropical Agriculture (IITA), Ibadan, Nigeria, c/o, L.W. Lambourn & Co., Carolyn House, 26 Dingwall Road, Croydon CR93EE, UK

Pérombelon, Michel, Scottish Crop Research Institute, Invergowrie, Dundee DD2 5DA, Scotland, UK

Rizk, A., Agricultural Economic Research Institute, Charbia Governorate, Egypt

Simko, Ivan, University of Maryland, Dept. Natural Resource Sciences and Landscape Architecture, College Park, MD 20742, USA

Solomos, Theophanes, University of Maryland, Dept. Natural Resource Sciences and Landscape Architecture, College Park, MD 20742, USA

Suherman, R. 2, Lembang Horticultural Research Institute, Lembang, Indonesia

Tai, George C.C., Potato Research Centre, Agriculture & Agri-Food Canada, P.O. Box 20280, Fredericton, NB, Canada E3B 4Z7

Thieme, R., Bundesanstalt für Züchtungsforschung an Kulturpflanzen Institut für Landwirtschaftliche Kulturen Rudolf-Schick-Platz 3a 18190 Gross Lüsewitz, Germany, *e-mail: r.thieme@bafz.de*

Thieme, T., Bio-Test Labor (GmbH) Sagerheide Birkenalle 19, 18184 Sagerheide, Germany

Veilleux, Richard E., Virginia Polytechnic Inst. & State University, Dept. Horticulture, Blacksburg, VA, USA 24061-0327, *e-mail: potato@vt.edu*

Walker, T.S., CIP, Av. La Molina 1895, Lima 12, Peru, *e-mail: t.walker@cgiar.org*

Watanabe, Junko A., University of Tsukuba, Gene Research Center, 1-1-1 Tennoudai, Tuskuba, Ibaraki 305-8572, Japan

Watanabe, Kazuo N., University of Tsukuba, Gene Research Center, 1-1-1 Tennoudai, Tuskuba, Ibaraki 305-8572, Japan, *e-mail: nabechan@gene.tsukuba.ac.jp*

Xiong, Xingyao, Potato Research Centre, Agriculture & Agri-Food Canada, P.O. Box 20280, Fredericton, NB, Canada E3B 4Z7

Zimnoch-Guzowska, Ewa, Plant Breeding and Acclimatization Institute, Mlochow Research Center, Platanowa 19, 05-831, Mlochow, Poland, *e-mail: E.Zimnoch-Guzowska@ihar.edu.pl*

Abbreviations Used Throughout the Book

acyl-HSL:	N-acyl derivatives of homoserine lacton
ADH:	Alcohol dehydrogenase
ADP:	Adenosinediphosphate
ADPase:	Adenodiphosphatase
AFGC:	*Arabidopsis* Functional Genomics Consortium
AFLP:	Amplified fragment length polymorphism
AGPase:	ADP-glucose pyrophosphorylase
AGPB:	Catalytic subunit of AG Pase
ANOVA:	Variance analysis
APIC:	Association of Potato Intergenebank Collaborators
ATP:	Adenosinetriphosphate
ATPase:	Adenosinetriphosphatase
BA:	Benzyladenine
BAC:	Bacterial artificial chromosome
BC1, BC2, BC3:	Backcross first, second or third
BIBAC:	BAC transformed vectors namely BIBAC vectors
BIO-PCR:	Enrichment PCR
BOX:	Repetitive extragenic palindromic sequence
BSA:	Bulked segregant analysis
BW:	Bacterial wilt
CAPS:	Cleaved amplified polymorphic sequence
cDNA:	Complementary DNA
CEPESER:	Central Peruana de Servicios (Peru)
cfu:	Colony-forming unit
CGIAR:	Consultative Group on International Agricultural Research
CIMMYT:	International de Mejoamiento de Maiz Y Trigo, Mexico
CIP.	International Potato Center, Lima, Peru
CIPC:	Chloroisoprophyl-N-phenylcarbamate
cM:	Centimorgan
CMV:	Cucumber mosaic virus
Ca MV:	Cauliflower mosaic virus
CO:	Crossover

COS:	Conserved orthologue set (i.e., conserved regions of functional genes)
cp:	Chloroplast
CP:	Coat protein
CPMR:	Coat-protein-mediated resistance
cpDNA:	Chloroplast DNA
CPRI:	Central Potato Research Institute (Shimla, HP, India)
cpSSR:	Chloroplast single sequence repeat
CRN:	Columbia root-knot nematode
CRS:	Corky ringspot disease
CV. (or cv.):	Cultivar
2,4-D:	2, 4-dichlorophenoxyacetic acid
DD:	Degree days
DEFRA:	Department of Environment, Food and Rural Affairs
DET:	Double exchange tetrad
DFID:	Department of International Development
DNA:	Deoxyribonucleic acid
DR:	Double reduction
DTS:	Distribution of tuber size
EAPR:	European Association for Potato Research
EB:	Early blight
EBN:	Endosperm balance number
ELISA:	Enzyme-linked immunosorbent-assay
EPG:	Electrical penetration graph
EPPO:	European and Mediterranean Plant Protection Organization
ER:	Extreme resistance
ERIC:	Enterobacterial repetitive intergenic consensus sequence
ESTs:	Expressed sequence tags
FAMEs:	Fatty acid methyl esters
FAO:	Food and Agriculture Organization
FAP:	Fatty acid profiles
FCM:	Flow Cytometry
FDR:	First division restitution
FISH:	Fluorescence *in situ* hybridization
GA:	Gibberellic acid
GCA:	General combining ability
gDNA:	Genomic DNA
GFG:	Gene for gene resistance
GISH:	Genomic insitu hybridization
GM:	Genetically modified
GPT:	Glucose phosphate translocator
GS:	Genetic similarity

ha:	Hectare
HR:	Hypersensitive response
IAA:	Indole acetic acid
ICTVdB:	International Committee on Taxonomy of Viruses Database
IFAS:	Indirect immunofluorescent assay/Immunofluorescence antibody staining
IHAR:	Institut Hodowli i Aklimatyzacji Roslin
IHN:	Internal heat necrosis
IITA:	International Institute for Tropical Agriculture
INCAGRO:	Innovacion y Competitividad para el Agro-Peruano
INCRISAT:	International Crop Research Institute for the Semi-arid Tropics
INRA:	Institut Nationale des Research Agronomiques
IPC:	Isopropyl-N-phenylcarbamate
ISSR:	Intersimple sequence repeat
ITS:	Intergenic transcribed spacer
JRN:	Javanese root-knot nematode
Kin:	Kinetin
LAR:	Local acquired resistance
LB:	Late blight
LI:	Lesion index
LRR:	Class/motif of a plant resistance gene (Leucine rich repeats)
MAPK:	Mitogen activated protein kinase
MAS:	Marker-assisted selection
MT:	Methyltryptophan
MS:	Murashige and Skoog Medium
mtDNA:	Mitochondrial DNA
NAA:	Naphthalene acetic acid
NASBA:	Nucleic acid sequence based amplification
NBS-LRR:	Nucleotide budding site-leucine rich repeat
NCM-ELISA:	Nitrocellulose membrane-ELISA
NCO:	Non crossover
NET:	No exchange tetrad
Non-DR:	Nondouble reduction
NPS:	National Park Service (USA)
NPT II:	Neomycin phosphotransferase II
NRA:	Newly reclaimed areas
NRC:	National Research Council (USA)
NRN:	Northern root-knot nematode
PAL:	Phenylalanine ammonia lyase
PARC:	Pakistan Agricultural Research Council

PCN: Potato cyst nematode
PCR: Polymerase chain reaction
PEBV: Pea early browing virus
PEG: Polyethylene glycol
PEP: Phosphoenyl pyruvate
PGA: Phosphoglyceric acid
PGIP: Polygalacturonase-inhibiting proteins
PGM: Phosphoglucomutase
PGRFA: Plant Genetic Resources for Food and Agriculture (FAO)
PL: Pectate Lyase
PLRV: Potato leafroll virus
PNW: Pacific North West of USA
PPG: Price per grade
PPO: Polyphenol oxidase
PRN: Peanut root-knot nematode
PRC: Potato Research Center (New Brunswick, Canada)
PRSV: Pepper ringspot virus
PS: Price seasonality
PTGS: Post-transcriptional gene silencing
PTM: Potato tuber moth
PVA: Potato virus A
PVM: Potato virus M
PVS: Potato virus S
PVV: Potato virus V
PVX: Potato virus X
PVY: Potato virus Y
QRTs: Quantitative resistance traits
QTA: Quantitative trait allele
QTL: Quantitative trait loci
RAPD: Randomly amplified polymorphic DNA
rDNA: Ribosomal DNA
recA PCR-RFLP: Restriction analysis of amplified *recA* gene fragment
rep-PCR: Repetitive element sequence PCR
rt-PCR: Reverse transcriptase PCR
RELP: Restriction fragment length polymorphism
REP: Repetitive extragenic palindromic sequence
RGL: Resistance gene like
RH: Relative humidity
RKN: Root-knot nematode
RNA: Ribonucleic acid
RT-PCR: Reverse transcriptase-polymerase chain reaction
SAI: Surface area infected
SAR: Systemic acquired resistance

SCA: Specific combining ability
SCAR: Sequence characterized amplified region
SCRI: Scottish Crop Research Institute
SDR: Second division restitution
SEERAD: Scottish Executive Environment and Rural Affairs Department
SET: Single exchange tetrad
SG: Specific gravity
SH: Schenk and Hildebrandt medium
SGT: Solanidine glucosyltransferase
SI: Scab index/Self-incompatible
SNPs: Single nucleotide polymorphisms
SPS: Sucrose phosphate synthase
SRN: Southern root-knot nematode
SSR: Single sequence repeat
ST: Seedling tuber
STEM: Symbiosis of Technology Environment and Management
STS: Silver thiosulfate
SuSy: Sucrose synthase
t: Ton
TaqMan PCR: fluorogenic PCR, real-time PCR
T-DNA: Ti plasmid DNA segment
TDZ: Thidiazuron
TEV Tobacco etch polyvirus
TMV: Tomato mosaic virus
TP: Transplants/TPS transplants
TPNI: Tachyplesin I protein
TPS: True potato seed
TRN: Thames' root-knot nematode
TRV: Tobacco rattle virus
TZC: Tetrazolium chloride
UDP: Uridine diphosphate
UDPase: Uridine diphosphatase
UGPase: UDP-Glucose pyrophosphorylase
UHD: Ultrahigh density
USDA: United States Department of Agriculture
UTP: α-D-glucose phosphate uridyl transferase
VIR: Vavilov Research Institute (St. Petersburg, Russia)
WIEWS (FAO): World Information and Early Warning System on Plant Genetic Resources of FAO
WRKY: Consensus sequence involved in elicitor induction of pathogenesis-related genes

Contents

1

Conservation of Potato Genetic Resources

J. Bamberg[1] and A. del Rio[2]

[1]USDA, Agricultural Research Service, Vegetable Crops Research Unit, Inter-Regional Potato Introduction Station, 4312 Highway 42, Sturgeon Bay, WI 54235, USA. (email: nr6jb@ars-grin.gov).

[2]Department of Horticulture, University of Wisconsin-Madison, 1575 Linden Drive, Madison WI 53706, USA

I. HISTORY, VALUE AND NEED FOR POTATO GERMPLASM

The greatest service which can be rendered by any country is to add a useful plant to its culture. Thomas Jefferson, 1790

Of all the improving and ameliorating crops, none in my opinion, is equal to potatoes. George Washington, in a letter to Thomas Jefferson, 1795

Origins of Potato Germplasm

Potato is a New World crop that was unknown to the rest of the world until the 1500's. The most obvious domestication originated in the Andes mountains of South America, which, in a sense, were named for potato agriculture ("Andes" is derived from a Quechua word for field terraces). Closely related wild species grow in diverse places from Chile through Central America and into North America as far as Utah. Correll (1962) says little or no domestication occurred in Central and North America because other foods were readily available. However, in the American Southwest, ostensibly *wild* potatoes growing among Puebloan ruin sites of the Anasazi culture suggest that these ancient peoples used potato for food. It is certainly plausible that an empire sophisticated enough to trade with the Pacific coast and Mexico may have also cultivated, selected and distributed the wild potatoes native to central Arizona and New Mexico.

The potato of commerce today was first domesticated in present-day Peru and Bolivia, and played an important role in that society, as is

Corresponding author: Dr J. Bamberg

witnessed by many representations of potato in ceramic artwork from the area. Did the Incas' extensive agricultural empire include centers for keeping and evaluating germplasm? It has been suggested that Machu Picchu, with its many microclimates and tiny field plots may have been an ancient prototype of modern genebanks (Weatherford 1988).

The genetic status of potato has had major influences on world history (Salaman 1985, Weatherford 1988), a famous example being the case of the Irish Potato Famine in 1845-50 that resulted in the population decreasing from 8 to 3 million due to starvation and emigration. Besides the social effects, the threat of losing potato as a food crop due to fungal and virus diseases instigated early importation of new germplasm from Latin America, shaped potato breeding and the genetic makeup of today's cultivars and set the stage for the development of genebanks. Several excellent reviews are available for the reader interested in details (e.g., Brown 1990 and 1993, Stevenson et al. 2001, Wilson 1993).

Importance of the Potato Crop

From the utilitarian perspective, the importance of a genebank is related to the importance of the food crop it represents. Potato is the world's 4th most important food crop (after rice, wheat and maize) in terms of production and area cultivated. The nutritive value of potato per unit of land is 2-3 times that of cereals (Shekhawat et al. 1999). These four crops provide a greater part of the world's diet than the next 26 ranked crops combined.

World potato production is increasing at a much faster rate than other leading crops. Production in the Asian region is growing at over twice the rate elsewhere— about 6% annually. Twenty years ago, only about 1/8 of the world crop was grown in Asia, but today, it is greater than 1/3, with China alone producing over 1/5 of the world's potato production. India has similar goals to sustain its dramatic increases in production over the past several decades through a "Brown Revolution" (Shekhawat et al. 1999), which is realistic considering the projected annual world growth rate of potato at about 2.8% over the next two decades (CIP 1999). World consumption need not plateau soon as per capita consumption in Asia is still less than 1/6 that of Europe and North America.

Potato exceeds all other crops in providing calories, protein and several other important nutrients (Neiderhauser 1993) and is a very good source of vitamin C. It can be cultivated in a broad variety of environments and produces an edible crop very rapidly (Rhoades 1982). Average yields are still a fraction of record yields, indicating that great potential still exists. A sometimes overlooked factor of great importance is that potatoes are both affordable and can be prepared in a greater variety of ways than most foods, so people are quite happy eating them as a major component of their diets (Kolasa 1993, Myers 1983). R. L. Sawyer, former Director

General of CIP said, "If one had to live on one food alone, the potato would be better by far than any other major food crop available today." (quoted in Shekhawat et al. 1999). High consumption also means that even small improvements in nutrition have a greater overall impact on the health of the population than larger gains in relatively minor foods. A wealth of detailed information and statistics about potato as a food resource is available at CIP's "Potato Facts" (Appendix 1.C).

Value and Ownership of Potato Germplasm

Potato's germplasm value stands out among all other crops. Assuming that a crop is an important one, the value of its germplasm will be determined by its genetic diversity, availability and utility. Some estimates have been made of the economic return from germplasm utilization. About 50% of the four-fold advance in potato yields during 1930-1980 have been due to genetic improvement (Myers 1983, Reid and Miller 1989) and about 1% of annual value of all crops may be credited to exotic germplasm. Pro-rated, this is a total of $10-25 million per year for potatoes in the USA over the past two decades (see also Thrupp 1998). No other crop approaches the existence, collection, research, and use of related wild species seen in potato. Current treatments (Spooner and Hijmans 2001) recognize a remarkable 197 tuber-bearing potato species, most of which are available to breeding with relative ease. Wild species represent only 2% of *ex situ* germplasm over all crops (Astley 1991), but for potato 50% of world genebank holdings are wild species (NRC 1993). For potato, it has been estimated that representative germplasm for 95% of related cultivated species and 40% of wild species have been collected. However, since species names have their origins in variable taxonomic concepts (Spooner et al. 2003) a quantitative assessment of genetic diversity based on species may not be very reliable.

No comprehensive empirical field studies of diversity in the wild are available to guide us in knowing how to compare what was accomplished in past sampling with the totality of available diversity, but some insights are being gathered with recent research (see research Section III following). Spooner (1999) indicates that when Peru collections are complete, most of the species will be available *ex situ*, making potato among the world's most completely collected crops. Williams (1984) concluded that potato is the *only* major crop with adequate representation of wild species in genebanks. It has been estimated that pressure for more resistant, higher quality cultivars requires an 8% annual influx of new genetics. This is hard to find in the breeding pool of many crops. For potato, however, breeders are readily finding this "new blood" in exotic species. A recent survey of breeders revealed that 19% of potato breeding research involves wild species from genebanks, far more than the averages of 1.2% for cereals and 1.4% for other vegetables (Swanson 1997).

Goodman (1990) notes that workers in all major crops have been hurrying to acquire disappearing exotic *in situ* germplasm, but only rice, potato and tomato workers have been successful in preserving, characterizing and using what they have collected. Engels and Wood (1999) suggest that conservation is no security against crop vulnerability until genetic diversity is deployed. More than 20 wild potato species have contributed genes to named cultivars (Reid and Miller 1989), and an estimated 10% of the genetics of the crop is derived from wild species—more than any other crop (Cox et al. 1988).

The idea that only cultivated species are of any value for potato breeding (Simmonds 1997) does not seem justified. The remarkable utilization of exotic potato germplasm has been attributed not only to the biology of the potato, but also to the approach taken by potato genebanks. For example, Cox et al. (1988) note that while genebank staff typically specialize in conservation physiology, the US Potato Genebank at Sturgeon Bay, Wisconsin has pursued a model more conducive to germplasm deployment. In this approach, curators specialize in genetics and enhancement breeding, thus doing the same kind of research as the clients to whom they supply germplasm. For additional information, the reader is referred to detailed reviews of the use of exotic germplasm for the development of improved potato cultivars (Hanneman 1989, Love 1999, Pavek and Corsini 2001, Plaisted and Hoopes 1989, Ross 1986, Spooner and Bamberg 1994).

Germplasm Ownership

Who owns germplasm, and if benefits are returned to countries where it naturally occurs, how is the amount calculated? This is a complicated and contentious issue for which policy is evolving. A detailed review of the history and philosophy of this topic is beyond the scope of presentation here, and is available elsewhere (Brown 1990, Eberhart et al. 1998, NRC 1993, Petit et al. 2001, Querol 1993, Reid and Miller 1989, Ryan 1992, Spooner 1999, Swanson 1997, Tisdell 1999). The most current developments are available on the Internet, e.g. from FAO. Some of the general considerations in the debate over international germplasm rights and ownership are discussed below. The convention of using "North" to signify the countries that are relatively rich in commercial breeding and poor in native germplasm, and "South" to signify countries that are relatively poor in commercial breeding and rich in native germplasm is used.

(1) The prevailing attitude about germplasm ownership has changed from a "common heritage of all mankind" to "sovereign ownership" (Ryan 1992).
(2) The North bred advanced varieties using germplasm freely collected from the South, then made profit from the products (even from

sales to the South), increasingly using legal protection to sell not only the physical propagules, but also the rights to their use.

(3) The above resulted in international treaties that regulate germplasm exchange and require "benefit sharing," meaning that profits from cultivars that use germplasm will be shared (in a variety of ways) with the country that originally provided the germplasm. Part of the benefits when returned to the donor country are to support conservation of genetic resources, thus benefiting everyone.

(4) *In situ* conservation of threatened diversity in countries of the South may have lower priority there if such germplasm is assigned no monetary value.

(5) How much of the value of the improved variety should be attributed to the exotic parental materials, and how much to the finesse of the breeder in manipulating those materials?

(6) The North has already contributed enormous benefits by bearing the costs of safely preserving hundreds of thousands of germplasm stocks from the South over many years. Spooner (1999) mentions this and several other contributions of the North, such as the funding of CGIAR centers and training of international graduate students, arguing for open access to plant genetic resources.

(7) Value is usually determined by the price that the market will bear, but this is a new paradigm for germplasm. It is difficult for those who have it and those who want it to agree on the price based on subjective philosophical considerations alone.

(8) Although the objective *monetary* return expected from any particular unit of germplasm is extremely unpredictable, most scientists agree that germplasm in general has great investment value. Wilson (1984) argues that value is attributed to that which is very hard to replace. Russian potato curators, for example, followed this ideal, enduring peril and starvation to protect the collection at VIR during the Siege of Leningrad (Alexanian and Krivchenko 1991). Germplasm, potato in particular, does have tangible value. The economic and political fortunes of Old World countries were radically changed when gold and silver were brought there from the New World (Weatherford 1988), but the sum value of those precious metals was less than the *annual* value of the world potato crop (Rhoades 1982).

(9) The term "genetic resources" is misleading if it suggests germplasm is a finite resource like minerals. The amount of physical material sampled from the South as seeds and plants is trivial, and the physical elements in Northern genebanks and the crops derived from them originated in the North. Thus, the only significant resource obtained from abroad was the genetic *information*. That genetic information, and the opportunity for the donor country's breeders to

use it have been in no way diminished, particularly since Northern genebanks will freely return it to the donor country if it is no longer accessible from native *in situ* or genebank sources.

(10) We sometimes deem benefit sharing for use of unique information as appropriate, sometimes not, and the differentiation is rather subjective. For example, if a breeder uses unique genetic *information* from a potato species of foreign origin to produce a commercial variety, benefit sharing may seem appropriate. But, in an apparently parallel activity, unique information could be used to produce illustrations of native species that are copyrighted and sold without any sharing of profits with the country from which those species originated (e.g., CIP 1990).

(11) It may be very difficult to gain consensus on the philosophical and subjective considerations in point 10. Similarly, trying to right alleged wrongs of the past can hardly be done objectively, since no group is innocent of exploitation at some point in its history. However, one objective, unequivocal reality exists in this debate: sovereign nations have the *power* to control their borders and thus restrict future sharing of their germplasm regardless of any consensus on "fairness."

(12) Studies have documented that no country, North or South, is presently nutritionally self-sufficient with only its native crops. Southern countries currently receive many more units of germplasm per year for breeding than they donate to the North. Because of these factors, if exchange becomes restricted, the South will have to commit much more spending to germplasm preservation and development in their own countries to maintain the current level of benefits they now enjoy (Fowler et al. 2001).

Plant Rights

A final form of value has nothing to do with the sovereignty of nations or ownership benefits derived from the utilization of their germplasm. Agricultural genebank scientists tend to put high value on collecting and using food-related genes, because there is general agreement that having abundant food is a good thing. But many other botanists and citizens take the converse approach, regarding protection of the *intrinsic value* of natural populations as top priority. Under this philosophy, an *ex situ* "genebank" may have a much different definition: a temporary place to preserve and bulk plants for the sole purpose of re-establishment of threatened natural populations. Although *intrinsic value* is an oxymoron (ascribing *value* is subjective and necessarily implies a human judgment), this view of nature has a significant following. The US National Park Service web page on biodiversity (NPS 1999) notes practical uses of native plants as well as this spiritual, non-anthropomorphic explanation for why natural flora should be the object of government protection:

If there is to be change for the better, it must begin with changed thinking, and that must be based on understanding Buddhists teach "each thing has its own intrinsic value, and is related to everything else in function and position." Tribal groups on all continents have lived this belief. The chief of the Kuna Indians in Panama recently expressed it this way: "The land is our mother, and all living things that live on her are brothers. We must live in a harmonious manner on her; because the extinction of one living thing is also the end of another."

Some suggest that an *anti*-anthropomorphic perspective is necessary if society is to improve conservation (Noss and Cooperridge 1994, Wilson 1984). Others contend that the "intrinsic value" approach is counterproductive because it disqualifies any rational judgments from a process that, in practice, *must* include prioritization (Brown 1990). Also, germplasm with a purely philosophical valuation does not generate any funds for its own conservation (Reid and Miller 1989, Tisdell 1999).

Genebank managers may increasingly encounter restrictions that reflect the attitude that plants have an intrinsic right to be undisturbed, and that the burden of proof is on the genebank to show why collection is an overriding consideration (Fox 1993). Regardless of their philosophies, governmental and private organizations that manage *in situ* diversity are important partners that preserve the future possibility of agricultural applications of native plants (Bamberg 1999, Becker et al. 1998, McGraw 1998).

Problems with Diseases, Pests, Stresses and Quality

Germplasm value can also be measured by the continuing needs of the crop for improvement. Despite the merits of the current potato crop, improvement is indeed needed. Jansky's review (2000) notes that disease problems are particularly severe for potato, reducing the world crop by an estimated 22%, not counting the enormous costs associated with chemical and cultural inputs used to prevent further losses (Martin 1988, Stevenson et al. 2001). The same is true with respect to a variety of insect pests. Similarly, susceptibility to stresses like frost and heat limits growing area and productivity (Levy et al. 2001, Richardson and Estrada 1971). Profitability is enhanced not only by increased yield and reduced losses, but also by reducing needs for high levels of inputs like water and fertilizer (Neiderhauser 1993). Finally, great increases in the value of the crop could be realized by improving tuber quality with respect to appearance, texture, cooking quality, storability and processing. Stocks with extreme expression of desirable traits such as these are often found in the exotic germplasm in potato genebanks.

Problems with Breeding Methods

Breeding has been a cumbersome task due to inherent factors in the cultivar breeding pool, such as tetrasomic inheritance, inbreeding depression and yield stasis (Douches, et al., 1996). Diversity with respect to ploidies, breeding behavior, reproductive mutants and basic heterozygosity in genebank stocks have the potential to mitigate these problems and improve the efficiency of breeding methods (see Hanneman 1999). Advances are also being made in more traditional approaches to potato breeding (e.g., Brown and Dale 1998).

II. STATUS OF GENETIC RESOURCES OUTSIDE GENEBANKS

Genebanks exist to distribute disease-tested germplasm that is available for breeding and research, and to document associated characterization and evaluation data. These stocks originate from the wild, and those domesticated originate from farmer's fields, markets, or public breeding programs. Considering the great number of species, broad geographic distribution, and varieties of habitats, germplasm in genebanks probably will always contain only a small sample of the total diversity that exists (Allard 1970).

The status and dynamics of genetic resources outside the genebank is important to consider because they determine the...

— Goals for adding diversity to the genebank
— Extent to which genetic diversity outside the genebank is being lost
— Extent to which genebank stocks represent plants in nature

Goals for Adding Diversity to the Genebank

Cultivated Species

Modern potato cultivars are highly heterozygous genotypes typically selected from hundreds of thousands of seedlings over 10-20 years, and thereafter preserved clonally. So, relative breeding value of cultivars is not as obvious as, for example, in inbred crops.

Potato is a major world crop with many professional breeders who develop parental stocks known to have basic adaptation to their regional growing conditions and also to possess the consumer and processing qualities in current demand. Sharing among such breeding programs fulfills a genebank role with respect to advanced breeding materials.

In contrast to genebanks for some other crops, potato genebanks rarely provide propagules for the crop or even the direct parents of a selected cultivar. Thus, potato genebanks emphasize being a bank for *genes*, rather than *genotypes*. Although a certain unique combination of genes may have once had enough merit to be a named cultivar, it will probably not be of much interest to researchers, breeders or to the genebank unless it contains genes of particular rarity and value (Allem 2001). One exception is when preservation of obsolete cultivars makes it possible to conduct studies on historic trends in breeding (Douches et al. 1996, Hosaka et al. 1994). Because only about 5% of the genetic variation in *Solanum* is being used in commercial cultivars (Stevenson et al. 2001), when a new type of resistance or quality is sought, the most potent expression of the desired trait is more likely to be found in exotic species.

In light of the above factors, strategies to expand the useful diversity in potato genebanks usually do not emphasize survey and preservation of modern cultivars. On the other hand, primitive (i.e., non-*tuberosum*) cultivated species' germplasm diversity is a major emphasis of world potato genebanks. The best example of a deliberate attempt to survey and capture the diversity of cultivated species is taking place at the International Potato Center (CIP) at Lima, Peru. Over 10,000 samples were collected from across the Andean growing area, characterized by morphology and evaluated for economic traits. Isozyme tests eliminated clonal duplicates and indicated which types were relatively rare in the collection (Huaman et al. 2000a). Other large collections of primitive cultivated species (mostly *S. andigena*) are also kept at the Vavilov collection (Russia) and the Sturgeon Bay collection (Wisconsin, USA).

Wild Species

The consortium of genebank leaders inaugurated in 1991 at the US Potato Genebank, Sturgeon Bay, Wisconsin, USA noted that future collection of wild stocks could be planned to increase diversity in genebanks with respect to *taxonomy, known economic traits,* and *geography/habitats* (Bamberg 1990). Engels and Wood (1999) add *rescue* and *research* as motives for collecting. Similarly, Reid and Miller (1989) list criteria for collecting as *distinctiveness, utility* and *threat*. Hawkes et al. (2000) include *biological importance,* in which case a species plays a special role in the integrity of an ecosystem, *cultural importance,* and *relative cost and sustainability of conservation*. For details of collecting techniques, the reader is referred to the several sources that cover this topic in detail (Astley 1991, Hawkes et al. 2000, Querol 1993).

Collection for Genetic Diversity

Predicting how to capture more wild species diversity usually begins with considering the number of populations representing a given species in the genebank, and the relative distinction of the species in question. For example, *S. leptosepalum* is very rare and has had no germplasm representation in genebanks until very recently. However, *leptosepalum* probably is not a distinct species from *S. fendleri,* for which many genebank populations exist. This illustrates the importance of taxonomic research and the species concept (for details see Spooner et al. 2003). But for taxonomic research to reveal the relative diversity of species, those populations must first be collected and brought to the genebank for analysis. Often the existence of those reputed novel species is documented in herbarium specimens by botanists making floras, not germplasm collectors. In some cases, germplasm from a site discovered and documented by collectors has been lost from the genebank, so re-collection may be a priority.

Collection for Habitat and Geographic Diversity

Habitat and geography may also provide clues for how wild resources compare to those in the genebank. Potatoes predominantly grow at higher elevations (Hijmans and Spooner 2001) where there is sufficient moisture, so such ostensibly appropriate habitats that have not been sampled are good candidates for exploration. Within that broad description, however, the specific habitats that favor potato may not be very predictable (Bamberg 1999). Within species, genetic separation is often assumed to be related to geographic separation. Thus, genebanks set goals for collecting across the reported range of a species, and are particularly interested in disjunct populations. However, while one can easily look at a map and set goals for maximizing geographic representation, the time and effort required to get to a site usually plays a significant role in determining where collections are made (Hijmans et al. 2000). Eco-geographic data are increasingly available and able to be mapped. But does such information predict genetic diversity? The answer to this question is "no" according to evidence gathered in some model potato species. Within the species *S. jamesii* and *S. fendleri* (from the USA) and *S. sucrense* (from Bolivia), variation in characteristics of the site of collection were very poor indicators of general genetic diversity. Even distance between collection sites did not associate with genetic distances according to DNA markers (del Rio et al. 2001). Similarly, when Spooner et al. (2001) collected and analyzed a population of *S. fendleri* on the tip of the Baja California peninsula, they found that it was not genetically distinctive, despite its remarkable physical isolation from other *S. fendleri* in Mexico. These results must be qualified, however. They pertain to differences *within* species. Geographic separation certainly is a predictor of genetic diversity *among* species, since we know that most species are narrow endemics, with only 39 occurring in more than one country (Hijmans and Spooner, 2001). Also, del Rio and Bamberg's results pertain to *general* genetic relationships. Associations between *specific* genes/traits and eco-geo parameters are certainly possible (Flanders et al. 1997).

Extent to Which Genetic Diversity Outside the Genebank is Changing

Cultivated Species

This kind of germplasm is lost because farmers *intentionally* replace it with modern cultivars. However, there are mitigating factors. Primitive potato cultivar clones have shapes, colors, textures and adaptation to particular growing conditions that continue to be prized as the subjects of local preference and traditions of indigenous farmers and consumers (Brown 1993 and 1999, Brush et al. 1995). Potato is a clonal crop, but occasional sexual cycles that enhance diversity are promoted by the proximity of wild species and the fact that most systemic viruses are eliminated in botanical seed (Quiros et al. 1992). It is also simplistic to think that potato crop diversity in the Andes was ever static, or that the availability of modern high yielding varieties will have an impact on native potato farming greater than other upheavals like the arrival of Europeans. Rather than a progressive decline to complete loss, the attrition of native varieties in potato seems to have come to a level of dynamic equilibrium (Brush 1991). Potato may have an unusual advantage in this regard too, in having many close relatives growing wild in habitats far from production zones (Wood and Lenné 1999).

Spooner et al. (1991) observed that Chiloe Island farmers and gardeners have a tradition of maintaining many different types of potato simply for the sake of variety. It is also encouraging to note that genetic analysis of ostensibly different collections of cultivated species revealed extensive duplication of genotypes across Latin America (Huaman, et al. 2000a). Habitat changes and the resulting possible genetic bottlenecks, drift, extinction, loss of pollinators, etc., are not pertinent to the extent that these are intentionally cultivated and clonally maintained. Collection of primitive cultivars of potato is estimated at 95% (Reid and Miller 1989).

Monocultures

Genetic specialization in crops has paralleled social specialization in manufacturing and careers in the Industrial Revolution. Cultivation of large areas of a single crop with a narrow genetic base is often presented as a recent dangerous and misguided trend (Reid and Miller 1989, Swanson 1997, Thrupp 1998). However, uniformity certainly has been a desirable and *intentional* means to efficient production in which the crop is optimally fitted to growing conditions and markets (Ryan 1992, Goodman 1990) and farmers have always selected some varieties and eliminated others (Engels and Wood 1999). The system is undesirable and vulnerable only if conditions change, and the genetic diversity that is then needed has been either discarded, allowed to go extinct or otherwise not accessible. Frequent replacement of uniform elite cultivars avoids vulnerability by providing diversity *over time*, if not within the crop of any given season (Duvik and Brown 1989), but this works only if new genetic inputs are rapidly available. This illustrates why strong genebanks with an infrastructure of enhancement and breeding are essential to efficient, specialized agriculture (Allem 2001, Pavek and Corsini 2001).

Losing Wild Populations

Much literature is available about the status of biodiversity and its impact on world food. Like the loss of crop diversity in the field, loss of wild species populations is also primarily due to the impact of man, but *unintentionally* through habitat degradation. It seems reasonable and prudent to assume that the factors causing the staggering pace of extinction worldwide (but particularly in the tropics), estimated at 100 to 1000 times that of pre-human levels (Allem 2001, Pimm 1995, Wilson 1984), are also impacting wild potato, making the threat of genetic loss greater in the wild than in the genebank (Lenné and Wood 1999). Carlos Ochoa, renowned potato collector and taxonomist from CIP, says that the chance of finding many more wild potato species is remote-that most of the rare, uncollected ones are highly threatened and may have already been destroyed by man or natural changes (Rhoades 1982, see also CIP web page, Appendix 1.C.). Correll (1962, p. 16) noted that the effect of man and his domestic animals can be both positive and negative: potato foliage is a popular food for both native and non-native herbivores, but potatoes are colonizers that thrive where man has disturbed the natural vegetation, and probably were intentionally distributed by ancient cultures. Johns and Keen (1986) note evidence of dispersal of potato by domestic animals. However, the overall status of wild potato species populations is complicated to assess, since little quantitative empirical evidence is available.

As already noted, specific favorable habitats for potato are not easy to define. Similarly, it may be difficult to define the robustness of populations at a given site. They may appear to be barely surviving in one season and thriving in the next. Little is known about the status of the reserve of dormant tubers and seeds in the soil, but observations by the authors suggest they must have the capacity to very rapidly capitalize on favorable conditions. It would be good to know this for certain, however, since collectors aim to sample without harming the *in situ* population, which might happen if the population really is quite small. Little is known about reproductive behavior and population structures in the wild, or the factors that are influencing them. The above considerations make it difficult to identify populations that are at risk.

Loss of Diversity within Populations

The above concepts relate to the reduction in survival of germplasm in nature, or, at the extreme, total extinction. However, populations that are apparently thriving may still be losing diversity for certain loci (Ryan 1992, p. 32). Since agricultural genebanks exist primarily for the economic enhancement of the crop, it is appropriate to consider the risk with respect to loci of economic importance. Clearly, the alleles most vulnerable to loss are those that are rare. So, whether alleles of agricultural importance tend to be rare in nature is an important question. Cases for which this is true are often mentioned because they so powerfully exemplify why germplasm conservation needs careful attention (and funding). However, there appears to be no empirical support for the idea that genes of agricultural value are usually rare (Lawrence et al. 1995). On the contrary, it may be more reasonable to assume that rare alleles are likely to have *less* evolutionary or agricultural significance (Allard 1996). It also seems true that traits conferring natural fitness for the plant can be desirable in domestication (e.g., disease resistance, fertility). Nevertheless, some have argued that genes valuable in cultivation are likely deselected in nature (Gale and Lawrence 1984, p. 78) and thus are rare. We cannot predict the abundance of valuable alleles that are yet to be discovered, but the abundance of currently-recognized valuable alleles may be instructive. Table 1.1 illustrates the fact that, in potato, valuable traits are often characteristic of whole species, fixed in several populations, or otherwise quite abundant. So there appears to be no reason to assume that yet unknown agriculturally valuable traits typically exist in a rare and vulnerable state in nature.

Table 1.1 *Abundance of economic traits in potato**

Trait	*No. species tested*	*No. populations tested*	*No. populations resistant*	*Percent of Populations tested exhibiting resistance*
Potato leafroll virus	63	2477	404	16
Potato virus S	15	115	19	17
Potato virus X	20	430	79	18
Potato virus Y	27	388	101	26
Tobacco Mosaic Virus	4	24	4	17
Tobacco Rattle Virus	19	42	22	52
Early Blight	43	676	77	11
Fusarium Dry Rot	1	209	39	19
Late Blight	19	687	156	23
Rhizoctonia	12	210	18	9
Verticillium Wilt	81	1239	659	53
Wart	44	200	107	54
Bacterial Wilt	13	1862	58	3
Blackleg	73	423	268	63
Ring Rot	75	1853	454	25
Columbia Root-Knot Nematode	28	390	65	17
Golden Potato Cyst	21	481	48	10
Root-Knot Nematode	48	197	135	69
Colorado Potato Beetle	59	1064	323	30
Potato Flea Beetle	49	1305	266	20
Potato Aphid	69	1618	310	19
Potato Leaf Hopper	76	1360	378	28

(Contd.)

(*Contd.*)

Green Peach Aphid	33	1290	120	9
Cold Stress	39	2761	744	27
Heat Stress	32	318	72	23

*Selected and adapted from Bamberg, JB, MW Martin and JJ Schartner. 1994. Elite selections of tuber-bearing Solanum species germplasm. Univ. Wisconsin, Madison/US Potato Genebank, 4312 Highway 42, Sturgeon Bay, WI 54235, USA. 56 pp.

Extent to Which Genebank Stocks Represent Plants in Nature

Getting a Good Original Sample

Knowing about the status of populations in the wild may also impact how one collects them in order to maximize sample quality within a location. At the extreme, a single genotype might be collected from a site during a poor growing season. Such accessions will require the genebank to maintain and distribute the germplasm clonally (if the species is self incompatible). This is more costly and time consuming, and a single genotype is likely to be a poor representation of the genetics that really exists at that location in an optimal growing season. In this case, re-collection would have a high priority. Unfortunately, we do not know much about the potential variation in sample quality or its causes. The number and size of plants, size of the colony and fruiting can vary greatly for a given site over seasons (Bamberg et al. 1996 and 2003).

Genebank plants may not represent their counterparts in the wild in reproductive behavior, a characteristic that has impact on many aspects of germplasm use. Most potato species can reproduce clonally as tubers or sexually as true seeds. Little is known about the situation in nature, but it seems likely that the prevalence of one or the other of these options varies with species and over different local growing conditions and seasons. In contrast, sexual reproduction and recombination is usually *enforced* at the genebank because of the advantages of botanical seeds in the genebank (see *Preservation* section, p. 19).

Unlike nature, regeneration at the genebank tries to avoid selection of seedling progeny to minimize genetic drift and loss of diversity. Thus, the genebank sample may have been "domesticated" to have a population structure and particular genotypes that never occur in the original population in nature. Similarly, reproductive factors in nature might make it such that the most representative sample of the population would be a clonal one, but a seed sample was taken (as is conventional). The authors are currently investigating this possibility.

Changes Occurring in the Genebank

Most germplasm research is done using genebank stocks. Conclusions that relate to the status of potato germplasm in the wild are usually made with the unstated assumption that materials from the genebank represent the wild germplasm growing at the natural sites of origin. Little direct evidence is available on how genebank germplasm differs from that in the wild, but some inferences can be made. Neither genebank germplasm nor natural germplasm is fixed. On one hand, there is a risk of preserving worthless mutants that can build up in the genebank but would be purged in nature (Schoen et al.1998), so some selection in the genebank may be beneficial. Schoen and Brown (2001) discuss several factors that might cause *ex situ* samples to diverge from their source populations in the wild, and become less diverse: selection for genebank conditions, lack of selection for dynamic biotic pressures in the wild, genetic drift, and inbreeding. On the other hand, there are several mechanisms by which *ex situ* germplasm may spontaneously *increase* in diversity (Weissinger 1990).

Re-collections of USA potato species (del Rio et al. 1997b) illustrated that the genebank sample, collected several decades earlier, may be much different from the material currently growing in the wild at the original collection site. We also know that collection tends to be done in the places easiest to get to, and therefore potentially unrepresentative (Hijmans et al. 2000).

Environmental Effects on Phenotype

How does one compare plants in the genebank and wild? Comprehensive comparison at the DNA level is currently impractical, so we must rely largely on phenotype. But growth habit may be very different in the genebank setting than in the wild. Herbarium specimens made from wild plants growing in shaded, stressed conditions look much different than the same population cultivated in full sun with optimal growing conditions (Correll 1962). Consider the importance of this in light of the fact that taxonomy is based partly on the examination of herbarium specimens from the wild and partly on the appearance of plants grown in artificial conditions. We also know that daylength sensitive species originating near the equator will exhibit different maturity-related features when grown in the field at genebanks at higher latitudes. A striking example of this is that hardly any of the tuber-bearing *Solanum* species normally produce tubers in the field at the US Potato Genebank in northeastern Wisconsin. Another very important aspect of daylength sensitivity is its impact on apparent disease resistances. Juvenility enhances resistance to many important diseases (like late blight and *Verticillium*). So, part of the resistance attributed to natural populations may actually be due to their response to the artificial genebank environment. Those artificial resistances will disappear in a cultivar breeding program where selected progeny must have early maturity. This illustrates why it is difficult to make a reliable estimate of the genetic status of genebank samples as compared to the potential in the wild.

Practical Realities

Of course, the problem of genetic changes in the genebank noted above assume that the optimum known technology is being employed. In practice, lack of funds, adequate facilities and trained workers can also undermine genetic integrity of germplasm in genebanks or put it at risk (Chang et al. 1989).

III. *EX-SITU* POTATO GERMPLASM COLLECTIONS: THEIR DATA, GENERAL OBJECTIVES AND TECHNICAL RESEARCH

Collections and Their Data

Hawkes (1990) provides a history of potato collections and the genebanks that house them. Detailed, up-to-date information about potato germplasm resources in various world genebanks is now available on the Internet on a scale impractical for paper publication. Principal genebank addresses are given in Appendix 1.A. If desired, free paper catalogs and reports are often available by directly contacting these genebanks. Appendix 1.A also lists other resources such as an Internet database available from WIEWS with which one can search all reported world sources for a given species name. Explicit contact information for the listed institutions is provided. However, only the number of accessions at each site is given, without other information about the populations themselves. In 1990, world potato genebank leaders formed APIC, the Association of Potato Intergenebank Collaborators (Bamberg et al 1995, Huaman et al. 1997), to address common problems and objectives of genebank managers. One product has been the Intergenebank Potato Database (Huaman et al. 2000b). This Internet-accessible database cross references all wild potato germplasm in major world genebanks, with details of passport and evaluation data.

The status of cultivated germplasm is more difficult to document. A large number of seed certification and breeding programs have their own extensive collections of named or numbered clonal selections at various stages of development. Various sources of such information are available as catalogs and Internet databases that list cultivar names, pedigree, health status, sources and availability, evaluation data with respect to disease, pests and stress; botanical, agronomic and quality characteristics, color and shape (often pictured), and suitability for various uses. Appendix 1.B lists some of these sources. As already mentioned, the International Potato Center, CIP at Lima Peru has had a particularly strong emphasis on primitive cultivated species, of which *S. andigena* predominates. Other genebanks also have many collections of this species. One of the goals of APIC is to produce a merged database of cultivated germplasm in world genebanks to complement that of the wild species.

The services agricultural genebanks provide are similar and relatively stable over time, regardless of crop or location: acquisition, classification, preservation, evaluation, and distribution. Others may be specific to the conditions under which individual genebanks operate or are in the realm of unpublished experience of the staff. Thus, one of the resources genebanks provide is the advice on how to germinate, grow, cross and store exotic germplasm. Following are the concepts and philosophies underlying the genebank goals listed above. Detailed descriptions of specialized considerations and techniques are readily available elsewhere (see Appendix 1.C). All of the following points pertain to working genebanks, as opposed to "base collections" focused on long term backup, for example, USDA's National Center for Genetic Resources Preservation (NCGRP— formerly known as the National Seed Storage Laboratory, NSSL) at Ft. Collins, Colorado (see Allem 2001).

Acquisition

Also referred to as *introduction*, acquisition takes two forms: incorporation of stocks already formally preserved by man (e.g., other genebanks) or novel capture of wild or cultivated material. In either case, acquisition is often a slow process. In most countries, especially where potato is an important crop, imported germplasm must undergo 1-2 years of quarantine testing to detect and eliminate systemic pathogens. Only small samples are imported, so at least one more year of seed bulking and processing at the genebank may be needed before the material is available for distribution. In the case of new collections, additional time and effort are often required because the precise source of such germplasm is not known, and access may be dependent on seasonal weather conditions or political restrictions (plant *exploration* is an apt term). But when the germplasm is finally established at the genebank, it can be delivered very rapidly within-country, within multi-country consortiums like the European Union (Appendix 1.A), or to foreign countries without quarantine restrictions. Thus, acquisition by genebanks does far more than safeguard germplasm. In most cases it is the key to making genetic materials available to researchers and breeders within a practical timeframe.

Classification

Identifying Plants Pertinent to the Potato Crop

Genebanks must classify their germplasm, so genebank staff are often involved with taxonomic research. What is the purpose? Species historically collected and conserved by public agricultural institutions are those cultivated, known to hybridize with cultivated species, or species with apparent similarity by virtue of their morphology and crossability. Thus, the first and most easily met practical objective of a potato germplasm classification scheme is to circumscribe the germplasm entities that could conceivably contribute to the potato crop or associated research. This initial goal is relatively easy to accomplish, since most species accessible to potato breeding when genebanks began collecting germplasm had the obvious characteristics of tubers.

Times are changing, however, and accessible genes are no longer restricted to those that can be incorporated by conventional crossing. One might argue that the ability to transform potatoes with genes from unrelated organisms reduces the need to focus on gathering close relatives in a potato genebank. However, the opposite may be true. Some question whether combining the genetics of widely separated organisms is safe with respect to health and the environment — and in the marketplace, perception is reality. The availability of valuable genes from close relatives may be the acceptable alternative that allows the potential of genetic engineering to be realized. Thus, germplasm of species that have always been classified as *potatoes* may be increasingly valuable. Certainly, developing tools that allow selection at the gene (versus phenotype) level (Witcombe 1999), and promote better management of resources in genebanks hold great potential for improving the potato crop.

Classification to Guide Collecting, Breeding and Evaluating "Core Subsets"

Once the taxonomic limits of potato germplasm have been established, there are at least two reasons for proceeding with additional studies focused on defining species boundaries and other subgeneric associations within the genebank. First, classification is a basis for assessing genetic diversity in the genebank. Surveys are made to plan new germplasm collections for the expansion of the genebank's total diversity, and a small fraction of the genebank's total holdings may be selected for a "core collection" with a disproportionately high amount of diversity. Such a sample should facilitate more efficient evaluation for useful traits. Core collections also eliminate the inefficiency of maintaining exact or near duplicates (Allem 2001). Both of these genebank management activities rely heavily on the species designations of populations.

With respect to utilization of the germplasm through breeding, Hawkes (1994) offers a second benefit of research to define species: such classification is also assumed to be grouping sources of useful traits, and thus is "...of value in choosing initial plant breeding material." How successful has taxonomy been as a germplasm management guide? Hawkes (1994) provides a detailed review of the advances made in our understanding of species relationships with the advent of modern techniques (e.g., based on DNA markers), but also notes that this new evidence has largely confirmed previous schemes based on visual assessment of gross morphology. This is good news. It implies that past breeding efforts that have used exotic germplasm were generally well guided by potato taxonomy, and that the value of classification as an aid to germplasm exploitation is likely to increase.

Preservation

This encompasses several concepts. At the most basic level, it means keeping *something* of the original acquisition alive. It also implies maximizing the conservation of the original genetic traits. Finally, it implies making the germplasm available for evaluation and distribution by bulking a sufficient number of disease-free propagules. Unless each of these aspects of preservation is accomplished, a working collection will not be a *bank* in the usual sense of the term (from which one can readily make withdrawals). At best, it is a closed bank with only the hope of someday being accessible, at worst a museum of poorly-understood curiosities. Unfortunately, assessments of the status of a genebank do not always sufficiently emphasize the critical question of whether its germplasm is freely available.

Clonal

This method completely preserves the genetics (the risk of culture-induced mutation seems trivial within the scope of the task for which genebanks are responsible) in the germplasm. Tuber reproduction is expensive and must be done every year, but its greatest drawback is poor control of systemic diseases. *In vitro* culture allows for control of diseases, but also requires expensive equipment and expertise. While the genetics and disease status are secure *in vitro*, the plants themselves are more vulnerable in storage and in transit to users. *In vitro* methods also entails a delay in delivery of germplasm while subcultures are prepared, since keeping inventory on hand of all clonal holdings is usually not practical. Clonal cryopreservation is long-term, but has the other disadvantages of *in vitro* preservation. While great advances have been made, cryopreservation is probably not yet practical for routine use in working collections. All forms of clonal preservation are too tedious and expensive to apply to germplasm other than selected genotypes of known merit such as named cultivars or breeding stocks (i.e., not populations).

Botanical Seeds

Preservation by botanical seeds is low-tech, long-term (20+ years) and eliminates most disease concerns other than some systemic viruses (notably PSTV). Most wild species have been collected as seed samples from populations. The main problem with seeds relates to the fact that most potato species are heterogeneous outbreeders. This introduces the threat of genetic drift through inadequate sampling or other inadvertent selection. Seeds also must be produced indoors by hand pollinations, or field plots spaced apart far enough such that bumblebees will pollinate only within (not among) cross-compatible populations that are being increased (Schittenhelm and Hoekstra 1995).

Toward a Utopian Scheme

Seeds are nearly ideal for the purposes of potato genebanks. They are long-lived, inexpensive in storage and are relatively imperishable and light for shipping to germplasm users. But some problems remain. Conventional seed production by hand pollination may be limited by plant growth, flowering, fertility and berry set. Bee pollination in caged field plots sometimes helps (per unpublished tests at US Potato Genebank, Sturgeon Bay, Wisconsin), but is expensive. Genebanks are constantly testing various modifications in growing conditions by mechanical and chemical means to promote production of high quality seeds. However, the ideal in technology would be to obviate routine pollination and recombinant seeds through apoximis (i.e., agamospermy). This would be a great advance for genebanks' evaluation and breeding (Hermsen 1980) as it would combine all the advantages of clonal and seed propagation.

The Aversion to Discard

It has already been noted that preserving all the existing germplasm of potato is an unattainable goal. In all likelihood, the genebank's capacity will eventually be reached, because it costs less to acquire new germplasm than to maintain it. What happens if capacity is reached but stocks more valuable than those already in the genebank become available for acquisition? Unfortunately, aversion to any discarding makes it seem more honorable for curators to either refuse to accept additional materials (a *de facto* choice to "discard" the opportunity to acquire superior new germplasm) or agree to take new germplasm, thus fail to adequately preserve more than the genebank capacity (discarding by default, probably in random fashion). In these circumstances it seems wiser to carefully and deliberately choose some items to discard.

Preservation Research

Readers concerned by the suggestion of intentionally discarding even the lowest priority germplasm because of limited capacity will be relieved to know that genebanks work very hard to avoid this by increasing capacity. Obviously, capacity can be increased by increasing budgets, expanding facilities and staff. But in most realistic economies, expanding capacity by increasing the efficiency at the current level of resources will also need to be pursued. One way to achieve increased efficiency is through studies that identify the most effective propagation techniques. Another approach involves assessment of genetic diversity itself:

What is the status of diversity, and how is it affected by standard genebank techniques? When empirical evidence is available to answer these questions, genebank managers know how to apportion fixed resources in a way that will maximize diversity in the genebank. Williams (1989) notes that while the basic functions of germplasm conservation are well understood, the "knowledge base on which the work is founded is extremely slender." Other than numerous studies measuring the extent of diversity in collections, very few have been done on aspects that may have an impact in the quality of conservation, such as how vulnerable the genetic diversity of the collections has been over years of conservation. Potato curators emphasized this need in the early stages of the development of the Association of Potato Intergenebank Collaborators (APIC) in 1991.

APIC emphasized the need to examine whether potato genebanks are taking the correct measures to preserve genetic diversity that has been deposited in their collections. In the past, there were technical limitations to detecting small genetic changes, loss of genes in populations or the amount of diversity. However, the use of new DNA-based markers (RFLP, RAPD, AFLP, SNPs, etc.) has now enabled effective study and comparison of plant materials (Karp et al. 1997). Presented below is a summary of selected results.

Effects of Seed Increase on the Genetic Integrity of Potato Collections

A question in the conservation of *ex situ* genetic resources is whether genebank management has altered the genetic structure of the collections. Ellstrand and Elam (1993) indicated that there are no precise assessments of how well diversity has been preserved over many years in genebanks. There are a number of factors that could generate negative effects in the genetic structure of *ex situ* populations. For instance, a chance of unintentional selection exists due to small samples, different germination rates, pathogen and disease attacks, seed contamination, etc. Unfortunately, little research has been done in this regard. A study by Widrlechner et al. (1989) reported that conditions for genetic drift do exist at genebanks because unintentional selection could take place during seed increases. Breese (1989) and Ellis et al. (1985) pointed out that in order to reduce chances of genetic drift in the collections, some guidelines for handling seed multiplication must be followed. Cross and Wallace (1994) suggested that loss of genetic diversity in self-pollinated species is likely to occur after a number of generations. However, most of these fears have been based upon theoretical calculations and simulations with no real evaluation of plant material. Thus, there was need to complete studies gauging the actual extent and dynamics of diversity in germplasm collections.

In del Rio et. al. (1997a), RAPD markers were used to compare populations regenerated from their parent populations at different periods in the genebank. Two potato species native to the United States,

Solanum jamesii and *S. fendleri*, were used as model plant systems. *Solanum jamesii* is a diploid (2n=2x=24) that reproduces by outcrossing, while *S. fendleri* is a disomic tetraploid (2n=4x=48) that can outcross or self. These differences made it possible to see if reproductive behavior and vulnerability to genetic drift are associated (Loveless and Hamrick 1984). Genetic differentiation was small and insignificant with an average genetic similarity (GS, the simple matching coefficient) in *S. jamesii* of 96.3% and GS in *S. fendleri* of 95.9%. Breeding systems did not have any obvious impact on genetic vulnerability. These results suggest that genetic diversity of *ex situ* potato germplasm is being rather well preserved.

Comparison of ex situ and in situ Populations

Another important concern in the conservation of potato genetic resources is whether samples in the genebank, although genetically stable (del Rio et al. 1997a), still are representative of the populations taken from the wild during explorations, sometimes many decades ago. This has consequences with respect to the value of *ex situ* collections as "back-ups" and the worth of re-collecting "new" samples at the exact original sites. Population dynamics in the wild are very complex. Various factors influence the distribution, extinction and expression of genetic diversity (see Loveless and Hamrick 1984, Slatkin 1987 for reviews). For curators, however, the practical issue is whether genetic differentiation has been taking place in the wild over time, and if so, how much.

Thus, an investigation was conducted to assess how populations in the genebank compare to their counterparts at the original sites in the wild (del Rio et al. 1997b). RAPD markers indicated that these two groups were often significantly different. The differences between samples collected 34 years apart were nearly as great as those of samples collected from different locations. Therefore, the wild does not provide an adequate "backup" for the genebank, and re-collections might be valuable for collecting additional diversity.

Comparison of Original Populations and their First Seed Multiplication

Samples of populations in the wild were shown to be different than those in the genebank. Which samples changed? Did the genebanks generate artificial differentiation by reducing (genetic bottleneck) or increasing (promoting allelic recombination) the genetic diversity? Is new diversity evolving in the wild or are these populations suffering genetic erosion? Genebank samples were shown to be quite stable (del Rio et al 1997a), but could they undergo a significant genetic change when they are first incorporated into the genebank? For one thing, the materials in question were collected as tubers and conversion to sexual seeds by artificial means at the genebank might have imposed selection.

To examine this possibility, populations of plants gathered directly from the wild were compared to their offspring generated by methods common in potato genebanks (del Rio and Bamberg 2003). The assessment of genetic similarity using DNA markers showed that the wild parents and their genebank offsprings were virtually identical. Therefore, large genetic changes do not result while incorporating these plants into the genebank from the wild. Alternate explanations for the differences are being investigated, among which the most obvious is that significant genetic changes are occurring in the wild over time.

Value of Germplasm with Unreliable or Missing Origin Data—The Problem of "Mystery" Samples

There is concern regarding the accumulation of accessions in the genebank and how effectively they can be maintained and evaluated. At the extreme, unrecognized exact duplicate samples use up twice the resources needed to maintain one unit of genetic diversity. When resources are limited, duplication may result in a net loss of diversity in the genebank. Huaman et al. (2000a, 2000c) using isozymes to select a core collection, identified redundancy in the collection of *S. tuberosum* ssp. *andigena* held at CIP. It was found that more than 75% of the original collection could be eliminated as redundant material. Such data give an idea of the magnitude of the problem and also about the opportunities for improvement.

Even if samples are not exact duplicates, incomplete or inaccurate data regarding their geographic origin make it difficult to estimate their value in the collection, since geographic origin is often assumed to be correlated with genetic diversity. Spooner et al. (1992) reported that the origin of many of the wild potato populations could not be geographically mapped because they lack precise data on site of origin. Similarly, Waycott and Fort (1994) emphasized that large numbers of collections at genebanks lack descriptive and/or pedigree information.

A model case for examination of "mystery" accessions occurred at the US Potato Genebank. A certain seed lot, according to the original seed packet, was species *S. sucrense* from Mexico. But *S. sucrense* is not known to occur in Mexico (the geographical range of this species is southern Bolivia). Since the documentation for this population was contradictory, these seeds were not increased, advertised or distributed by the genebank for over 23 years. Although this case was extreme, many other populations in the genebank have limited collection data, e.g., only country of origin.

del Rio and Bamberg (2000) conducted a genetic analysis based on RAPD markers , comparing this "mystery" population and each of 30 other *S. sucrense* populations in the genebank (and known related species). All populations within this species, including the mystery population, were significantly different from being duplicates, and were therefore worthy of separate conservation. RAPD markers also distinguished the mystery population from closely related tetraploid species *S. oplocense, S. gourlayi* and *S. andigena* suggesting that it is also not a duplicate of a population of these species. One major outcome from this research was that the DNA markers used clearly differentiated populations within a highly heterogeneous species like *S. sucrense* (Hosaka and Hanneman 1991), thus, RAPDs appear to be a powerful technique in determining germplasm organization within potato species (Quiros et al. 1993).

Vulnerability of Alleles in the Potato Genebank

Once diversity is captured in the genebank, the task is to keep it. Potato genebanks typically have many heterogeneous populations. Clonal maintenance for all of these is impractical. Seed populations that need periodic regeneration become subject to changes associated with sampling error: samples may be altered in a certain direction by unintentional selection, or simply drift randomly to a condition unrepresentative of the original parental source. In the case of random mis-sampling, low frequency alleles are most vulnerable to loss due to exclusion from the sample of seed increase parents by chance. A number of studies have been conducted dealing with mathematical models and recommendations that follow from hypothetical calculations on vulnerable alleles within single populations (Crossa et al. 1993, Gale and Lawrence 1984, Lawrence et al. 1995).

A study was undertaken to empirically detect the prevalence of low frequency alleles in the US Potato Genebank by extrapolation from RAPD markers. Two outcrossing species known to be very heterogeneous were selected, *Solanum jamesii* and *Solanum sucrense* (del Rio et al. 1997a, Hosaka and Hanneman 1991). The marker frequencies of 83 RAPD loci within each of 15 populations were assessed. For both species, about 10% of loci within-populations had vulnerable alleles (banded plant frequency <40%) considering just the particular population in question. However, about half of these were fixed in at least one of the other population (making that marker invulnerable to loss). The highest within-population banded plant frequency for loci not fixed in any population was, with one exception, always >40%. Thus, while several alleles may be vulnerable within a given population, these were nearly always fixed or nearly fixed (invulnerable) in another population. This suggests that the safest strategy for maximizing conservation of alleles is to apportion available resources over many separate Mendelian populations, rather than adopt a preservation protocol so rigorous that it can only be practically applied to a limited number of populations (Bamberg and del Rio 1999).

Unintentional Seedling Selection

Even relatively common alleles could be eliminated if a strong, unintentional, selection pressure is imposed by standard genebank protocols. For example, since germination cannot be precisely predicted, excess seeds are sown to produce plants for use as parents in a seed increase. Often, more seedling plants than needed are available. It seems counterproductive to include less vigorous seedlings, but does seedling selection at transplanting impose genetic shifts in the population? Recent evidence suggests that it certainly can for some species (Bamberg et al. 2003).

Analysis of Genetic Parity Between Potato Germplasm from Different Genebanks

As mentioned above, world potato genebanks have a formal network (APIC) to exchange information and techniques and to work on problems of mutual interest. As a result, a comprehensive database of passport and evaluation data has been synthesized for wild potato species that is available on the Internet (Appendix 1.A). This database shows that, in many cases, several genebanks have samples of the progeny from a single original germplasm collection (Huaman et al. 2000). The assumption has been that these samples are genetically equivalent, so all the characterization and evaluation data gathered on a seedlot from one genebank can be applied to all the other "duplicate" seedlots in other genebanks.

This assumption was tested by comparing pairs of reputed duplicates between potato genebanks: VIR (St. Petersburg, Russia) and US (Sturgeon Bay, USA) and, between CIP (Lima, Peru) and US, with RAPD markers. In the majority of the cases of the VIR vs. US comparisons (Bamberg et al. 2001) and in some of the CIP vs. US comparisons, reputed duplicates among genebanks had significantly less similarity than replicate samples taken from a single population. Thus, users of germplasm should be aware that while samples under the same identifier in different genebanks are usually quite similar, they might not be identical. Work is in progress to measure how much divergence may exist with respect to the expression of economic traits.

Studies on the Relationship of Genetic Diversity to Natural Habitats

Altieri and Merrick (1987) and Brown et al. (1997) recognized that there are problems associated with inadequate sampling procedures during explorations, which has an impact on the representation of the total gene pool. However, genebanks have limited resources to address these types of problems. Ferguson et al. (1998) suggested that sampling would be more successful if the exploration trips had clearly defined targets (e.g. geographical areas and habitats). Different experts have advised that an approach to identify these areas is to undertake eco-geographic studies preceding explorations, because plant populations exhibit structured genetic organization along their geographical ranges (Antonovics 1971, Loveless and Hamrick 1984). The general and historic practice in explorations has been to sample as many different niches as possible (Brown 1978, Marshall and Brown 1975). Sampling methods have been set for collecting germplasm, but they are based on probability schemes combined with theoretical predictions of population genetics (Crossa et al. 1993, Zoro et al. 1999). Such recommendations can offer only imprecise predictions when empirical information about species reproductive biology is unknown. For example, Hamrick et al. (1991) emphasize that there is extensive variation among plant species for ecological traits that influence the distribution of genetic variation, so a genetically effective management strategy for one species may not be effective for another. Therefore, comprehensive and empirical studies on ecology, population biology, genetics and reproductive biology are essential to predict proper sampling strategies (Allard 1970, Yonezawa and Ichihashi 1989).

Wild potato species have a wide range of ecological and geographic distributions. Is it possible to correlate genetic variation observed in natural wild potato populations with certain ecological, geographical or reproductive variables? There are two ways that this might occur, for example, using the parameter of altitude: 1) collections might be equally variable within altitudes, and group genetically by altitude (e.g., collections at higher altitudes can be differentiated from those at low altitude); or, 2) diversity itself might be correlated with altitude (e.g., collections from high altitude may be more variable than those from low altitudes).

Since reproductive system plays a major role in shaping genetic structure, the wild populations examined represented four different breeding systems: *S. fendleri* (disomic polyploid, partially selfer), *S. jamesii* (diploid outcrosser), *S. sucrense* (polysomic polyploid, outcrosser) and *S. verrucosum* (diploid selfer). This study evaluated 151 wild potato populations from different geographical origins. Three of the breeding systems studied exhibited no significant associations between genetic and eco-geographic variation. Even populations from sites in close proximity could be quite different genetically. Only for *S. verrucosum* (the inbred) were ecogeographic parameters predictive of genetic differences, which is in agreement with studies in other inbred wild populations (Nevo et al. 1988). Inbreeding diploids like *S. verrucosum* expose all of their alleles to the environment for adapting selection. But other potato species are outcrossing, so can have recessive alleles in heterozygous state that are not subject to environmental selection pressure. Since potatoes can reproduce clonally by tubers, these plants can avoid the loss of reproductive fitness that would occur in their segregating sexual progeny, many of which would be unadapted to their environment. Some polyploid species are inbreeding but with disomic chromosome pairing. In this case as well, recessive or unexpressed alleles may be present that do not reflect adaptation to the environment of origin. For such species, both tuber and seed reproduction maintains heterozygosity in uniform, adapted progeny.

There are several other possible explanations for a lack of association between ecogeographic traits and genetics in the more heterogeneous species: poor sampling, excessive within-population diversity, population subdivision, small population size, gene flow caused by humans or animals, poor resolution of the climatic, geographical or ecological variables used, etc. In the future, these problems may be overcome. But for now, the practical and prudent approach indicated for heterogeneous species is to collect as many populations as possible and study their genetic potential empirically.

The Impact of Breeding System on the Preservation and Use of Germplasm
As noted above, seed populations imply sampling. Sample quality depends on sample variation, and sample variation depends largely on sample size and variation among the units. In many instances, practical limits of resources restrict genebanks to the use of relatively small sample sizes. For example, the theoretical solution to the risk of drift using only 20 parents for a seed increase is easy: multiply by several fold the number of parents used (with a corresponding multiplication of the resources required). A more practical approach would be to increase the sample size of more heterogeneous populations and decrease the sample size of more uniform populations. The heterogeneity of populations also effects sample size needed to representatively collect and evaluate germplasm. Thus, another thrust of genebank research has been to characterize the with-population heterogeneity of species, which can vary greatly (del Rio and Bamberg 2001). Other studies demonstrate a related effect of within-population heterogeneity: the reduced ability to precisely and accurately assess genetic differences among populations using DNA markers (del Rio and Bamberg 1998, Bamberg and del Rio 2004).

Evaluation Research

Details of assessment and utilization of diseases and pest resistance are covered elsewhere in this volume. The following discussion focuses on evaluation for better use and management of germplasm in the genebank.

Whether planning new acquisitions or deciding what can be discarded, efficient management of the genebank requires an empirical basis for prioritizing by assessing relative value—that is, *evaluating* the germplasm. Information is needed about the history of collection, general diversity (i.e., taxonomic and intra-specific characterization based on phenotypic and neutral DNA markers), presence of economic traits, and prospects for utilization (e.g., status of crossing barriers). For example, a certain species may have been heavily collected only because it is widespread and a prolific seed producer. Perhaps an accession from within this species was a re-collection from the same location and determined empirically to be a near genetic duplicate of the original. Finally, if the accession has been thoroughly evaluated and exhibits no unique useful traits and is very difficult to incorporate into the breeding pool, it would have low priority for retention in the genebank. These examples illustrate why evaluation research is critical for efficient genebank management. The considerations are similar for the germplasm user who has a particular use for the germplasm in mind, and wants to know whether a given stock has strong expression for the trait in the form of novel genetics and can be easily incorporated into a research or breeding program.

There is a large body of literature detailing evaluation of potato germplasm for various traits. Potato genebanks often summarize and format evaluation information in catalog and database forms. The usual purpose of evaluation data is to identify germplasm directly useful for breeding. However, the extreme expression of traits found in the breadth of diversity may also be useful to breeding indirectly, when they are useful as standards or research tools and models that elucidate the genetics and physiology of the trait (e.g. Bamberg et al. 1998, Singh 1984).

Another type of evaluation for traits indirectly related to breeding is referred to as *characterization*. Characterization is any innate feature of the plant pertinent to its culture and use (for example: vigor in greenhouse, field or *in vitro*; flowering, pollen dehiscence, crossability, seed longevity, etc.). As for evaluation, there is a large body of research literature on various potato germplasm characterization topics, in some cases done by genebank personnel themselves. Genebank staff also typically summarize and format such information for the benefit of their clients.

Distribution

Delivery and use of potato germplasm are relatively easy. It has small, long lived botanical seeds that are typically not difficult to germinate. In exceptional cases, requesters can be advised of needed technology, or it can be provided. The standard distribution unit at the US Potato Genebank is 50 botanical seeds. Clonal items are usually distributed in the form of *in vitro* plantlets or minitubers. *In vitro* plantlets may be preferred over minitubers because plantlets can be rapidly subcultured as ordered, so there is no expiring unused inventory. When shipping *in vitro* stocks, using tubes of smaller diameter, a more solid medium and padding in the box helps to keep the cultures intact despite rough handling. Explicit labeling, insulating packaging, or waiting for milder weather may be necessary to avoid freezing in transit.

Other Forms of Germplasm Distribution

The expectation in a typical distribution is that the recipient will propagate the germplasm pursuant to particular interests. However, since genebanks specialize in propagation of exotic species, they can provide a very valuable service in partnering with requesters to supply customized forms of germplasm or associated products. For example, seedling tubers may be generated for the germplasm recipient. Although tubers present disease risks mentioned above, this is not germane in some cases in which the tubers are not intended for propagation (e.g., tests of tuber constituents, tubers for display or educational purposes). In such cases, the recipient may be able to get a special permit to avoid the delays of quarantine testing. Nodal cuttings are relatively easy to produce in potato, and are a convenient way to replicate genotypes without having to wait for tubers to be produced and break dormancy. Rooted cuttings in moist paper survive well in transit for several days, and rapidly establish a research plant for the recipient. Herbarium specimens are not subject to quarantine testing, and their study is sometimes a convenient alternative to growing a live plant. Some genebanks have associated herbaria and loan specimens (Bamberg and Spooner 1994, Spooner and Bamberg 1998). In some applications, emailed digital photographs are a convenient and quick substitute for herbarium specimens. Other atypical items distributed include pollen, DNA, or even custom hybrids. Even if the genebank is not able to provide these extended services *per se*, genebank personnel can often share advice and technology that will make a significant contribution to the success of the requester's experiment. These atypical forms of distributions may be wholly at the recipient's request, or may be the genebank's contribution to a research partnership with a specialist in some aspect of germplasm evaluation. For example, genebank personnel might contribute their knowledge of the genetics, propagation and reproduction of germplasm to select and prepare materials for disease resistance screening in collaboration with a pathologist having the appropriate knowledge, experience and facilities.

Promoting Distribution

As suggested above, the genebank should pursue every opportunity to enhance distribution. The other supporting activities of genebanks are only of *potential* value until germplasm is delivered into the hands of researchers and breeders. Unused germplasm is actually a liability since keeping it requires ongoing outlays of resources. In addition to meeting special needs of requesters, genebank personnel should be promoting creative new uses for the germplasm to a broader audience of specialists who may not be aware that potato genebanks exist and can provide tools that may be ideal for their research. One paradigm in this regard is encouraging the testing of exotic species for traits that have been researched extensively only in commercial cultivars.

Distribution of Information

Like germplasm that is not available for distribution, data that is available for distribution only has *potential* value. Without specific data, there will be few requests, since potentially interested parties will not know what is available, its characteristics or condition. Thus, a major function of genebanks is to gather, organize and distribute data. As is true for many fields, computers have caused a revolution of efficiency in data management in genebanks. Databases and on-line access have also made it easier for germplasm users to obtain and manipulate up-to-date evaluation data. This change in the dynamics of information availability raises some questions for genebank managers. For example, online databases, with all their advantages, are now in place and accessible to most germplasm customers. Is there any further need for genebanks to expend the time and expense of making paper catalogs, or provide custom searches and summaries for germplasm users?

REFERENCES

Alexanian, S. M. and V. I. Krivchenko. 1991. Vavilov Institute scientists heroically preserve world plant genetic resources collections during World War II Siege of Leningrad. Diversity 7(4):10-13.

Allard, R. W. 1970. Population structure and sampling methods. In: Frankel, O. H. and E. Bennett (eds.), Genetic resources in plants, IBP Handbook No. 11. F. A. Davis Co. Philadelphia. P. 97-109.

Allard, R. W. 1996. Genetic basis of the evolution of adaptedness in plants. Euphytica 92:1-11.

Allem, A. C. 2001. Managing genebanks: Seed base collections examined. Genetic Resources and Crop Evolution 48:321-328.

Altieri, M. A. and L.C. Merrick. 1987. *In situ* conservation of crop genetic resources through maintenance of traditional farming systems. Eco Bot 41: 86-96.

Antonovics, J. 1971. The effects of heterogeneous environment on the genetics of natural populations. Am Scientist 59:593-599.

Astley, D. 1991. Exploration: Methods and problems of exploration and field collecting. In: Hawkes, J. G. (ed.), Genetic conservation of world crop plants. Academic Press, London. 87 pp.

Bamberg, J. B. 1990. Intergenebank collaboration launched for global potato collections. Diversity 6(3&4):6-7.

Bamberg, J. B. 1999. Wild potatoes on public lands of the Southwest. NRSP-6 genebank circular. US Potato Genebank, Sturgeon Bay, WI. 6pp.

Bamberg, J. B. and A. H. del Rio. 1999. Vulnerability of alleles in the US Potato Genebank extrapolated from RAPDs. Amer J Potato Res 76:363.

Bamberg, J. B., A. H. del Rio, Z. Huaman, S. Vega, M. Martin, A. Salas, J. Pavek, S. Kiru, C. Fernandez, and D. Spooner. 2003. A Decade of Collecting and Research on Wild potatoes of the Southwest USA. Amer J. Potato Res. 80:159-172.

Bamberg, J. B. and A. H. del Rio. 2004. Genetic heterogeneity estimated by RAPD polymorphism of four tuber-bearing potato species differing by breeding structure. Amer J. Potato Res. (in press).

Bamberg, J. B., A. Salas, S. E. Vega, R. Hoekstra, and A. Huaman. 1996. Notes on wild potato populations in Arizona and New Mexico. Am Potato J 73:342.

Bamberg, J. B., S. D. Kiru and A. H. del Rio. 2001. Comparison of reputed duplicate populations in the Russian and US potato genebanks using RAPD markers. Amer J Potato Res 78: 365-369.

Bamberg, J., J. Palta, L. Peterson, Max Martin and A. Krueger. 1998. Fine screening potato (*Solanum*) species germplasm for tuber calcium. Amer J Potato Res 75:181-186.

Bamberg, J.B. and D.M. Spooner. 1994. The United States Potato Introduction Station herbarium. Taxon 43:489-496.

Bamberg, J.B., Z. Huaman and R. Hoekstra. 1995. International Cooperation in Potato Germplasm. In: R. Duncan, (ed.), International Germplasm Transfer: Past and Present, CSSA Special Publication #23. CSSA/ASA/SSSA Madison, WI. pp 177-182.

Becker, H., L. McGraw and K. Barry Stelljes. 1998. Why *in situ*?— Finding and protecting wild relatives of American crops. Agric Res, (12/1998):4-8.

Breese E.L. 1989. Regeneration and multiplication of germplasm resources in seed genebanks: The scientific background. IBPGR, Rome.

Brown A. H. D., C. L. Brubaker and J. P. Grace. 1997. Regeneration of germplasm samples: Wild versus cultivated plant species. Crop Sci 37:7-13.

Brown, A. H. D. 1978. Isozymes, plant population genetic structure and genetic conservation. Theor Appl Genet 52:145-157.

Brown, C. R. 1990. Modern evolution of the cultivated potato genepool. In: Vayda, M. E. and W. D. Park (eds.). The Molecular and Cellular Biology of the Potato. CAB International, Wallingford, UK. pp 1-11.

Brown, C. R. 1993. Origin and history of the potato. Am. Potato J 70:363-373.

Brown, C. R. 1999. A Native American technology transfer: The diffusion of potato. HortScience 34:817-821.

Brown, G. M., Jr. 1990. Valuation of Genetic Resources. In: The preservation and valuation of biological resources, G. H. Orians et al., (eds.), University of Washington Press, pp. 203-228.

Brown, J. and M. F. B. Dale. 1998. Identifying superior parents in a potato breeding program using cross prediction techniques. Euphytica 104:143-149.

Brush, S. B. 1991. Farmer conservation of new world crops: The case of Andean potatoes. Diversity 7(1&2):75-79.

Brush, S., R. Kesseli, R. Ortega, P. Cisneros, K. Zimmerer and C. Quiros. 1995. Potato diversity in the Andean center of crop domestication. Conservation Biol 9:1189-1198. Blackwell Scientific Publications, Cambridge, MA.

Chang, T., S. M. Dietz and M. N. Westwood. 1989. Management and use of plant germplasm collections. In: L. Knutson and A. Stoner (eds.), Biotic diversity and germplasm preservation, global imperatives. Kluwer. Dordrecht, pp. 127-160.

CIP (International Potato Center). 1990. The Potatoes of South America: Bolivia. A portfolio of fine botanical prints. (derived from: Ochoa, C. M. 1990. The Potatoes of South America: Bolivia. Cambridge University Press, 512 pp.).

CIP (International Potato Center). 1999. Annual Report. CIP. Lima, Peru. p. 12.

Correll, D. S. 1962. The potato and its wild relatives. Texas Research Foundation, Renner, TX, 606 pp.

Cox, T. S., J. P. Murphy and M. Goodman. 1988. The contribution of exotic germplasm to American agriculture. In: J. R. Kloppenburg, Jr. (ed.), Seeds and sovereignty: The use and control of plant genetic resources. Duke University Press, Durham, NC, 368 pp.

Cross R. J. and A. R. Wallace. 1994. Loss of genetic diversity from heterogeneous self-pollinating genebank accessions. Theor Appl Genet 88:885-890.

Crossa, J., C. M. Hernandez, P. Bretting, S. A. Eberhart and S. Taba. 1993. Statistical genetic considerations for maintaining germplasm collections. Theor Appl Genet 86:673-678.

del Rio, A. H. and J. B. Bamberg. 1998. Effects of sampling size and RAPD marker heterogeneity on the estimation of genetic relationships. Amer J Potato Res 75:275.

del Rio, A. H. and J. B. Bamberg. 2000. RAPD markers efficiently distinguish heterogeneous populations of wild potato (*Solanum*). Genet Resources and Crop Evol 47:115-121.

del Rio, A. H. and J. B. Bamberg. 2001. Genetic heterogeneity among breeding systems of potato species and its ramifications in germplasm conservation. Amer J Potato Res 78:452.

del Rio, A. H. and J. B. Bamberg. 2003. The effect of genebank seed increase on the genetics of recently collected potato (*Solanum*) germplasm. Amer J Potato Res 80:215-218.

del Rio, A. H., J. B. Bamberg and Z. Huaman. 1997a. Assessing changes in the genetic diversity of potato genebanks. 1. Effects of seed increase. Theor Appl Genet 95:191-198.

del Rio, A. H., J. B. Bamberg, Z. Huaman, A. Salas and S. E. Vega. 1997b. Assessing changes in the genetic diversity of potato genebanks 2. *In situ* vs *ex situ*. Theor Appl Genet 95:199-204.

del Rio, A. H., J. B. Bamberg, Z. Huaman, A. Salas and S. E.Vega. 2001. Association of eco-geographical variables and genetic variation in native wild US potato populations determined by RAPD markers. Crop Sci 41:870-878.

Douches, D. S., D. Maas, K. Jastrzebski and R. W. Chase. 1996. Assessment of potato breeding progress in the United States over the last century. Crop Sci 36:1544-1552.

Duvik, D. and W. Brown. 1989. Plant Germplasm and the economics of agriculture. In: L. Knutson and A. Stoner (eds.). Biotic diversity and germplasm preservation, global imperatives. Kluwer, Dordrecht, pp. 499-513.

Eberhart, S., H. Shands, W. Collins and R. Lower (eds.). 1998. IPR Global Genetic Resources: Access and property rights. Crop Science Society of America, Madison Wisconsin, USA. 176 pp.

Ellis, R. H., T. D. Hong and E. H. Roberts. 1985. Handbook of seed technology for genebanks No. 2. Vol. I: Principles and methodology. IBPGR, Rome, 210 pp.

Ellstrand N. C. and D. R. Elam. 1993. Population genetic consequences of small population size: Implications for Plant Conservation. Ann Rev Ecol Syst 24:217-242.

Engels, J. M. M. and D. Wood. 1999. Conservation of agrobiodiversity. In: D. Wood and J.M. Lenné (eds.), Agrobiodiversity: Characterization, utilization, and management. CAB International, Wallingford, UK, 490 pp.

Ferguson, M. E., B. V. Ford-Lloyd, L. D. Robertson, N. Maxted and H. J. Newbury. 1998. Mapping the geographical distribution of genetic variation in the genus *Lens* for the enhanced conservation of plant genetic diversity. Mol Ecol 7:1743-1755.

Flanders, K. L., E. B. Radcliffe and J. G. Hawkes. 1997. Geographic distribution of insect resistance in potatoes. Euphytica 93:201-221.

Fowler, C., M. Smale and S Gaiji. 2001. Unequal Exchange? Recent transfers of agricultural resources and their implications for developing countries. Developmental Policy Review, June 2001, pp 1-26.

Fox, Warwick. 1993. What does the recognition of intrinsic value entail? Trumpeter 10:3. ISSN: 0832-6193

Gale, J. and M. Lawrence. 1984. The decay of variability. In: J. H. W. Holden and J. T. Williams (eds.), Crop genetic resources: Conservation and evaluation. Allen and Unwin, London, pp. 77-101.

Goodman, M. M. 1990. What genetic and germplasm stocks are worth conserving? In: P. E. McGuire and C. O. Qualset (eds.), Genetic resources at risk: Scientific issues, technologies, and funding policies. Symposium Proceedings, American Association for the Advancement of Science, Annual meeting, San Francisco, January 16, 1989. Genetic Resources Conservation Program. University of California, Oakland, 42 pp.

Hamrick, J. L., M. J. W. Godt, D. A. Murawski and M. D. Loveless. 1991. Correlations between species traits and allozyme diversity: Implications for conservation biology. In: Falk, D.A., K.E. Holsinger (eds.), Genetics and Conservation of Rare Plants. Oxford Univ Press, Oxford, UK.

Hanneman, R. E. Jr. 1989. The potato germplasm resource. Amer Potato J 66:655-667.

Hanneman, R. E. Jr. 1999. The reproductive biology of the potato and its implication for breeding. Potato Res 42:283-312.

Hawkes J. G., 1990. The potato: Evolution, biodiversity, and genetic resources. Belhaven Press, London. 259 pp.

Hawkes, J. G. 1994. Origins of cultivated potatoes and species relationships. In: J. E. Bradshaw and G. R. Mackay (eds.). Potato Genetics. CAB International. Wallingford, UK, p. 3-42.

Hawkes, J. G., N. Maxted and B.V. Ford-Lloyd (eds.). 2000. The *ex situ* conservation of plant genetic resources. Kluwer Academic Publishers, Boston, 250 pp.

Hermsen, J. G. T. 1980. Breeding for apomixis in potato: pursuing a utopian scheme. Euphytica 29:595-607.

Hijmans, R. J. and D. M. Spooner. 2001. Geographic distribution of wild potato species. Amer J Bot 88:2101-2112.

Hijmans, R. J., K. A. Garrett, Z. Huaman, D. P. Zhang, M. Schreuder and M. Bonierbale. 2000. Assessing the geographic representativeness of genebank collections: The case of bolivian wild potatoes. Conservation Biol 14:1755-1765.

Hosaka, K. and R. E. Hanneman, Jr. 1991. Seed protein variation within accessions of wild and cultivated potato species and inbred *Solanum chacoense*. Potato Res 34:419-428.

Hosaka,K., M. Mori and K. Ogawa. 1994. Genetic relationships of Japanese potato cultivars assessed by RAPD analysis. Am Potato J 71:535-546.

Huaman, Z., R. Ortiz, D. Zhang and F. Rodriguez. 2000a. Isozyme analysis of entire and core collections of *Solanum tuberosum* subsp. *andigena* potato cultivars. Crop. Sci 40:273-276.

Huaman, Z., A. Golmirzaie and W. Amoros. 1997. The Potato. In: D. Fuccillo, et al. (eds.), Biodiversity in trust: Conservation and use of plant genetic resources in CGIAR centres. Cambridge University Press, Cambridge, pp. 21-28.

Huaman, Z., R. Hoekstra and J. B. Bamberg. 2000b. The inter-genebank potato database and the dimensions of available wild potato germplasm. Amer J Potato Res 77:353-362.

Huaman, Z., R. Ortiz and R. Gomez. 2000c. Selecting a *Solanum tuberosum* subsp. *andigena* core collection using morphological, geographical, disease and pest descriptors. Amer J Potato Res 77:183-190.

Jansky, S. 2000. Breeding for disease resistance in potato. Plant Breeding Reviews 19:69-155.

Johns, T. and S. Keen. 1986. Ongoing evolution of the potato on the Altiplano of western Bolivia. Eco Bot 40:409-424.

Karp, A., S. Kresovich, K. Bhat, W. Ayad and T. Hodgkin. 1997. Molecular tools in plant genetic resources conservation: A guide to the technologies. IPGRI Tech Bull No. 2. IPGRI. Rome, 47 pp.

Kolasa, K. M. 1993. The potato and human nutrition. Am Potato J 70:375-384.

Lawrence, M. J., D. F. Marshall and P. Davies. 1995. Genetics of genetic conservation. I. Sample size when collecting germplasm. Euphytica 84:89-99.

Lenné, J.M. and D. Wood. 1999. Optimizing biodiversity for production agriculture. In: D. Wood and J.M. Lenné (eds.). Agrobiodiversity: Characterization, utilization, and management. CAB International. Wallingford, UK, pp. 447-470.

Levy, D., Y. Itzhak, E. Fogelman, E. Margalit and R. E. Veilleux. 2001. Ori, Idit, Zohar and Zahov: Tablestock and chipstock cultivars bred for adaptation to Israel. Amer J Potato Res 78:167-174.

Love, S. L. 1999. Founding clones, major contributing ancestors, and exotic progenitors of prominent North American potato cultivars. Amer J Potato Res 76:263-272.

Loveless, M. D. and J. L. Hamrick. 1984. Ecological determinants of genetic structure in plant populations. Ann Rev Ecol Syst 15:65-95.

Marshall, D. R. and A. D. Brown. 1975. Optimum sampling strategies in genetic conservation. In: O. H. Frankel and J.G. Hawkes (eds.), Crop genetic resources for today and tomorrow. Cambridge Univ Press, pp. 53-80

Martin, M. 1988. Potato production and chemical dependency. Am Potato J Symp suppl.: 1-4.

McGraw, L. 1998. Gene Bank at the Park. News from the USDA Agricultural Research Service Information Staff, Agricultural Research Service. November 4, 1998. http://www.ars.usda.gov/is/pr/1998/981104.htm

Myers, N. 1983. A wealth of wild species: Storehouse for human welfare. Westview Press, Boulder, CO, 274 pp.

Neiderhauser, J. S. 1993. International cooperation and the role of the potato in feeding the world. Am Potato J 70:385-403.

Nevo, E., B. Baum, A. Beiles and D. A. Johnson. 1998. Ecological correlates of RAPD DNA diversity of wild barley, *Hordeum spontaneum*, in the Fertile Crescent. Genet Resources and Crop. Evol. 45:151-159.

Noss, R. F. and A. Y. Cooperrider. 1994. Saving nature's legacy: Protecting and restoring biodiversity. Island Press, Washington, D.C., 416 pp.

NPS (US National Park Service). 4/1999. Webpage on Biodiversity "The Eroding Foundation of Life" http://www.nature.nps.gov/wv/biodiv.htm

NRC (US National Research Council). 1993. Agricultural Crop Issues and Policies / Committee on Managing Global Genetic Resources, National Academy Press, 449 pp.

Pavek, J. J. and D. L. Corsini. 2001. Utilization of potato genetic resources in variety development. Amer J Potato Res 78:433-441.

Petit, M., C. Fowler, W. Collins, C. Correa and C. Thornström. 2001. Why governments can't make policy: The case of plant genetic resources in the international arena. International Potato Center (CIP). Lima, Peru, 80 pp.

Pimm, S. L., G. J. Russell, J. L. Gittleman, and T. M. Brooks. 1995. The future of biodiversity. Science (Washington) 26:347-350

Plaisted, R. L and R. W. Hoopes. 1989. The past record and future prospects for the use of exotic potato germplasm. Amer Potato J 66:603-627.

Querol, D. 1993. Genetic resources: A practical guide to their conservation. Third World Network, London, 252 pp.

Quiros, C. F., A. Ceada, A. Georgescu and J. Hu. 1993. Use of RAPD markers in potato genetics: segregations in diploid and tetraploid families. Am Potato J 70:35-42.

Quiros, C.F., R. Ortega, L. van Raamsdonk, M. Herrera-Montoya, P. Cisneros, E. Schmidt and S. B. Brush. 1992. Increase of potato genetic resources in their center of diversity: the role of natural outcrossing and selection by the Andean farmer. Genet Resources and Crop Evol 39:107-113.

Reid, W. and K. R. Miller. 1989. Keeping options alive: The scientific basis for conserving biodiversity. World Resources Institute, Washington, DC, 128 pp.

Rhoades, R. 1982. The incredible potato. National Geographic 161:668-694.

Richardson, D. G. and N. Estrada-Ramos. 1971. Evaluation of frost resistant tuber-bearing *Solanum* hybrids. Am Potato J 48:-339-343.

Ross, H. 1986. Potato breeding—Problems and perspectives. Advances in plant breeding. Supplement 13 to Journal of Plant Breeding. Verlag Paul Parey. Berlin and Hamburg, 132 pp.

Ryan, J. C. 1992. Life support: Conserving biological diversity. Worldwatch Institute, Washington, DC, 62 pp.

Salaman, R. N. 1985. The history and social influence of the potato. Revised impression by J. G. Hawkes (ed.), Cambridge Univ. Press. Cambridge, UK, 685 pp.

Schittenhelm, S. and R. Hoekstra. 1995. Recommended isolation distances for the field multiplication of diploid tuber-bearing *Solanum* species. Plant Breed 114:369-371.

Schoen, D. J. and A. D. H. Brown. 2001. The conservation of wild plant species in seed banks. Bioscience 51:960-966.

Schoen, D. J., J. L. David and T. M. Bataillon. 1998. Deleterious mutation accumulation and the regeneration of genetic resources. Proc Nat Acad Sci (USA), 95:394-399.

Shekhawat, G. S., A. K. Sharma, and S. Uppal (eds.). 1999. CPRI (Central Potato Research Institute, Shimla, India), Malhotra Publishing, New Delhi, 24 pp.

Simmonds, N. W. 1997. Letter to the Editor. Diversity 13(1):2.

Singh, R. P. 1984. *Solanum Xberthaultii*, a sensitive host for indexing potato spindle tuber viroid from dormant tubers. Potato Res 27:163-172.

Slatkin, M. 1987. Gene flow and the geographic structure of natural populations. Science 236: 787-792.

Spooner, D. M and R. J. Hijmans. 2001. Potato systematics and germplasm collecting, 1989-2000. [Erratum: Sept/Oct 2001, v. 78 (5), p. 395.]. Amer J Potato Res 78:237-268.

Spooner, D. M. 1999. Exploring options for the list approach. Proceedings, Intl. Workshop: Inter-dependence and food security: Which list of PGRFA for the future multilateral system? Instituto Agronomico per L'Oltremare, Florence, Italy. p. 133-164.

Spooner, D. M. and J. B. Bamberg. 1994. Potato genetics resources: Sources of resistance and systematics. Am Potato J 71: 325-338.

Spooner, D. M. and J. B. Bamberg. 1998. Potato Herbarium. In: Lichtenfels, J. R., J. H. Kirkbride Jr., and D. J. Chitwood, eds. Systematic collections of the Agricultural Research Service. USDA, ARS Misc pub 1343, pp. 53-56.

Spooner, D. M., A. Contreras and J. B. Bamberg. 1991. Potato germplasm collecting expedition to Chile, 1989, and utility of the Chilean species. Amer Potato J 68:681-690.

Spooner, D. M., R .G. van den Berg, W. L. Hetterscheid and W. Brandenburg. 2003. Plant nomenclature and taxonomy: An horticultural and agronomic perspective. In: J. Janick (ed.), Hortic Rev 28:1-60.

Spooner, D. M., R. G. van den Berg, and J. T. Miller. 2001. Species and series boundaries of *Solanum* series Longipedicellata (Solanaceae) and phenetically similar species in ser. Demissa and ser. Tuberosa: Implications for a practical taxonomy of sect. Petota. Amer J Bot 88:113-130.

Spooner, D.M., D. S. Douches, and M. A. Contreras. 1992. Allozyme variation within Solanum sect. Petota, ser. Etuberosa (Solanaceae). Amer J Bot 79: 467-471.

Stevenson, R. W., R. Loria, G. D. Franc and D. P. Weingartner (eds.). 2001. Compendium of potato diseases, 2nd edition. Amer Phytopath Soc Press, 106 pp.

Swanson, T. M. 1997. Global action for biodiversity: An international framework for implementing the convention on biological diversity. Earthscan Publications. London, 191 pp.

Thrupp, L. A. 1998. Cultivating diversity: Agrobiodiversity and food security. World Resources Institute, Washington, DC, 80 pp.

Tisdell, C. A. 1999. Biodiversity, conservation, and sustainable development: Principles and practices with Asian examples. Edward Elgar, Northamptom, MA, 263 pp.

Waycott, W. and S. B. Fort. 1994. Differentiation of nearly identical germplasm accessions by a combination of molecular and morphological analyses. Genome 37:577-583.

Weatherford, J. 1988. Indian givers. Fawcett Columbine, New York, 272 pp.

Weissinger, A. K. 1990. Technologies for germplasm conservation *ex situ*. In: G. H. Orians et al. (eds.). The preservation and valuation of biological resources. University of Washington Press, p 3-31.

Widrlechner, M. P., L. D. Knerr, J. E. Staub and K. Reitsma. 1989. Biochemical evaluation of germplasm regeneration methods for cucumber, *Cucumis sativa* L. FAO/IBPGR Plant Genet Res Newslr 88/89:1-4.

Williams, J. T. 1984. A decade of crop genetic resources research. In: J. H. W. Holden and J. T. Williams (eds.). Crop genetic resources: Conservation and evaluation. Allen and Unwin, London, 296 pp.

Williams, T. J. 1989. Plant germplasm preservation: A global perspective. In: L. Knutson and A. Stoner (eds.), Biotic diversity and germplasm preservation, global imperatives. Kluwer, Dordrecht, pp. 81-96.

Wilson, Alan. 1993. The story of the potato. Balding & Mansell, London, 120 pp.

Wilson, E. O. 1984. Biophilia. Harvard University Press. Cambridge, MA, 157 pp.

Witcombe, J. R. 1999. Does plant breeding lead to loss of genetic diversity? In: D. Wood and J.M. Lenné (eds.), Agrobiodiversity: Characterization, utilization, and management, CAB International, Wallingford, UK, pp. 245-272.

Wood, D. and J. M. Lenné. 1999. Why agrobiodiversity? In: D. Wood and J.M. Lenné (eds.). Agrobiodiversity: Characterization, utilization, and management. CAB International. Wallingford, UK, pp. 1 5.

Yonezawa, K. and T. Ichihashi. 1989. Sampling size for collecting germplasm from natural populations in view of the genotypic multiplicity of seed embryos borne in a single plant. Euphytica 41:91-97.

Zoro, I., A. Maquet, J. Degreef, B. Wathelet and J. P. Baudoin. 1998. Sample size for collecting seeds in germplasm conservation: The case of the Lima bean (*Phaseolus unatus* L.). Theor Appl Genet 97:187-194.

APPENDIX 1
INFORMATION SOURCES

A. Potato Genebanks and Related Sites

CGIAR System-wide Information Network for Genetic Resources
http://www.singer.cgiar.org/
CIP, International Potato Center collection, Lima, Peru
http://www.cipotato.org/index.asp?bhcp=1
Commonwealth Potato Collection, Dundee, Scotland, UK
http://www.scri.sari.ac.uk/cpc/
CORPOICA Potato Collection, Santafe de Bogota, D.C., Colombia
http://www.corpoica.org.co
European Union joint potato project RESGEN (initiated 1996)
http://www.plant.dlo.nl/about/Biodiversity/Cgn/research/eupotato/
FAO Plant Genetic Resources for Food and Agriculture
http://www.fao.org/ag/agp/agps/PGRFA/Home.htm (including "State of the World's Plant Genetic Resources - 1997")
Dutch-German Potato collection, Wageningen, Netherlands (prior to 1995 in Braunschweig, Germany)
http://www.plant.wageningen-ur.nl/cgn/potato/
INTA-Balcarce Potato collection, Balcarce, Argentina
http://www.inta.gov.ar/crbsass/balcarce
IPK Genebank Gatersleben North Branch, Gross Lüsewitz, Germany
http://www.ipk-gatersleben.de/en/02/02/06/
Potato Collection of the University of Cusco Dr. Ramiro Ortega, CERRGETYR, University of Cusco, P.O. Box 295, Cusco, Peru.
PROINPA potato collection, Cochabamba, Bolivia
http://www.condesan.org/socios/proinpa/proinpa.htm
Search WIEWS-FAO database for ex-situ plant genetic resources collections Home page:
http://www.fao.org/ag/AGP/AGPS/default.htm
The Inter-genebank Potato Database
(http://www.potgenebank.org).
US Potato Genebank, Sturgeon Bay, Wisconsin, USA
http://www.ars-grin.gov/ars/MidWest/NR6/
USA National Plant Germplasm System
http://www.ars-grin.gov/npgs/
Vavilov Institute potato collection, St. Petersburg, Russia
http://www.vir.nw.ru/

B. Information about Named *S. tuberosum* Cultivars

Cultivar database with pedigree and traits
http://www.europotato.org/

Hamester, W. and U. Hils. 1998. World catalogue of potato varieties, 1999. World Potato Congress. Buchedition Agrimedia GmbH, Bergen, Germany. 208 pp. (See also 2003 updated version)

Kehoe, H. W. 1986. Inventory of potato variety collections in EC countries. A. A. Balkema, Rotterdam. 129 pp.

Potato Information Exchange (http://www.css.orst.edu/potatoes) links to many variety description web sites.

Stegemann, H. and V. Loeschcke. 1976. Index of European potato varieties. Paul Parey, Berlin. 213 pp.

Search WIEWS-FAO database for ex-situ plant genetic resources collections Home page: http://www.fao.org/ag/AGP/AGPS/default.htm

C. General Germplasm Conservation Information

Ellis, R. H., T. D. Hong and E. H. Roberts. 1985. Handbook of Seed Technology for Genebanks No. 2. Vol. I: Principles and Methodology. IBPGR, Rome, 210 pp.

Engels, J. M. M. 2002. Genebank Management: An essential activity to link conservation and plant breeding. Plant Genetic Resources Newsletter (No. 129):17-24.

FAO Report on the State of the World's Plant Genetic Resources (*in situ* and *ex situ*)
http://www.fao.org/WAICENT/FAOINFO/AGRICULT/cgrfa/PGR.htm

Frankel, O. H. and E. Bennett (eds.). 1970. Genetic resources in plants, IBP Handbood No. 11. F. A. Davis Co., Philadelphia, PA. 554 pp.

Fuccillo, D., L. Sears and P. Stapleton (eds). 1997. Biodiversity in Trust: Conservation and use of plant genetic resources in CGIAR Centres. Cambridge University Press, Cambridge, 371 pp.

Hawkes, J. G. (ed). 1991. Genetic Conservation of World Crop Plants. Academic Press, London. 87 pp.

Hawkes, J. G., N. Maxted and B.V. Ford-Lloyd (eds.). 2000. The *ex situ* Conservation of Plant Genetic Resources. Kluwer Academic Publishers, Boston, 250 pp.

Holden, J.H.W. and J. T. Williams. 1984. Crop Genetic Resources: Conservation and Evaluation. Allen & Unwin, London, 296 pp.

Huaman Z 1998. Collection, maintenance and evaluation of potato genetic resources. Plant Var Seeds 11 (1): 29-38.

IPGRI Home page: http://www.ipgri.cgiar.org/

Knutson, L. and A. Stoner (eds.). 1989. Biotic Diversity and Germplasm Preservation: Global Imperatives. Kluwer, Dordrecht, pp. 81-96.

Orians, G. H., et al. (eds.). 1990. The Preservation and Valuation of Biological Resources, University of Washington Press, 301 pp.

Plucknett, et al. (eds). 1987. Gene Banks and the World's Food. University Press, Princeton, 247 pp.

Querol, D. 1993. Genetic Resources: A Practical Guide to their Conservation. Zed Books, Third World Network., London, 252 pp.

Reid, W. V. and K. R. Miller. 1989. Keeping Options Alive : The Scientific Basis For Conserving Biodiversity. World Resources Institute, Washington, DC, 128 pp.

Swanson, T. M. 1997. Global Action for Biodiversity: An international framework for implementing the convention on biological diversity. Earthscan Publications, London, 191 pp.

Thrupp, L. A. 1998. Cultivating diversity: Agrobiodiversity and food security. World Resources Institute, Washington, DC, 80 pp.

Wood, D. and J.M. Lenné (eds.). 1999. Agrobiodiversity: Characterization, utilization, and management. CAB International, Wallingford, UK, 490 pp.

2

On-farm Profitability and Prospects for True Potato Seed (TPS)

A. CHILVER[1], T.S. WALKER[2], V. KHATANA[3], H. FANO[4], R. SUHERMAN[5], and A. RIZK[6]

[1]*DFID, c/o FCO, Kampala, Uganda*

[2]*International Potato Center (CIP), Lima, Peru*

[3]*STEM, New Delhi, India*

[4]*Proyecto INCAGRO, Ministry of Agriculture, Lima, Peru*

[5]*Lembang Horticultural Research Institute, Lembang, Indonesia*

[6]*Agricultural Economic Research Institute, Charbia Governorate, Egypt*

INTRODUCTION

Poor seed quality is often cited as the most important factor limiting potato productivity in developing countries (Rasco 1994). Improving tuber seed quality and availability is institutionally complex in the tropics and subtropics. Progress of public sector (government-financed) seed programs has usually been slow and disappointing even in countries where sites for seed multiplication are agroecologically attractive (Crissman 1987). True potato seed (TPS) is a captivating technological alternative that offers farmers an option to overcome the aforesaid weaknesses of clonally propagated tubers as a source of planting materials. Side-by-side comparisons of a jar of TPS with twenty 100 kg bags of clonal planting material—the amounts needed to plant one hectare (ha)—are visually impressive evidence of the potential of the technology.*

Although the investment in agricultural research and extension of TPS technologies is small compared with the amount spent on clonal seed technologies and selection systems, sufficient experience has accumulated, largely over the past 20 years, to review achievements, shortcomings, and future prospects (Almekinders et al. 1996; Simmonds 1997). "At

Corresponding author: T.S. Walker, e-mail: walkerts@msu.edu; present US mailing address: Michigan State University, Department of Agriculture Economics, 207 Agriculture Hall East Lansang, MI 48824-1039, USA

*A jar of TPS as planting material is equivalent to 2,000 kgs of clonal planting material; i.e., a common seedling rate/ha for potatoes.

present, there are few potato-producing countries in the developing world where TPS has not been tried, or suggested as a means to alleviate seed problems" (p. 290, Almekinders et al. 1996). One of the deficiencies in this experience has been the lack of documentation of on-farm performance with farmer management. For example, Simmonds (1997) in his excellent review of TPS posed the key economic issue: "Can clean (-ish) seedlings deliver better profits than infected clones?" (p. 202). In reviewing the evidence, he did not definitively answer this question and noted that "there is still, it seems, plenty of room for good costing of practical, real life rather than theoretical production systems" (pp. 202-203).

This large gap in the literature is addressed here by reports of the results of on-farm research in Egypt, India, Indonesia, and Peru conducted in the mid-1900s. The research design is the same in the first three countries where farmers are experimenting with TPS technologies in the initial stages of acceptance or rejection. The Peruvian results are based on farmer-managed, on-farm trials.

The focus is on variables hypothesized to influence the comparative profitability of TPS technologies in farmer circumstances. Areas for improvement in the design of TPS technologies are identified and a rule of thumb to guide the targeting of these technologies was generated from the empirical evidence.

TPS TECHNOLOGIES

TPS can be grown by farmers, in a number of ways. TPS (botanical seed) can be sown directly in the field or in a nursery bed and the seedlings in the nursery bed may be left in place or transplanted to the field (Fig. 2.1). The output of TPS is called seedling tubers or tuberlets as distinct from seed tubers or tuber seed, which is used to propagate a clone. Seedling tubers can be consumed by the farm household, sold as ware, or saved for seed (TPS) for the next growing season. Most on-farm experience with TPS comes in the form of planting and replanting of seedling tubers, and this is the technology focused on here. Substituting seedling tubers for tuber seed is a relatively small step for most farmers who cultivate seedling tubers without making significant changes in their production practices. Farmers replace a clone with a family of individuals, but the mode of propagation is still vegetative. Seedling tubers may be supplied by governmental organizations (India), nongovernmental organizations (India and Peru), the private sector (Egypt), or raised from TPS by farmers themselves (Indonesia and India). Seedling tubers, similar to tuber seed, may be replanted over several seasons, but the on-farm experience reported in this paper is concentrated on seedling tubers in their first or second use as a source of planting material.

The economics of raising TPS via the transplant route in Fig. 2.1 are also appraised. With this option, potatoes are cultivated from the tiny

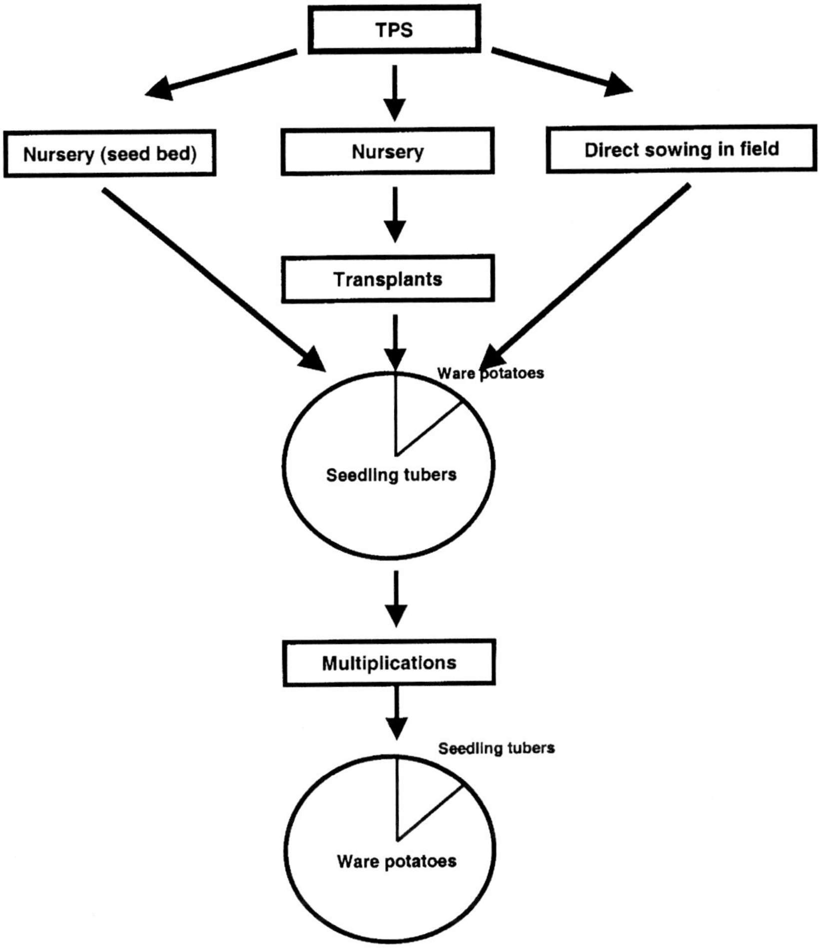

Fig. 2.1 Potential TPS-related technologies. (Adapted from Almekinders et al., 1996).

botanical seed similar to tomatoes or lettuce in a garden or small plot setting. In its most commercialized form, seed is repurchased every year as tuber output is not saved for replanting in subsequent years. Sufficient data are available in only two of the nine multilocational sites to make meaningful statements about the comparative on-farm profitability of transplants; therefore, the seedling tuber stages of Fig. 2.1 receive most of our attention. This focus is not unduly restrictive because the demand for TPS is derived from the demand for seedling tubers.

Direct sowing is another TPS technology (Fig. 2.1). True seed is usually pelleted and sown directly into the field with a mechanical planter roughly in the same manner that cereals are planted in developed countries. Seed is repurchased and sown mechanically each year. This option imposes higher specifications on technology design and implies major changes in small farmer practice. With direct seedling, elimination of the seedling tuber stage for further propagation places TPS in the genre of newly developed farming systems that imply a radical departure from existing practice (Merrill-Sands 1986). Examples include ICRISAT's (International Crop Research Institute for the Semi-Arid Tropics) Vertisol Technology Options and IITA's (International Institute for Tropical Agriculture) hedgerow intercropping of annuals and leguminous perennials. Although these new farming systems score high marks on environmental grounds, diffusion has been negligible. Direct sowing of TPS may eventually become a reality, but it is not tailored to the conditions of labor-intensive potato production in a small-field setting characteristic of developing-country agriculture. Because of its potential for displacing labor, the direct sowing mode of producing potatoes with TPS has not been emphasized at CIP and is not addressed in this chapter.

METHODS AND EXPECTATIONS

Partial budgeting is used for economic evaluation. With partial budgeting, the focus is on relative profitability. One is not interested in whether or not the farmer should grow potatoes, but rather how well one technology compares with another in an economic dimension. Several cost components, such as land preparation, are the same irrespective of whether the farmer uses tuber seed or seedling tubers. The emphasis is squarely on those costs that change and on the attributes of output that make one technology more or less profitable than another. This focus prompts collection of data on the key input and output variables which are expected to vary in switching from one technology to another.

The partial budgeting approach is straightforward. Net benefits and total varying costs of the two technologies are compared. Usually, the prospective technology implies an additional investment by the farmer; therefore, total varying costs of the prospective technology are higher than total varying costs of farmer technology. For this conventional case, the prospects for farmer acceptance of the prospective technologies are bright if the marginal rate of return on investment, i.e., the change in net benefits divided by the change in total varying costs expressed as a percent, exceeds 100% (CIMMYT 1988). This high 100% threshold level for economic gain to translate into adoption applies to developing country agriculture where capital is scarce. But this decision rule is not hard and fast. For example, the earlier version of the CIMMYT manual (Perrin et al. 1979) used 40% as a cut-off criterion for establishing economic performance synonymous with good prospects for technology acceptance.

Because TPS-related technologies are widely touted as cost saving, they may not imply additional investment; hence, the marginal rate of return criterion would not apply. In our case, the change in net benefits would have to be positive for the farmer to switch to TPS, but how large the level has to be is arbitrary. We set the threshold level of profitability at $100 per hectare which approaches the lower limit of per hectare benefits of nine CIP-related success stories of technological change documented in Walker and Crissman (1996). For cereals, an improvement of $100 ha^{-1} is a relatively large amount, but for a high value crop like potatoes $100 ha^{-1} is equivalent to less than the value of one ton or (5-15%) of production in typical developing country conditions.

The threshold value also depends on the complexity of the technology and the amount of effort required for adoption. For example, purchasing seedling tubers from a commercial source in lieu of planting the saved tuber seed would not seem to imply a significant change in farming practice. Thus the $100 threshold for adoption would seem to be appropriate. On the other hand, planting TPS and growing transplants is tantamount to a large change in production practice for most farmers (with the exception of specialized vegetable growers) and a threshold level higher than $100 ha^{-1} would appear to be needed to induce adoption. Therefore, a higher threshold of around $200 ha^{-1} may be required to ensure the competitiveness of transplant technologies. Intuitively, these adoption thresholds are equivalent to 5-10% of the value of production with the farmer's technology.

Net benefit and components that change in switching from tuber seed (CP) to TPS technologies are defined below:

(1) Net benefit = ((price × yield) – varying costs)
(2) Net benefit (TPS) – net benefit (CP) > $100 ha^{-1}
(3) Price = f(DTS, PPG, PS)
(4) Varying costs = Seed cost + fungicide cost + other varying costs
(5) Seed cost = seed price × seed quantity

Several cost-and output-related variables are identified in equations (1) to (5). At a higher level of aggregation, we have total yield, per unit harvest value, seed cost, fungicide cost, and other varying costs for TPS and clonal propagation technologies. Per unit harvest value is mainly a function of the graded distribution of tuber size (DTS). Price per grade (PPG) and price seasonality (PS) at harvest can also influence the per unit harvest value. Seed cost is the product of the price of seed and the seed rate. Hence, at a lower level of aggregation, we have eight key input and output determinants of relative profitability.

Our interpretation of conventional wisdom on how well TPS technologies will perform vis-à-vis tuber seed technologies is presented in Table 2.1. Starting with yield, comparative productivity depends on the seed quality of the two competing technologies, particularly the cleanliness and physiological correctness of the farmer's clonally propagated seed. Yield advantages or disadvantages should be location specific. But

Table 2.1 *Expectations on variables conditioning comparative profitability*

Variable	*Expectation*[a]
Yield	The same or positive
Per unit harvest value	Strongly negative
—Per grade	Negative
—Distribution of tuber size	Negative
—Price seasonality	Negative
Seed costs	Strongly positive
—Seed rates	Positive
—Seed prices	Positive
Fungicide costs	Positive
Other variable costs	The same

[a]Positive denotes an advantage of TPS technology compared with clonal propagation.

there is a presumption that yields of clonal material are low; therefore, seedling tubers should display a productivity advantage. Simmonds (1997) takes stock of the literature:

> It would be rash indeed to predict yield; authors seem commonly to agree that current yields are 10 t/ha or less of crops raised from usually infected clones but that competent seedling production by way of tuberlets of good hybrids might be in the order of 20-30 t/ha. But experimental results and agricultural achievements are very different things (p. 204).

All three components of our per-unit-harvest-value variable point to higher price expectations for output grown from clonally propagated seed than for output produced from seedling tubers. Firstly, botanical seed generates relatively large numbers of small-size tubers. When these are planted in the next generation as seedling tubers, they are usually smaller than the farmers' seed tubers; thus, the tendency exists to generate relatively more tubers in the smaller-size categories. Planting smaller-sized seedling tubers may also increase selection pressure for progenies with high tuber number and small tuber size (E. Chujoy, pers. commu. 1998; van Hest 1994). Secondly, TPS output is comprised of many individuals from the same family which are less uniform than clonal seed of the same individual. Lack of uniformity can be associated with a price penalty for tubers of the same size and shape harvested at the same time. Indeed, when people discuss TPS, lack of uniformity is usually foremost in their minds as a disadvantage of the technology (Chilver 1997). Heterogeneity is cited as a reason why potato scientists in India stopped working on TPS in 1949 (Almekinders et al. 1996). Lack of uniformity is expected to be more problematic when the processing of high-quality products looms large as a destination for production.

Lastly, TPS technologies may reach maturity later than earlier maturing clonally propagated material. Later harvesting often translates into a price discount especially in regions where potatoes are not cultivated throughout the year.

Although price discounts are often perceived as a minus for TPS technologies, cost savings are viewed as a plus. Indeed, lower seed cost and increased availability of cleaner planting material is the raison d'etre for TPS. Planting material to sow the same area in TPS is measured in grams per hectare and seedling tubers may weigh several hundred pounds less than clonally propagated material. Farmers may be able to produce and store seedling tubers at a cost that is substantially lower than the production and storage of clonally propagated material or the purchase of such material in the market. For seedling-tuber technologies, less and lower priced tuber material potentially translates into a large savings in the cost of planting materials. For TPS transplant technologies, a 20,000-fold reduction in the quantity of material to be planted more than compensates for a 150-fold increase in per unit cost. Summing up the literature, Simmonds (1997) reported:

> Several calculations suggest that seedling production should indeed be cheap . . . On balance, calculations seem to suggest that seed might cost as little as 20-50 percent of tubers for planting but there is the difficult question of the quality of the seed (p. 202).

Fungicide costs are also expected to be less with TPS than with clonally propagated material because a heterogeneous family of individuals carrying different segregating genes (both major and minor) for resistance to late blight should be able to withstand pressure better than a uniform clone (Upadhya and Thakur 1998). Most clones grown by farmers in both developed and developing countries are susceptible to late blight, therefore an advantage should accrue to TPS technologies when late blight infestation is high. In regions where chemical control is ineffective, this advantage should manifest itself in heavier yields. In regions where chemical control is effective, this advantage should translate into cost savings in the use of fungicide.

Other varying costs are included in Table 2.1 to incorporate information on systematic changes in farmer production practices when they plant TPS technologies vis-à-vis clonally propagated material. A priori, there is no reason to think that farmers will consistently apply more inputs such as fertilizer or manure to potatoes grown from TPS or seedling tubers than to potatoes raised from clonally propagated tubers.

Summing up, the trade-off between cost and output price is the main message emerging from Table 2.1. Where yields are not significantly different between the two technologies, the important economic question is: Can lower seed and fungicide costs compensate for an expected decline in output price?

SAMPLING AND DATA COLLECTION

Based on farmer experience with TPS technologies, sites were selected in Egypt, India, and Indonesia. An inventory was taken of all farmers planning to cultivate TPS transplants (TP) and seedling tubers in the upcoming cropping season. TPS farmers were stratified by wealth group based on asset holdings, mainly land, and a random stratified sample was taken.

The unit of observation was the field. One field planted using a TPS technology was monitored during the growing season. The point of reference for comparative inference was a neighboring or nearby field planted using conventional seed tuber technology. For those TPS farmers without a tuber-seed field available for a with-and-without comparison, a nearby field of another farmer was chosen.

An integrated production questionnaire combining survey responses from the farmer and field measurements was canvassed five times during the growing season. Field visits were made at planting, 45 days after planting, harvesting (twice), and at post-harvest marketing. Seed rate, yield, and tuber size were measured. Local market data were collected weekly on seed and ware prices.

Data were collected across consecutive cropping seasons in India and Indonesia. In India the survey was conducted between March 1995 and May 1996. In Indonesia, the survey included both the wet and dry seasons and was implemented between June 1994 and April 1995. In Egypt, the farmer survey was carried out during the 1994 autumn season.

The format of data collection in Peru was in the spirit but not the letter of data collected in the other three countries. Data in Peru refer to more structured on-farm trials with two treatments in fields of approximately one hectare in size. Fields in 11 village locales were monitored during the main growing season in the 1997-1998 cropping year.

Comparative evaluation was carried out in nine sites for TPS seedling tubers (ST) and in three agroecologies for TPS transplants (TP). Sites are represented by their agroecologies in Table 2.2. Almost all sites contained multiple locations comprising several or more villages.

Table 2.2 *Description of recent on-farm research comparing TPS technologies with clonal seed tubers*

Country	*Agroecology*	*Fields*			
		Farmers	*CV (Cultivar-clonal Seed)*	*ST (Seedling tubers)*	*TP (Transplants)*
India	Northeastern plains	119	99	99	34
India	Northeastern hills	24	20	15	8
India	Northcentral plains	22	17	17	0
India	Western arid zone	11	10	13	3

(*Contd.*)

(Contd.)

India	Deccan Plateau	14	11	14	0
Egypt	Desert	7	3	13	0
Egypt	Delta	40	26	42	0
Indonesia	Highlands	67	32	45	0
Peru	Highlands	11*	11	11	0
Total plots			229	269	45

*Groups comprised of 197 farmers.

Although a minimum of 15 to 25 plots per site was the goal, the number of TPS fields ranged from 13 to 99. Experience with and availability of TPS in the region significantly influenced sample size.

RESULTS

The economic assessment in Table 2.3 for India and Table 2.4 for Egypt, Indonesia, and Peru supports the view that TPS has mixed prospects.[1] Of the nine generalized locations, TPS seedling tuber technology was spectacularly profitable in Chacas, Peru, robustly profitable in the northeastern hills of India and the Nite Delta of Egypt, and marginally profitable in the northeastern plains of India (Fig. 2.2). In the other five sites, farmers lost more than US\$100 ha^{-1} (relative to clonally propagated material) by investing in seedling tubers.

The comparative profitability of TPS transplants hinges on whether or not the output is destined for ware or seed use. Production from TPS transplants was only competitive if the smaller size output was retained or sold for seed (see the hollow point estimates in Fig. 2.2). Destining all output for ware consumption in the same year was a decidedly uneconomical proposition (the black point estimates in Fig. 2.2). Demand for TPS as transplants should be strong in India's northeastern hills because of the competitiveness of seedling tubers which generated about \$500 more per hectare than clonally propagated material. The fate of transplants on the northeastern plains is decidedly more problematic because the economic advantage of seedling tubers was only about \$60 per hectare. At this lower level of profitability, it is unlikely that the derived demand for transplants could be sustained.

[1]In Egypt and Indonesia, more than one TPS field was often sampled. Usually, the comparative "without" plot belonged to the same farmer, but focusing the analysis only on pairwise differences would mean that the sample size would be reduced by about 30%. A pairwise analysis generates most of the same findings as averaging across all observations (Chilver, 1997). Results based on the mean values for each technology are presented below.

Table 2.3 *Partial budgets comparing TPS seedling tuber and transplant technologies (ST & TP) to clonal seed tuber technology (CV) for India.*

Item		India											
		Northeastern plains			NEH[a]			NCP[b]		WAZ[c]		DP[d]	
	Units	CV	ST	TP	CV	ST	TP	CV	ST	CV	ST	CV	ST
Seed quantity	t ha^{-1}	2.00	1.20	0.34	2.30	1.60	0.21	3.10	3.00	2.90	1.50	0.91	0.84
Seed price	US$ t^{-1}	219	219	625.0	344	250	625.0	119	119	275	275	216	188
Seed cost	US$ ha^{-1}	438	263	210.0	791	400	129.4	368	356	798	413	196	158
Manure cost	US$ ha^{-1}	134	128	181.3	125	131	137.5	75	81	228	338	144	175
Fertilizer cost	US$ ha^{-1}	288	272	200.0	141	134	134.4	75	84	434	306	141	172
Spray cost	US$ ha^{-1}	16	13	20.0	29	27	21.9	3	2	9	10	11	11
Fungicide cost	US$ ha^{-1}	30	27	39.7	47	37	31.3	4	3	16	21	13	16
Insecticide cost	US$ ha^{-1}	7	4	4.1	21	16	20.6	1	1	13	17	12	13
Total varying costs	US$ ha^{-1}	912	707	767.5	1,153	745	587.5	525	525	1,498	1,104	517	545
Yield	t ha^{-1}	27.7	26.8	19.8	27.5	28.3	17.5	30.9	33.3	45.2	43.8	12.6	12.1
Harvest grade ABCD[e]	%	11:54:26:8	9:46:35:9	0:10:85:4	8:68:23:0	4:72:23:0	0:56:38:6	47:25:23:4	51:13:34:1	28:66:2:4	18:49:24:9	0:84:7:8	0:24:55:20
Harvest value	US$ t^{-1}	78	75	62.7	105	105	98.2	72	63	113	104	119	103
Value Production	US$ ha^{-1}	2,157	2,012	1,241.8	2,892	2,968	1,719.1	2,232	2,093	5,093	4,567	1,504	1,245
Net revenue	US$ ha^{-1}	1,245	1,305	474.3	1,739	2,223	1,131.6	1,707	1,565	3,596	3,463	987	700
Net economic benefit	US$ ha^{-1}		60	–770.9 222.5[f]		484	–607.7 767.6[f]		–142		–133		–287

Source: Constructed from Chilver (1997) and Fano and Tello (1998).

a = Northeastern hills

b = North-central plains

c = Western arid zone

d = Deccan plateau

e = Percentage of harvested tubers in grade A, B, C or D (percent damage not included). CV = cultivar; ST = seedling tuber; TP = transplants

f = Medium, small, and very small tubers from transplant harvest valued as seed (1.8 times ware prices).

Table 2.4 *Partial budgets comparing TPS seedling tuber technology (ST) to clonal seed tuber technology (CV) for Egypt, Indonesia and Peru.*

Item	*Units*	*Delta-Egypt*		*NRA-Egypt*		*Indonesia*		*Peru*	
		CV	*St*	*CV*	*ST*	*CV*	*ST*	*CV*	*ST*
Seed quantity	t ha^{-1}	3.54	1.74	3.64	1.8	1.29	1.15	2.38	1.1
Seed price	US$ t^{-1}	377	599	377	599	933	800	342	418
Speed cost	US$ ha^{-1}	1,335	1,042	1,368	1,078	1,209	916	813	460
Manure cost	US$ ha^{-1}	129.0	129.0	245.5	245.5	0	0	183	109
Fertilizer cost	US$ ha^{-1}	251.0	289.0	245.0	245.0	164	218	294	226
Spray cost	US$ ha^{-1}	16.7	16.7	20.9	20.9	142	102		
Fungicide cost	US$ ha^{-1}	46.7	46.7	79.6	79.6	333	258	106	124
Insecticide cost	US$ ha^{-1}	[a]	[a]	[a]	[a]	[a]	[a]	85	65
Total varying costs	US$ ha^{-1}	2,246	1,968	2,365	2,077	2,196[b]	1,871[b]	1,893	1,463
Yield	t ha^{-1}	27.5	28.7	34.6	32.4	18	11.8	24.1	35.5
Harvest grade ABCD[c]	%	62:26:10:2	60:28:11:1	83:14:4:0	74:19:8:0	59:22:15:4	54:23:17:6	[d]	[d]
Harvest value	US$ t^{-1}	213	213	228	222	369	356	201	201
Value Production	US$ ha^{-1}	5,865	6,127	7,827	7,167	6,671	4,204	4,857	7,154
Net revenue	US$ ha^{-1}	3,619	4,158	5,452	5,090	4,476	2,333	2,964	5,691
Net economic benefit	US$ ha^{-1}		539		(372)		(2,142)		2,727

Source: constructed from Chilver (1997) and Fano and Tello (1998).

a = Disaggregate costs are not presented, but are included in total varying costs.

b = Includes a land cost charged to the longer duration TPS progenies.

c = Percentage of harvested tubers in grade A, B, C, or D (percent damage not included). CV = cultivar; ST = seedling tuber.

d = Data were not taken on the distribution of output by quality grade.

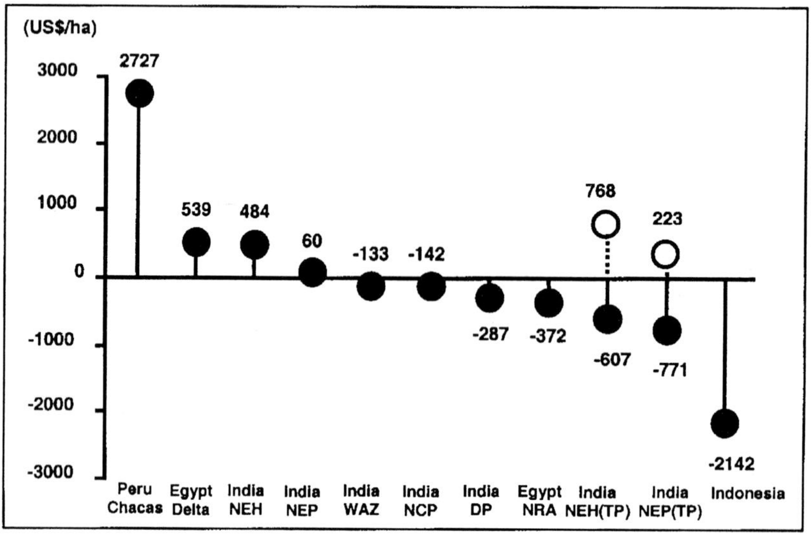

Fig. 2.2 Difference in net benefit (US$ ha^{-1}) of TPS technologies from clonal tuber seed.

Source: Constructed from Chilver (1977), and Fano and Tello (1998).

With a few notable exceptions, the conventional wisdom expressed in Table 2.1 was confirmed. The symmetrical pattern of point observations in Fig. 2.3 indicates that productivity behavior on aggregate was pretty much the same. Average seedling tuber yields were markedly higher in the Peru site, significantly lower in the highlands of Indonesia, and not significantly different from average yields of clonally propagated material in the other seven agroecologies. The Indonesian results show higher rates of yield deterioration over time for seedling tubers compared to the dominant clone Granola, which demonstrated a slower rate of productivity decay. Farmers in Indonesia commonly plant Granola for six to seven cycles before they renew their seed.

The large yield advantage of the progeny Chacasina in Peru shown in Fig. 2.3 was attributed to loss of stand in clonally propagated material in a dry year. Chacasina matures about six weeks earlier than Yungay, the clonal check. Per plant yields were similar, but Yungay had a significantly lower stand establishment than Chacasina which greatly favored the latter over the former. But the superiority of Chacasina over Yungay was not nearly as surprising as the high productivity level of 24 t in a dry year in a remote poverty-stricken region where potato is cultivated only for subsistence consumption and where seed quality of clonally propagated material is reportedly a severe constraint to production. In the following year, these yield differences disappeared in a wet year in a more commercialized setting for production across 14 locations (Appendix Fig. 2.1).

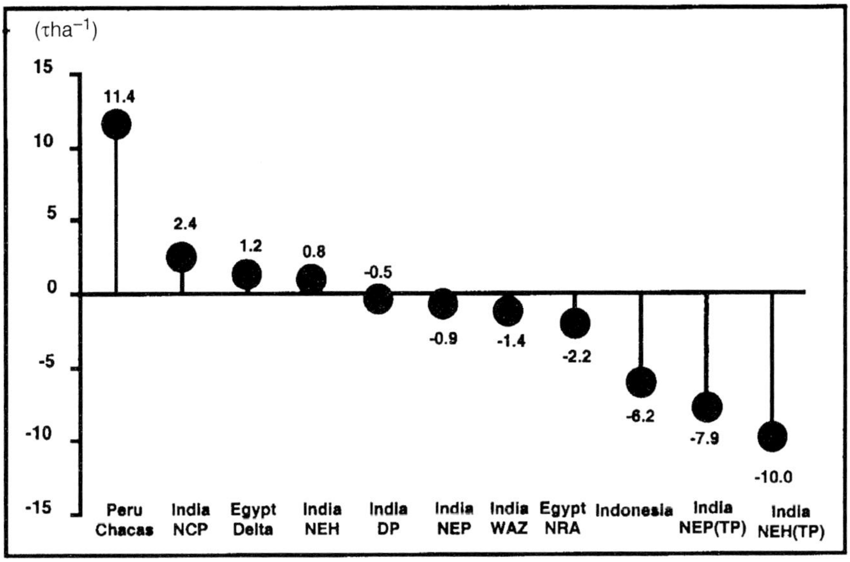

Fig. 2.3 Difference in yields (t ha^{-1}) of TPS technologies from clonal tuber seed.

Source: Constructed from Chilver (1997), and Fano and Tello (1998).

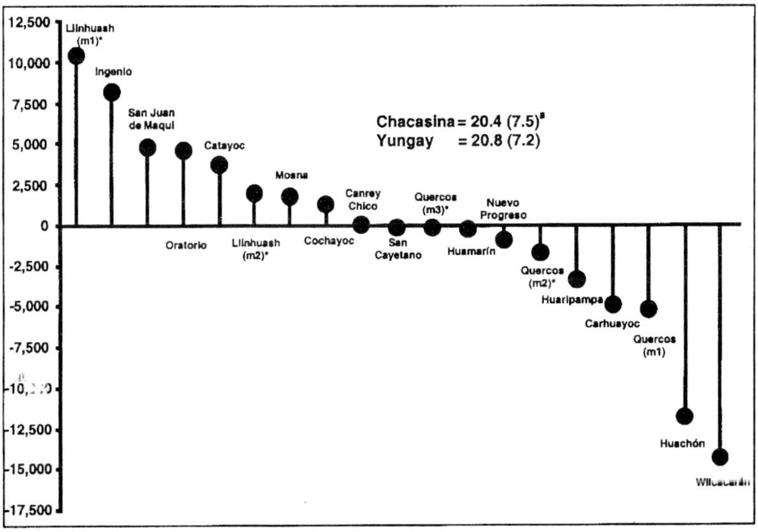

Appendix Fig. 2.1 Yield of Chacasina (ST) *vs.* Yungay (CV) in the Callejón de Huaylas in 1998.

Source: Constructed from Fano and Tello (1998). *a* = Mean (standard deviation).

TPS transplants produced fairly good yields, between 15 and 20 t per hectare in both the northeastern plains and the northeastern hills. However, farmers in the same regions were producing 25-30 t ha^{-1} with clonally propagated material. Cost savings would have to be huge or output prices substantially higher to compensate for a relative loss of 10 t of yield.

Considering Simmonds (1997) assessment of the literature on the productivity differences between seedling tubers and clonal materials, the results strongly confirm that seedling tubers of good hybrids are high yielding at levels exceeding 20 t ha^{-1}, but the same is true of competing clones. The estimated high yields of competing clones is the most surprising finding in our study. None of the average yields of clonal material fell below 10 t ha^{-1}. One can choose among three competing hypotheses to explain this simple but amazing fact: (1) seedling tubers were poorly targeted to regions where seed quality was not a major problem, (2) yield sampling grossly overestimated per hectare productivity, or (3) scientists and research administrators have severely overestimated the size of the seed-quality constraint.

As expected, the savings in seed costs had a marked advantage for the seedling tuber technology. The average savings in seed cost ranged from about US\$10 ha^{-1} on the north-central plains of India to about US\$400 ha^{-1} in the northeastern hills (Fig. 2.4). The difference in these two regions largely reflects differences in seed prices. Tuber seed in western

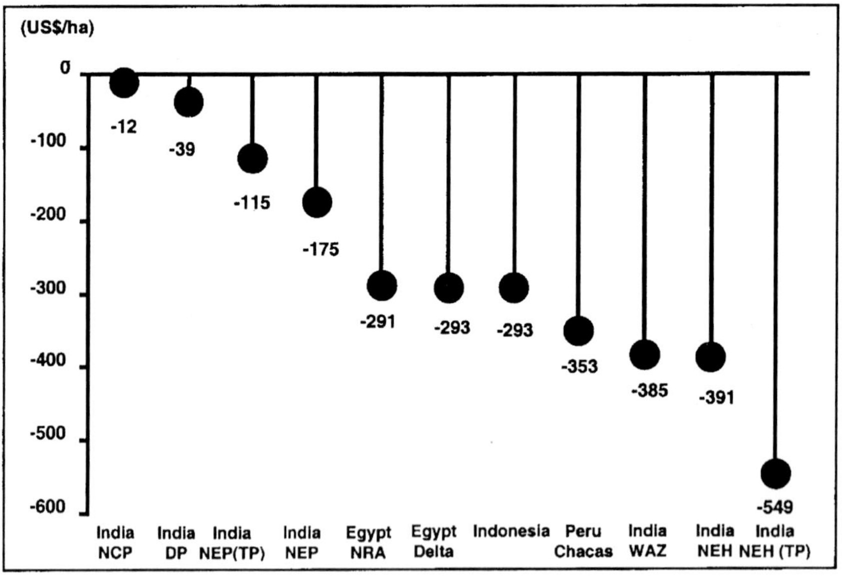

Fig. 2.4 Difference in seed cost (US\$ ha^{-1}) of TPS technologies from clonal tuber seed.

Source: Constructed from Chilver (1997), and Fano and Tello (1998).

Uttar Pradesh on the north-central plains is readily available from nearby Punjab and Haryana, the main sources of clonally propagated material in India. That material has to be trucked across the plains up into the mountains of Assam, one of the states in the northeastern hills. The remoteness of Assam and transport difficulties make good-quality tuber seed a dear commodity.

Lower seed rates contributed proportionally more to the savings in seed costs than lower seed prices. The seed rate declined across all the cited agroecologies, but lower per unit prices for seedling tubers were only notable in three settings, the northeastern hills and the Deccan Plateau of India and the highlands of Indonesia. The commercialization of seedling tubers in Egypt by large dealers led to substantially high prices per unit of material for propagation; so much so that the advantage of a 1.8 t ha^{-1} lower seed rate with seedling tubers was not fully reflected in savings in seed cost.

Summing up, the cost savings with seedling tubers did approach 50% of the cost of tuber seed in three of the nine sites. A 50% estimate appears to be realistic for the potential for cost saving at this time.

Partial budgeting results also support the hypothesis that replacing clonal tuber seed with seedling tubers is accompanied by a fall in output price (Fig. 2.5). But output price discrimination against seedling tubers was not as widespread as expected.

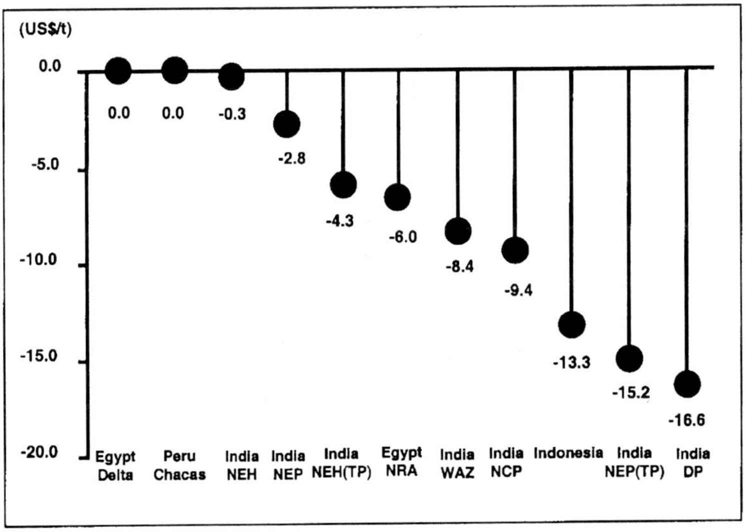

Fig. 2.5 Difference in price (US$ ha^{-1}) of TPS technologies from clonal tuber seed.

Source: Constructed from Chilver (1997) and Fano and Tello (1998)

Two of the expected considerations pointing to lower prices for the output of TPS seedlings did not materialize. Except for Peru where the TPS progeny matured earlier than the clonal check, farmers harvested the production from seedling tuber and the conventional seed technology at the same time; therefore, price differences associated with seasonality were not a factor. Only in Indonesia did the later maturity of TPS progenies result in somewhat lower (4%) prices relative to the early maturing variety Granola. Nor did we find tangible differences in prices between TPS and clonal output for production classified in the same grade. This positive finding for TPS indicates that the progenies are sufficiently homogeneous to compete in the marketplace in these countries where fresh table consumption is the dominant form of utilization. Indeed, farmers commented that the glossier offspring of seedling tubers were even slightly preferred in some agroecologies.

The cause of price disparities between TPS and clonal seed technologies centers on the potential for a smaller distribution of tuber size of the former relative to the latter. This disadvantage was most salient in the Deccan Plateau in Karnataka where the output of seedling tubers fetched about \$15 per ton less than the production of clonally propagated material. The bulk (or 85%) of the production of the dominant clone Kufri Joythi belonged to the Grade B category; 75% of the output of seedling tubers was destined for the smaller size Grades C and D categories. Farmers in Karnataka display a strong preference for fewer, but larger size, more marketable tubers. An inferior distribution of tuber size was also a disadvantage of seedling tuber technology in Gujarat in the western arid zone and in Uttar Pradesh in the north-central plains of India. Price differences of 8-12% may not seem that large, but they are equivalent to differences in net benefits of US\$300 to US\$400 ha^{-1} in these high-yielding regions.

Changes in other costs did not unduly affect the outcomes of the comparative profitability of the two types of potato propagation technologies. Mean savings in fungicide cost of US\$115 ha^{-1} were large only in Indonesia. TPS progenies were more resistant to late blight than Granola, but their longer duration increased the time for application; thus, the scope for cost savings was somewhat less than anticipated.

The cost of soil amendments did not vary much between the two propagation systems. The western arid zone in India where farmers applied relatively more fertilizer to clonal material and relatively more manure to seedling tubers was the exception. In general, other varying costs did not influence the analysis by more than US\$50 ha^{-1} (Fig. 2.6). This suggests that farmers did not modify their production practices in experimenting with seedling tuber technology.

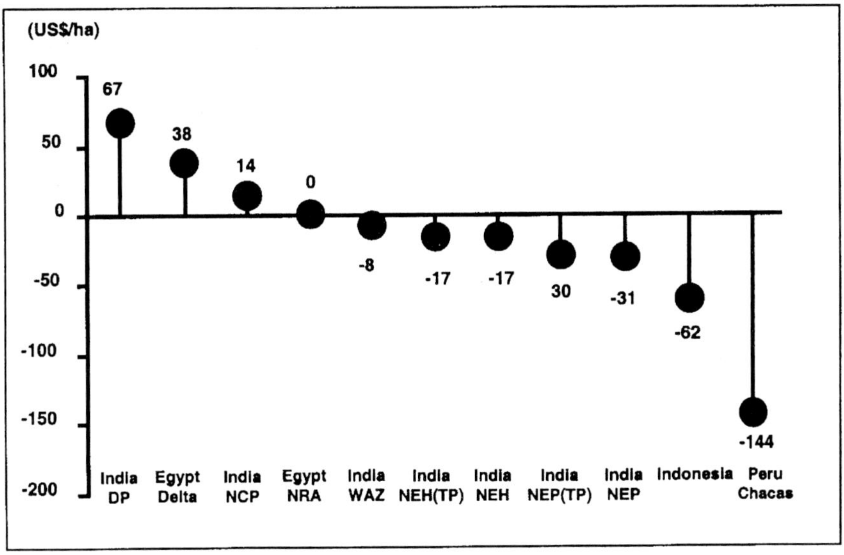

Fig. 2.6 Difference in other costs (US$ ha^{-1}) of TPS technologies from clonal tuber seed. *Source:* Constructed from Chilver (1997), Fano and Tello (1998).

SENSITIVITY ANALYSIS

A simple sensitivity analysis was conducted to illustrate the most important variables in conditioning the comparative profitability of TPS seedling tuber technologies. Three possible improvements are envisaged in Table 2.5: (1) a reduction of 10% in the cost of seedling tubers, (2) a 10% higher yield, and (3) a 10% increase in size distribution of output from small to large-size categories. Proportional changes in yield and in tuber size distribution of yield are much more effective in increasing TPS benefits than proportional changes in seed costs. In terms of hybrid TPS family selection, the level and distribution of output should receive pride of place. The scope for leveraging adoption via reduced input costs is limited. A premium is placed on the ability to generate larger size tuber distributions in both transplant and seedling tuber technologies.

Table 2.5 *Effect on TPS benefits of improvements to TPS technology*

Possible improvement to TPS technology, center is paribus	*Increase in TPS benefits (%)*		
	Indonesia	*Egypt*	*India*
10% lower seedling tuber seed costs per ha	4	39	44
10% higher seedling tuber yields per ha	20	230	333
10% more large, 10% less small tubers	12	78	154

Source: Chilver, 1997.

EMPIRICAL RULES OF THUMB

The above results suggest that TPS technologies are not presently sufficiently well developed to replace clonal propagation systems in developing countries. Although definitely not a universal solution, some places at some times would appear to be ripe for their deployment. Are there any empirical rules of thumb which would be informative about prospects for success to encourage more effective targeting of research and extension effort? In other words, can we find a good predictor of TPS profitability solely with information on the economic performance of clonal propagation systems?

The response to the above questions appears to be yes: the cost of propagation material divided by the value of production is a reasonably good predictor of TPS performance when significant yield effects between the two types of technologies are absent. This is shown in Fig. 2.7 for the seven agroecologies. Whereas yield differences in Fig. 2.3 are not substantial, the scatter of the average values in Fig. 2.7 suggests a linear association between the quotient of seed cost to value of production and change in net benefits between TPS and clonal propagation systems. The break-even point for the economic profitability of TPS technologies is

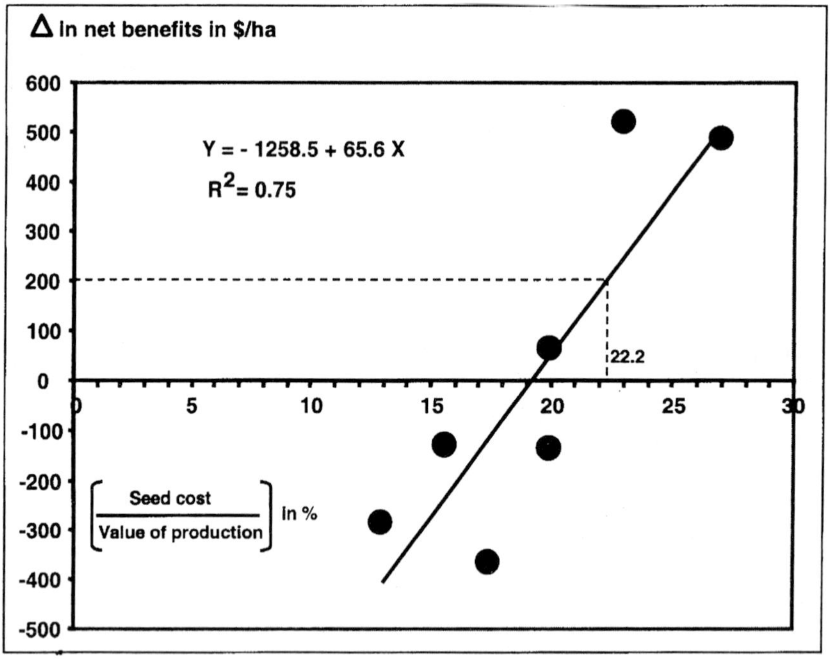

Fig. 2.7 Relationship between comparative profitability of TPS utilization technologies and seed cost as percent of value production in clonal tuber seed.

equivalent to seed cost comprising 19% of the value of clonal production. A value of 22% equates to an economic gain of \$200 ha^{-1}, which should be more than sufficient to stimulate adoption.

It takes some audacity to base a decisive rule on only seven observations, but the intuition behind the relationship depicted in Fig. 2.7 is appealing. The quotient comprises two ratios; inverse of the multiplication ratio and the seed: output price ratio. Clonal propagation systems with higher seed to output price ratios and lower multiplication ratios will generate higher values for seed costs in terms of production. Neither of these rules separately correlates as closely with economic performance as their combination in the suggested quotient.

As a point of reference, in the Red River Valley of the United States the value for this quotient is about 8-10%, indicating a very effective seed system. In the Netherlands, the value is about 12%. Higher values signal more inefficiencies in the clonal seed system. Advances in TPS technology will shift the relationship depicted in Fig. 2.7 downward; improvements in the clonal seed system is tantamount to observations sliding down the line.

Applying even this very simple empirical rule is difficult but the problems are not insurmountable. Estimates are needed on farmer yield, seed rate, price of tuber seed, and price of output at harvest. There is no substitute to investing in well-focused, albeit limited, diagnostic research in farmer fields and in local markets to obtain these estimates. The temptation to use national or FAO data on yield levels and import data on seed costs should be resisted. Yield sampling in farmer fields show that officially published estimates often underreport per hectare potato productivity by 30-100% (Pakistan-Swiss Potato Development Project 1991; Terrazas et al. 1998). Imported propagation material is usually multiplied several times before it becomes cost effective to produce potatoes destined for the fresh market.

The Indonesian highlands is an apt example where the use of the wrong data can give spurious results on the prospects for TPS-related technologies. Combining national yield levels at 15 tha^{-1} with an imported seed price at \$2,000 per hectare gives an estimate of 46% for our empirical rule of thumb. The farm-level data in Tables 2.3 and 2.4 show a "true" estimate of 18%, which is less than our 20% threshold value, indicating that TPS is not nearly as promising a venture as published data would suggest.

DISTRIBUTION OF BENEFITS

TPS has always been advocated as a technology favoring the rural poor, particularly poorer producers and landless laborers (Malagamba and

Monares 1988). Optimism about differential benefits to the poorer segments of society is based mainly on the potential for increasing labor use and on the extent of discrimination against poorer producers in the market for vegetatively propagated material. Growing TPS to produce seedling tubers is a very labor-intensive horticultural activity, considerably more so than cultivating potatoes from vegetatively propagated material or from seedling tubers. If material for clonal propagation is limited and allocation is rationed, smaller farmers may have to pay more for tuber seed of the same quality or may end up purchasing poorer quality seed for the same price.

The pattern of these and other differential benefits is likely to be heavily context-specific. The sample sizes used in this study are generally too small to detect significant differences. But because empirical information on these issues is so scant—even anecdotal evidence is in short supply—discussion is warranted of those results that are available. We therefore confine our remarks to three agroecologies where TPS has economic promise: the Nile Delta of Egypt, the northeastern hills, and the plains of India.

In the northeastern hills and plains of India, raising one hectare of transplants generated an incremental demand of labor for 100 days per hectare over potato production either from seedling tubers or vegetatively propagated material. In this context, wages in the casual labor market rarely exceed one dollar per day in a land scare economy. Under these conditions, any technology that increases labor demand will be socially beneficial. Agricultural laborers benefit directly when hired for the production of seedling tubers from TPS transplants or indirectly when small-farm family members, who engage in this production, become less likely to participate in the casual labor market.

The potential for increased labor use from TPS did not materialize in the Nile Delta of Egypt. When seedling tubers were first produced in Egypt in the mid-1980s, they were grown from TPS sown in beds with two techniques, one recommended by researchers and the other adapted by farmers (See Fig. 2.1). Both were extremely labor intensive, equivalent to about 2,000 days of labor per hectare with the researcher's method and 3,000 days with the farmer's (Chilver et al. 1997). Yields were also very high, 60 t ha^{-1} with the farmer's method and 45 t ha^{-1} with the researcher's, but scientists believed that seedling tubers produced from the bed method would be more expensive than tuber seed multiplied from imports. They also felt that farmers were not capable of cost-effectively engaging in the transplant option chalked out in Fig. 2.1. Interest from a private venture appeared decisive in selecting the direct sowing option. This highly mechanized and capital-intensive system entailed sowing TPS directly in the sandy fields of newly reclaimed areas (NRA). Producing seedling tubers by direct sowing, using this commercial company's method, drastically cut labor demand to about 50% of that used in growing potatoes from tuber seed or seedling tubers. However, productivity was low at only about 7 t ha^{-1}, but seedling tubers were purportedly produced more cheaply than under the bed method. The fact that mechanized direct sowing emerged as the best candidate for producing seedling tubers illustrates how a distorted policy and production environment can influence the choice of technique. Fortunately, this capital-intensive method of production has not been successful.

Are poorer potato producers more likely to raise transplants than wealthier growers? This is a moot question for Egypt because the trajectory in Fig. 2.1 excluded farmers from the initial production of seedling tubers. For India, answers are not as clear on the direction of benefits for agricultural labor. On the one hand, the opportune labor costs should be lower for poorer farm households than for richer producers whose family members may not engage in manual labor even on their own farm. Because of the delicate nature of transplants and skill involved in growing them, we would expect family labor to have an advantage in this activity since monitoring hired labor is difficult. However, the survey data indicated that poorer farmers had lower transport establishment rates and higher transplant densities than richer farmers. Wealthy farmers established a hectare of transplants with 240 g of TPS, medium-wealthy farmers with 400 g of TPS, and poor farmers with 610 g of TPS. We have no satisfactory explanations for why seed consumptions are higher for poorer producers, but we do know that they negate any benefits from differential labor costs between large and small producers.

Producing seedling tubers from true seed is a delicate, knowledge-intensive operation requiring considerable extension effort complemented with a small but critical level of investment. Private sector attempts to distribute seed packets in the hope that farmers will produce seedling tubers sufficient to generate a sustained demand for TPS have not been successful in India. Similar to the experience in the successful *in vitro* multiplication of potatoes in the Dalat highlands of Vietnam, it is likely that a smaller number of specialist producers will emerge to supply the needs of the larger population of farmers growing seedling tubers (Van Uyen et al. 1996). Experience in Nicaragua suggests the need for specialists (Torres 1997), but whether this specialist minority will be drawn from the ranks of the poorer or the richer potato producers is uncertain.

One other consideration that merits attention is assessing who will capture the benefits from the production of transplants. Because transplants produce small potatoes that are good for seed but not for market ware, they can generate cash-flow problems unless the output is sold as seed at harvest. Postponing sales to the next planting season can result in cash-flow problems in particular for poorer producers.

The hypothesis for price discrimination by wealth category was supported by data on the northeastern plains where the sample size was sufficiently large to test for significant differences. On average, farmers in the poor stratum paid about 20-25% more for tuber seed than farmers in the wealthy category. Higher unit costs were associated with the purchases of small quantities of seed and the tendency of poor farmers to purchase tuber seed from wealthy farmers on credit. This price premium meant that poor farmers gained more from planting seedling tubers than rich farmers and the difference in cost savings in seed was equivalent to about US$120 ha^{-1}.

In the Nile Delta in Egypt, the unit price of tuber seed did not vary systematically across wealth categories. But wealthier farmers appeared to have higher tuber seed quality. The 10 farmers in the wealthy stratum had yields that were about 10t ha^{-1} higher than the 19 farmers in the medium and poor strata. About 70% of the richer farmer production was classified as Grade A; about half of the poorer farmer output was placed in the same grade. The seed rate was also about 0.5 t higher for the richer than for the poorer farmers.

These wealth-wise disparities in productivity were almost the same with seedling tuber technology. Richer farmers had higher yields than poorer farmers and a higher percent of production in Grade 1. Nonetheless, poorer farmers received about a three-ton per hectare yield advantage with seedling tuber technology. For the richer farmer, clonal yields were marginally higher than seedling tuber yields. These productivity differences translated into an average gain of about US$700 ha^{-1} for switching to seedling tuber technology for the 19 poorer farmers compared to a US$250 ha^{-1} change in net benefits for the 10 richer farmers in the Nile Delta. These differences in net benefit are not statistically significant but they are indicative that seedling tuber technology has some potential for correcting for seed-quality differentials between richer and poorer farmers.

Differential gains hinge on sustained equality of access to seedling tubers by all farmers. The opportunity for farmers to produce their own seedling tubers from TPS transplants implies seedling tuber access should be more equitable than access to cultivar seed.

The above discussion is by no means exhaustive. For example, the attractiveness of seedling tubers in Egypt was expected to be further enhanced by the propensity of poorer producers to concentrate on fresh ware production for the local market, where seedling tuber output is acceptable, during the autumn growing season. Potatoes harvested from the winter crop, compared to the autumn one, provide lucrative off-season markets in Europe where market quality is high and richer farmers tend to be more dominant producers. Poor consumers are unlikely to benefit noticeably from increased spring crop production because it is exported. Seasonally targeting seedling tuber technology for the autumn season increases the odds that cost savings and increased production will be passed on in the form of lower food prices to poor rural and urban consumers. Recent poverty profiles in Upper and Lower Egypt indicate that poor rural and urban households consume on average 50 to 80 grams of potato daily (Datt and Joliffe 1998).

CONCLUDING COMMENTS

We conclude that earlier ex-ante assessments and priority—setting exercises, such as Khatana et al. (1996) and Walker and Collion (1997), were overly optimistic about the potential areal adoption and large increments in net benefits per hectare replacing vegetative propagation of potato with TPS technologies. In India, incremental net benefits above our cutoff criterion of \$100 ha^{-1} appear to be restricted to the northeastern hills. Our results reject widespread geographic applicability of TPS, even in India, which has benefited from a sustained breeding effort since the early 1980s. Selective targeting is needed to adjust expectations to the economic reality of the state of technological progress. Penetration of TPS technologies into the northeastern hills of India or the southwestern mountains of China would undoubtedly be a tremendous achievement. Both India and China have sufficient potato production to justify investment in a small but sustained TPS breeding and selection effort to complement their conventional clonal crop improvement programs.

Several positive findings also emerged from this latest investigation. For a given distribution of tuber size, price differences between TPS progenies and clones did not differ statistically. Indeed, at times the glossy (waxy) skinned seedling tubers fetched more attractive prices than the output of clonal seed. The very fact that these evaluations could be carried out indicates that the supply of TPS technologies is not a problem or should not be considered a constraint in technology assessment. Sufficient material is available internationally for this technology to take off.

Say's[2] law that "supply creates its own demand" does not apply to TPS (or to many other prospective technologies for that matter). The immediate problems related to TPS technologies are on the demand side. Cost savings are not sufficient to leverage adoption. TPS technologies have to produce as much or more than clonally propagated tuber seed. Seedling tubers more or less meet this production condition, but significantly lower yields of transplants from botanical seed are a binding con-

[2]Say, Jean Baptiste (1767-1832) French economist, who propounded the principle that the supply of goods is always matched by the demand for them.

straint to technology acceptance. Moreover, the distribution of tuber size has been highlighted as a critical variable for improving the design of both transplant and seedling tuber technologies.

The results of these recent evaluations again confirm that TPS technologies will only be economically viable in regions such as the Red River Delta of Vietnam where clonal seed availability and quality are severely limiting. We have derived an empirical rule for the identification of such regions. Potential for TPS is greatest when seed cost exceeds 20% of the value of production. But even in these identified regions one has to remember that the potential for clonal propagation systems is also changing over time. For example, in the Red River Delta of Vietnam, liberalizing the imports of clonally propagated material from China or investing in cold storage could substantially erode the attractiveness of TPS technologies.

Egypt is another case where the economic fundamentals are rapidly changing (Chilver et al. 1997). TPS research began in Egypt under a highly regulated agricultural policy context with a bias toward labor-saving mechanization. Potatoes and other vegetable crops were comparatively free from direct regulation, but tuber seed imports of clonal material were tightly controlled by the government both in terms of the quantity imported and the sales price. Egypt recently, lifted the quota on these seed imports and deregulated their price. In 1996, seed imports doubled and tuber seed prices fell to 40% of the level used in this study. At these price levels, the $500 advantage in net benefit ha^{-1} of seedling tubers in the Egyptian Delta vanishes. Fortunately, for the viability of TPS technology, 1996 may have been a glut year when importers overestimated the demand for tuber seed. Nonetheless, this example demonstrates how a changing policy context can markedly alter the prospects for TPS-related technology.

Of the nine multilocational sites in Egypt, India, and Indonesia, TPS appears to have the brightest prospects in the northeastern hills of India. Thus far the experience with TPS is heavily concentrated in Tripura. Extension of TPS technologies from Tripura to other states, particularly Assam, in the northeastern hills is a priority for monitoring.

Acknowledgments

We thank Mahesh Upadhya and Keith Fuglie for their comments.

REFERENCES

Almekinders, A. Chilver and H. Renia. Current status of the TPS technology in the world. Potato Res 39: 289–303.

Chilver, A. 1997. Innovation paths in developing country agriculture: True potato seed in India, Egypt, and Indonesia. PhD thesis, University East Anglia, 229 pp.

Chilver, A., R. El-Bedewy, and A. Rizk. 1997. True potato seed: Research, diffusion, and outcomes in Egypt. Intl Potato Center, Lima, Peru, 29 pp.

CIMMYT. 1988. Formulation of recommendations from agronomic data. An Economics Training Manual, Economics Program. CIMMYT, Mexico D.F., Mexico.

Crissman, C. 1987. Identifying strengths and weaknesses of seed programs in developing countries. Rept. Third Social Science Planning Cong., September 7–10. CIP: Lima, Peru.

Datt, G. and D. Joliffe. 1998. Determinants of poverty in Egypt: 1997. FCND Discussion Paper No. 75. Washington, D.C.: International Food Policy Research Institute.

Fano, H. and C. Tello. 1998. Informe Parcelas Demostrativas. Avances de Evaluactión. CIP, Lima, Peru (unpubl.).

Khatana, V.S., M.D. Upadhya, A. Chilver, and C.C. Crissman. 1996. Economic impact of true potato seed on potato. *In:* T. Walker and C. Crissman (eds.). Case Studies of the Economic Impact of CIP-related Technology. CIP, Lima, Peru, pp. 139–156.

Malagamba, P. and A. Monares. 1988. True potato seed: Past and present uses. CIP: Lima, Peru, 40 pp.

Merrill-Sands, D. 1986. Farming systems research: Clarification of terms and concepts. Experim. Agric. 22: 87–104.

Pakistan-Swiss Potato Development Project. 1991. Potato yield estimation survey of autumn crop of Punjab-January, 1991: A study carried out jointly by Directorate of Agriculture, Crop Reporting Services, Punjab, Ministry of Food, Agriculture, and Cooperatives, Government of Pakistan, and the Pakistan-Swiss Potato Development Project, PARC. Islamabad, Pakistan.

Perrin, R., D. Winkelmann, R. Moscardi, and J. Anderson. 1979. From Agronomic Data to Farmer Recommendations. An Economics Training Manual. Inform Bull 27. CIMMYT, Mexico D.F., Mexico, iv + 54 pp.

Rasco, E.T. 1994. SAPPRAD on the Third Year of Phase III. Coordinator's Report July 1993-June 1994. SAPPRAD, Manila, Philippines.

Simmonds, N. 1997. A review of potato propagation by means of seed, as distinct from clonal propagation by tubers. Potato Res 40: 191–214.

Terrazas, F., V. Suárez, G. Gardner, G. Thiele, A. Devaux, and T. Walker. 1998. Diagnosing potato productivity in farmers' fields in Bolivia. Social Science Dept, working paper no. 1998-5. Intl Potato Center (CIP), Lima, Peru.

Torres, F. 1997. Semilla sexual de papa (SSp) en Piura. Una nueva posibilidad de desarrollo. Central Peruana de Servicios (CEPESER), Piura, Peru, 48 pp.

Upadhya, M. and K.C. Thakur. 1998. True potato seed—A piece of the late blight puzzle? Paper presented Natl Conf Veg Improv, Nov. 1998, Varanasi, India.

van Hest, P. 1994. Production of potatoes through true botanic seed. Univ. Reading, UK, 196 pp.

Van Uyen, N., Troung Van Ho, Pham Xuan Tung, P. Vander Zaag, and T. Walker, 1996. Economic impact of the rapid multiplication of high-yielding, late blight-resistant varieties in Dalat, Vietnam. In: CIP T. Walker and C. Crissman (eds.). Case Studies of the Economic Impact of CIP-related Technologies. CIP, Lima, Peru, pp. 127–138.

Walker, T. and Collion. 1997. Priority Setting at CIP for the 1988-2000, Medium term Plan. CIP, Lima, Peru, 48 pp.

3

Propagation by Traditional Breeding Methods

GEORGE R. MACKAY

[1]*Scottish Crop Research Institute, Invergowrie, Dundee DD2 5DA, Scotland, UK*
Tel: +44 1382 562731; Fax: +44 1382 568587; Email: gmacka@scri.sari.ac.uk

INTRODUCTION

More than 200 tuber-bearing species of genus *Solanum* are known, of which seven are cultivated forms (see Hawkes 1990; Dodds 1965 for reviews). However, the predominant cultivated form worldwide is *Solanum tuberosum* ssp. *tuberosum* (Hawkes 1990) or *Solanum tuberosum* Group Tuberosum (Dodds 1965). Tuberosums are tetraploids, $2n = 4x = 48$, believed to have evolved in Europe from introductions of the primitive cultivated form *S. tuberosum* ssp. *andigena*, or Group Andigena, via the Canary Islands during the late 16th century (Hawkes and Francisco-Ortega 1993). The means by which Andigena cultivars were bred and selected by the indigenous people of the Andes is a matter of conjecture. However, this tetraploid species is believed to have arisen several thousand years ago, either by spontaneous doubling of the somatic chromosome number of a diploid cultivated form, *S. stenotomum*, $2n = 2x = 24$, or the doubling of chromosome number of a hybrid(s) between Stenotomum and a wild diploid species, *S. sparsipilum* (Hawkes 1990) (Fig. 3.1). Either way, Tuberosum is assumed for most purposes to be an autotetraploid and usually demonstrates tetrasomic inheritance. Early cytological studies of meiosis (Swaminathan and Howard 1953), plus the fact that modern breeders have introgressed genetic variation from several wild species of varying degrees of ploidy (Ortiz 1998), suggest that modern cultivars may in fact be segmental allopolyploids. This could, in part, explain the disturbance of segregation ratios observed in modern molecular studies of tetraploid populations (Bradshaw et al. 1998).

Nevertheless, genetics aside, "deterioration" of clonally reproduced cultivars was a commonly observed, but initially little understood, phenomenon in Europe (Knight 1807; Wilson 1907). The fact that high-yield-

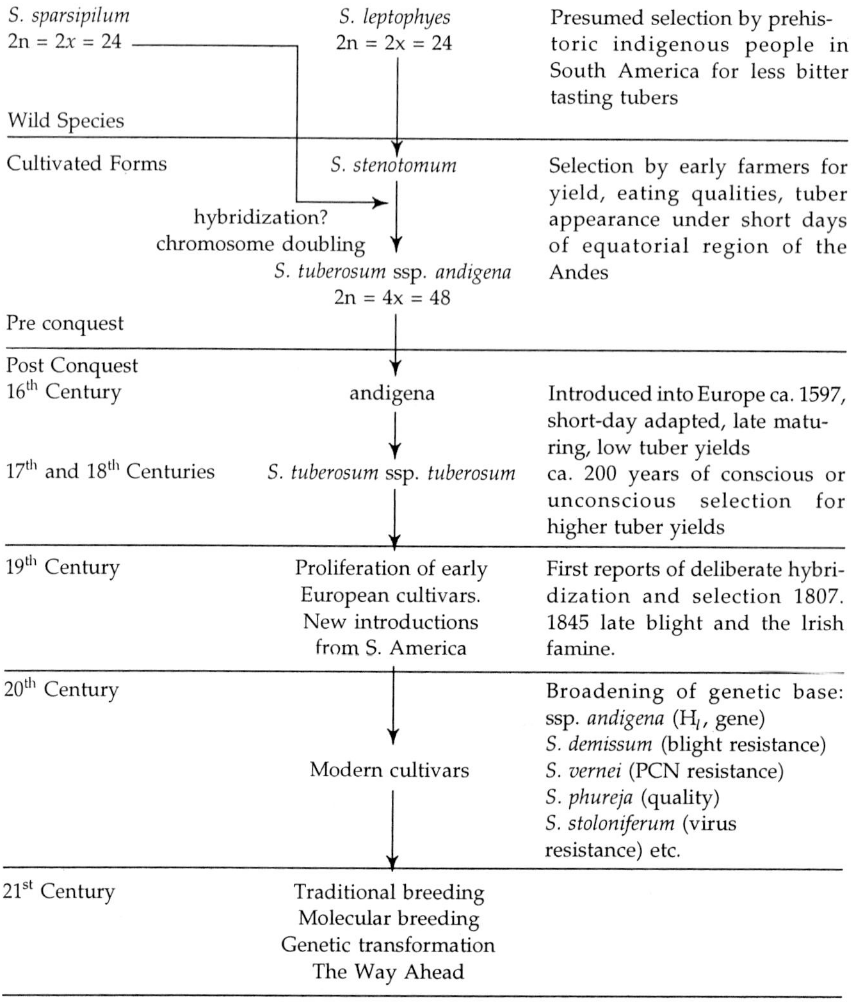

Fig. 3.1 Outline of the origin of the European form of the cultivated potato, *Solanum tuberosum* L. ssp. *tuberosum* (after Hawkes 1990).

ing clones, apparently restored in vigor by passage through sexual hybridization, could be selected from seedlings lends credence to the hypothesis that in years past farmers of South America selected new cultivars from seedlings as and when their clonally reproduced stocks declined in vigor. This process, which can be described as cyclical phenotypic recurrent selection, was perhaps not too dissimilar to the means by which Tuberosum evolved and early European cultivars were bred. Although the cause of "degeneration" of stocks was eventually recognized as due to systemic viral infection (Salaman 1924), the fact that common viruses are not transmitted through botanic seed supports the idea that virus-free and hence new vigorous clones could be and were selected from seedlings from berries produced by sexual means. In Europe, a substantial increase in number of new cultivars selected from seedlings, was reported during the late 18th and 19th centuries (Glendinning 1983). The precise means by which early breeders produced berries is not well documented. Knight (1807) reported deliberate crossing but Glendinning (1983) expressed the view that such crossing did not become common practice until the latter half of the 19th century. This being so, the tetraploid potato (if fertile), was self-compatible, berries produced by uncontrolled open pollination were very likely selfs, supporting the view that inbreeding may have been an additional factor in reducing the gene pool of the European potato. With the rediscovery of Mendel's work, which laid the foundation of modern genetics at the start of the 20th century (Mendel 1865), crop breeding—including that of potato—began to be based on a sounder understanding of the mechanisms of inheritance and the development of more scientifically based breeding methods. Nevertheless, it took several decades before Tuberosum was recognized as a tetraploid (Lunden 1937; Cadman 1942), and potato breeding methods differed little, except perhaps in scale and objectives, from those of Knight (1807) and Wilson (1916) or perhaps even the early farmers of Peru. Modern breeders, particularly during the latter half of the 20th century, have added substantially to knowledge of potato genetics and the technologies for its improvement by breeding in the broadest sense (see Bradshaw and Mackay 1994a for review). It is now, however, becoming increasingly difficult to draw a distinction between what is meant by traditional breeding of potatoes and less conventional approaches, though some of the latter have yet to be proven in practice by the successful production of new cultivars.

TRADITIONAL BREEDING OF CULTIVARS VS. GERMPLASM ENHANCEMENT

The narrow genetic base of the European potato and its extreme susceptibility to diseases, in particular late blight (*Phytophthora infestans*), allied to the fact that Tuberosum can, with varying degrees of difficulty, be hybridized directly or indirectly by bridging crosses with many of its wild relatives and primitive forms (see Ross 1986 for review); complicates the distinction between breeding cultivars and germplasm enhancement (Muller and Black 1951). For the purpose of this review, it is assumed that the traditional method most commonly deployed in breeding of potato cultivars *per se* has been by sexual hybridization of selected tetraploid clones, themselves often cultivars, followed by selection among the ensuing segregating populations for superior individuals. The parental clones used in such a traditional program may themselves have been the product of generations of prebreeding by a variety of methods, including interspecific hybridizations, backcrossing and selection, or even by asexual means such as somatic cell fusion (see Louwes et al. 1989 for reviews).

Like other crop breeding programs, the techniques adopted by potato breeders are largely determined by the breeding system of their crop and the genetics of the traits of interest. Practically all diploid members of genus *Solanum* are obligate outbreeders with a gametophytic self-incompatibility system. Tetraploid potatoes, if fertile, are self-fertile but suffer the deleterious effects of inbreeding. Consequently, breeders treat Tuberosum as an outbreeder for the purpose of maintaining vigor and parental clones as well as cultivars are therefore highly heterozygous. In terms of traits of interest, the potato breeder is faced with a multiplicity of objectives, varying substantially in their genetic complexity. However, in so far as potato breeding is basically a matter of selecting from among clonally reproduced individuals of a heterogenous F_1, from crosses between heterozygous parents, it is possible to exploit all the genetic variability. The problem is prioritizing the objectives, which are numerous.

OBJECTIVES OF TRADITIONAL BREEDING PROGRAMS

The primary objective of traditional breeders is to achieve the highest yield per unit area of utilizable product. This is complicated by the fact that potato tubers have numerous end uses. These can probably be best partitioned into three major categories: (a) table use (or direct consumption), (b) processing, and (c) industrial uses. As the fourth most important food crop in the world, most of the global crop is grown for direct consumption. However, an increasing proportion of the crop is now grown for processing into chips (crisps, UK) and French fries (chips, UK), or other processed food products such as prepared or semiprepared meals (Anon 1995). The cooked potato tuber constitutes an almost complete diet for humans, with an adequate balance of carbohydrate, fiber, minerals and vitamins, as well as a protein component superior in quality to that of cereals, deficient only in the sulfur-bearing amino acids, methionine and cysteine (Storey and Davies 1992). Thus, when selecting for table use, the breeder principally must pay heed to characters that could detract from eating quality, such as after-cooking blackening, and select positively for apparently cosmetic features such as skin and flesh color, tuber morphology, shape, size, and depth of eyes, suitable to the intended market place of a cultivar. It is also necessary to select negatively (discard) for traits which would render tubers unmarketable, such as secondary growth problems and internal defects (Ortiz and Huaman 1994;

Dale and Mackay 1994), traits which, unfortunately, are often intermittent in expression and of low heritability. Where wild species may have featured in the pedigree of a potential cultivar, it is also necessary to ensure that tuber glycoalkaloids do not exceed certain statutory or advisory limits (Dale and Mackay 1994). Yield is a function of tuber number and tuber size but, in the increasingly sophisticated markets of North America and Europe, speciality potatoes are presented to the consumer, tightly graded, washed and ready to cook, so the total yield of tubers per plant within a fairly broad size range is becoming less important. Whether the breeder selects for a few large tubers per plant or numerous small ones depends on his intention to produce a cultivar suitable for producing a higher proportion of "baker" or "baby" potatoes during harvest.

Furthermore, to some extent tuber number and tuber size can be modified by the grower's choice of chitting regime and/or agronomic practices (Allen and Wurr 1992). Moreover, in countries such as the UK, where the bulk of the main crop is harvested in autumn (fall) and stored for many months, growers of early or new (immature) or fresh ripe potatoes in spring and summer can realize a premium price so that potential total yield at maturity is less important than timely marketing. In short, while selecting for table use, the breeder's objective will be to produce cultivars capable of producing the highest yield of marketable tubers per unit area, although "marketable" itself may encompass a multiplicity of traits of complex and varying genetic architecture (Bradshaw and Mackay 1994b).

Catering to the needs of the processing industry, the breeder's objectives are somewhat more precisely defined than those for table use. In common with table potatoes, tubers for processing need to be free of major secondary growth and internal defects, whereas "cosmetic" features are of lesser importance. For the manufacture of both chips and French fries, as well as other processed products, tuber dry matter, hexose sugar levels (particularly post-storage) and to some extent tuber size and shape are more critical. When the bulk of the crop has to be stored for extended periods, as in Europe and North America, the control of sugar levels in tubers is paramount. Storage of tubers at low temperature (ca. 4°C), though desirable to prevent sprouting, results in low-temperature sweetening. Even low levels of the reducing sugars, fructose and glucose, will result in a raw product that produces a processed product that is dark colored and bitter tasting due to the Maillard reaction (Schallenberger et al. 1959). To avoid this, tubers for processing are stored at relatively high temperatures (>6°C) and sprout suppressant chemicals used to prevent sprouting (Storey and Davies 1992). In recent years, much effort by breeders has been devoted to producing cultivars which will store at low temperature without accumulating sugar, and this is meeting with some success (Mackay et al. 1990; Hayes and Till 2002).

Nevertheless, this adds yet another additional trait to the traditional breeder's already extensive list.

Breeding for industrial uses is less encumbered by the need to consider many cosmetic features. Total yield of dry matter and hence starch per unit area, is the prime objective. Potato starch is preferred for certain end uses, but it is price competitive against alternative sources (Bachelor et al. 1996). Attempts to commercialize potatoes with novel and higher value starches, modified *in planta* using GM technology, was not approved in Europe due to use of an antibiotic marker gene in their production (Anon 2001a). GM plants therefore have not been used in traditional methods of potato breeding to date. In traditional breeding for industrial use, the objective is still to produce the highest yield of dry matter (starch) per unit area in varieties which are resistant to biotic constraints such as late blight and cyst nematodes.

Biotic Constraints

The list of tuber traits that the breeder has to take cognizance of is long but pales into insignificance when considering the fact that potatoes are subject to attack by many pests, and pathogens, putting constraints on productivity and deleteriously affecting quality. Furthermore, the maladies caused by pests and pathogens are costly to control.

In a recent compendium, 61 pests and pathogens of potato were described, in addition to 26 abiotic disorders which can affect yield, storability, and quality (Stevenson et al. 2001). While many of these constraints are of less economic importance than others, extreme susceptibility to a single pathogen or pest can lead to failure of a new cultivar in the marketplace. Fortunately, not all pests and pathogens are present in all environments in which potatoes are grown. For example, both the golden cyst nematode (*Globodera rostochiensis*) and the white cyst nematode, (*Globodera pallida*) are major pests in the UK and Europe, but only the former is found in the USA and is limited in distribution (Marks and Brodie 1998). Conversely, the UK is free of Colorado beetle, an endemic pest in North America and mainland Europe. Late blight is, of course, a global menace (Trognitz et al. 2001). It is difficult to perceive of a traditional breeding program that attempts to produce cultivars with resistance to all pests and pathogens, to which cultivars from that program may be exposed. Therefore, most traditional, practical, breeding programs tend to prioritize parental choice and selection of clones for the most important traits and resistances. These may be designed to enable the breeder to discard clones during routine testing and trials, which are too susceptible or intolerant of lesser constraints. In a few instances, paucity of heritable variation may suggest that an alternative, such as integrated pathogen management, may be a more successful option than breeding for resistance, for example, to bacterial wilt (French 1994; Mackay 2000).

Typical Traditional Breeding Programs

Several publications in recent decades have reviewed the genetics and breeding of potatoes, which include examples of typical, traditional breeding programs (Howard 1970; Mackay 1987; Tarn et al. 1992; Caligari 1992; Bradshaw and Mackay 1994b; Poehlman and Sleper 1995). Descriptions of other programs are now available on the Internet. Though differing in emphasis, all share common factors. Parental clones may be selected on the basis of their phenotype, on the assumption that complementary phenotypes will produce recombinants with the desirable traits of both parents in ensuing populations. Mid-parental values may provide good indicators of the likely average value of their progeny and thus serve a useful means of cross prediction in some circumstances (Brown and Caligari 1988). Some breeders choose to cross a potentially new source of parental clones with tester clones whose value as parents is already known (Tarn, pers. comm. 2002). It is also possible, by the use of progeny tests, to estimate breeding value or combining ability of parents (Wastie et al. 1993). Whatever be the method of identifying potentially useful parents, the probability of finding recombinants in ensuing progenies possessing all the desirable attributes of an improved cultivar is very low. Consequently, traditional breeders tend to raise many thousands of seedlings from among which individual clones are selected (Fig. 3.2). For practical reasons, selection in the early stages of such a breeding program is usually based on fairly cursory, visual appraisal of seedling tubers, produce of single hill plants and small, often unreplicated plots. Once the population reaches a manageable size, the clones selected may be grown in small-scale replicated ware trials, or observation plots, from which quantifiable data on yield, quality and some disease resistances are collected, analyzed and used to select preferred clones for more sophisticated testing and trials as potential varieties. During these final stages of selection, breeders may collaborate with potential end users to more widely assess advanced clones on the basis of independent pre-variety trials at multiple sites. Having identified potential varieties, progress to varietal release of newly named cultivars varies according to particular country's regulatory system. In the European Union, for example, before a new cultivar can be commercialized and its seed certified, it has to be entered into the European Common Catalogue of varieties. This statutory hurdle has to be overcome by the new cultivar satisfying the differing criteria of an EU member state as to its value for cultivation and use (Anon 2001b).

In attempting to increase the efficiency of breeding and selection of new cultivars, traditional breeders have adopted several complementary strategies. Some of these have been successfully applied and thus may constitute components of traditional breeding in future.

Year	*Activity*	*No. of clones*
1	Select parents, clones and cultivars with complementary phenotypes, make crosses	ca 200-300 pair crosses
2	Raise seedlings from previous year's crosses and resowings of progenies based on performance of those progenies in routine trialling system. Select single tubers from individuals	ca 100,000 seedlings
3	Grow singles at seed site, select by visual appraisal of tubers at harvest	ca 50,000
4	Selected clones grown as 3-4 plant plots, unreplicated with controls (cultivars), select by visual appraisal at harvest. Some limited disease testing (for PCN resistance)	ca 4,000
5 6 7	Two sites, seed and ware, yield trials, quality and disease resistance testing, assess maturity types etc.	1,000 500 200
8 9 10	Regional trials, 6-7 sites in UK, 2-3 sites overseas, samples to collaborators for independent assessments. Produce Approved Seed stocks. Submit potential cultivars to statutory trials	60 10 5
11 12	Two years National List Trials (in UK). Breeders maintain Approved Stocks, pathogen tested prebasic seed initiated by certifying authority (SEERAD* and DEFRA† in UK)	1 to 5
13	Named cultivar(s) National Listed, if deemed of value for cultivation and use. Commercial seed production initiated by private sector	1 to 5

*SEERAD = Scottish Executive Environment and Rural Affairs Department
†DEFRA = Department of Environment, Food and Rural Affairs

Fig. 3.2 Outline of the SCRI potato breeding program prior to 1982: example of a typical traditional breeding program.

EXAMPLES OF COMPLEMENTARY STRATEGIES IN TRADITIONAL BREEDING METHODS

Parental Breeding

The choice of parents in a cultivar breeding program is critical. Though the phenotype and mid-parental value of the phenotypes of any two parents may provide a guide to their selection, the complexity of tetrasomic inheritance and low heritability of some traits, renders this only a partially effective predictor of performance. However, there are a number of useful traits governed by simply inherited major dominant genes. For example the H_I gene (group Andigena) conveys total resistance to strains Ro_1 and Ro_4 of *Globodera rostochiensis* (Toxopaeus and Huijsman 1953) and the genes Ry_{sto} and Rx_{adg} confer almost total resistance to viruses

PVY and PVX respectively (Cockerham 1970). A parental clone simplex for one such gene will guarantee that 50% of its progeny from a cross to a susceptible clone will inherit that resistance. If the parental clone is duplex, the probability is raised to more than 80%. Selective breeding and test-crossing has enabled breeders to produce clones which are triplex or quadruplex for genes such as these (Mackay 1989; Mendoza et al. 1996; Fig. 3.3). The use of these clones as parents then virtually guarantees that all the progeny will inherit that particular resistance, thus guaranteeing that any cultivar from such a program will possess the desired resistance to that trait, thereby relieving the breeder of the need for selecting this trait. This frees him or her to focus on other characters. A newly released cultivar, namely, cv. Spey, from the Scottish Crop Research Institute (SCRI) is triplex of the H_l gene, a fact that lends credence to the belief that such an approach need not be at the expense of other desirable agronomic and quality traits (Anon 1997).

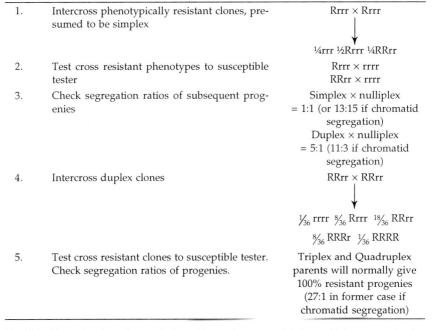

1.	Intercross phenotypically resistant clones, presumed to be simplex	Rrrr × Rrrr ↓ ¼rrr ½Rrrr ¼RRrr
2.	Test cross resistant phenotypes to susceptible tester	Rrrr × rrrr RRrr × rrrr
3.	Check segregation ratios of subsequent progenies	Simplex × nulliplex = 1:1 (or 13:15 if chromatid segregation) Duplex × nulliplex = 5:1 (11:3 if chromatid segregation)
4.	Intercross duplex clones	RRrr × RRrr ↓ $\frac{1}{36}$ rrrr $\frac{8}{36}$ Rrrr $\frac{18}{36}$ RRrr $\frac{8}{36}$ RRRr $\frac{1}{36}$ RRRR
5.	Test cross resistant clones to susceptible tester. Check segregation ratios of progenies.	Triplex and Quadruplex parents will normally give 100% resistant progenies (27:1 in former case if chromatid segregation)

Fig. 3.3 Example of a scheme designed to produce parental clones triplex or quadruplex for a single dominant resistance gene (after Mackay 1989, Mendoza et al. 1996).

What is true in simple inheritance of major dominant genes is also true for traits of more complex inheritance and has been attempted by use of seedling progeny tests (Mackay and Wastie 1981; Caligari et al. 1985;

Wastie et al. 1990; Wastie et al. 1992; Bradshaw et al. 1995). The use of these and similar progeny tests for other traits permits identification of genetically superior parents whose application in breeding significantly raises the probability of their progeny inheriting some important traits. This likewise frees the breeder to target other characters during routine testing and trial procedures.

Targeted Breeding Approach

Large-scale potato breeding programs are often aimed at a multiplicity of potential end uses. Consequently, clones which fail to satisfy one criterion may be retained as they might satisfy some other. For instance, clones with yellow flesh might be acceptable in one market but not in another. Hence breeding programs tend to retain a larger proportion of segregating populations than a program targeted to a specific end user's needs. The publicly funded breeding program of SCRI was, as a matter of fact, historically designed to produce cultivars for all manner of end uses (Holden 1977). This later also came to include breeding cultivars for export as seed to countries in the Mediterranean and North Africa, where diseases such as *Alternaria* and *Verticillium*, are important though not in the UK (Mackay et al. 1994). However, by using proven progeny tested parents and, more precisely, plants targeting the needs of particular end users, it has been possible to notably accelerate the breeding process from seedling to cultivar (Mackay et al. 1997). This approach relies heavily on discarding entire progenies and subsequently all clones, which fail to meet very specific criteria and select/retain only those which possess the particular trait desired, say, the capability of producing acceptable fried products after storage at low temperature (Fig. 3.4). A similar approach to accelerated breeding of "cold chipping" genotypes by crossing conventional or experimental tetraploid parents with diploid ones, was developed in the USA and is also proving successful (Hayes and Till 2002). The targeted approach is presently being implemented by SCRI in association with various commercial companies who have very specific markets for their products. However, such a focused approach will continue to rely on access to parental material of proven genetic worth with respect to major biotic and abiotic constraints, if the cultivars so produced are to satisfy both the end user's needs and the consumer's desire for less reliance on agrochemicals.

Multitrait Selection

In contrast to the targeted approach, the objective of most breeding programs remains to produce cultivars which combine resistance to several major pests or pathogens with good yield and agronomic performance. At SCRI, as progeny tests were developed and validated, they were

Year/Site	*Activities*
1991 Glasshouse	43 progenies of c. 72 seedlings per progeny. total ca. 3,100 seedlings sown to produce tubers. No selection
1992 Seed site, Scotland	43 progenies planted as 4 replicates of 15 clones ca. 2,500 clones. One plant per clone. No selection at harvest, samples taken for storage and fry tests from every clone.
1992/1993 Store/laboratory	Fry tests on samples at 4°C and 10°C storage. Data analyzed, superior progenies identified
1993 Seed site, Scotland	17 superior progenies (960 clones) plus controls in 4-plant plots × 1 replicate. Visual appraisal* of tubers. Samples from clones placed in storage at 4°C and 10°C.
1993/1994 Store/laboratory	Superior clones identified on basis of tuber traits, fry tests and dry matter
1994 Scotland and England	136 superior clones plus controls field trialled, as ware, at two sites and maintained as seed. Yield, visual appraisal of tubers in store, tuber size graded, fry tests post-storage etc.
1995 Scotland and England	28 clones and controls subjected to replicated ware trials as in 1994 plus some screening for disease resistance (late blight and PCN).
Glasshouse	Clones micropropagated to enhance seed production.
1996 Scotland and England	16 'best' clones subjected to repeated replicated trials, ware traits, storage quality, and disease resistances rested.
Seed site	Products of micropropagation grown as 'Approved Stocks'. Three clones submitted to official National List Trials as potential varieties.
1997/98	Breeders' scale up seed production, continue testing and trialling with commercial collaborators | Two years statutory National List Trials
1999	All three clones deemed of satisfactory value for cultivation and use: Golden Millennium, Harborugh Harvest and Montrose entered on UK National List as new processing cultivars with enhanced (low temperature) storage characteristics

*Visual appraisal = visual assessment of tubers; size, shape, internal or external defects etc.

Trial sites

Seed site: High grade seed farm Peeblesshire, Scotland

Scotland: Ware site at SCRI, Dundee

England: Ware sites in Lancashire and Yorkshire, proximal to processing plants.

Fig. 3.4 Examples of successfully targeted and accelerated breeding programs for processing potato varieties.

initially applied independently to the core breeding program so that, for example, clones from progenies with resistance to late blight but susceptible to cyst nematodes, or vice versa, went into routine testing and trials as potential varieties. By 1990, SCRI breeders had sufficient confidence in the application of their progeny test-based system that they undertook simultaneous selection for resistance to late blight (in foliage and tubers), resistance to cyst nematodes, and other satisfactory agronomic traits (yield, tuber characteristics, including cooking and processing quality). To maximize the opportunity for recombination events, this "multitrait" scheme operates on a three-year cycle, wherein the "best" clones from the "best" families are crossed to produce a new segregating population, upon which to practice selection at a much earlier stage than hitherto. In a traditional scheme, clones would seldom be used as parents until they successfully reached the final stage of selection as potential varieties or, indeed, were declared varieties. The SCRI multitrait selection approach is not yet concluded although the indications are encouraging. Moreover, based on appropriate crossing schemes, it has been possible to determine the heritable potential of various traits for selection and breeding values or combining abilities of the parents, thus enabling more emphasis on genotypic vs. phenotypic selection (Bradshaw and Mackay 1994b; Bradshaw et al. 1995; Bradshaw et al. 1999).

SOME PRACTICAL CONSIDERATIONS

In order to pursue any of these approaches or, indeed, to create the segregating populations upon which to practice selection in a more traditional breeding program, it is, of course, necessary to make crosses between selected parents. Without sexual hybridization and meiosis, recombination is not possible. The fact that fertile tetraploid potatoes are self-compatible has already been alluded to above. Unfortunately, many cultivars and potential parental clones are "shy flowerers". There is a tendency for buds and/or flowers to drop before berries can be secured. Some prolific flowering varieties are male sterile, producing little or no pollen, or pollen which is nonfunctional. The possible causes for these difficulties are numerous (Howard 1970). Anecdotal evidence even suggests that some of the early breeders selected clones that did not berry freely in the mistaken belief that sexual fertility might be negatively correlated with clonal propagation and thus yeild. The fact that many modern cultivars produce prolific berry yields as well as high tuber yields tends to negate this notion (Lawson 1983). Nevertheless, it was recognized in earlier studies that inhibiting or precluding production of daughter tubers would encourage flowering while not necessarily affecting fertility *per se*. This offered breeders an enhanced probability of making the crosses they desired (Knight 1807). Similar approaches are still followed today and irrespective of whether a breeder chooses to grow parental clones "on a brick" or graft their scions to the stocks of other non-tubering solanaceous species, e.g., tomato, is largely a matter or choice or convenience. However, the sterility problems associated with tetraploid potato are an impediment in breeding and could well be one of the factors which stimulated interest in the potential of such techniques as induced somaclonal variation (Kumar 1994), somatic cell fusion (Wenzel 1994), and genetic transformation. There are many potential benefits in the latter approach (Mackay 1999) but despite improved potato varieties being produced and commercialized (Davies 2002), genetic transformation or other modern methods such as molecular marker-assisted selection, have yet to be fully exploited even though they are likely to provide very useful supplementary tools for traditional breeders in the foreseeable future.

SUMMARY

It may be assumed that the ancient indigenous people of South America first "bred" the early cultivated forms of potatoes by selection of less bitter tasting, hence less toxic tubers from the wild tuber-bearing species of the genus *Solanum*. When these hunter-gatherers became farmers, they deliberately selected high-yielding clones with more attractive tubers from seedlings of natural berries, to replace clonal stocks that "degenerated" due to systemic viral infections. This primitive form of cyclical, phenotypic, recurrent selection was possibly aided by spontaneous hybridization between the primitive diploid cultivated species, *S. stenotomum* Juz. et Buk with *S. leptophyes* Bitt. and subsequent spontaneous doubling of the somatic chromosome number, to give rise to the tetraploid cultivated form of potato, *S. tuberosum* L. ssp. *andigena* Hawkes. Cultivated in its center of origin for thousands of years, *S. tuberosum* ssp. *andigena* was introduced into Europe at the time of Spanish conquests in South America. By a process of recurrent selection for larger tubers and higher yields from seedlings, these early introductions were evolved into the European or Irish potato (*S. tuberosum* L. ssp. *tuberosum*), which became a staple in the European diet within 200 years. By the early 19th century, selectors were making controlled hybridizations and practicing deliberate selection for new cultivars among ensuing seedling populations. After the 1845 blight epidemic and with the rediscovery of Mendel's laws of inheritance in the early 20th century, new introductions were imported from South America, resulting in proliferation of new varieties. During the 20th century, the tetraploid status of Tuberosum was confirmed and the genetics of some important traits understood. This led to basing traditional breeding on sounder, more scientific technology. However, the complexity of tetrasomic inheritance and the great number of traits the breeder must take into account led to the development of large programs generating many thousands of seedlings, upon which selection could be practiced each year, but remained based largely on phenotype. As the genetic base of Tuberosum was considerably broadened during the 20th century, mainly by introgression of disease and pest resistances from related wild species, many modern cultivars benefited from this genetic enhancement in the latter half of the century. Steps to increase the efficiency of selection in the early stages of traditional breeding programs were taken and more emphasis given to genotypic vs. phenotypic selection, both of parental clones and their seedling progenies. With the advent of DNA-based methods, the possibility of marker-assisted selection using molecular techniques is being pursued, despite the complications of tetrasomic inheritance. Subsequently, identification, localization, and cloning of genes has facilitated genetic transformation of potatoes, leading already to the production and commercialization in North America of genetically modified (GM) versions of existing cultivars. If and when GM technology surmounts the opposition to it, especially in Europe, then during the 21st century molecular biology and genetic transformation may augment the skills of the traditional breeder and thus provide the means to produce new, improved cultivars whose yield, quality, and disease as well as pest resistances are significantly improved.

Acknowledgments

The author gratefully acknowledges the support and help of Mrs Sheena Forsyth, without whose computer skills the typescript of this chapter would have been seriously delayed, Dr John Bradshaw for his constructive comments, and the Scottish Executive Environment and Rural Affairs Department for financial assistance.

REFERENCES

Allen, E.J. and D.C.E. Wurr. 1992. Plant density. *In:* Paul Harris (ed.), The Potato Crop, the Scientific Basis for Improvement. Chapman & Hall, London, pp. 292–333.

Anon. 1995. Potatoes in the 1990s: Situation and Prospects of the World Potato Economy. FAO, Rome.

Anon. 1997. New potato cultivars. *In:* Scottish Crop Research Institute Annual Report 1996-97, pp. 54–55.

Anon. 2001a. Amylose-free potatoes. Genet Eng Newsl, sp. issue 8, October 2001, p. 2.

Anon. 2001b. Guide to National Listing of Varieties of Agricultural and Vegetable Crops in the UK. Dept Environment, Food & Rural Affairs, UK, and Scottish Executive Environment and Rural Affairs Dept., November 2001.

Batchelor, S., E. Booth, G. Entwistle, K. Walker, T. Rees, A. Hacking, G. Mackay, and I. Morrison. 1996. Industrial markets for UK-grown crop polysaccharides. Res Rev no. 32. Home Grown Cereals Authority, London.

Bradshaw, J.E. and G.R. Mackay (eds.) 1994a. Potato Genetics. CAB Intl, Wallingford, UK.

Bradshaw, J.E. and G.R. Mackay, 1994b. Breeding strategies for clonally propagated potatoes. *In:* J.E. Bradshaw and G.R. Mackay (eds.), Potato Genetics. CAB Intl, Wallingford, UK, pp. 467–498.

Bradshaw, J.E., H.E. Stewart, R.L. Wastie, M.F.B. Dale, and M.S. Phillips. 1995. Use of seedling progeny tests for genetical studies as part of a potato (*Solanum tuberosum* ssp. *tuberosum*) breeding programme. Theor Appl Genet 90: 899–905.

Bradshaw, J.E., C.A. Hackett, R.C. Meyer, D. Milbourne, J.W. McNicol, M.S. Phillips, and R. Waugh. 1998. Identification of AFLP and SSR markers associated with quantitative resistance to *Globodera pallida* (Stone) in tetraploid potato (*Solanum tuberosum* subsp. *tuberosum*) with a view to marker assisted selection. Theor Appl Genet 97: 202–210.

Bradshaw, J.E., I.M. Chapman, M.F.B. Dale, G.R. Mackay, R.M. Solomon-Blackburn, M.S. Phillips, H.E. Stewart, G.E.L. Swan, D. Todd, and R.N. Wilson. 1999. Applied potato genetics and breeding: potato improvement by multitrait genotypic recurrent selection. *In:* Scottish Crop Research Institute Annual Report, 1998-99, pp. 92–96.

Brown, J. and P.D.S. Caligari. 1988. Cross prediction in a potato breeding programme by evaluation of parental material. Theor Appl Genet 77: 246–252.

Cadman, C.H. 1942. Autotetraploid inheritance in the potato: some new evidence. Gene 44: 33–52.

Caligari, P.D.S. 1992. Breeding new varieties. *In:* Paul Harris (ed.), The Potato Crop, the Scientific Basis for Improvement. Chapman & Hall, London, pp. 334–372.

Caligari, P.D.S., G.R. Mackay, H.E. Stewart, and R.L. Wastie. 1985. Confirmatory evidence for the efficacy of a seedling progeny test for resistance to potato foliage blight [*Phytophthora infestans* (Mont.) de Bary]. Pot Res 28: 439–442.

Cockerham, G. 1970. Genetical studies on resistance to potato viruses X and Y. Heredity 25: 309–348.

Dale, M.F.B. and G.R. Mackay. 1994. Inheritance of table and processing quality. *In:* J.E. Bradshaw and G.R. Mackay (eds.), Potato Genetics. CAB Intl, Wallingford, UK, pp. 285–318.

Davies, H.V. 2002. Commercial developments with transgenic potato. In: Victorianu Valpuesta (ed.), Fruit and Vegetable Biotechnology. Woodhead Pub Ltd, Cambridge, England, pp. 222–249.

Dodds, K.S. 1965. The history and relationships of cultivated potatoes. In: Sir Joseph Hutchinson (ed.), Essays on Crop Plant Evolution. Cambridge Univ Press, UK, pp. 123–141.

French, E.R. 1994. Integrated control of bacterial wilt of potatoes. C.I.P. Circ 20(2): 8–11.

Glendinning, D.R. 1983. Potato introductions and breeding up to the early 20th century. New Phytol 94: 479–505.

Hawkes, J.G. 1990. The Potato, Evolution, Biodiversity and Genetic Resources. Bellhaven Press, London.

Hawkes, J.G. and J. Franciso-Ortega. 1993. The early history of the potato in Europe. Euphytica 70: 1–7.

Hayes, R.J. and C.A. Till. 2002. Selection for cold chipping genotypes from three early generations in a potato breeding programme. Euphytica 128(3): 353-362.

Holden, J.H.W. 1997. Potato breeding at Pentlandfield. Re Scott Pl Breeding Sta 1976-77, pp. 66–97.

Howard, H.W. 1970. Genetics of the Potato, *Solanum tuberosum*. Logos Press Ltd, London.

Knight, T.A. 1807. On raising new and early varieties of the potato (*Solanum tuberosum*). Trans Hortic Soc London 1: 57–59.

Kumar, A. 1994. Somaclonal variation. *In:* J.E. Bradshaw and G.R. Mackay (eds.), Potato Genetics. CAB Intl, Wallingford, UK, pp. 197–212.

Lawson, H.M. 1983. True potato seeds on arable weeds. Pot Res 26: 237–246.

Louwes, K.M., H.A.J.M. Toussaint, and L.M.W. Dellaert (eds.), 1989. Parental Line Breeding and Selection in Potato Breeding. Proc Joint Conf EAPR Breeding Sect and EUCARPIA Potato Sect, 11–16 Dec. 1988. Pudoc, Wageningen, Netherlands.

Lunden, A.P. 1937. Arvelighetsundersokelser i potet [Inheritance studies in the potato]. Meldinger fra Norges Laudrukshøishkole 17: 1–156.

Mackay, G.R. 1987. Selecting and breeding for better potato cultivars. *In:* A.J. Abbott and R.K. Atkin (eds.), Improving Vegetatitively Propagated Crops. Academic Press, London, pp. 183–193.

Mackay, G.R. 1989. Parental line breeding at the tetraploid level. *In:* K.M. Louwes, H.A.J.M. Toussaint and L.M.W. Dellaert (eds.), Parental Line Breeding and Selection in Potato Breeding. Pudoc, Wageningen, Netherlands, pp. 131–135.

Mackay, G.R. 1999. Potential benefits of genetically-modified potatoes. *In:* The Proceedings Crop Protection in Northern Britain 1999, Dundee Scotland, 23-25 March 1999, pp. 203–210.

Mackay, G.R. 2000. Basic and applied research needs for sustained crop health of the potato. Amer J Potato Res 77: 334–338.

Mackay, G.R. and R.L. Wastie. 1981. Problems and prospects on progeny testing for disease and pest resistance in a commercial potato breeding programme. *In:* Abstracts Conference Papers, 8th Triennial Conference EAPR 1981. EAPR, Wageningen, Netherlands, pp. 30–31.

Mackay, G.R., J. Brown, and C.J.W. Torrance. 1990. The processing potential of tubers of the cultivated potato, *Solanum tuberosum* L. after storage at low temperature, 1. Fry colour. Pot Res 33: 211–218.

Mackay, G.R., R.L. Wastie, and H.E. Stewart. 1994. Breeding potatoes for warm climates. *In:* Scottish Crop Research Institute Annual Report, 1993, pp. 20–23.

Mackay, G.R., D. Todd, J.E. Bradshaw, and M.F.B. Dale. 1997. The targeted and accelerated breeding of potatoes. *In:* Scottish Crop Research Institute Annual Report, 1996-97: 40–45.

Marks, R.J. and B.B. Brodie (eds.). 1998. Potato Cyst Nematodes: Biology, Distribution and Control. CABI Publ, Oxon, Wallingford, UK.

Mendel, G. 1865. Experiments in Plant Hybridisation, with an Introduction By R.A. Fisher, 1965. Centennial reprint, Olivery & Boyd, Edinburgh.

Mendoza, H.A., E.J. Mihovilovich, and F. Saguma. 1996. Identification of triplex (YYYy) potato virus Y (PVY) immune progenitors derived from *Solanum tuberosum* ssp. *andigena*. Amer Pot J 73: 13–19.

Muller, K.O. and W. Black. 1951. Potato breeding for resistance to blight and virus diseases during the last hundred years. Zeitschrift für Pflanzenzüchtung 31: 306–318.

Ortiz, R. 1998. Potato breeding via ploidy manipulations. Plant Breeding Reviews 16: 15–86.

Ortiz, R. and Z. Huaman. 1994. Inheritance of morphological and tuber characteristics, *In:* J.E. Bradshaw and G.R. Mackay, (eds.), Potato Genetics. CAB Intl, Wallingford, UK. pp. 263-284.

Poehlman, J.M. and D.A. Sleper. 1995. Breeding potato. In: Breeding Field Crops, Iowa State Univ Press, IA (4th ed), pp. 419–433.

Ross, H. 1986. Potato Breeding— Problems and Perspectives. Advances in Plant Breeding. Verlag Paul Parey, Berlin.

Salaman, R.N. 1924. Potato Varieties. Camb Univ Press, London.

Schallenberger, R.S., O. Smith, and R.H. Treadaway. 1959. Role of sugars in the browning reaction in potato chips. J Agric Fd Chem 7: 274.

Stevenson, W.R., R. Loria, G.D. Franc, and Weingartner (eds.). 2001. Compendium of Potato Diseases. APS Press, St Paul, MN USA (2nd ed).

Storey, R.M.J. and H.V. Davies. 1992. Tuber quality. *In:* P. Harris (ed.). The Potato Crop, the Scientific Basis for Improvement. Chapman & Hall, London, pp. 507–552.

Swaminathan, M.S. and H.W. Howard, 1953. The cytology and genetics of the potato (*Solanum tubersum*) and related species. Bibliographica Genetica XVI: 1–192.

Tarn, T.R., G.C.C. Tai, H. De Jong, A.M. Murphy, and J.E.A. Seabrook. 1992. Breeding potatoes for long-day temperate climates. Plant Breeding Reviews 9: 217–332.

Toxopaeus, H.J. and C.A. Huijsman. 1953. Breeding for resistance to potato root eelworm. Euphytica 2: 180–186.

Trognitz, B.R., M. Bonierbale, J.A. Landeo, G. Forbes, J.E. Bradshaw, G.R. Mackay, R. Waugh, M. Huarte, and L. Colon. 2001. Improving potato resistance to disease under the global initiative on late blight. In: H.D. Cooper, C. Spillane and T. Hodgkin (eds.), Broadening the Genetic Base of Crop Production. CABI Publ, Oxon, Wallingford, UK.

Wastie, R. L., J.E. Bradshaw, and H.E. Stewart. 1993. Assessing general combining ability for late blight resistance and tuber characteristics by means of glasshouse seedling tests. Pot Res 36: 353–357.

Wastie, R.L., G.R. Mackay, P.D.S. Caligari, and H.E. Stewart. 1990. A glasshouse progeny test for resistance to gangrene (*Phoma foveata*). Pot Res 33: 131–133.

Wastie, R.L., J.E. Bradshaw, M.F.B. Dale, G.R. Mackay, M.S. Phillips, and R.M. Solomon-Blackburn. 1992. Progeny testing for resistance to diseases and pests of potatoes. In: Scottish Crop Research Institute Annual Report, 1991, pp. 13–16.

Wenzel, G. 1994. Tissue Culture. In: J.E. Bradshaw and G.R. Mackay (eds), Potato Genetics. CAB Intl. Wallingford, UK, pp. 173–192.

Wilson, J.H. 1907. Experiments in crossing potatoes. Trans Highland and Hortic. Soc Scotland, 5th series, 19: 74–92.

Wilson, J.H. 1916. Further experiments in crossing potatoes. Trans Highland and Hortic Soc Scotland, 5th series, 28: 33–55.

4

Breeding Potential and Transmission of Traits in 4x-2x Crosses

KAZUO N. WATANABE[1], RODOMIRO ORTIZ[2], AND JUNKO A. WATANABE[1]

[1]*Gene Research Center, University of Tsukuba, 1-1-1 Tennoudai, Tsukuba, Ibaraki, 305-8572, Japan. Tel: +81-298-53-4663, Fax: +81-298-53-7723, e-mail: nabechan@gene.tsukuba.ac.jp*

[2]*International Institute of Tropical Agriculture (IITA), Ibadan, Nigeria c/o L.W. Lambourn & Co., Carolyn House, 26 Dingwall Road, Croydon CR93EE, U.K. e-mail: R.Ortiz@cgiar.org*

INTRODUCTION TO TETRASOMIC POTATO GENETICS

Tetrasomic Inheritance

The genetics of tetrasomic inheritance in tetraploid (cultivated) potato is very unique and far more complex than diploid genetics (Wricke and Weber 1986). At a diallelic locus, a diploid can have three classes of genotypes, e.g. *AA*, *Aa*, and *aa*; in contrast, a tetrasomic tetraploid may have five classes, *AAAA*, *AAAa*, *AAaa*, *Aaaa*, and *aaaa*.

Polysomic tetraploids may have two alternative segregation patterns in gamete formation depending on the chromosomal position of a locus in relationship to recombination events during the early stages of meiosis: they are called chromosome and chromatid segregations respectively (Peloquin et al. 1989; Wricke and Weber 1986). Segregation of gametes in heterozygotes at the tetrasomic tetraploid are comparable to diploid due to polyploidy and also the two possible segregation patterns: e.g., assuming a diallelic locus, gametic output of the *AAaa* genotype is *AA* : *Aa* : *aa* = 1 : 4 : 1 for chromosome segregation and 3 : 8 : 3 for chromatid segregation respectively, although while intermating the tetrasomic tetraploid, the resultant genotypic segregation in the progeny happens to be more complex than that at diploid even at a locus. For example, a cross between duplex *AAaa* and simplex *Aaaa* genotypes results in segregation ratios with *AAAa*, *AAaa*, *Aaaa* and *aaaa* 1 : 5 : 5 : 1 at chromosome segregation and *AAAA* : *AAAa* : *AAaa* : *Aaaa* : *aaaa* = 3 : 44 : 144 : 156 : 45 for chromatid segregation respectively. Emphasis is given here to the

[1]Corresponding author: Kazuo N. Watanabe

following points: (1) when a nulliplex such as *aaaa* is preferred for specific trait expression at the tetrasomic tetraploid crosses, the required population size to select the target genotype could be theoretically much larger than that in selecting a homozygote at diploid; and (2) if one of the alleles in a diallelic locus in the tetraploid, e.g. "*A*" allele, were to have an additive effect over one phenotypic expression, more quantitative segregation could be expected even on a single locus compared with diploid (Bever and Felber 1992; Ronfort et al. 1998). Molecular marker studies over many loci have suggested that chromatid segregation could take place generally at the chromosomal regions apart from their centromeres and relatively close to the end of the chromosomes (Barone et al. 2002; Yamada et al. 1998); thus, both the possible segregation types should be considered for estimation of gametic output in conjunction with location of the locus with respect to the centromere and the site of maximum crossover.

Double Reduction

While efforts have been made to estimate the level of chromosome and chromatid segregations at specific loci, still there is need to elucidate the question of double reduction. The coefficient of double reduction can be estimated for inference of the level of chromatid segregation and inbreeding effects to specific chromosomal regions and loci (Tai 1982a, b). It can be obtained in tetrasomic tetraploids such as potatoes in two ways: (1) crossing heterozygotes with homozygote tester at a locus and (2) extraction of a number of (di)haploids from a tetraploid with heterozygous for the loci of interest (Tai 1982a, b; Wricke and Weber 1986). An elaborative multivalent pairing model was proposed by Wu et al. (2001) for estimating relative proportions of different modes of gamete formation that generate gametes with identical genotypes due to multivalent pairing and consequent separation of the sister-chromatid segments in meiosis. However, more experimental data shall be provided to support these estimations for use as a breeding parameter.

Findings on the level of occurrence of double reduction in tetrasomic tetraploid potatoes differ. Haynes and Douches (1993), using several multiallelic isozyme loci found it to be sporadic. In contrast, Barone et al. (2002) claimed that double reduction at the end of chromosomes was more frequent using more than 100 molecular marker loci covering the potato genome. Thus, it seems likely that double reduction and chromatid segregation occur generally at chromosomal regions apart from their centromeres. However, more specific case reports are required on this aspect.

Allelic Diversity and Allelic Interactions

Potato is an outbreeder with crossbreeding, in the range 1-74% and therefore does not exhibit the characteristics of inheritance and breeding of an autogamous species. This outbreeding nature of potato causes difficulties in fixing genotypes for making pure breeding lines in both tetraploids and diploids.

When multiple alleles exist, many genotypes are found even on a single locus. The number is quite enormous in tetrasomic tetraploid potatoes compared with autogamous diploid species. Moreover, if a trait is controlled by more than one locus, its expression may be further complicated. Consequently, large variations in intralocus interaction and interlocus interaction should be considered, especially on various levels of allelic interactions (Mendiburu and Peloquin 1977a, b).

Multiallelic loci have been reported for various traits such as tuber shape (van Eck et al. 1994), starch metabolic pathways (van de Wal et al. 2001), late blight resistance (Meyer et al. 1998) and, needless to say, many molecular marker loci (reviewed in Watanabe and Watanabe 2000). Thus, the models previously proposed with multiallelic cases indeed occur and can be applied in explaining many cases of quantitative traits in potatoes.

Inbreeding Depression and Heterosis

On selfing, decline in shape, vigor and consequent quality fitness of individual tetraploids may be expected; this is associated with inbreeding depression in the digenic locus of a tetrasomic tetraploid (Bennett 1976; Wricke and Weber 1986). Golmirzaie et al. (1998a) demonstrated strong inbreeding depression in selfed progeny of tetraploids with respect to pollen viability, number of flowers and berries/plant, number of seeds/berry, pollen, seed production and tuber yield/plant. Among these traits, Golmirzaie et al. (1998b) further added pollen viability ($r = -0.912$, $P < 0.01$) and tuber yield ($r = 0.837$, $0.01 < P < 0.05$) as strongly influenced by inbreeding depression.

Mendoza and Haynes (1974) and Mendoza (1989) considered heterozygosity components important for the yield and the basis for heterosis in tetrasomic tetraploid potatoes. Ortiz et al. (1993) further indicated overdominance/heterozygosity an important genetic component in productivity of tetrasomic potatoes. But, the importance of maximum heterozyogosity for breeding traits has not proven universal and depends on the genetic background of the genetic materials under specific environments (Bonierbale et al. 1993). When tetraploid offsprings resulting from crosses between adapted materials were tested, positive correlations on tuber yield were observed, but when filial generations from crosses between adapted and unadapted germplasm were examined, no effect on any agronomic trait was observed. However, only three populations were tested to derive this conclusion, and no negative correlation between heterozygosity and yield components was found. While the objectives of the experiments based on application of molecular markers were helpful in elucidating the question, many cases favoring the aforesaid hypothesis must be adduced before it can be accepted.

Estimation of Genetic Variation and Covariances

Theoretical models, sometimes validated by experimental results, are available to determine the significance of allelic diversity and 2n gametes for approaching heterozygosity in tetraploid potatoes (Werner and Peloquin 1991), or to describe the genetic value of tetraploid-diploid hybrids (David et al. 1995). Likewise, other models help to establish associations between genetic markers and quantitative traits (Ortiz and Peloquin 1992), and assist in marker-based analysis of tetrasomic inheritance of quantitative traits (Tai 1994).

Covariances between diploid parent-tetraploid offspring were derived from noninbred parents by computing the coefficient of coancestry and double coancestry (Haynes 1990). The covariance between a haploid species hybrid and its derived tetraploid hybrid offspring, when the same Tuberosum parent was used for haploid induction, and of sexual polyploidization were also investigated (Haynes 1992). This covariance depends on the mechanism of 2n gamete formation and the frequency of single exchange tetrads, and is a function of the ploidy levels involved. These models are becoming important tools for gaining insights into the potato genome.

General and Specific Combining Ability in Tetraploid Potatoes

In plant breeding, determination of combining abilities is an important parameter in selecting parents to be crossed. The GCA of a parental clone provides an assessment of its gametic input, as judged by the mean performance of its progenies from crosses with other clones (Rowe 1967; Bradshaw and Mackay 1994). Specific combining ability (SCA) is the variation in the progeny mean from the expected on the basis of the GCAs of the parents. The methods employed for estimation of these combining abilities somehow become biased due to deviation from random mating populations in equilibrium and due to gametic disequilibrium (Bradshaw 1994; Rowe 1987). Thus, it was cardinal to elaborate methods for more precisely estimating the combining abilities, and also specific mating designs needed for quantitative analysis of tetrasomic tetraploids. Ortiz and Golmirzaie (2002) examined the comparative advantages of hierarchical and factorial mating designs of, namely, North Carolina mating designs I and II respectively, and indicated that North Carolina mating design I is more appropriate for tuber yield. They also suggested that increase in number of female parents from two to four per male is preferred in determining additive variance in design I for the tetrasomic tetraploid potato.

Bradshaw and Mackay (1994) compiled a comprehensive list of studies on the general and specific combining abilities of various traits in tetraploid potatoes. The potential contribution of GCA is relatively higher in many traits, especially tuber yield and quality components than that of SCA in crossing schemes involving closely related parents. However, tuber yield-related components often have relatively low GCA or heritability in 4x-4x crosses and vary depending on matings and experimental conditions, such as locations (for reviews, see Bradshaw and Mackay 1994; Ortiz and Golmirzaie 2002). On the other hand, GCA or heritability of tuber quality, such as internal heat necrosis (IHN) and specific gravity (SG), can be rather high, ranging from 0.83 to 0.88 for IHN and 0.77 to 0.92 for SG (Henninger et al. 2000). For general information on combining ability and/or heritability, see Dale and Mackay (1994).

Ortiz (1997) discussed combining ability tests and their implications in potato breeding, hypothesizing that hybrids from crosses with both parents showing high and positive GCA and SCA may exhibit high yield and stable performance across various environments. Selected clones of such a cross could also assist in genetic gains from traits associated with additive gene variance. On the contrary, clones from crosses having poor SCA may exhibit unstable and poor performance across diverse environments, particularly for traits associated with heterozygosity.

Recent combining ability analyses of potato suggest that loci which mainly affect tuber yield occur predominantly between centromeres and crossovers/chiasmata (Tai and De Jong 1997; Buso et al. 1991). This chromosome segregation provides a means for reducing recombination for this trait in potato (Peloquin et al. 1999).

Complexity of Quantitative Traits in Tetrasomic Tetraploid Potato

Most of the agronomic and pest-resistant traits targeted in breeding programs are quantitatively inherited. Expression of quantitatively inherited traits in potato is markedly complicated due to the nature of tetraploids and genotype—environment interactions. Screening procedures for quantitatively inherited traits often require repeated trials to confirm results.

The dicot potato has cytogenetic characteristics that differ from the monocot wheat, especially in expression of microsporogenesis. Chromosome manipulation, a common feature in wheat breeding, does not appear feasible in potato. At present, distinguishing each chromosome of the potato is not readily accomplished. Recombination of chromosomes originating from various genetic backgrounds is not simple. Thus, the addition of extra chromosome(s) for creation of aneuploids is not readily achievable in potato (Watanabe et al. 1995a).

Conventional potato breeding methods have taken decades to incorporate variable traits from various genetic resources. However, some progress was made in coping with constraints (Peloquin et al. 1989; Ross 1986). Incorporation of knowledge of polyploid genetics, haploidization techniques (Kotch et al. 1992), breaking interspecific barriers by the concept of EBN (Ehlenfeldt and Hanneman 1988), and use of 2n gametes, made access possible to the diverse range of wild and cultivated potato genetic resources, offering a spectrum of resistances to various biological constraints (Peloquin et al. 1989). This has been demonstrated by ploidy manipulation or analytical breeding schemes (reviewed by Peloquin et al. 1989). Nonetheless, despite integration of such knowledge and new concepts, the complicated genetics of potatoes continues to slow down progress in potato breeding (Watanabe 1994; Watanabe et al. 1999a).

Besides the aforesaid impediments, variable traits may include deleterious traits such as high glycoalkaloid content (Ross 1986). Glycoalkaloids (solanidine in *S. tuberosum* and solasodine in *S. berthaultii*) were quantitatively inherited in potatoes and presumably their levels are controlled by recessive genes (Yencho et al. 1998). Modern tools are now available to alleviate pitfalls in conventional germplasm enhancement through incorporation of diverse genetic resources (Watanabe et al. 1997).

Generally, the expression of quantitatively inherited traits depends on the collective interaction of numerous genes, i.e., a polygenic system (Stoskopf et al. 1993). Therefore, many traits, including the classical example of corolla length of tobacco (East 1915), can be explained quantitatively instead of simply by dominant or recessive genes. Such inheritance is common for economically important traits in domesticated animals as well as cultivated plants, and those traits that affect the survival and fertility in wild species (Bradshaw 1994).

GENETICS OF 2n GAMETES WITH 4x-2x BREEDING SCHEMES

Occurrence of 2n Gametes

While both 2n eggs and 2n pollen generally occur in tuber-bearing *Solanum*, it is pertinent that 2n pollen is observed more often in various taxa (Watanabe and Peloquin 1989, 1991, 1993; Werner and Peloquin 1990). Furthermore, the selection process of 2n pollen is easier than that for 2n eggs in a germplasm enhancement program (Watanabe et al. 1996a, b). Occurrence of 2n pollen in an individual potato genotype varies from less than 5% to 80% (Watanabe and Peloquin 1989); however, improvement and selection are more easily done on diploid genotypes with a high frequency of 2n pollen (Qu et al. 1996; Watanabe et al. 1995b; Yerk and Peloquin 1990).

Frequency of 2n eggs in a genotype is low compared with 2n pollen, while frequency of number of individuals having 2n pollen is much more common than that of individuals having 2n eggs (Jongedijik et al. 1991a, b). While breeding can be done by recurrent selection of diploid genotypes with 2n eggs, the selection of 2n pollen is far more convenient.

Genetic Mode and Genetic Consequence of 2n Gametes

The genetic mode of 2n gamete formation is classified into two categories: first division restitution (FDR), which is equivalent to the occurrence of genetic restitution at the first division of meiosis; and second division restitution (SDR), which is equal genetically to the genetic restitution event at the second division of meiosis (Hermsen 1984; Mendiburu and Peloquin 1977a, b; Mok and Peloquin 1975 a, b, c; Peloquin et al. 1989, 1999). It should be noted that the genetic mode and cytological event at meiosis must be considered separate entities. For example, the recessive meiotic mutant gene *ps* (parallel spindles), which controls the formation of 2n pollen, influences the second division at microsporogenesis. However, the genetic mode is equal to an FDR (Mok and Peloquin 1975 a, b).

While considering the higher frequency of SDR over FDR 2n eggs (Werner and Peloquin 1990; Werner et al. 1991), FDR 2n pollen may be more advantageous for transmitting overall heterozygosity. Overall, 4x × 2x with 2n pollen may be preferred to 2x with 2n eggs × 4x crosses for operation in potato breeding. However, when an elite diploid clone is available without conferring 2n pollen producing capacity, then evaluation of occurrence of 2n eggs should definitely be considered.

Genetic Consequences

Models and genetic experiments have been reported by various authors (Carputo et al. 2000; David et al. 1995; Haynes 1990, 1992; Peloquin et al. 1989, 1999; Tai 1994 and Chapter 7 this volume; Watanabe et al. 1991a).

Transmission of the heterozygosity of the diploid parental genome by 2n gametes has been predicted as 80% and 40% respectively for FDR and SDR (Hermsen 1984; Peloquin et al. 1989, 1999). Douches and Quiros (1988) and Jongedijik et al. (1991a) confirmed these estimates using several isozyme markers with known gene-centromere distances. Barone et al. (1995) used RFLP markers on thirteen loci distributed at five chromosomes in the 4x progenies derived from 2x × 2x crosses and estimated 71.4% and 31.8% respectively for FDR and SDR. This was also supported by an independent study conducted by Carputo et al. (2000) in 4x progenies derived from 2x × 2x crosses. They reported a high level of heterozygosity in RFLP marker loci and up to 90% of such heterozygosity was transmitted in the bilateral sexual polyploidization.

Uniformity of genotypes could be expected on 2n gametes over n gametes due to less segregation during meiotic events and, consequently, the mean values in tetraploid progeny from 4x-2x crosses are higher with smaller variation than those from 4x-4x matings (Ortiz et al. 1991). It was also reported that FDR 2n pollen-producing diploid potato clones had progenies with a significantly higher pollen stainability (i.e., inference for a viable pollen) in 4x × 2x matings than those derived from crosses between tetraploids (Ortiz and Peloquin 1994). The authors attributed this observation to the transmission of a high level of heterozygosity and epistasis by FDR 2n pollen to the progeny. The high frequency of pollen stainability correlated significantly with the fruit and seedset in the breeding process and hence may be considered a key factor in conducting breeding with a 4x × 2x scheme.

Preferential Pairing in 4x × 2x Hybridization

It was postulated that preferential pairings between homologous chromosomes derived from the original tetraploid or diploid parent occur in the tetraploid × diploid hybrid progenies, which vary among genetic modes of 2n gamete formation in the diploids (Haynes et al. 1991). Occurrence of random multiple exchange tetrads is unlikely (Wagenvoort and Zimnoch-Guzowska 1992) and segregation distortion can be observed in 4x-2x crosses. Indeed, recent models on chromosome configuration (Wu et al. 2001) and population structure (Ronfort et al. 1998) in tetrasomic tetraploids do not fit with the expected performance in progenies obtained from 4x-2x matings. These progenies often perform far better than plants with normal agronomic traits (reviewed in Ortiz 1998) and exhibit a high frequency of quantitative disease resistances (Watanabe et al. 1999a; Watanabe and Watanabe 2000).

Somatic Chromosome Doubling Vs. 2n Gametes

Somatic chromosome doubling and cell fusion have been considered an alternative method to sexual hybridization, using 2n gamete-producing diploids for ploidy manipulation in tetraploid potatoes (Watanabe et al. 1991a, 1995a). However, the genetic consequences and practical breeding value of the progeny manipulated by somatic chromosome doubling appear less valuable than the progeny obtained from 4x-2x mating using 2n gametes. Somatic chromosome doubling and cell fusion of genetically closely related individuals would conclude in drastic increase in inbreeding with proportionate decrease in heterozygosity. Consequently, the progeny could not perform for traits expressed by heterotic gene loci, such as tuber yield and quality (Maris 1990; Rowe 1967; Tai and De Jong 1997). However, when cases wherein diploid genotype with 2n gametes conferring specific quantitative trait(s) of interest are not available for tetraploid breeding, highly heterozygous individuals shall be selected from somatically doubled diploids to transfer the quantitative trait(s). In spite of overall loss of heterozygosity and vigor, transfer of resistance and high heterozygosity may be possible by crossing with the original diploid in order to reduce the chance of relative loss of general vigor in somatically doubled diploids (Watanabe et al. 1999b).

TRANSMISSION OF TRAITS AND HERITABILITY IN 4x-2x SCHEME

Advantages in General Combining Ability and Heritability

It has been reported that the estimated heritability of tuber yield was not high in 4x × 4x potato breeding: 0.13 (Killick 1977), 0.22 (Plaisted et al. 1962), 0.26 (Tai 1976) and 0.14-0.29 (Ortiz and Golmirzaie 2002), depending on the experimental environments. Only Henninger et al. (2000) indicated extremely high heritability for tuber yield: broad-sense $h^2 = 0.86$. Bradshaw et al. (2000) also pointed out that the offspring-midparent regression coefficient was low for tuber yield and attributed this to significant SCA. On the contrary, the 4x-2x mating scheme could provide high frequency of progeny-individuals with traits of interest (Peloquin et al. 1989, 1999). Peloquin and coworkers at the Univ. of Wisconsin, Madison, WI(USA) demonstrated the potential of the 4x-2x breeding scheme, by obtaining the offspring desired with high frequency. Ortiz et al. (1991) further demonstrated high transmission and heritability of specific traits, e.g. tuber yield, using the same scheme.

Selection of elite diploid clones was the key to successful progeny selectivity under the 4x-2x scheme, particularly for diploid progenitors of early selection cycles (Rouselle and Rouselle 1995). Progress with respect to improvement of diploid progenitors appears far easier than with tetraploids as various combined traits could be obtained in the elite diploid clones in a short period (Watanabe et al. 1994, 1995a, b, 1996a, b).

FDR Vs. SDR in 4x-2x Crosses for Agronomic Traits

Various discussions and experiments have centered on the choice of FDR vs. SDR 2n gametes in 4x-2x matings (Carputo et al. 2000; Peloquin et al. 1989, 1999). Tai (1994) and David et al. (1995) provided some genetic considerations on differences in FDR and SDR mechanisms for modeling the inbreeding effects in 4x-2x crosses.

No differences in total tuber yield and SG were observed in 2x × 4x between FDR and SDR 2n eggs (Douches and Maas 1998). However, due to more inbreeding and lack of genetic diversity, FDR-derived progeny appears to transmit greater heterozygosity than SDR-derived progeny. Hutten et al. (1994) reported that 4x-2x progeny from FDR producing diploid progenitors was superior to progeny resulting from SDR with reference to mean yield, mean underwater weight, and mean tuber size, although vine maturity and chip color between FDR- and SDR-derived offspring were almost the same. However, SDR-derived progeny had a higher variation in progeny mean except for vine maturity. Hence, FDR gametes may possess an overall advantage in general agronomic traits. Buso et al. (1999a) suggested that genes for tuber yield could be between the centromeres and the chromosomal sites exhibiting maximum recombination, so that FDR 2n gametes would be preferable to SDR 2n gametes because of maximum transmission of genetic components for tuber yield. Indeed, Buso et al. (1999b, c, 2000) emphasized 4x-FDR 2x mating as the scheme preferred for unilateral sexual hybridization in the interploidy breeding scheme of potato.

Breeding value of both types of 2n gametes has been examined in interploidy crosses 4x × 2x and 2x × 2x (Concillio and Peloquin 1991; Darmo and Peloquin 1990; Mendiburu and Peloquin 1977a, b; Mok and Peloquin 1975c; Ortiz and Peloquin 1991, 1992; Ortiz et al. 1991). In these reports also the breeding value of 2n gametes was superior for FDR over SDR in tuber yield of the tetraploid progeny resulting from 4x × 2x crosses. Heterotic responses from 4x-FDR 2x matings have shown progeny yield that outperformed midparent values and 4x cultivars in yield trials. Enhanced yield production of the 4x × 2x populations was attributed to the high level of heterozygosity transmission via FDR 2n gametes vis-à-vis SDR. Mendiburu and Peloquin (1977b) explained the lower yield in progenies from SDR 2x × 4x crosses by the fact that SDR 2n eggs only transmit on average 40% of the overall genomic heterozygosity from the diploid parent to the offspring.

The relative advantages of using FDR over SDR 2n gametes have been reported for many other agronomic traits (Buso et al. 1999a; Peloquin et al. 1999). Also, the genetic variances of these agronomic traits in 4x-2x progenies were smaller than those observed in 4x-4x crosses (Buso et al. 1999a, b, c, 2000; Ortiz et al. 1991; Peloquin et al. 1999; Tai and De Jong 1997).

FDR-NCO (Non-Crossover) Vs. FDR-CO (Crossover) in Transmission of Agronomic Traits

It appears that FDR-NCO gametes are superior to FDR-CO gametes (Peloquin et al. 1989). However, Buso et al. (1999a, b, c, 2000) demonstrated that there were no significant differences for several traits, such as total tuber yield, haulm maturity, plant vigor, plant uniformity, eye depth, number of tubers per mound, and commercial yield/total yield index between offspring from 4x × 2x crosses in which both 2x parents were either FDR-NCO or FDR-CO. They also indicated that FDR-CO and FDR-NCO gametes were expected to be genetically equivalent for all loci between centromeres and chromosal sites exhibiting maximum recombination.

Transmission of Quantitative Resistance Traits in 4x-2x Matings in Potato

As shown above, FDR 2n gametes have proven to be a powerful tool for transmission of specific desirable traits from diploid germplasm directly into 4x cultivated potatoes. Indeed, 2n pollen has distinct advantages over 2n eggs in potato improvement due to ease of selection (Iwanaga et al. 1989; Peloquin et al. 1989).

The following pest-resistance traits in potatoes appear to be controlled by quantitative loci: (1) field resistance to late blight (LB) [*Phytophthora infestans* (Meyer et al. 1998)]; (2) early blight (EB) [*Alternaria solani* (Herriott et al. 1986; Herriott and Haynes 1990; Pavek and Corsini 1994)]; (3) common scab [*Streptomyces scabies* (Cipar and Lawrence 1972; Howard 1978; Pfeffer and Effmert 1985)]; (4) bacterial wilt (BW) [*Pseudomonas solanacearum* (Tung et al. 1990a, b; Watanabe et al. 1992; Watanabe et al. 1999c)]; (5) blackleg and bacterial soft rot [*Erwinia carotovora* (Lellbach 1978)]; (6) potato leaf roll virus (PLRV) (Peters and Jones 1981); (7) cucumber mosaic virus (CMV) (Valkonen and Watanabe 1999); (8) potato cyst nematodes (PCN) [*Globodera rostochiensis* and *G. pallida* (Dale and Phillips 1982)]; (9) root-knot nematodes (RKN) [*Meloidogyne* spp. (Gomez et al. 1983; Janssen et al. 1995)]; and (10) potato tuber moth (PTM) [*Phthorimaea operculella* (Ortiz et al. 1990; Raman et al. 1981; Watanabe et al. 1999b)]. Types A and B glandular trichomes on leaves are also controlled by quantitative loci; they confer resistance to a broad range of insects and mites (Bonierbale et al. 1994).

Previous research has demonstrated transmission of resistance to biotic stresses in 4x cultivars via FDR 2n pollen, e.g. RKN, race 3 (Iwanaga et al. 1989; Watanabe et al. 1991b), BW (Watanabe et al. 1992), EB (Herriott and Haynes 1990), PCN (Ortiz et al. 1997), and common scab (Murphy et al. 1995). Since in these studies, the resistance transmitted showed varying degrees of frequency or heritability, prediction in the progeny is very important for effective deployment of such resistances through 4x × 2x crosses.

Transmission of Combined Traits Related to Pest Resistance with 4x-2x

An example of combining resistance to BW and RKN best explains the process of transmission. Potato cultivars resistant to BW became susceptible when RKN and BW infection occurred simultaneously, especially in hot climates (Ortiz et al. 1994). Hence breeding for combined resistance to these two pests is of paramount importance, especially in tropical and subtropical regions in developing countries. Watanabe et al. (1999a) reported high frequency of combined resistance to RKN and BW in 4x-2x crosses. About 85 clones from 557 progeny individuals (15.3%) were obtained from these crosses. Although no specific mating design was applied in this experiment, the diploid clone "381348.7" with resistance to both BW and RKN, showed a high frequency of combining resistance. These results suggested that transmission of combined resistance from the 2x clones to a 4x population via 2n pollen should be effective. In a previous study at the diploid level, 22 clones with combined resistance to at least two different pests (equivalent to 2.2%) were identified among 990 seedlings tested (Watanabe et al. 1996b). The much higher frequency of combined resistance (15.3%) detected later (Watanabe et al. 1999a). was probably due to more advanced stages of selection in the diploid population. It was also found that combined resistances with LB and BW, or BW plus RKN and glandular trichomes A, were likewise readily selected from relatively small populations emanating from 4x-2x crosses.

Efficiency of Multiple Quantitative Pest Resistance Traits in 4x × 2x Crosses Via FDR 2n Pollen

Watanabe et al. (1998, 1999a, c) indicated that (1) transmission of combined quantitative traits to pest resistance was possible via FDR 2n pollen and (2) quantitative traits for LB resistance as well as the presence of glandular trichomes (type A) could be transmitted to progenies with a high frequency (Watanabe et al. 1999a). Transmission rate for combining quantitative pest resistance traits in conventional 4x × 4x breeding is often below 5% and selection in such a mating scheme may require a far larger population size (Watanabe et al. 1999a). Furthermore, the desirable combination of quantitative resistance traits, such as BW + RKN in the tropics and subtropics, and LB + BW + RKN in the lowland tropics is likewise worth pursuing (Ortiz et al. 1994; Ross 1986).

CONCLUSION

Breeding cultivated tetraploid potatoes has been a challenge given the complexity of their polysomic genetic nature but steady advances have been, and are being made, by talented breeders concomitant with elucidation of trait genetics. However, more systematic research is needed for genetic analysis of the variations observed across diverse experimental environments to ascertain whether some general principles regarding these variations can be drawn. The 4x-2x matings have proven useful in the transmission of specific traits and even combining quantitative traits located on different genomic chromosomes. With integration of functional genomics into polyploid genetic research, further development of theory and resultant strategies for better utilization of 2n gametes can be expected. This would be a golden opportunity for enhancing available genetic resources, in particular from wild species, to alleviate the pitfalls entrained by shortage of resources in potato breeding and to ensure greater efficiency in the overall breeding process.

Acknowledgment

KNW acknowledges financial support from the Japan Society for the Promotion of Sciences, Research for Future Programs (grants RFTF 96-L603 and RFTF-00-L01602).

REFERENCES

Barone, A., C. Gebherdt, and L. Frusciante. 1995. Heterozygosity in 2n gametes of potato evaluated by RFLP markers. Theor Appl Genet 91: 98–104.

Barone, A., A. Sebastiano, T. Cardi, and L. Frusciante. 2002. Evidence for tetrasomic inheritance in a tetraploid *Solanum commersonii* (+) *S. tuberosum* somatic hybrid through the use of molecular markers. Theor Appl Genet 104: 539–546.

Bennett, J. H. 1976. Expectation for inbreeding depression on self-fertilization of tetraploids. Biometrics 32: 449–452.

Bever, J. D. and F. Felber. 1992. The theoretical population genetics of autopolyploidy. Oxford Surv Evol Biol 8: 185–217.

Bonierbale, M. W., R.L. Plaisted, and S. D. Tanksley. 1993. A test of the maximum heterozygosity hypothesis using molecular markers in the tetraploid potatoes. Theor Appl Genet 86: 481–491.

Bonierbale, M. W., R. L. Plaisted, O. Pineda, and S. D. Tansley. 1994. QTL analysis of trichomemediated insect resistance in potato. Theor Appl Genet 87: 973–987.

Bradshaw, J. E. 1994. Quantitative genetics theory for tetrasomic inheritance. *In:* J. E. Bradshaw and G. R. Mackay (eds.), Potato Genetics. CAB Intl, Wallingford, UK, pp. 71–100.

Bradshaw, J. E. and G. R. Mackay. 1994. Breeding strategies for clonally propagated potatoes. *In:* J. E. Bradshaw and G. R. Mackay (eds.), Potato Genetics. CAB Intl, Wallingford, UK, pp. 467–497.

Bradshaw, J.E., D. Todd, and R. N. Wilson. 2000. Use of tuber progeny tests for genetical studies as part of a potato (*Solanum tuberosum* subsp. *tuberosum*) breeding programme. Theor Appl Genet 100: 772–781.

Buso, J.A., L. S. Boiteux, G. C. C. Tai, and S. J. Peloquin. 1999a. Chromosome regions between centromeres and proximal crossovers are the physical sites of major effect loci for yield in potato: Genetic analysis employing meiotic mutants. Proc Natl Acad Sci USA 96: 1773–1778.

Buso, J. A., L. S. Boiteux, and S. J. Peloquin. 1999b. Multitrait selection system using populations with a small number of interploid (4x-2x) hybrid seedlings in potato: degree of high-parent heterosis for yield and frequency of clones combining quantitative agronomic traits. Theor Appl Genet 99: 81–91.

Buso, J. A., F. J. B. Reifschneider, L. S. Boiteux, and S. J. Peloquin. 1999c. Effects of 2n-pollen formation by first meiotic division restitution with and without crossover on eight quantitative traits in 4x-2x potato progenies. Theor Appl Genet 98: 1311–1319.

Buso, J. A., L. S. Boiteux, and S. J. Peloquin. 2000. Heterotic effects for yield and tuber solids and type of gene action for five traits in 4x potato families derived from interploid (4x-2x) crosses. Plant Breeding 119: 111–117.

Carputo, D., A. Barone, and L. Frusciante. 2000. 2n gametes in the potato: essential ingredients for breeding and germplasm transfer. Theor Appl Genet 101: 805–813.

Cipar, M. S. and C. H. Lawrence. 1972. Scab resistance to haploid from two *Solanum tuberosum* cultivars. Amer Potato J 49: 117–120.

Concillio, L. and S. J. Peloquin. 1991. Evaluation of the 4x-2x breeding scheme in a potato breeding program adapted to local conditions. J Genet Breed 45: 13–18.

Dale, M. F. B. and M. S. Phillips. 1982. An investigation of resistance to the white potato cyst nematode. J Agric Sci 99: 325–328.

Dale, M. F. B. and G. R. Mackay. 1994. Inheritance of table and processing quality. *In:* J. E. Bradshaw and G. R. Mackay (eds.), Potato Genetics. CAB Intl, Wallingford, UK, pp. 285–318.

Darmo, E. and S.J. Peloquin. 1990. Performance and stability of nine 4x clones from 4x-2x crosses and four commercial cultivars. Potato Res 33: 357–365.

David, J.L., P. Boudec, and A. Gallais. 1995. Quantitative genetics of 4x-2x populations with first-division restitution and second-division restituion 2n gametes produced by diploid parents. Genetics 139: 1797–1803.

Douches, D. S. and C. F. Quiros. 1988. Genetic recombination in a diploid synaptic mutant and a *Solanum tuberosum* x *S. chacoense* diploid hybrid. Heredity 60: 183–191.

Douches, D. S. and D. L. Maas. 1998. Comparison of FDR- and SDR-derived tetraploid progeny from 2x × 4x crosses using haploids of *Solanum tuberosum* L. that produce mixed modes of 2n eggs. Theor Appl Genet 97: 1307–1313.

East, E. M. 1915. Studies on size inheritance in *Nicotiana*. Genetics 1 : 164–176.

Ehlenfeldt, M. K., and R. E. Hanneman, Jr. 1988. Genetic control of Endosperm Balance Number (EBN): three additive loci in a threshold-like system. Theor Appl Genet 75: 825–832.

Golmirzaie, A. M., R. Ortiz, G. N. Atlin, and M. Iwanaga. 1998a. Inbreeding and true seed in tetrasomic potato. I. Selfing and open pollination in Andean landraces (*Solanum tuberosum* Gp. Andigena). Theor Appl Genet 97: 1125–1128.

Golmirzaie, A. M., K. Bretschneider, and R. Ortiz. 1998b. Inbreeding and true seed in tetrasomic potato. II. Selfing and sib-mating in heterogeneous hybrid populations of *Solanum tuberosum*. Theor Appl Genet 97: 1129–1132.

Gomez, P. L., R. L. Plaisted, and H. D. Thurston. 1983. Combining resistance to *Meloidogyne incognita, M. javanica, M. arenaria,* and *Pseudomonas solanecearum* in potatoes. Amer Potato J 60: 353–360.

Haynes, K. G. 1990. Covariances between diploid parent and tetraploid offspring in tetraploid × haploid-species hybrid in 4x-2x crosses of *Solanum tuberosum*. J Hered 81: 208–210.

Haynes, K. G. 1992. Covarince between haploid species hybrid and *tuberosum* × haploid species hybrid in 4x-2x crosses of *Solanum tuberosum* L. J Hered 83: 119–122.

Haynes, K. G. and D. S. Douches. 1993. Estimation of the coefficient of double reduction in the cultivated tetraploid potato. Theor Appl Genet 85: 857–862.

Haynes, K. G., W. E. Potts, and M. J. Camp. 1991. Estimation of preferential pairing in tetraploid × diploid hybridizations. Theor Appl Genet 81: 504–508.

Henninger, M. R., S. B. Sterrett, and K. G. Haynes. 2000. Broad-sense heritability and stability of internal heat necrosis and specific gravity in tetraploid potatoes. Crop Sci 40: 977-984.

Hermsen, J. G. Th. 1984. Mechanisms and genetic implication of 2n gamete formation. Iowa State J Res 58: 421–434.

Herriott, A. B. and F. L. Haynes. 1990. Inheritance of resistance to early blight disease in tetraploid × dipiold crosses of potatoes. HortSci 25: 224–226.

Herriott, A. B., F. L. Haynes, Jr., and P. B. Shoemaker. 1986. The heritability of resistance to early blight in diploid potatoes (*Solanum tuberosum* subspp. *phureja* and *S. stenotomum*). Amer Potato J 63: 229–232.

Howard, H. W. 1978. The production of new varieties: *In:* P. M. Harris (ed.), The Potato Crop. Chapman & Hall, London, UK, pp. 607–646.

Hutten, R. C. B., M. G. M. Schippers, J. G. Th. Hermsen, and M. S. Ramanna. 1994. Comparative performance of FDR and SDR progenies from reciprocal 4x-2x crosses in potato. Theor Appl Genet 89: 545–550.

Iwanaga, M., P. Jatala, R. Ortiz, and E. Guevara. 1989. Use of FDR 2n pollen on transfer resistance to root-knot nematodes into cultivated 4x potatoes. J Amer Hort Sci 114: 1108–1013.

Janssen, G. J. W., A. van Norel, B. Verkerk-Bakker, and R. Janssen. 1995. Detecting resistance to the root-knot nematodes *Meloidogyne hapla* and *M. chitwoodi* in potato and wild potato *Solanum* spp. Potato Res 38: 353–362.

Jongedijik, E., R. C. B. Hutten, J. M. A. S. A. van der Wolk, S. I. J. Sxhuurmans- Stekhoven. 1991a. Synaptic mutants in potato, *Solanum tuberosum* L. III. Effects of the *Ds-1/ds-1* locus (desynapsis) on genetic recombination in male and female meiosis. Genome 34: 121–130.

Jongedijik, E., M. S. Ramanna, Z. Sawor, and J. G. Th. Hermsen. 1991b. Formation of first division restitution (FDR) 2n-megaspores through pseudohomotypic division in *ds-1* (desynapsis) mutants of diploid potato: routine production of tetraploid progeny from 2x FDR × 2x FDR crosses. Theor Appl Genet 82 : 645–656.

Killick, R. J. 1977. Genetic analysis of several traits in potato by means of diallel cross. Ann Appl Biol 86: 279–289.

Kotch, G. P., R. Ortiz, and S. J. Peloquin. 1992. Genetic analysis by use of potato haploid population. Genome 35: 103–108.

Lellbach, H. 1978. Estimating genetic parameters derived from diallel crosses in cases of susceptibility to soft rot in the potato. Archiv für Zuchtungsforchung, 8: 193–199.

Maris B. 1990. Comparison of diploid and tetraploid potato families derived from *Solanum phureja* × dihaploid *S. tuberosum* hybrids and their vegetatively doubled counterparts. Euphytica 46: 15–33.

Mendiburu, A. and S. J. Peloquin. 1997a. Bilateral sexual polyploidization in potatoes. Euphytica 26: 573–583.

Mendiburu, A. and S. J. Peloquin. 1977b. The significance of 2n gametes in potato breeding. Theor Appl Genet 19: 53–61.

Mendoza, H. A. 1989. Population breeding as a tool for germplasm enhancement. Amer Potato J 66: 639–653.

Mendoza, H. A. and F. L. Haynes. 1974. Genetic basis of heterosis for yield in the autotetraploid potato. Theor Appl Genet 45: 21–25.

Meyer, R. C., D. Milbourne, C. A. Hackett, J. E. Bradshaw, J. W. McNichol, and R. Waugh. 1998. Linkage analysis in tetraploid potato and association of markers with quantitative resistance to late blight (*Phytophthora infestans*). Mol Gen Genet 259: 150–160.

Mok, D. W. S. and S. J. Peloquin. 1975a. Three mechanisms of 2n pollen formation in diploid potatoes. Can J Genet Cytol 17: 217–225.

Mok, D. W. S. and S. J Peloquin. 1975b. The inheritance of three mechanisms of diplandroid (2n pollen) formation in diploid potatoes. Heredity 35: 295–302.

Mok, D. W. S. and S. J. Peloquin. 1975c. Breeding value of 2n pollen (diplandroids) in tetraploid × diploid crosses in potatoes. Theor Appl Genet 46: 307–314.

Murphy, A. M., H. de Jong and G. C. C. Tai. 1995. Transmission of resistance to common scab from the diploid to the tetraploid level via 4x-2x crosses in potatoes. Euphytica 82: 227–233.

Ortiz, R. 1997. Breeding for potato production from true seed. Plant Breed Abst 67: 1355–1360.

Ortiz, R. 1998. Potato breeding via ploidy manipulation. Plant Breed Rev 16: 15–85.

Ortiz, R. and S. J. Peloquin. 1991. Breeding for 2n egg production in haploid × species 2x potato hybrids. Amer Potato J 68: 691–703.

Ortiz, R. and S. J. Peloquin. 1992. Recurrent selection for 2n gamete production in 2x potatoes. J Genet Breed 46: 383–390.

Ortiz, R. and S. J. Peloquin. 1994. Effect of sporophytic heterozygosity on the male gametophyte of the tetraploid potato (*Solanum tuberosum* L.). Ann Botany 73: 61–64.

Ortiz, R. and A.M. Golmirzaie. 2002. Hierarchical and factorial mating designs for quantitative genetic analysis in tetrasomic potato. Theor Appl Genet 104: 675–679.

Ortiz, R., M. Iwanaga, and S. J. Peloquin. 1994. Breeding potatoes for developing countries. J Genet Breed 48: 89–98.

Ortiz, R., F. Javier, and M. Iwanaga. 1997. Transfer of resistance of potato cyst nematode (*Globodra pallida)* into cultivated potato *Solanum tuberosum* through first division restitution 2n pollen. Euphytica 96: 339–344.

Ortiz, R., M. Iwanaga, K. V. Raman, and M. Palacios. 1990. Breeding for resistance to potato tuber moth *Phthorimaea operculella* (Zeller). Euphytica 50: 119–126.

Ortiz, R., S. J. Peloquin, R. Freyre, and M. Iwanaga. 1991. Efficiency of potato breeding using FDR 2n gametes for multitrait selection and progeny testing. Theor Appl Genet 82: 602–608.

Ortiz, R., D. S. Douches, G. P. Kotch, and S.J. Peloquin. 1993. Use of haploids and isozyme markers for genetic analysis in the polysomic polyploid potato. J Genet Breed 47: 283–288.

Pavek, J. J. and D. L. Corsini, 1994. Inheritance of resistance to warm-growing season fungal diseases. *In:* J. E. Bradshaw, and G. R. Mackay (eds.), Potato Genetics. CAB Intl, Wallingford, UK, pp. 403–409.

Peloquin, S. J., L. Boiteux, and D. Carputo. 1999. Meiotic mutants in potato: valuable variants. Genetics 153: 1493–1499.

Peloquin, S.J., G.L. Yerk, J.E. Werner, and E. Darmo 1989. Potato breeding with haploids and 2n gametes. Genome 31: 1000-1004.

Peters, D. and R. A. C. Jones 1981. Potato leaf roll virus. *In:* W. J. Hooker (ed.), Compendium of Potato Disease. Amri Phytopath Soc, St. Paul, MN. USA, pp. 68–70.

Pfeffer, C. and M. Effmert. 1985. Die Züchtung homozygoter Eltern für Resistenz gegens Kartoffelschorf, verursacht durch Sreptomyces scabies (Thaxt.) Waksman & Henrich. Archiv für Züchtungsforschung, 15: 325–333.

Plaisted, R. L., L. Sanford, W. T. Federer, A. E. Kehr, and L. C. Peterson 1962. Specific and general combining ability for yield in potato. Amer Potato J 61: 395–403.

Qu, D. G., Z. Dewei, M.S. Ramanna, and E. Jacobsen. 1996. A comparison of progeny from diallelic crosses of potato diploid with regard to the frequencies of 2n-pollen grains. Euphytica 92: 313–320.

Raman, K. V., M. Iwanaga, M. Palacio, and R. Egusquiza. 1981. Breeding for resistance to potato tuber worm *Phthorimaea operculella* (Zeller). Amer Potato J 58: 516.

Ronfort, J., E. Jenczewski, T. Bataillon, and F. Rousset. 1998. Analysis of population structure in autotetraploid species. Genetics 150: 921–930.

Ross, H. 1986. Potato Breeding—Problems and Perspectives, 132. Verlag Paul Parey, Berlin, Germany.

Rouselle, B. F. and P. Rouselle. 1995. Agronomic and technological evaluation and selection of tetraploid clones of potato (*Solanum tuberosum* L.) originating from diploid populations. Agronomie 15: 284–293.

Rowe, D. E. 1987. Theoretical value of estimates of general combining ability in the autotetraploid crop. Theor appl Genet 73: 537–541.

Rowe, P. R. 1967. Performance of diploid and vegetatively doubled clones of *Phureja*-haploid *tuberosum* hybrids. Amer Potato J 44: 195–203.

Stoskopf, N., C. Tomes, T. Dwight, and B. R. Christie. 1993. Genetic Variation in Plants, *In:* Plant Breeding: Theory and Practice. Westview Press, Boulder, CO. USA, pp. 87–120.

Tai, G. C. C. 1976. Estimation of general and specific combining ability in potato. Can J Genet Cytol 18: 463–470.

Tai, G. C. C. 1982a. Estimation of double reduction and genetic parameters of autotetraploids. Heredity 49: 63–70.

Tai, G. C. C. 1982b. Estimation of double reduction and genetic parameters in autotetraploids based on 4x-2x and 4x-4x matings. Heredity 49: 331–335.

Tai, G. C. C. 1994. Use of 2n gametes. *In:* J. E. Bradshaw and G. R. Mackay (eds.). Potato Genetics. CAB Intl, Wallingford, UK, pp. 109–132. ISBN 0-85198-869-5.

Tai, G. C. C. and H. De Jong. 1997. A comparison of performance of tetraploid progenies produced by diploid and their vegetatively doubled (tetraploid) counterpart parents. Theor Appl Genet 94: 303–308.

Tung, P. X., E T. Rasco, Jr., P. Vander Zaag, and P. Schmiediche. 1990a. Resistance to *Pseudomonas solanacearum* in the potato: I. Effect of source of resistance and adaptation. Euphytica 45: 203–210.

Tung P. X., E T. Rasco, Jr., P. Vander Zaag and P. Schmiediche, 1990b. Resistance to *Pseudomonas solanacearum* in the potato: II. Aspects of host—pathogen—environment interaction. Euphytica 45: 211–215.

Valkonen, J. P. T. and K. N. Watanabe. 1999. Autonomous cell death, temperature-sensitivity and the genetic control associated with resistance to cucumber mosaic virus (CMV) in diploid potato (*Solanum* spp.) Theor Appl Genet 99: 996–1005.

van Eck, H. J., J.M. E. Jacobs, P. Stam, J. Ton, W. J. Stiekma, and E. Jacobsen. 1994. Multiple alleles for tuber shape in diploid potato detection by qualitative and quantitative genetic analysis using RFLPs. Genetics 136: 303–309.

van de Wal, M. H. B. J., E. Jacobsen, and R. G. Visser. 2001. Multiple allelism as a control mechanism in metabolic pathways: GBSSI allelic composition affects the activity of granule-bound starch synthease I and starch composition in potato. Mol Genet Genomics 265: 1011–1021.

Wagenvoort, M. and E. Zimnoch-Guzowska. 1992. Gene-centromere mapping in potato by half-tetrad analysis: map distance of H_l, Rx, and Ry and their possible use for ascertaining the mode of 2n-pollen formation. Genome 35: 1–7.

Watanabe, J. A. and K. N. Watanabe. 2000. Pest resistance traits controlled by quantitative loci and molecular breeding strategies in tuber-bearing *Solanum*. Plant Biotec 17(1): 1–16.

Watanabe, J. A., R. Ortiz, and K. N. Watanabe. 1998. Resistance to Potato Late Blight (*Phytophthora infestans* [Mont.] de Bary) in crosses between resistant tetraploids and susceptible diploids. Mem Fac Biol Sci Tech, Kinki Univ 4: 65–72.

Watanabe, J. A, M. Orrillo, and K. N. Watanabe. 1999a. Frequency of multiple quantitative pest resistance traits in 4x × 2x crosses of potato. Breed Sci 49(2): 53–62.

Watanabe, J. A., M. Orrillo, and K. N. Watanabe. 1999b. Evaluation of in vitro chromosome-doubled regenerates with resistance to potato tuber moth (*Phthorimaea opercullella* (Zeller)). Plant Biotec 16(3): 223–228.

Watanabe J. A., M. Orrillo, and K. N. Watanabe. 1999c. Resistance to bacterial wilt (*Pseudomonas solanacearum*) of potato evaluated by survival and yield performance at high temperatures. Breed Sci 9: 63–68.

Watanabe, K. N. 1994. Potato molecular genetics. In J. E. Bradshaw and G. R. Mackay (eds.), Potato Genetics. CAB Intl, Wallingford, UK, pp. 213–235.

Watanabe, K. N. and S. J. Peloquin. 1989 Occurrence of 2n pollen and ps gene frequencies in cultivated groups and their related wild species in tuber-bearing Solanums. Theor Appl Genet 78: 329–335.

Watanabe, K. N. and S. J. Peloquin. 1991. The Occurrence and frequency of 2n pollen in 2x, 4x, and 6x wild tuber-bearing *Solanum* species from Mexico, and Central and South Americas. Theor Appl Genet 82: 621–626.

Watanabe, K. N. and S. J. Peloquin. 1993. Cytological basis of 2n pollen formation in a wide range of 2x, 4x, and 6x wild taxa of tuber-bearing *Solanum*. Genome 36: 8–13.

Watanabe, K. N., S. J. Peloquin, and T. Endo. 1991a. Genetic significance of mode of polyploidization: Somatic doubling or 2n gametes? Genome 34: 28–34.

Watanabe, K. N., M. Iwanaga, H. El-Nashaar, and P. Jatala. 1991b. Transmission of resistances on root-knot nematodes and bacterial wilt by 2n pollen in potatoes. Proc 2nd. Intl Symp Chrom Eng Plant, Univ. Missouri, Columbia, MO, USA, August 13–15, 1990, pp. 128–132.

Watanabe, K. N., H. EL-Nashaar, and M. Iwanaga. 1992. Transmission of bacterial wilt resistance by FDR 2n pollen via 4x × 2x crosses in potatoes. Euphytica 60: 21–26.

Watanabe, K. N., M. Orrillo, and A. M. Golmirzaie. 1995a. Potato germplasm enhancement for resistance to biotic stresses at CIP. Conventional and biotechnology-assisted approaches using a wide range of *Solanum* species. Euphytica 85: 457–464.

Watanabe, K. N., M. Orrillo, S. Vega, M. Iwanaga, R. Ortiz, R. Freyre, G. Yerk, S. J. Peloquin, and K. Ishiki. 1995b. Selection of diploid potato clones from diploid haploid-species F_1 families for short day conditions. Breed Sci 45(3): 341–348.

Watanabe, K. N., A. M. Golmirzaie, and P. Gregory 1997. Use of biotechnology tools in potato genetic resources management and breeding. *In:* K. N. Watanabe and E. Pehu (eds.). Plant Biotechnology and Plant Genetic Resources for Sustainability and Productivity. R. G. Landes Company, Austin, TX, USA, pp. 143–154.

Watanabe, K. N., M. Orrillo, S. Perez, J. Crusado, and J. A. Watanabe. 1996a. Testing yield of diploid potato breeding lines for cultivar development. Breed Sci 46(3): 245–250.

Watanabe, K. N., M. Orrillo, S. Vega, S. Perez, J. Crusado, A. M. Golmirzaie, and J. A. Watanabe. 1996b. Generation of pest resistant, diploid potato germplasm with short day adaptation from diverse range of genetic stocks. Breed Sci 46(4): 327–336.

Watanabe, K. N., M. Orrillo, M. Iwanaga, R. Ortiz, R. Freyre, and S. Perez. 1994. Diploid germplasm derived from wild and landrace genetic resources. Amer Potato J 71: 599–604.

Werner, J. E. and S. J. Peloquin. 1990. Inheritance and two mechanisms of 2n egg formation in 2x potatoes. J Hered 81: 371–374.

Werner, J. E. and S. J. Peloquin. 1991. Yield and tuber characteristics of 4x progeny from 2x × 2x crosses. Potato Res 34: 261–267.

Werner, J. E., D. S. Douches, and R. Freyre. 1991. Use of half-tetrad analysis to discriminate between two types of 2n egg formation in a potato haploid. Genome 35: 471–475.

Wricke, G. and W. E. Weber. 1986. Quantitative Genetics and Selection in Plant Breeding. Walter de Gruyter Publ, Berlin.

Wu, S.S., R. Wu, C.-X. Ma, Z.-B. Zeng, M. C. K. Yang, and G. Casella. 2001. A multivalent pairing model of linkage analysis in autotetraploids. Genetics 159: 1339–1350.

Yamada, T., K. Hosaka, K. Nakagawa, S Misoo, and O. Kamijima. 1998. Cytological and molecular characterization of BCI progeny from two somatic hybrids between dihaploid *Solanum acaule* and tetraploid *S. tuberosum.* Genome 41: 743–750.

Yencho, G. C., S. P. Kowalski, R S. Kobayashi, S. L. Sinden, M. W. Bonierbale, and K. L. Deahl. 1998. QTL mapping of foliar glycoalkaloid aglycones in *Solanum tuberosum* × *S. berthaultii* potato progenies: quantitative variation and plant secondary metabolism Theor Appl Genet 97: 563–574.

Yerk, G.L. and S. J. Peloquin. 1990. Performance of haploid × wild species hybrids (involving five newly evaluated species) in 4x × 2x families. Amer Potato J 67: 405–417.

5

Improvement at the Diploid Species Level

Kathleen G. Haynes[1] **and Wenhe Lu**[2]

[1] *USDA/ARS, Vegetable Laboratory, Plant Sciences Institute, 10300 Baltimore Ave., Building 010A, Room 312 Beltsville, MD 20705 (USA).*

[2] *Northeast Agricultural University, Dept. Agronomy, Harbin, Heilongjiang 150030, People's Republic of China*

In South America, ten *Solanum* species and/or subspecies, including diploids, triploids, and tetraploids, are still cultivated (Spooner and Hijmans 2001). However, commercial potatoes are primarily tetraploid in the rest of the world. Objectives in most tetraploid cultivar breeding programs include: improving yield, reducing susceptibility to biotic and abiotic stresses, and enhancing those qualities important to the end-use market. Thus, high specific gravity (SG) and low reducing sugar content are important traits for a cultivar destined for the processing market, whereas more moderate SG, nutritional factors, taste and culinary characteristics are important traits for a cultivar destined for the fresh market. With over 200 potato species from Central and South America, the majority of them diploid (Hawkes 1990), there is a wealth of germplasm that can potentially contribute allelic diversity for traits of economic importance and for the development of new commercial cultivars. Some of the genetic manipulations necessary to incorporate genetic material from these diverse species into a tetraploid form have been discussed in other chapters of this book and are not included here, namely: introgression of genes from wild species; breeding potential and combining ability in 4x-2x crosses; and applications of cell and tissue culture. The purpose of this chapter is to review recent progress in population improvement at the diploid species level *per se*.

Until recently, little information was available on the genetics of the potato (Tarn et al. 1992). Most quantitative genetic theory was developed for either diploid breeding populations resulting from random mating

Corresponding author: Dr. Kathleen G. Haynes. Tel: 01-301-504-7405, fax: 01-301-504-5555, e-mail: haynesk@ba.ars.usda.gov and in linkage equilibrium or for inbred lines and generations subsequently derived from them (Nyquist 1991). Since few diploid potato populations which fulfill these requirements exist in nature, in breeders' germplasm bases, or in potato germplasm collections, it is not surprising that genetic information has been scarce. Although a quantitative genetic theory for tetrasomic inheritance was developed (Wricke and Weber 1986), it is much more complex than disomic inheritance and based on the same assumptions held for diploid populations. For these reasons, tetraploid potato populations which fulfill the necessary requirements are even fewer.

Many factors contribute to the lack of random mating and linkage equilibrium in potatoes. The primary factor is that cultivars are vegetatively propagated clones. In addition, all of the following, which occur to some extent in all *Solanum* species, greatly reduce the chances for establishing a random-mating population in linkage equilibrium: low female fertility (Trognitz 1995), male sterility (Birhman and Kaul 1989), cytoplasmic male sterility (Amoah and Grun 1988), self-incompatibility (Abdalla and Hermsen 1971), and cross-incompatibility (Camadro and Peloquin 1981; Fritz and Hanneman 1989).

Interspecific hybridizations produce plants that are typically not sterile since the chromosomes of the different tuber-bearing *Solanum* species are not sufficiently structurally differentiated to preclude chromosome pairing during meiosis (Ramanna and Hermsen 1979). However, the Endosperm Balance Number (EBN) must be considered when interspecific crosses between *Solanum* species are attempted. Diploid potato species have an EBN of 1 or 2 (Hawkes and Jackson 1992; Hanneman 1994). Most of the South American diploid *Solanum* species are 2EBN, whereas, many of the Mexican diploid *Solanum* species are 1EBN. The EBN hypothesis is based on preserving a female: male EBN ratio of 2:1 in the endosperm (Johnston et al. 1980). Species with the same EBN can be readily sexually hybridized; in species with different EBN numbers the endosperm aborts when hybridizations are attempted, although a few exceptions to this general rule have been reported (Jackson and Hanneman 1999). Since tetraploid potato species have an EBN of 2 or 4, it might appear at first glance that most of the diploid species could not be crossed with tetraploids. However, the production of unreduced gametes in diploid species (discussed elsewhere, this volume) can allow for the successful crossing of tetraploid (4EBN) potato species with diploid (2EBN) potato species, which produce 2n gametes (4x-2x), or tetraploid (2EBN) potato species with diploid (1EBN) potato species which produce 2n gametes. (4x – 2x).

Nevertheless, there is considerable interest in breeding potatoes at the diploid level for a number of reasons; foremost among them is that disomic inheritance is much simpler than tetrasomic inheritance. Of almost equal importance is the recognition of the narrow genetic base existing among commercially available potato cultivars (Mendoza and Haynes 1974; Simmonds 1969) which limits its breeding progress. Nowhere is this more apparent than the documented lack of progress in improving potato yield in North American cultivars during the 20th century, despite years of public and private breeding efforts (Douches et al. 1996). This is even more amazing given the fact that 18 of 22 North American potato breeding programs listed yield and quality as their top breeding priority in a survey conducted in 1986 (Pavek 1987). Commercial potato cultivars are also markedly deficient in resistance to many fungal, bacterial, and viral pathogens as well as numerous insect and nematode pests. Adequate levels of resistance to pests are lacking in *S. tuberosum* germplasm. Worldwide, an estimated 22% of the potato yield is lost each year due to these biotic stresses (Ross 1986). This lack of resistance has made potato production in the developed countries one of the most intensely managed crops ever grown. Numerous traits of economic importance have been identified through screening accessions within existing germplasm collections, the results of which are available to researchers (Hanneman and Bamberg 1986; Hoekstra and Seidewitz 1987).

Germplasm improvement can be defined as transferring useful genes from exotic or wild types into agronomically acceptable backgrounds (Roath 1989). The need for a comprehensive improvement program depends on the role of the donor species (Tarn et al. 1992). If the primary role of the donor is to contribute only one specific trait from among those which it possesses, then the need for a comprehensive amelioration program is less than if the primary role of the donor is the addition of genetic variability for many traits (Tarn et al. 1992). In potatoes, these two distinctly different roles for donor species have given rise to different breeding strategies to capture and exploit the genetic diversity and valuable economic traits that exist in diploid species, such as utilizing (di)haploid species hybrids and breeding diploid species *per se*.

Worldwide, the (di)haploid-species breeding approach is practiced much more than breeding at the diploid species level. For a more detailed discussion of the first approach, see other chapters of this volume; it is briefly summarized here for the purpose of comparison of the outcomes of the two approaches given later.

Haploids ($2n = 2x = 24$) can be isolated from tetraploid *S. tuberosum* (Peloquin et al. 1996) by a number of different techniques but the one preferred is pollination of tetraploid potatoes with a dihaploid inducer clone from *S. phureja* (Hougas and Peloquin 1957). These (di)haploids are then crossed with diploid wild species, which is why the progeny have been termed "haploid-species hybrids". The haploids in haploid-species hybrids contribute adaptation and tuberization ability to the progeny of the cross (Hermundstad and Peloquin 1986). The wild species contributes the genetic trait(s) of interest lacking in the cultivated species to the progeny of the cross. Research on improving the population of haploids has mostly focused on tuberization, ability to transmit tuberization to the progeny (Hermundstad and Peloquin 1986), tuberization combined with number of fruits per pollination, seeds per fruit, and gene(s) for parallel spindles (Yerk and Peloquin 1990).

Little research has been undertaken on improving wild *Solanum* species since they are generally short-day adapted species that fail to tuberize under long-day conditions or produce tubers which are small and few in number. These tubers exhibit a range of dormancy from very long to none, a wide variety of shapes and colors, have deep eyes, and deep apical and stem ends. They may adhere to the stolon, sometimes reaching a meter or more in length, with such tenacity that they literally have to be pulled off the plant at harvest. Jacobsen and Jansky (1989) reported that selection of wild species for crossing with haploids for tuberization was not at all beneficial.

The role of haploid-species hybrids falls primarily under the category of the donor possessing a desirable specific trait(s) to contribute to the breeding program, although in a narrower sense it could be seen as a means of broadening the genetic base. With one set of chromosomes from the diploid species parent and three sets of chromosomes from *S. tuberosum*, progeny from such tetraploid × haploid-species hybridizations are reported to exhibit heterosis for yield (Bani-Aameur et al. 1991; Tai 1994; Clulow et al. 1995; Buso et al. 1999). Another advantage to utilizing haploid-species hybrids in 4x-2x crosses has recently been highlighted by several researchers: it is more effective in generating heterotic clones with a combination of desirable traits using smaller population size than hybridizations among 4x-4x *S. tuberosum* (Buso et al. 1999; Zimnoch-Guzowska et al. 1999). The Polish diploid breeding program at Rozalin, Poland, reported use of haploid species in 4x-2x crosses to be five times more effective than conventional 4x-4x crosses in development of parental lines and cultivars (Zimnoch-Guzowska et al. 1999). It may be noted that the Polish diploid breeding program, which generated these haploid-species hybrids, has been ongoing for more than 30 years (Jakuczun et al. 1997).

The advantage of the haploid-species approach is that the haploid is extracted from parental material which already has good adaptive value and good results have already been obtained using this approach. Nonetheless, there are several disadvantages in the haploid-species approach. Using probablistic methods, Haynes (1993) showed that haploids can be more inbred than the tetraploids from which they were extracted. Furthermore, *S. tuberosum* haploids are male sterile (Peloquin and Hougas 1960).

Recent research also suggests that for at least one cultivar ("Pentland Crown"), use of the haploid inducer results in the elimination of *S. phureja* chromosomes from triploid zygotes instead of the parthenogenic development of unfertilized eggs (Clulow et al. 1991). Besides, haploid extraction is time consuming and costly (Neele and Louwes 1986). The crossability of haploids with different diploid species is unpredictable and must be tested by test crossing (Hermundstad and Peloquin 1985, 1986). Most haploid-species hybrids are also male sterile (Amoah and Grun 1988). Hence in exploiting genetic diversity, approximately 75% of the genes in the tetraploid × haploid-species progeny will still originate from *S. tuberosum*. The exact percentage of genes originating from *S. tuberosum* will depend on the frequency of single exchange tetrads in the haploid-species parent and the mechanism of 2n pollen formation (Haynes 1992b).

Relatively little research has been done on improving diploid populations for traits of interest. Large-scale programs involved in enhancing the diploid germplasm base exist in Poland, Scotland, the United States, and the International Potato Center (CIP), Lima, Peru. A number of other places are also working with smaller diploid populations. To improve the diploid germplasm base, however, Poland, Scotland, and the USA are confronted with an additional challenge, namely adapting short-day diploid species to long-day growing conditions. Only CIP (Peru) is involved in enhancing the diploid germplasm base for pest-resistant traits under short-day growing conditions.

At CIP, Watanabe et al. (1994, 1995, 1996a) have been involved in creating pest-resistant, diploid, short-day populations using a number of *Solanum* species—*S. phureja*, *S. stenotomum*, *S. bukasovii*, *S. chacoense*, *S. multidissectum*, and *S. sparsipilum*—as well as haploids of *S. tuberosum* × *S. andigena*. Interestingly enough, the 26 parental lines (clones) that made up the base of their breeding population originated from Poland (2), CIP (15), the NRSP-6 germplasm bank in Wisconsin (1), and haploid-species hybrids from the Univ. of Wisconsin and CIP materials (8). The CIP team has attempted to combine resistance to bacterial wilt, white cyst nematode, potato leaf roll virus, potato tuber moth, potato virus M, potato virus X, potato virus Y, and root-knot nematode. They have reported one clone with resistance to four pests, four clones with resistance to three pests, and 147 clones resistant to one or two pests. Watanabe et al. (1996b) believe that some of their adapted diploids with improved levels of disease resistance may have a niche as landrace cultivars in local ethnic markets.

In Poland, a diploid breeding program using *S. phureja*, *S. verrucosum*, *S. microdontum*, *S. gourlayi*, *S. chacoense*, *S. yungasense* and haploids of *S. tuberosum*, has been underway for over 30 years (Jakuczun et al. 1997).

A primary goal of this breeding program is the development of diploid populations with resistance to several viruses (Dziewonska 1986; Muchalski et al. 1997). This has resulted in the release of DW 84-1457, a diploid clone with resistance to potato leaf roll virus, potato virus X and potato virus M, along with good table and processing quality (Dziewonska and Was 1994). This diploid population has also been reported to have high starch content and resistance to soft rot (Zimnoch-Guzowska and Lojkowska 1993; Zimnoch-Guzowska et al. 1999) and variation in levels of reducing sugars before and after cold storage (Jakuczun et al. 1995; also, this volume).

In Scotland, a population of *S. phureja* and *S. stenotomum* was adapted to long-day conditions by Carroll (1982). Unfortunately, this diploid breeding work has been discontinued in favor of focusing on genetic studies within their diploid populations by developing molecular markers for various diseases and pests (George Mackay, pers. comm.).

In 1966, F.L. Haynes (1972, 1980) of North Carolina State University (NCSU) began a program aimed at adapting the two short-day Andean cultivated diploid species *S. phureja* Juz. and Bukasov subspecies *phureja* (*phu*) and *S. stenotomum* Juz. and Bukasov subspecies *stenotomum* (*stn*) to the long-day conditions of North Carolina, USA. Starting with approximately 30 plant introductions of each species, he initiated a program of recurrent mass selection for adaptaton as measured by the number of tubers initiated and their size. Although a few of the original plant introductions failed to tuberize, the breeding effort was expanded so that eventually 72 plant introductions formed this interbreeding population. These 72 plant introductions gave rise to the 72 families that still formed the backbone of this population in 2002. Each family can trace its ancestry to one of the original founding plant introductions. Progeny from each of these 72 families were produced and evaluated using a two-year recurrent mass selection scheme. In the first year, seedlings would be transplanted to the field in North Carolina and at harvest the four "best-looking" genotypes selected. These four clones were then brought into storage and evaluated for the trait(s) of interest. The best clone from each family was then planted the following year in a randomized complete block design and open-pollinated seed was collected to start the selection cycle all over again. Since these species are self-incompatible, any fruit produced was the result of cross-pollination. In all, ten cycles of recurrent mass selection were made from 1966 to 1986: the first five to six selection cycles for general adaptation (Haynes, 1980), followed by four to five selection cycles for high specific gravity. Unfortunately, no further reports documenting the progress obtained by this method of selection were published. However, Plaisted (1985), who was involved in a similar population improvement program utilizing tetraploid *S. andigena*, believed that population improvement efforts were a powerful tool for developing potato germplasm resources into useful breeding resources. He found population amelioration effective when based on good genetic material with selection pressure steadily applied over time. He also stated that a broader range of genetic diversity could be maintained in such population improvement efforts than would be possible in pedigree selection.

Concurrent with the diploid recurrent mass selection scheme, pedigree selection was also carried out utilizing individual clones from various cycles of selection (Haynes 1972, 1980). The *phu-stn* population was found to produce clones with high specific gravity (Ruttencutter et al. 1979, Haynes and Haynes 1990, Haynes et al. 1989), heat tolerance (Gautney and Haynes 1983), resistance to early blight disease (Herriott et al. 1986), soft rot *Erwinias* (Wolters and Collins 1995), and potato virus X and Y (Vallejo et al. 1994). Following F.L. Haynes retirement in 1986, remanent true seed from the tenth cycle of recurrent mass selection was transferred to the United States Department of Agriculture, Agricultural Research Service (USDA/ARS), in Beltsville, Maryland. The diploid *phu-stn* population work at NCSU was subsequently discontinued.

In 1988, K.G. Haynes began the USDA/ARS-Beltsville phase aimed at continual enhancement of this diploid *phu-stn* population. The first breeding technique to be changed was the length of each selection cycle from two years to four years (Haynes 2001; Fig. 5.1). In North Carolina, with irrigation available, it was possible to transplant seedlings directly to the field. In Presque Isle, Maine, the site of the USDA/ARS-Beltsville field plots, no irrigation was available, making transplantation of seedlings a risky proposition. Therefore, seedling generations were planted in the Beltsville greenhouses in the fall: a single tuber was harvested from each pot and tubers were bulked by family for planting the following May in Maine. The two-year recurrent mass selection scheme utilized in North Carolina was due to high virus pressure in the field. In order to have viable material to evaluate, it was necessary to produce new material every other year. In Presque Isle, Maine, this *phu-stn* population was grown on Chapman Farm, an isolated seed farm, with minimal virus pressure. Therefore, each selection cycle was grown and evaluated over the course of three years. The tuberling generation arising from the Beltsville greenhouses, consisting of approximately 100 clones from each of the 72 *phu-stn* families, was planted during the second year of this selection cycle and at harvest the four "best-looking" genotypes were selected for further evaluation. In each of the next two years (years three and four of the selection cycle), 288 clones (4 clones for each of 72 families) were planted in a randomized complete block design and evaluated for yield and specific gravity. In the third year of this selection cycle,

Year	*Breeding Scheme*
1.	Produce seedling generation in greenhouse in Beltsville, MD during the fall. ↓
2.	Grow tuberling generation in the field in Presque Isle, ME. (72 families × 100 genotypes/family) Select 4 "best" genotypes at harvest time. ↓
3.	Evaluate 288 (72 families × 4 clones/family) clones in randomized complete block design. Collect open-pollinated fruit from all 288 clones. Measure clones for trait(s) of interest. ↓
4.	Evaluate the same 288 clones in randomized complete block design. Measure clones for trait(s) of interest. Choose the "best" clone from each family for the trait(s) of interest to plant open-pollinated seed from that clone to start the cycle all over again.

Fig. 5.1 Scheme utilized in the USDA/ARS-Beltsville diploid potato breeding program

open-pollinated seeds were also collected from all 288 clones. At the end of the fourth year, data were analyzed and open-pollinated seed from the highest specific gravity clone in each family selected for planting the following fall in order to start the next selection cycle. This selection scheme was an improvement over the earlier one because data on specific gravity (SG) and yield became available from two years of replicated testing, instead of just one year of unreplicated testing, and four clones from each maternal half-sib family were evaluated, instead of only the "best" one based on selection in the first year. It also expanded the genetic base, since selection though still based on performance of the maternal plant, now had the paternal plant selected by random sampling of the possible gametes from 288 clones instead of 72. It was also possible to determine the importance of genotype—environment interactions after two years of testing.

This particular diploid population opened the door for several quantitative studies on the heritability of traits of interest at a population rather than a pedigree level. This diploid population, with a definite family structure, was randomly mated (i.e., open-pollinated) for more than 12 cycles of selection and the clones tested under replicated conditions over years. Thus, Haynes (2001) was able to demonstrate that the genotypic variation in this *phu-stn* population, from the first to the second selection cycle, decreased in Maine by 34% for SG and 81% for yield. This indicates that this population may be rapidly approaching its genetically imposed "yield limit", which in the second cycle of selection had only about half the yield of "Atlantic", the number one chipping cultivar in the USA, even though a substantial number of clones from this population expressed higher SG than 'Atlantic', one of the highest SG chipping cultivars in the USA.

Estimates obtained on narrow-sense heritability of SG from this diploid population, based on pedigree selection and recurrent mass selection, were similar over years in different parts of the country. The average of the three least biased estimates on narrow-sense heritability of SG reported in Ruttencutter et al. (1979) was 0.428. Ruttencutter's estimates were obtained using pedigree selection in clones from the fifth cycle of recurrent mass selection evaluated in the mountains of North Carolina. The estimates on heritability of SG obtained by Haynes (2001) after two cycles of recurrent mass selection were 0.37 and 0.43 respectively. Haynes' estimates were obtained from the first and second recurrent mass selection cycles of the USDA/ARS-Beltsville population (representing the 11th and 12th recurrent mass selection cycles overall for materials evaluated in northern Maine). The average SG of the 72 families constituting Haynes' material was 1.085 and 1.093 respectively for each selection cycle.

This diploid *phu-stn* population was also unique because either no correlation or a positive correlation was obtained between SG and yield in each of the four years of evaluation (Haynes 2001). In tetraploid *S. tuberosum*, a negative correlation was found between yield and SG in advanced selections from the USDA/ARS potato breeding program (Haynes et al. 1989) and in the tetraploid tuberling population selected for horticultural characteristics (Haynes and Wilson 1991).

Estimates obtained on narrow-sense heritability of yield from this diploid population, also based on pedigree selection and recurrent mass selection, were similar across years (except for the first selection cycle in Maine) in different parts of the country. Gautney and Haynes (1983) estimated narrow-sense heritability of yield under high temperature growing conditions in North Carolina from the fifth selection cycle parents of this *phu-stn* population using pedigree selection as 0.07. Haynes (2001) estimated narrow-sense heritability of yield in the USDA/ARS-Beltsville *phu-stn* population to be 0.60 and 0.06 respectively in the first and second cycles of recurrent selection in Maine (11th and 12th recurrent mass selection cycles overall). The high estimate of narrow-sense heritability for yield in the first selection cycle may have been due to the completely new environmental conditions under which this genetic material was grown. The rapid decrease in genetic variance for yield from the first to the second selection cycle in Maine might be indicative of rapid genetic equilibrium for yield attained by the population after just one cycle of selection. Any further marked increases in yield appear unlikely, considering this low estimate of heritability. But more tests on this aspect will be conducted as the population advances through several more rounds of recurrent mass selection for SG.

Estimates obtained of narrow-sense heritability of resistance to early blight (EB) from this diploid population, based on pedigree selection and recurrent mass selection, were also similar over years in different parts of the country. Herriott et al. (1986) estimated narrow-sense heritability for EB resistance in parents from the sixth selection cycle of this *phu-stn* population using pedigree selection in North Carolina as 0.83. This estimate is biased toward the high side because parents and offspring were evaluated in the same year. Christ and Haynes (2001) estimated narrow-sense heritability for EB resistance in this *phu-stn* population from the second USDA/ARS-Beltsville selection cycle (12th selection cycle overall) as 0.61. Based on this, they decided to establish another breeding population just for EB resistance similar to the one previously set up for SG.

Haynes and Christ (1999) likewise estimated narrow-sense heritability for resistance to late blight (LB) in the second USDA/ARS-Beltsville selection cycle (12th selection cycle overall) to be 0.78. This encouraged them to establish another breeding population, this one for LB resistance, similar to that previously set up for SG.

Future prospects for continuing use of this *phu-stn* population for recurrent selection look very promising. High levels of resistance to EB and LB, the major foliar diseases of potatoes in most of the world, have been found in this population; estimates of narrow-sense heritability for both traits are moderately high, indicating that it should be possible to continue to improve this population for these traits (Christ and Haynes 2001; Haynes and Christ 1999). Population amelioration efforts for these traits are underway, in addition to molecular mapping studies to locate the genes governing these resistances, which should hopefully lead to marker-assisted selection in the future.

The USDA/ARS-Beltsville diploid *phu-stn* population was developed to maintain maximum genetic diversity with a minimal amount of inbreeding, allow quantitative genetic studies of traits of interest, and provide long-day adapted material for future breeding efforts at both the diploid and tetraploid level. Clones taken from the first two selection cycles at the USDA were utilized directly in 4x-2x crosses and found to transmit high SG, and freedom from internal heat necrosis in their tetraploid offspring (Sterrett et al. 2003), and to develop more intense yellow flesh (K.G. Haynes; unpublished). In addition, it is envisioned that hybrids between these adapted *phu-stn* clones and wild species can be utilized in much the same manner that haploid-species hybrids have been utilized to capture the genetic diversity in wild diploid *Solanum* species for future improvement efforts, as proposed by Bani-Aameur et al. (1993). Use of adapted *phu-stn* clones in 4x-2x crosses would result in the tetraploid progeny having 50% of its genetic composition from the *phu-stn* germplasm base. This should ultimately contribute to less inbreeding in tetraploids derived from 4x-2x crosses than from 4x-haploid-species crosses (Haynes 1990, 1992a,b; Haynes and Potts 1993).

Lastly, greater effort needs to be expanded in the various potato breeding programs worldwide in utilizing the diploid *Solanum* species. Many breeding programs elsewhere have devoted a small portion of their resources to working with diploid species and have captured very little of the genetic diversity and potential available at present. Work with diploid species may not pay off in immediate short-term benefits, but as evidenced by the *phu-stn* population we now have at the USDA, there is a tremendous amount of valuable germplasm waiting to be exploited. This work is expensive in terms of land, labor, and time, but the potential rewards are large.

Acknowledgments

To those members of the eastern potato-growing states of the USA who faithfully make the journey to Presque Isle, Maine every year to assist in harvesting both advanced tetraploid selections for varietal evaluation and also the diploids, our sincerest thanks. The work we do depends on you: Steve O'Hair, Marion White, Pete Weingartner, Craig Yencho, Mark Clough, Rikki Sterrett, Mel Henninger, Barb Christ, Mike Peck, Joe Sieczka, and Don Halseth. Just as it depends on members of the USDA potato breeding team who also help to harvest the diploids: Karl DeLong, Merle Bragg, Bonnie Adams, Charles Legasse, Diane Fleck, and Karen Frazier. To those USDA/ARS administrators who have likewise journeyed to Presque Isle, Maine—we appreciate their encouragement and support.

REFERENCES

Abdalla, M.M.F. and J.G.Th. Hermsen. 1971. A two-locus system of gametophytic incompatibility in *Solanum phureja* and *S. stenotomum*. Euphytica 21: 345–350.

Amoah, V. and P. Grun. 1988. Cytoplasmic substitution in *Solanum*. I. Seed production, germination, and sterilities of reciprocal backcross generations. Potato Res 31: 113–119.

Bani-Aameur, F., F.I. Lauer, and R.E. Veilleux. 1993. Enhancement of diploid *Solanum chacoense* Bitt. using adapted clones of *Solanum phureja* Juz. et Buk. Euphytica 68: 169–179.

Bani-Aameur, F., F.L. Lauer, R.E. Veilleux, and A. Halali. 1991. Genomic composition of 4x-2x potato hybrids: influence of *Solanum chacoense*. Genome 34: 420–423.

Birhman, R.K. and M.L.H. Kaul. 1989. Flower production, male sterility and berry setting in Andigena potato. Theor Appl Genet 78: 884–888.

Buso, J.A., L.S. Boiteux, and S.J. Peloquin. 1999. Multitrait selection system using populations with a small number of interploid (4x-2x) hybrid seedlings in potato: degree of high parent heterosis for yield and frequency of clones combining quantitative agronomic traits. Theor Appl Genet 99: 81–91.

Camadro, E.L. and S.J. Peloquin. 1981. Cross-incompatibility between two sympatric polyploid *Solanum* species. Theor Appl Genet 60: 65–70.

Carroll, C.P. 1982. A mass-selection method for the acclimatization and improvement of edible diploid potatoes in the United Kingdom. J Agric Sci, Camb 99: 631–640.

Christ, B.J. and K.G. Haynes. 2001. Inheritance of resistance to early blight disease in a diploid potato population. Plant Breed 120: 169–172.

Clulow, S.A., J. McNicoll, and J.E. Bradshaw. 1995. Producing commercially attractive, uniform true potato seed progenies: the influence of breeding scheme and parental genotype. Theor Appl Genet 90: 519–525.

Clulow, S.A., M.J. Wilkinson, R. Waugh, E. Baird, M.J. DeMaine, and W. Powell. 1991. Cytological and molecular observations on *Solanum phureja*-induced dihaploid potatoes. Theor Appl Genet 82: 545–551.

Douches, D.S., D. Maas, K. Jastrzebski, and R.W. Chase. 1996. Assessment of potato breeding progress in the USA over the last century. Crop Sci 36: 1544–1552.

Dziewonska, M.A. 1986. Development of parental lines for breeding of potatoes resistant to viruses and associated research. *In:* A.G.B. Beekman, K.M. Louwes, L.M.W. Dellaert, and A.E.F. Neele (eds.), Potato Research of Tomorrow. Pudoc, Wageningen, Netherlands, pp. 96–100.

Dziewonska, M.A. and M. Was. 1994. Diploid genotype DW 84–1457, highly resistant to potato leaf roll virus. Potato Res 37: 217–224.

Fritz, N. and R.E. Hanneman, Jr. 1989. Interspecific incompatibility due to stylar barriers in tuber-bearing and closely related non-tuber-bearing *Solanums.* Sex Plant Reprod 2: 184–192.

Gautney, T.L. and F.L. Haynes. 1983. Recurrent selection for heat tolerance in diploid potatoes (*Solanum tuberosum* subspp. *phureja* and *stenotomum*). Amer Potato J 60: 537–5452.

Hanneman, R.E. Jr. 1994. Assignment of Endosperm Balance Numbers to the tuber-bearing *Solanums* and their close non-tuber-bearing relatives. Euphytica 74: 19–25.

Hanneman, R.E., Jr. and J.B. Bamberg. 1986. Inventory of tuber-bearing *Solanum* species. Univ Wisconsin, Madison, WI Bull 533.

Hawkes, J.G. 1990. The Potato: Evolution, Biodiversity and Genetic Resources. Belhaven Press, London, UK.

Hawkes, J.G. and M.T. Jackson. 1992. Taxonomic and evolutionary implications of the Endosperm Balance Number hypothesis in potatoes. Theor Appl Genet 84: 180–185.

Haynes, F.L. 1972. The use of cultivated diploid *Solanum* species in potato breeding. *In:* E.R. French, (ed.). Prospects for the Potato in the Developing World. Intl Potato Center Symp, 1972. Lima, Peru pp. 100–110.

Haynes, F.L. 1980. Progress and future plans for the use of Phureja-Stenotomum populations. *In:* O.T. Page, (ed.). Utilization of the Genetic Resources of the Potato. III. Report Planning Conference, Intl Potato Center, 1980. Lima, Peru, pp. 80–88.

Haynes, K.G. 1990. Covariances between diploid parent and tetraploid offspring in tetraploid × diploid crosses of *Solanum tuberosum* L. J Heredity 81: 208–210.

Haynes, K.G. 1992a. Covariance between haploid-species hybrid and *Tuberosum* × haploid-species hybrid in 4x-2x crosses of *Solanum tuberosum* L. J Heredity 83: 119–122.

Haynes, K.G. 1992b. Some aspects of inbreeding in derived tetraploids of potatoes. J Heredity 83: 67–70.

Haynes, K.G. 1993. Some aspects of inbreeding in haploids of tetraploid *Solanum tuberosum.* Amer Potato J 70: 39–344.

Haynes, K.G. 2001. Variance components for yield and specific gravity in a diploid potato population after two cycles of recurrent selection. Amer J Potato Res 78: 69–75.

Haynes, K.G. and F.L. Haynes. 1990. Selection for tuber characters can maintain high specific gravity in a diploid potato breeding population. HortSci 25: 227–228.

Haynes, K.G. and D.R. Wilson. 1991. Correlation of yield and specific gravity in a tetraploid potato tuberling population. Amer Potato J 68: 355–362.

Haynes, K.G. and W.E. Potts. 1993. Minimizing inbreeding in tetraploids derived through sexual polyploidization. Amer Potato J 70: 617–624.

Haynes, K.G. and B.J. Christ. 1999. Heritability of resistance to foliar late blight in a diploid hybrid potato population of *Solanum phureja* x *Solanum stenotomum*. Plant Breed 118: 431–434.

Haynes, K.G., F.L. Haynes, and W.R. Henderson. 1989. Heritability of specific gravity of diploid potato under high temperature growing conditions. Crop Sci 29: 622–625.

Haynes, K.G., R.E. Webb, R.W. Goth, and D.R. Wilson. 1989. The correlation of yield and specific gravity in the USDA potato breeding program. Amer Potato J 66: 587–592.

Hermundstad, S.A. and S.J. Peloquin. 1985. Germplasm enhancement with potato haploids. J Heredity 76: 463–467.

Hermundstad, S.A. and S.J. Peloquin. 1986. Tuber yields and tuber traits of haploid-wild species F_1 hybrids. Potato Res 29: 289–297.

Herriott, A.B., F.L. Haynes, and P.B. Shoemaker. 1986. The heritability of resistance to early blight in diploid potatoes (*Solanum tuberosum* subspp. *phureja* and *stenotomum*). Amer Potato J 63: 229–232.

Hoekstra, R. and L. Seidewitz. 1987. Evaluation data on tuber-bearing *Solanum* species. Institut für Pflanzenbau and Pflanzenzüchting der FAL-Stichting voor Plantenveredeling (SVP).

Hougas, R.W. and S.J. Peloquin. 1957. A haploid plant of the potato variety Katahdin. Nature 180: 1209–1210.

Jackson, S.A. and R.E. Hanneman, Jr. 1999. Crossability between cultivated and wild tuber-bearing *Solanums*. Euphytica 109: 51–67.

Jacobsen, T.L. and S.H. Jansky. 1989. Effects of pre-breeding wild species on tuberization of *Solanum tuberosum* haploid-wild species hybrids. Amer Potato J 66: 803–811.

Jakuczun, H., K. Zgorska, and E. Zimnoch-Guzowska. 1995. An investigation of the level of reducing sugars in diploid potatoes before and after cold storage. Potato Res 38: 331–338.

Jakuczun, H., E. Zimnoch-Guzowska, M. Osiecka, and D. Strzelczyk-Zyta. 1997. Diploid parental line breeding—summary and forecast. Biuletyn Instytutu Ziemniaka 48: 55–62.

Johnston, S.A., T.P.M. den Nijs, S.J. Peloquin, and R.E. Hanneman. 1980. The significance of genic balance to endosperm development in interspecific crosses. Theor Appl Genet 57: 5–9.

Mendoza, H.A. and F.L. Haynes. 1974. Genetic relationship among potato cultivars grown in the United States. HortSci 9: 328–330.

Muchalski, T., I. Wasilewicz-Flis, M.A. Dziewonska, and K. Ostrowska. 1997. Synteza diploidalnych ziemniakow o kompleksowej odpornosci na wirusy. Biuletyn Instytutu Ziemniaka 48: 63–70.

Neele, A.E.F. and K.M. Louwes. 1986. The analytic breeding method: possibilities for potato breeding. *In:* A.G.B. Beekman, K.M. Louwes, L.M.W. Dellaert, and A.E.F. Neele (eds.). Potato Research of Tomorrow Pudoc, Wageningen, Netherlands, pp. 107–114.

Nyquist, W.E. 1991. Estimation of heritability and prediction of selection response in plant populations. Crit Rev Plant Sci 10: 235–322.

Pavek, J.J. 1987. Some interesting aspects of recent and expected developments in potato breeding in North America. Acta Hortic 213: 61–65.

Peloquin, S.J. and R.W. Hougas. 1960. Genetic variation among haploids of the common potato. Amer Potato J 37: 289–297.

Peloquin, S.J., A.C. Gabart, and R. Ortiz. 1996. Nature of 'pollinator' effect in potato (*Solanum tuberosum*) haploid production. Ann Bot 77: 539–542.

Plaisted, R.L. 1985. Population breeding applied to improvement of unadapted *Solanum* cultivated species. *In:* Present and Future Strategies for Potato Breeding and Improvement. Report of the XXVI Planning Conference. Lima, Peru. December 12–14, 1983.

Ramanna, M.S. and J.G. Th. Hermsen. 1979. Genome relationships in tuber-bearing *Solanums*. *In:* J.G. Hawkes, R.N. Lester, and A.D. Skelding (eds.). The Biology and Taxonomy of the Solanaceae. Linn. Soc. Symp., Ser. 7, Academic Press, London, pp. 647–654.

Roath, W.W. 1989. Evaluation and enhancement. Plant Breed Rev 7: 183–211.

Ross, H. 1986. Potato Breeding: Problems and Perspectives. Verlag Paul Parey, Berlin, 132 pp.

Ruttencutter, G.E., F.L. Haynes, and R.H. Moll. 1979. Estimation of narrow-sense heritability for specific gravity in diploid potatoes (*S. tuberosum* Subsp.*phureja* and *stenotomum*). Amer Potato J 56: 447–453.

Simmonds, N.W. 1969. Prospects of potato improvement. Scottish Soc Res Plant Breed Ann Meet, Scottish Plant Breed Station, Pentlandfield, pp. 18–38.

Spooner, D.M. and R.J. Hijmans. 2001. Potato systematics and germplasm collecting, 1989-2000. Amer J Potato Res. 78: 237–268.

Sterrett, S.B., M.R. Henninger, G.C. Yencho, W. Lu, B.T. Vinyard, and K.G. Haynes. 2003. Stability of internal heat necrosis and specific gravity in tetraploid × diploid potatoes. Crop Sci 43: 790-796.

Tai, G.C.C. 1994. Use of 2n gametes. *In:* J.E. Bradshaw and G.R. Mackay (eds.). Potato genetics. CAB Intl. Wallingford, UK, pp. 109–132.

Tarn, T.R., G.C.C. Tai, H. DeJong, A.M. Murphy, and J.E.A. Seabrook. 1992. Breeding potatoes for long-day temperate climates. Plant Breed Rev 9: 217–332.

Trognitz, B.R. 1995. Female fertility of potato (*Solanum tuberosum* spp. *tuberosum*) dihaploids. Euphytica 81: 27–33.

Vallejo, R.L., W.W. Collins, R.D. Schiavone, S.A. Lommel, and J.B. Young. 1994. Extreme resistance to infection by potato virus Y and potato virus X in an advanced hybrid *Solanum phureja-S. stenotomum* diploid potato population. Amer Potato J 71: 617–628.

Watanabe, K., M. Orrillo, M. Iwanaga, R. Ortiz, R. Freyre, and S. Perez. 1994. Diploid potato germplasm derived from wild and landrace genetic resources. Amer Potato J 71: 599–604.

Watanabe, K.N., M. Orrillo, S. Vega, A.M. Golmirzaie, S. Perez, J. Crusado, and J.A. Watanabe. 1996a. Generation of pest resistant, diploid potato germplasm with short-day adaptation. Breed Sci 46: 329–336.

Watanabe, K.N., M. Orrillo, S. Perez, J. Crusado, and J.A. Watanabe. 1996b. Testing yield of diploid potato breeding lines for cultivar development. Breed Sci 46: 245–249.

Watanabe, K.N., M. Orrillo, S. Vega, M. Iwanaga, R. Ortiz, R. Freyre, G. Yerk, S.J. Peloquin, and K. Ishiki. 1995. Selection of diploid potato clones from diploid (haploid × wild species) F_1 hybrid families for short-day conditions. Breed Sci 45: 341–347.

Wolters, P.J.C.C. and W.W. Collins. 1995. Estimation of genetic parameters for resistance to *Erwinia* soft rot, specific gravity, and calcium concentration in diploid potatoes. Crop Sci 35: 1346–1352.

Wricke, G. and W.E. Weber. 1986. Quantitative Genetics and Selection in Plant Breeding. Walter de Gruyter, New York, NY.

Yerk, G.L. and S.J. Peloquin. 1990. Selection of potato haploid parents for use in crosses with 2x (2 Endosperm Balance Number) wild species. Crop Sci 30: 943–946.

Zimnoch-Guzowska, E. and E. Lojkowska. 1993. Resistance to *Erwinia* spp. in diploid potato with a high starch content. Potato Res. 36: 177–182.

Zimnoch-Guzowska, E., R. Lebecka, and J. Pietrak. 1999. Soft rot and blackleg reactions in diploid potato hybrids inoculated with *Erwinia* species. Amer J Potato Res 76: 199–207.

Zimnoch-Guzowska, E., M. Sieczka, H. Jakuczun, and L. Domansky. 1999. Diploid and tetraploid parental line breeding focused on resistance to pathogens and quality traits. Proc 14th Trien Conf Euro Assoc Potato Res, Sorrento, Itlay, May 2-7, 1999, pp. 329–330.

6

Molecular Markers in Identification of Genotypic Variation

F. Celebi-Toprak[1,2], Junko A. Watanabe[1], and Kazuo N. Watanabe[1,3]

[1]*Gene Research Center and Institute of Biological Sciences, University of Tsukuba, 1-1-1 Tennoudai, Tsukuba, Ibaraki, 305-8572, Japan*

[2]*Pamukkale University, Department of Biology, Fen Edebiyat Fakultesi, Kinikli, Denizli, 20010 Turkey. e-mail: fctoprak@pamukkale.edu.tr, fevziye26@hotmail.com*

[3]*International Plant Genetic Resources Institute (IPGRI), Rome, Italy*

INTRODUCTION: POTATO GENETIC MAPS AND MOLECULAR GENETICS

Potato genetics is very complicated because of high heterozygosity, polysomic polyploidy, and quantitative inheritance of key breeding traits (Watanabe et al. 2002; also Chapter 4). Therefore, introducing a desirable trait in cultivated potato is a time-consuming process (10 to 15 years) with traditional breeding (Ross 1986).

Selection of target traits on the basis of phenotypic evaluation is generally cumbersome with a number of limitations in terms of long timeframe, skills and experience for evaluation, and requires a relatively large scale of evaluation and repeated progeny testing. Conventional phenotypic markers such as those used in other crop plants—e.g. rice, wheat, and tomato—are not available for potatoes (Watanabe et al. 1997). However, recent developments in molecular biology and comparative genomics could provide molecular genetic markers to select genes of interests, including quantitative inherited traits as well as single dominant inherited traits in potato. These are recent advances in potato genetics especially when studies using genetic markers in this crop are far fewer than in other major crops.

To analyze plant genome and to better understand plant genome evolution, genetic markers are invaluable tools. Once plant chromosomes are well characterized by molecular markers, they can be used in breeding to

Corresponding author: K.N. Watanabe. e-mail: nabechan@gene.tsukuba.ac.jp monitor introgression of specific chromosomal regions for selection for closely linked target traits. Potato mapping techniques have been developed during the last ten years and many genes mapped. Some of these genes were cloned by using molecular markers such as restriction fragment length polymorphisms (RFLP), amplified fragment polymorphisms (AFLP), randomly amplified polymorphic DNA (RAPD), single sequence repeat (SSR), intersimple sequence repeat amplification (ISSR), cleaved amplified polymorphic sequence (CAPS), and sequence characterized amplified regions (SCARs) (Bendahmane et al. 1999; Bonierbale et al. 1994; Brigneti et al. 1997; Celebi-Toprak et al. 2002; Chen et al. 2001; Kasai et al. 2000; Marano et al. 2002; Meyer et al. 1998; Milbourne et al. 1998; Prevost and Wilkinson 1999; Rouppe van der Voort et al. 2000).

Most of the technical procedures for using molecular markers are individually referenced and also reviewed by Karp et al. (1997). However, a comparison with respect to characteristics of individual molecular markers is provided here to assist in deciding which specific type of marker be used in a particular breeding project. The objective of this chapter is therefore to review the progress made to date in using molecular markers as tools for potato improvement.

Potato Genetic Maps

The potato genetic maps available at present are summarized in Table 6.1. An interspecific cross between *Solanum phureja* × (*S. tuberosum* × *S. chacoense*) (2n = 2x = 24) was used by Bonierbale et al. (1988) to generate the first potato map. This map contains 134 DNA and isozyme markers, which cover approximately 606 cM of 12 linkage groups on 12 chromosomes. The second potato genetic map was generated by Gebhardt et al. (1989) and 141 markers were used to cover 690 cM of 12 linkage groups on 12 chromosomes of S. *tuberosum*. A plant population of 155 plants was obtained from *S. tuberosum* × *S. berthaultii*, backcrossed to *S. berthaultii* to generate the third potato genetic map by Tanksley et al. (1992). Genomic and cDNA probes were used to cover 684 cM of 12 linkage groups on 12 chromosomes (http://www.sgn.cornell.edu/cgi-bin/mapviewer/mapviewerHome.pl). The fourth potato genetic map with a total of 1,120 cM was published by Jacobs et al. (1995). They used molecular, phenotypic, and isozyme markers. Diploid potato populations were used for these four genetic maps. Recently, Meyer et al. (1998) constructed a partial linkage map in tetraploid potato by using AFLP markers. The map consists of 231 maternal and 106 paternal markers with total map lengths of 990.9 cM and 484.6 cM respectively. A linkage map was constructed between a *Solanum phureja* (*phu*) and (di)-haploid *Solanum tuberosum* (dih-*tbr*) population by Ghislain et al. (2001) using AFLP, RFLP, RAPD, and SSR markers. The total map lengths were 987.4 cM for *phu* and 773.7 cM for dih-*tbr*. Two important quantitative trait loci (QTLs) for the resistance to late blight (LB) were detected on chromosomes VII and XII as a contribution from both parents.

Table 6.1 *Summary of potato genetic maps in* solanum *species*

Solanum species	*Methods*	*Marker no.*	*Size*	*Reference*
S. phureja × (*S. tuberosum* × *S. chacoense*)	RFLP, isozyme	134	606 cM	Bonierbale et al. (1988)
S. tuberosum (diploid)	RFLP	141	690 cM	Gebhardt et al. (1989)
S. tuberosum × *S. berthaultii*, Backcross to *S. berthaultii* (diploid)	RFLP	260	684 cM	Tanksley et al. (1992)
S. tuberosum diploid	RFLP, transposon, isozyme		1120 cM	Jacobs et al. (1995)
USW5337.3 × 77.2102.37	AFLP	264	1170 cM	van Eck et al. (1995)
SCRI clone 1260 labl × cv. *Stirling* tetraploid	AFLP	231 (m) 106 (p)	990.9 cM 484.6 cM	Meyer et al. (1998)
SCRI clone 1260 1ab1 × cv. *Stirling* reciprocal cross (tetraploid)	AFLP, SSR	266 (m) 164 (p)	990.9 cM 484.6 cM	Bradshaw ėt al. (1998)
S. palustre × *S. tuberosum*	RFLP, RAPD, AFLP	102	720.4 cM	Perez et al. (1999)
Solanum phureja (phu) and dihaploid	AFLP, RFLP,		987.4 cM	Ghislain et al. (2001)
Solanum tuberosum (dih-*tbr*)	RAPD, SSR		773.7 cM	
S. cardiophyllum × *S. pinnatisectum* (BC1) diploid	RFLP	99	683.9 cM	Kuhl et al. (2001)
Solanum phureja clones 3704-76	AFLP	149	603 cM	Rouppe van der Voort et al. (2000)

(m) maternal; (p) paternal

Molecular Markers

The use of RFLP, RAPD, AFLP, SSR, and ISSR maps could help accelerate transfer of genes for resistance or other traits from wild to cultivated species or at the interspecies level (Brigneti et al. 1997; Celebi-Toprak et al. 2002; Meyer et al. 1998). A comparison on characteristic features of various molecular markers is given in Table 6.2. RFLP, RAPD, and AFLP potato maps highly saturated with molecular markers were constructed (Bonierbale et al. 1988; Gebhardt et al. 1989; Tanksley et al. 1992). These molecular potato maps could be employed to map many new genes for breeding purposes. For example, a diploid backcross population derived from a cross between *Solanum tuberosum* and *Solanum berthaultii* segregated for monogenic dominant hypersensitivity to *Potato virus Y* (PVY).

Table 6.2 *Comparison of the molecular markers*

Characteristic	*RFLPs*	*RAPDs*	*SSRs*	*AFLPs*	*PCR sequencing*	*ISSR*	*CAPS*	*SCARs*
Development cost of facilities	Medium	Low	High	Low	High	Low	High	High
Running cost	High	Low	Low	Medium	High	Low	Low	Low
Cost of automation	High	Medium	High	High	High	Medium	Medium	Medium
Radioactive used	Yes/No	No	Yes/No	Yes/No	Yes/No	No	Yes/No	No
Level of training required	Low	Low	Low	Medium	High	High	Medium	Medium
Time taken for completion	Long	Quick	Quick	Medium	Quick	Quick	Quick	Quick
Complexity of process	Complex	Simple	Simple	Complex	Complex	Simple	Simple	Simple
Information per sample	Low	Medium	Medium	High	High	Low	Specific	Specific
Reproducibility	High	Low	High	High	High	Medium	High	High
Level of polymorphism	Low	Medium	High	Medium	Medium	High	Medium	Medium
Marker types	Co-dominant	Dominant	Co-dominant	Dominant	Dominant	Dominant	Dominant	Dominant

Modified from Karp et al. (1997).

Celebi-Toprak et al. (2002) proposed the symbol Ny_{tbr} for this locus because plants carrying this gene developed necrosis after inoculation with PVY due to the allele originating from *S. tuberosum*. The gene mapped to chromosome IV between TG316 and TG208 at LOD = 2.72. This location corresponds to no other mapped resistance gene of potato such as Ry_{adg} on chromosome XI (Hämäläinen et al. 1997). The major dominant resistance genes and important agronomic traits are given in Tables 6.3 and 6.4. Because wild tuber-bearing *Solanum* species have genes resistant to diverse pathogens, many of them have been characterized with the help of molecular markers for the purpose of fingerprinting, which is useful for further molecular analysis.

Table 6.3 *Molecular markers related to pest resistance traits in potato*

Pest or related trait	*Target gene*	*Method*	*Markers for MAS*	*References*
Potato Virus X (PVX)	Nx_{phu}	RFLP	*TG424*	Tommiska et al. (1998)
	Nb	RFLP	*TG432*	De Jong et al. (1997)
	Nb	AFLP	*SPUD237*	De Jong et al. (1997)
	Nb	CAPS	*GP21, SPUD237*	De Jong et al. (1997)
	Rx2	RFLP	*TG432*	De Jong et al. (1997)
	Rx	RFLP	*GP34, CT100, CT129*	Bendahmane et al. (1997)
	Rx	AFLP	*PM4, PM3*	Bendahmane et al. (1997)
	Rx	CAPS	*IPM3, IPM4*	Bendahmane et al. (1997)
	Rx1, Rx2	RFLP	*CP60, GP21, GP213*	Ritter et al. (1991)
	Rx	CAPS	*IPM4-c, 77R*	Kanyuka et al., (1999)
Potato Virus Y (PVY)	Ry_{adg}	RFLP	*TG508*	Hämäläinen et al. (1997)
	Ry_{adg}	RFLP	*CP58, CT182, CD17, TG523, ADG1, ADG2*	Hämäläinen et al. (1998)
	Ry_{adg}	RFLP	*ADG2*	Shiranita et al. (1999a)
	Ry_{adg}	CAPS	*ADG2*	Sorri et al. (1999)
	Ry_{adg}	SCAR	*SYRC3*	Kasai et al. (1999)
	Ry_{sto}	RFLP	*CP58, TG523*	Brigneti et al. (1997)
	Ny_{tbr}	RFLP	*TG506, TG208*	Celebi-Toprak et al. (2002)

(*Contd.*)

(*Contd.*)

Potato Virus A (PVA)	Ra_{adg}	RFLP	*GP125, TG508*	Hämäläinen et al. (1998)
	Na_{adg}	RFLP	*TG523*	Hämäläinen et al. (2000)
Potato Virus S (PVS)	*Ns*	RFLP, ISSR		Marczewski et al. (2002)
Phytophthora infestans	*R1*	RFLP	*GP21, GP179*	Leonards-Schippers et al. (1992)
	R1	AFLP	*GP21, GP179*	Meksem et al. (1995)
	R1	RFLP	*GP21*	El-Kharbotly et al. (1996a)
	R1	RFLP	*TG432*	De Jong et al. (1997)
	R1	AFLP	*SPUD237*	De Jong et al. (1997)
	R2	AFLP		Li et al. (1998)
	R3	RFLP	*GP105(a), TG105 (a), GP185, GP35 (k)*	El-Kharbotly et al. (1994)
	R6, R7	RFLP	*GP185(a), GP250 (a)*	El-Kharbotly et al. (1996b)
	Pi (QTL)	RFLP		Leonards-Schippers et al. (1994)
	Pi (QTL)	SSR	*Stm3016*	Milbourne et al. (1998)
	Pi (QTL)	RFLP		Ghislain et al. (2001)
	QTL	AFLP		Meyer et al. (1998)
	Rpi1	RFLP		Kuhl et al. (2001)
	R_{blc}	RFLP	*CP53*	Naess et al. (2000)
Globodera rostochiensis	*H1*	RFLP	*CD78*	Pineda et al. (1993)
	H1	RFLP	*CP113*	Gebhardt et al. (1993)
	QTL	RFLP		Kreike et al. (1993, 1996)
	Gro1	RFLP	*CP51*	Barone et al. (1990)
	Gro1	RFLP	*CP56, GP516(c), CP51(c)*	Ballvora et al. (1995)
	Gro1	RAPD	$OPR10_{700}$	Ballvora et al. (1995)
	Gro1	AFLP	*AFLP1, AFLP2*	Ballvora et al. (1995)

(*Contd.*)

(*Contd.*)

Globodera pallida	*Gpa* (QTL)	RFLP		Kreike et al. (1994)
	Gpa2	RFLP	*GP34, CT100*	Rouppe van Der Voort et al. (1997)
	Gpa2	AFLP	*E + ATG/M + CTA-148*	Rouppe van Der Voort et al. (1997)
	Gpa2	CAPS	*77L, PM4c, 77R*	Rouppe van Der Voort et al. (1999b)
	Gpa4 (QTL)	AFLP, SSR		Bradshaw et al. (1998)
	Gpa5 (QTL)	AFLP		Rouppe van Der Voort et al. (2000)
		CAPS		
	Gpa6 (QTL)	AFLP		Rouppe van Der Voort et al. (2000)
		CAPS		
	QTL	AFLP, SSR		Bradshaw et al. (1998)
	QTL	AFLP		Bryan et al. (2002)
Globodera ssp.	*Grp1*	CAPS	*GP21, GP179*	Rouppe van Der Voort et al. (1998)
Meloidogyne chitwoodi	R_{mcl}	RFLP	*TG523*	Brown et al. (1996)
Meloidogyne spp.	R_{mcl}	AFLP	*E + AAC/M + CGA-170*	Rouppe van Der Voort et al. (1999a)
	R_{mcl}	CAPS	*CT182, M39b*	Rouppe van Der Voort et al. (1999a)
Glandular trichomes	QTL	RFLP		Bonierbale et al. (1994)
Erwinia carotovora ssp. *Eca*	(QTL)	AFLP,		Zimnoch-Guzowska et al. (2000)
atroseptica		RFLP		
Synchytrium endobioticum Sen1		RFLP	*CP58, GP125*	Hehl et al. (1999)

Molecular Markers with QTLs: Agronomic Traits and Disease Resistances

Most plant characteristics vary quantitatively under the influence of both environmental and genetic factors and are encoded by QTLs (Gelderman 1975). Mapping QTL is impossible by simple Mendelian analysis because discrete phenotypic classes cannot be identified. Many QTLs have been

Table 6.4 *Examples of molecular markers reported for important agronomic traits of potatoes*

Traits	*Target gene(s)*	*Method*	*Chromosome number*	*References*
Protein storage	PTN	RFLP	VIII	Bonierbale et al. (1988), Ganal et al. (1991)
Flower color	P	Morphological	XI	Van Eck (1995)
	F	Morphological	X	Van Eck (1995)
	D	Morphological	II	Van Eck (1995)
	R	Morphological	X	Van Eck (1995)
Tuber shape and skin color	Ro	RFLP	X	Gebhardt et al. (1989)
Tuber shape		AFLP	X	Van Eck et al. (1995)
		RFLP	X	Van Eck et al. (1994)
Tuber skin color		RFLP	X	Gebhardt et al. (1989, 1991)
Tuber flesh color		RFLP	III	Bonierbale et al. (1988)
Tuberization		RFLP (QTL)		Bonierbale et al. (1988)
		RFLP (QTL)	V	van den Berg et al. (1996)
Tuber dormancy		RFLP (QTL)	II	van der Berg et al. (1996)
Tuber starch content		RFLP (QTL)	12 linkage groups	Shäfer-Pregl et al. (1998)
Tuber yield		RFLP (QTL)	12 linkage groups	Schäfer-Pregl et al. (1998)
Leptine production		RAPD	I	Ronning et al. (1999)
Solanidin accumulation		RFLP (QTL)	I	Bonierbale et al. (1988)
		RFLP (QTL)	I	Yencho et al. (1998)

mapped in potato by using molecular markers. Paterson et al. (1988) published a resolution of QTL by using a complete linkage RFLP map. For example, cosegregation for quantitative resistance to LB and white potato cyst nematode (PCN) was observed in a tetraploid F_1 population with 94 plants derived from a cross between cultivar Stirling and advanced line SCRI breeding line 12601ab1. The QTL analysis revealed LB resistance genes mapped to chromosome VIII. This location corresponds to no other QTL-mapped resistance genes in potato (Meyer et al. 1998). The most important QTL maps have been shown in Tables 6.3 and 6.4.

DNA markers can be used for obtaining various types of information, such as: (1) introgression of foreign genes, (2) screening of qualitative and quantitative characters (number, effect, and chromosomal location of each gene affecting a trait, effect of multiple copies of individual genes (gene dosage), (3) interaction between/among genes controlling a trait (epistasis), (4) early generation selection by direct selection of DNA types, (5) detection of recessive genes, (6) analyzing the genetic structure of a polyploid for estimating the number of favorable alleles, (7) genotyping/fingerprinting, (8) prediction of success in hybridization in terms of involvement of self-or cross-incompatibility (Ehlenfeldt and Hanneman 1988; Fritz and Hanneman 1989), (9) estimation of occurrence of male sterility for hybridization (Iwanaga et al. 1991), and (10) diagnostics of various traits.

MODALITIES FOR CHOICE OF MARKERS

Without doubt molecular markers are invaluable tools for plant genetic studies and their use facilitates plant breeding programs. Various types of molecular markers have been used for potato molecular genetic studies (Bendahmane et al. 1999; Bonierbale et al. 1994; Brigneti et al 1997; Celebi-Toprak et al. 2002; Chen et al. 2001; Marano et al. 2002; Meyer et al. 1998; Milbourne et al. 1998; Prevost and Wilkinson 1999; Rouppe van der Voort et al. 2000). However, not all markers can be used in one study. The decision as to which should be employed in a particular study depends on many factors. It also depends on the objective, budget and time constraint of a project, database of the plant material, and availability of genetic materials, such as probe or primer.

Table 6.2 provides information on choice criteria for molecular markers suitable to a specific study vis-à-vis the major type of molecular markers. No single set of markers is suitable in terms of the technology used; however, recent advancement in functional genomics and COS (conserved regions of functional genes) markers, such as from tomato, can provide the additional components of choice criteria in selection of the markers to be used.

RFLP markers proved useful tools as the first generation of marker technology in potato because they are codominant and locus-specific, but preparatory steps take long time and require a large amount of high-quality DNA (Bonierbale et al. 1988; van Eck et al. 1995). RAPD markers are PCR based. If no information is available about a plant, and cost and time are critical for the project, then these markers are very useful but for reproducibility of results in different laboratories (Karp et al. 1997). Notwithstanding reproducibility problems, RAPDs, being dominant markers, are not commonly used for potato mapping studies (see Table 6.1). They are, however, used for other types of studies, such as phylogeny and systematics. AFLP markers are PCR based and fill the need for a more efficient tool to construct a dense linkage potato map (Rouppe van der Voort et al. 1997; van Eck et al. 1995). AFLPs are mapped as alleles, rather than loci, which is important for the alignment of genetic maps of plant species such as *S. tuberosum* having high levels of intraspecific variation. The usefulness of this technique is discussed by van Eck et al. (1995) and Rouppe van der Voort et al. (1997). AFLP markers have been used in potato linkage maps (see Table 6.1). Millbourne et al. (1998) developed a potato linkage map by using SSR markers which are highly polymorphic, codominant, and easier to handle. SSR markers are useful for studies on genetic diversity, phylogeny, and origin of potato (Bryan et al. 1999). The only disadvantages of SSR are initial investment and the technical expertise required to clone and sequence the loci, but once they are cloned they can be used for many applications. Ghislain et al. (2001) used RFLP, RAPD, AFLP, and SSR markers to construct two potato maps considering reproducibility, cost, and time. The present work and their studies conclude that a combination of AFLP and SSR markers is most appropriate for efficient genetic map construction in potato. Prevost and Wilkinson (1999) made a DNA fingerprint for selection of potato cultivars by using ISSR-PCR. ISSR markers are highly polymorphic, easy to produce, and reproducible for fingerprinting purposes.

FINGERPRINTING CULTIVARS AND MONITORING INTROGRESSION

Fingerprinting Genotypes

Molecular markers have also been used to identify and select specific genotypes for genetic studies (Meyer et al. 1993) and for genotyping diploid and tetraploid clones (Demecke et al. 1993; Gorg et al. 1992). Most traditional potato breeding took place at the tetraploid level with a relatively narrow genetic base (Ross 1986). Increasing genetic diversity in cultivar development is one of the important aspects in potato breeding and identification of the consequent cultivars cardinal for protection of intellectual property rights. Thus, cultivar identification and fingerprinting have been very important components of potato breeding dealing with aspects such as (1) increasing heterozygosity and avoiding inbreeding that could be crucial for the vigor and productivity of the resultant progeny; (2) production and distribution of seeds in commercial markets—the homogeneity of the clonal cultivars is paramount in assuring purity of the propagules; and (3) monitoring the correct use of the intellectual property rights associated with developed cultivars. It is estimated that more than 1,000 potato cultivars are actively used throughout the world and every year new cultivars are added to the modern cultivar genepool (Prevost and Wilkinson 1999). Genetic similarity and difference can be identified by using molecular markers besides the information provided from the pedigree. Thus, DNA fingerprinting methods should provide valuable information for the plant breeder concerned with enhancement of potato cultivars.

Several different molecular markers have been used for cultivar diagnosis, such as RFLP, RAPD, AFLP, SSR and ISSR (Gebhardt et al. 1989; Milbourne et al. 1997; Prevost and Wilkinson 1999). The molecular marker used for fingerprinting should be cost effective, reliable, highly informative and produce quick results in routine applications. A comparison of these aspects and analysis of the genetic relationships among cultivated potatoes from PCR-based marker systems are provided by Milbourne et al. (1997).

A new generation of techniques termed microarray is now emerging to understand gene functioning. It was originally used for human and animal genome studies. Now it has become available for plant genome studies such as in the model plant system, *Arabidopsis*. Microarray analysis allows the screening of thousands of identifiable genes in single experiment. Microarrays of *A. thalilana* are available both publicly (*Arabidopsis* Functional Genomics Consortium (AFGC) at http://afgc.stanford.edu/) and commercially (http://www.affimetrix.com). Principles of microarray are detailed by Wisman and Ohlrogge (2000). One of the major advantages of this method is that no prior information on gene function is needed and hence abundant information can be obtained by its application. However, DNA microarray is currently expensive because of DNA sequencing, and expensive instruments required, such as hybridization system, scanner, and software (Karp 2001).

Monitoring Introgression

Potato genetic resources involving wild and cultivated taxa, are genetically very diverse, in contrast to the narrow genetic base of modern cultivars, represented mainly by important traits, such as tuber size, shape, skin color, flesh quality, appearance, and pest resistance. However, introgression of such traits from wild germplasm into commercial cultivars requires many generations but, unfortunately, traditional potato breeding is time consuming (Ross 1986). Potato breeding programs can be accelerated by the use of molecular markers, facilitating the selection of desired characteristics on the basis of genotype rather than phenotype. Molecular markers allow early selection of traits linked to a market(s) (Watanabe et al. 1994, 1995). Molecular markers have further use in germplasm enhancement, which in combination with the knowledge of potato genetics (Peloquin et al. 1989, 1999), can accelerate the process of breeding and production of new cultivars.

Wild species of potato have valuable genes. For example, pest resistance genes do not exist in the modern cultivar genepool. The desirable resistance trait(s) from wild species can be sexually transferred to cultivars based on knowledge of potato genetics and reproductive biology, but sometimes there are limitations in sexual hybridization between specific wild and cultivated species due to lack of introgression of the specific genes or chromosomal regions of the donar germplasm. These limitations can be overcome by application of biotechnological methods, such as embryo rescue and somatic hybridization (Watanabe et al. 1997), which can cross the gene transfer barriers. Furthermore, introgression can be monitored by use of molecular markers to confirm the hybridity and introgression of specific traits or chromosome segments of interest into the cultivar genepool by looking at the donor and recipient specific markers positioned in the genome (Watanabe 1994). Where the linkage maps including the desirable trait are to be constructed in the population, molecular markers can target the introgression of specific traits. Molecular linkage maps have been used for targeting major genes and QTLs in potato and researchers have indicated that they proved very useful tools in monitoring introgression (Bradshaw et al. 1998; Brigneti et al. 1997; Celebi-Toprak et al. 2002; Ghislain et al. 2001; Hämäläinen et al. 1997; Meyer et al. 1998; Naess et al. 2001; Paterson et al. 1988).

GENETIC DIVERSITY, TAXONOMY AND PHYLOGENY

Tuber-bearing *Solanum*, consisting of cultivated potato and its wild relatives, is a highly diverse group with wild taxa distributed throughout the New World, from Nebraska, USA to the southern end of Chile, covering coastal deserts to highland Andes (Hawkes 1990). Most of the information on taxonomy and phylogenetic relationships, genetic diversity, origin of cultivated potato, and genebank management of germplasm, was based earlier on morphological data, and to some extent on biochemical data, until DNA markers dramatically increased this fundamental knowledge. This gave rise to some problems, viz: (a) while classical science constitutes the basis for new sciences, taxonomy and phylogeny based on phenotypic information has certain limitations due to altered morphology under diverse environmental conditions; the same genotype may express different phenotypes under changed ecological conditions; (b) change in genetic diversity both *in situ* and *ex situ* could be expressed far less efficiently with phenotypic variation; and (c) occurrence of low frequency of duplicates in genebank collections.

Maintaining duplicate potato germplasm in a genebank increases cost and labor. Traditionally, genebank accessions have been described using morphological characteristics, but environmental conditions can influence morphological features. Application of molecular markers in characterizing germplasm on the other hand, decreases germplasm redundancy and increases the efficiency in conserving a number of varieties in a genebank (Kresovich et al. 1994; Verma et al. 1999).

Various studies on molecular taxonomy and phylogeny of potato have been reported. RFLP (Hosaka and Spooner 1992), RAPD (del Rio et al. 1997), AFLP (McGregor et al. 2002), isozyme (Chandra et al. 2002), and chloroplast (cp) DNA (Kawagoe and Kikuta 1991) as molecular markers have been used to study genetic diversity, taxonomy, and phylogeny of potato even though each molecular marker has advantages and disadvantages. Use of molecular markers gives certain and reliable information about genetic diversity which cannot always be obtained by morphological characterization. Some reports on the use of molecular markers to study these aspects are cited here.

Spooner and colleagues (1995) compared molecular markers while studying relationships among wild potato relatives belonging to *Solanum* section Etuberosum. They found the RFLP marker good for study of interspecies but less so for intraspecies while RAPD, on the other hand, seemed best for intraspecific studies, albeit some reproducibility problems that were encountered. Milbourne et al. (1997) used AFLPs, RAPDs, and SSRs to analyze the genetic relationship in northwestern European cultivated potato genepools. RAPDs were good, easy, cheap, and gave quick results but proved unsuitable for tetraploid inheritance because of the dominant marker. AFLPs and SSRs were efficient for tetraploid studies because they detected allelic differences. Additional research needs to be done to find inexpensive markers. Potato cultivars constitute a very large group and need to be distinguished by reliable, reproducible, easy, and quick methods. Prevost and Wilkinson (1999) used ISSR-PCR methods to distinguish 34 tetraploid potato cultivars. They observed ISSR-PCR analysis to be quick, reproducible, and able to generate sufficient polymorphism to serve as a tool for large-scale DNA fingerprinting purposes.

Plants have three different types of DNA: nuclear or genomic (gDNA), mitochondrial (mtDNA), and chloroplast (cpDNA). The mtDNA is prone to recombination, which could alter the interpretation of evolutional studies. On the other hand, the sequencing of cpDNA conducted on many plant taxa appears to be highly conservative. Therefore, cpDNA is good for phlyogenetic or systematic studies, in particular those tracing maternal origins (Hancock 1992). Analysis of cpDNAs have been used for po tato phylogenetic studies and maternal origins of cultivated potatoes have been identified (Hosaka and Hanneman 1988; Hosaka 1995; Spooner and Systma 1992).

Previous knowledge about potato origin and phylogeny was based on morphological characteristics (Hawkes 1990). Use of nuclear (Debener et al. 1990) or chloroplast RFLP markers (Hosaka et al 1984; Spooner et al.

1991) proved uninformative. New cpSSR markers were developed and applied in different plant species for germplasm analysis (Powell et al. 1996; Proval et al. 1996). Bryan et al. (1999) developed polymorphic cpSSR loci from chloroplast genomes of solanaceous plants and demonstrated that cpSSR markers have a higher resolution level than RFLP markers. Thus, cpSSRs could be of great utility in population genetics, germplasm management, evolutional and phlyogenetic studies as well as in the analysis of material from introgression and somatic-fusion experiments.

Recently, genetic diversity was reported between European and Argentinian cultivated potatoes by Bornet et al. (2002) using ISSR. The authors found ISSR to be a very efficient, quick, and low-cost technique for potato programs. However, further case comparisons need to be undertaken to ascertain whether this approach would be suitable for prompt assessment of genetic diversity while comparing existing germplasm.

MARKERS FOR BREEDING DISEASE RESISTANCES

Markers for Linkage Analysis, Marker-Assisted Diagnosis, and Marker-Assisted Selection (MAS)

General aspects, advantages, and disadvantages in application of each type of molecular markers for analysis of specific traits were summarized by Karp and Edwards (1995). The genes for disease resistance and their genomic distribution were extensively reviewed by Gebhardt and Valkonen (2001). Here we provide examples of markers used for the selection of pest resistance and related traits in potatoes including the most recent reports (Table 6.3).

Major requirements for MAS should be accuracy, rapidity, ease of manipulation, and cost effectiveness. For this reason, PCR-based markers were developed after mapping the target loci using RFLP and/or AFLP markers. PCR-based markers are increasingly available for many simple inherited traits, such as resistance to PVX, PVY, LB, PCN, and RKN. Also, some simple markers have become increasingly useful for QTLs on resistance to LB, PCN, and the presence of glandular trichomes (Table 6.3).

In recent years, molecular markers have been used to localize resistance genes in the potato genome. Examples: detection of genes for extreme resistance of PVY based on Ry_{adg}, can be accurately done using CAPS (Sorri et al. 1999), SCAR (Kasai et al. 2000), and RAPD (Shiranita et al. 1999b); linkage of these markers with the trait appeared to be 100% (77.77 cultivars with diverse genetic background), 100% (103/103), 97% (31/32) respectively, while an RFLP marker (ADG2) gave 100% (54/54) correspondence (Hämäläinen et al. 1997). On chromosome XI, Ry_{sto} gene for extreme resistance to PVY, and Na_{adg} gene expressing hypersensitive resistance to PVA have been mapped (Brigneti et al. 1997; Hämäläinen et al. 2000) respectively. Ny_{tbr}, a hypersensitive resistance gene to PVY, was recently mapped by using RFLP (Celebi-Toprak et al. 2002). A single dominant *Ns* gene was mapped on chromosome VIII for PVS by using RFLP markers (Marczewski et al. 2002).

Resistance to Late Blight (LB)

Resistance genes to LB bere mapped on several chromosomal regions of the potato genome: *R1* on chromosome V (Leonards-Schippers et al. 1992) and *R3, R6,* and *R7* on chromosome XI (El-Kharbotly et al. 1994; El-Kharbotly et al. 1996a; Leister et al. 1996) in addition to possible existence of dominant suppressors to LB resistance (El-Kharbotly et al. 1996b). QTL analysis of LB resistance identified multiple chromosome segments on chromosomes II, III, IV, V, VI, VII, VIII, IX, XI, and XII (Leonards-Schippers et al. 1994; Meyer et al. 1998). Kuhl et al. (2001) mapped the *Rpi1* resistance gene to *Phytophthora infestans* using RFLP marker. Li et al. (1998) mapped the *R2* resistance gene on chromosome IV to *P. infestans* using AFLP markers. Resistance gene *Rblc* to *P. infestans* in *S. bulbocastanum* was mapped on chromosome VIII (Naess et al. 2000). *Pi* QTLs were mapped to LB using SSR (Milbourne et al. 1998). A new *Pi* QTL was also mapped to *P. infestans* using RFLP markers (Ghislain et al. 2001).

Resistance to Potato Cyst Nematode (PCN)

Molecular markers mapped *H1* [gene conferring resistance to *Globodera rostochensis* pathotypes Ro1 and Ro4 (Kort et al. 1977)] on chromosome V (Gebhardt et al. 1993, Ploneda et al. 1993); and *Gro1* (gene conferring resistance to *G. rostochensis* pathotype Ro1) corresponding to *Fb*, on chromosome VII (Barone et al. 1990; Ballvora et al. 1995). QTL analysis identified three loci involved in resistance to *G. rostochiensis* pathotype Ro1 on chromosomes III, X, and XI (Kreike et al. 1993, 1996).

With respect to resistance to *G. pallida*, the nematode resistance locus *Gpa2* (resistance to *G. pallida* pathotype Pa2) was mapped on chromosome XII using information on the genomic positions of known AFLP (Rouppe van der Voort et al. 1997) and CAPS markers (Rouppe van der Voort et al. 1998). One major locus, *Gpa*, conferring resistance to *G. pallida* pathotypes Pa2 and Pa3, was mapped on chromosome V, and two minor loci mapped on chromosomes IV and VII based on the QTL analysis of *S. spegazzinii* (Kreike et al. 1994).

Broad-specturm resistance to both *G. rostochiensis* and *G. pallida*, regarded as polygenetically inherited, could be ascribed to the action of locus *Grp1* (Rouppe van der Voort et al. 1998). QTL analysis allocated *Grp1* (resistance to *G. rostochiensis* pathotype Po5 and *G. pallida* pathotype Pa2) on chromosome V (Rouppe van der Voort et al. 1998). Interestingly,

Grp1 was mapped on a genomic region harboring other resistance factors for viral, fungal, and nematodal pathogens (Rouppe van der Voort et al. 1998). Bradshaw et al. (1998) mapped the *Gpa4* (QTL) using AFLP and SSR markers.

Recently, Rouppe van der Voort et al. (2000) mapped the *Gpa5* on chromosome V and *Gpa6* on chromosome IX (QTL) of potato by using AFLP, CAPS markers. Bradshaw et al. (1998) mapped the QTL for *G. pallida* by using AFLP and SSR markers. Bryan et al. (2002) mapped the QTL on chromosome V and IX for *G. pallida* by using AFLP markers.

Resistance to Root-Knot Nematode (RKN)

In linkage analysis of RKN resistance, only resistance to *meloidogyne chitwoodi* has been mapped on the potato genome although many resistance genes have been identified for various *Meloidogyne* species. The *Rmc1* (resistance to *M. chitwoodi*, race 1) locus originates from the wild species *S. bulbocastanum* and is localized on chromosome XI (Brown et al. 1996). It was subsequently demonstrated that the resistance of *Rmc1* includes not only *M. chitwoodi* and the related species *M. fallax*, but also genetically distinct populations of *M. hapla* (Rouppe van der Voort et al. 1999b). The present authors have achieved some progress in mapping and cloning the resistance genes to *M. pallida*, which are patent protected.

Glandular Trichomes

Based on the QTL map constructed by RFLP markers and phenotypic data of glandular trichomes, two quantitative trait loci, one on Chromosome VI and the other on Chromosome X, were identified as highly associated with the indication levels of oxidase, phenolic structure, and trichome density of type A (Bonierbale et al. 1994). For type B trichomes, five QTLs were found for fatty acid sucrose ester levels (Bonierbale et al. 1994). These QTLs were experimentally used for marker-assisted selection and at least selection of Type A trichomes was efficient (Watanabe 1994 and unpubl. data).

MARKERS FOR BREEDING AGRONOMIC TRAITS

Some important agronomic traits studied by means of molecular markers are given in Table 6.4 together with examples. For potato plant amelioration, breeders have to deal with both qualitative and quantitative traits. Nevertheless, quantitative traits still remain agronomically the most important (Ross 1986), e.g. yield quality, tuber shape, tuber size, tuber skin color, tuber flesh color, gravity, and flower color. Molecular markers help in analyzing some quantitative traits to increase efficiency in selection and relevant combination of those traits cardinal for germplasm conservation and cultivar development. But in spite of this information, breeders still know very little about the chromosomal locations of these genetic components, which directly affect their expression and heritability, because quantitative trait loci are under the control of both environmental and genetic factors (Gelderman 1975). To analyze QTL, the quickest technique would be inbreeding homozygous isogenic lines. But the cultivated potato and its wild relatives are highly heterozygous, polyploid, and/or self-incompatible. Therefore, establishing homozygous inbred lines genetically is just not feasible.

Many potato linkage maps were constructed for noninbred segregation populations using molecular markers (Bonierbale et al. 1988; Gebhardt et al. 1989; Jacobs et al. 1995; van Eck et al. 1995). These maps were then used for pinpointing major loci for tuber flesh color on chromosome III and tuber skin color on chromosome. X (Bonierbale et al. 1988; Gebhardt et al. 1989, 1991). A few studies on QTL mapping of major agronomic traits have also been done, for instance tuber shape (Van Eck et al. 1994) and tuberization (Bonierbale et al. 1988) using RFLP linkage maps, QTL for tuberization mapped on chromosome V in different segregating potato populations (van den Berg et al. 1996a), and another for tuber dormancy mapped mainly on chromosome II (van der Berg et al. 1996b). Leptine production controlled by QTL has likewise been mapped on chromosome I and *S. chacoense.* Bitter population using RAPD I (Ronning et al. 1999) and solanidin accumulation QTL on chromosome I using RFLP (Bonierbale et al. 1988; Yencho et al. 1998).

Schäfer-Pregl and colleagues (1998) studied QTLs and quantitative trait alleles (QTAs) for potato tuber yield and starch content using RFLP markers on two different crosses among (di)haploid breeding lines. QTLs for tuber starch content and tuber yield have been identified and localized in the potato genome. The effects of QTAs at specific QTLs have been analyzed. The stability of QTLs across various genetic backgrounds and environments has also been analyzed. From the results eighteen putative QTLs for tuber starch content and eight putative QTLs for tuber yield were identified and all twenty-six proved reproducible at two different environments, while a few major QTLs for tuber starch content appeared highly stable under various environments.

Chen et al. (2001) introduced a new approach for using the molecular marker map for potato. They constructed such a map based only on genes with known functional roles, such as carbohydrate metabolism and carbon transport using CAPS, SCAR, and RFLP. Comparison of this molecular-function map (Chen et al. 2001) with QTL maps for tuber starch content (Schäfer-Pregl et al. 1998), indicated a number of candidate genes having alleles that might differentially influence the starch content of potato tubers. This kind of map may also be useful for comparing the structure and function of distantly related plant genomes. To conclude, it may be noted that although some advances have been made in QTL mapping of important agronomic traits in potato, there are many important agronomic QTL traits waiting to be investigated.

GENERAL CONCLUSIONS

We have presented an overview of the potential use of molecular markers in potato genetic studies and breeding. Most importantly, there is need for advancement in scientific knowledge and technology to achieve better understanding of potato genetics, especially with relevance to the important agronomic traits in terms of genomics, gene expression, and practical applications.

Comparative genomic studies have been done in genera of family Solanaceae (Doganlar et al. 2002a,b; Livingstone et al. 1999; Tanksley et al. 1992) and family Gramineae (Chen et al. 1997). This type of study may facilitate the transfer of information among solanaceous genera and actual use of genopools in germplasm enhancement (Watanabe et al. 1995). Thus, attention should also be paid to ongoing comparative and functional genomic research concerning related taxa in order to incorporate this information for maximum utility and benefit.

One major point that requires elucidation is that cost assessments of molecular marker technology are not available in a comprehensive manner to ascertain its practicability in potato science and production. In particular, there should be decisive criteria for selection and employment of marker methods in evaluation of specific traits for potato breeding.

Information is scant on the use of QTLs with respect to agriculturally important traits. Yet technology must be standardized and simplified to make it acceptable for general use. Of particular consideration is the fact that information based on various trials using this technology be incorporated for dissemination throughout the world, especially in countries where demand for more potato production has been clearly expressed (see Fig. 6.1 and Appendix 6.1).

Acknowledgments

The authors thank the Japan Society for the Promotion of Sciences for financial support in the form of grants under Research for Future Programs (RFTF-96-L603 and RFTF-00L01602). FCT acknowledges financial support by the visiting faculty/scholarship program at University of Tsukuba. The authors also thank Prof. Jari P.T. Valkonen, SLU, Uppsala, Sweden, for his helpful comments on reviewing the manuscript.

REFERENCES

Ballvora, A., J. Hesselbach, J. Niewöhner, D. Leister, F. Salamini, and C. Gebhardt. 1995. Marker enrichment and high-resolution map of the segment of potato chromosome VII harbouring the nematode resistance gene *Gro1*. Mol Gen Genet 249: 82–90.

Barone, A., E. Ritter, U. Schachtschabel, T. Debener, F. Salamini, and C. Gebhardt. 1990. Localization by restriction fragment length polymorphism mapping in potato of a major dominant gene conferring resistance to the potato cyst nematode *Globodera rostochiensis*. Mol Gen Genet 224: 117–182.

Bendahmane, A., K. Kanyuka, and D.C. Baulcombe, 1997. High-resolution genetical and physical mapping of the *Rx* gene for extreme resistance to potato virus X in tetraploid potato. Theor Appl Gent 95: 153–162.

Bendahmane, A., K. Kanyuka, and D.C. Baulcombe. 1999. The *Rx* gene from potato controls separate virus resistance and cell death responses. Plant Cell 11: 781–791.

Bonierbale, M., R.L. Plaisted, and S.D. Tanksley. 1988. RFLP maps based on a common set of clones reveal modes of chromosomal evolution in potato and tomato. Genetics 120: 1095–1103.

Bonierbale, M.W., R.L. Plaisted, O. Pineda, and S.D. Tanksley. 1994. QTL analysis of trichomemediated insect resistance in potato. Theor Appl Genet 87: 973–987.

Bornet, B., F. Goraguer, G. Joly, and M. Branchard. 2002. Genetic diversity in European and Argentinian cultivated potatoes (*Solanum tuberosum* subsp. *tuberosum*) detected by inter-simple sequence repeats (ISSRs). Genome 45: 481–484.

Bradshaw, J.E., C.A. Hackett, R.C. Meyer, D. Milbourne, J.W. McNicol, M.S. Philips, and R. Waugh. 1998. Identification of AFLP and SSR markers associated with quantitative resistance to *Globedera pallida* (Stone) in tetraploid potato (*Solanum tuberosum* supsp. *tuberosum*) with a view to marker-assisted selection. Theor Appl Genet. 97: 202–210.

Brigneti, G., J. Garcia-Mas, and D.C., Baulcombe. 1997. Molecular mapping of the potato virus Y resistance gene *Rysto* in potato. Theor Appl Genet 94: 198–203.

Brown, C.R., C.P. Yang, H. Mojtahedi, G.S. Santo, and R. Masueli, 1996. RFLP analysis of resistance to Columbia root-knot nematode derived from *Solanum bulbocastanum* in a BC-2 population. Theor Appl Genet 92: 572–576.

Bryan, G.J., J. McNicoll, G. Ramsay, R.C. Meyer, and W.S. De Jong. 1999. Polymorphic simple sequence repeat markers in chloroplast genomes of Solanaceous plants. Theor Appl Genet 99: 859–867.

Bryan, G.J., K. McLean, J.E. Bradshaw, W.S. De Jong, M. Phillips, L. Castelli, and R. Waugh. 2002. Mapping QTLs for resistance to the cyst nematode *Globodera pallida* derived from the wild potato species *Solanum vernei*. Theor Appl Genet 105: 68–77.

Celebi-Toprak, F., S.A. Slack, and M.M. Jahn. 2002. A new gene, *Nytbr*, for hypersensitivity to potato virus Y from *Solanum tuberosum* maps to chromosome IV. Theor Appl Genet 104: 669–674.

Chandra, S., Z. Huaman, S. Hari Krishna, and R. Ortiz. 2002. Optimal sampling strategy and core collection size of Andean tetraploid potato based on isozyme data—a simulation study. Theor Appl Genet. Published on-line, 26 April 2002.

Chen, M., P. Sanmiguel, A.C.D. Oliveria, S.S. Woo, H. Zhang, R.A. Wing, and J.L. Bennetzen 1997. Microcolinearity in *sh*-2-homologous regions of the maize, rice, and sorghum genomes. Proc Natl Acad Sci USA 94: 3431–3435.

Chen, X., F. Salamini, and C. Gebhardt. 2001. A potato molecular-function map for carbohydrate metabolism and transport. Theor Appl Genet 102: 284–295.

De Jong, W., A. Forsyth, D. Leister, C. Gebhardt, and D.C. Baulcombe. 1997. A potato hypersensitive resistance gene against potato virus X maps to a resistance gene cluster on chromosome 5. Theor Appl Genet 65: 246–252.

Debener, T., F. Salamini, and C. Gebhardt. 1990. Phylogeny of wild and cultivated *Solanum* species based on nuclear restriction fragment length polymorphisms (RFLPs). Theor Appl Genet 79: 360–368.

del Rio, A.H., J.B. Bamberg, and Z. Huaman. 1997. Assessing changes in the genetic diversity of potato gene banks, 1. Effects of seed increase. Theor Appl Genet 95: 191–198.

Demecke, T., L.H. Kawchuk, and D.R. Lynch. 1993. Identification of potato cultivars and clonal variants by random amplified polymorphic DNA analysis. Amer Potato J 70: 561–570.

Doganlar, S., A. Frary, C.M. Daunay, R.N. Lester and S.D. Tanksley. 2002a. A comparative genetic linkage map of eggplant (*Solanum melongena*) and its implication to genome evolution in the Solanaceae. Genetics 161: 1697–1711.

Doganlar, S., A. Frary, C.M. Daunay, R.N. Lester and S.D. Tanksley. 2002b. Comparative mapping of domestication traits in eggplant reveals conservation of gene function in the Solanaceae. Genetics 161: 1713–1726.

Ehlenfeldt, M.K. and R.E. Hanneman, Jr. 1988. Genetic control of Endosperm Balance Number (EBN): three additive loci in a threshold-like system. Theor Appl Genet. 75: 825–832.

El-Kharbotly, A., A. Pereira, W.J. Stiekema, and E. Jacobsen. 1996a. Race specific resistance against *Phytophthora infestans* in potato is controlled by more genetic factors than only R-genes. Euphytica, 90: 331–336.

El-Kharbotly, A., C. Palomino-Sánchez, F. Salamini, E. Jacobsen, and C. Gebhardt. 1996b. *R6* and *R7* alleles of potato conferring race-specific resistance to *Phytophthora infestans* (Mont.) de Bary identified genetic loci clustering with *R3* locus on chromosome XI. Theor Appl Genet 92: 880–884.

El-Kharbotly, A., C. Leonards-Schippers, D.J. Huigen, E. Jacobsen, A. Pereira, W.J. Steikema, F. Salamini, and C. Gebhardt. 1994. Segregation analysis and RFLP mapping of *R1* and *R3* alleles conferring race-specific resistance to *Phytophthora infestans* in progenies of dihaploid potato parents. Mol Gen Genet 242: 749–754.

Fritz, N.K. and R.E. Hanneman, Jr. 1989. Interspecific incompatibility due to stylar barriers in tuber-bearing and closely related non-tuber-bearing *Solanums*. Sex Plant Reprod 2: 184–192.

Ganal, M.W., M.W. Bonierbale, and M.S. Roeder et al. 1991. Genetic and physical mapping of the patatin genes in potato and tomato. Mol Gen Genet 225: 501–509.

Gebhardt, C. and J.P.T. Valkonen. 2001. Organization of genes controlling disease resistance in the potato genome. Annu Rev Phytopathol. 39: 79–102.

Gebhardt, C., D. Muginery, E. Ritter, F. Salamini, and E. Bonnel. 1993. Identification of RFLP markers closely linked to the *H1* gene conferring resistance to *Globodera rostochiensis* in potato. Theor Appl Genet 85: 541–544.

Gebhardt, C., E. Ritter, T. Debener, U. Schachtschabel, B. Walkemeier, H. Uhrig, and F. Salamini. 1989. RFLP analysis and linkage mapping in *Solanum tuberosum*. Theor Appl Genet 78: 65–75.

Gebhardt, C., E. Ritter, A. Barone, T. Debener, B. Walkemeier, U. Schachtschabel, H. Kaufmann, R.D. Thompson, M.W. Bonierbale, M.W. Ganal, S.D. Tanksley, and F. Salamini. 1991. RFLP maps of potato and their alignment with the homoeologous tomato genome. Theor Appl Genet 83: 49–57.

Geldermann, H. 1975. Investigations on inheritance of quantitative characters in animals by gene markers, I. Methods. Theor Appl Genet 46: 319–330.

Ghislain, M., B. Trognitz, R. Ma. del Herrera, J. Solis, G. Casallo, C. Vásquez, O. Hurtado, R. Castillo, L. Portal, and M. Orrillo. 2001. Genetic loci associated with field resistance to late blight in offspring of *Solanum phureja* and *S. tuberosum* grown under short-day conditions. Theor Appl Gent 103: 433–442.

Gorg, R., U. Schachtschabel, E. Ritter, F. Salamini, and C. Gebhardt. 1992. Discrimination among 136 tetraploid potato varieties by fingerprinting using highly polymorphic DNA markers. Crop Sci 32: 815–819.

Hämäläinen, J.H., V.A. Sorri, K.N. Watanabe, C. Gebhardt, and J.P.T. Valkonen, 1998. Molecular examination of a chromosome region that controls resistance to potato Y and A polyviruses in potato. Theor Appl Genet 96: 1036–1043.

Hämäläinen, J.H., T. Kekareinin, C. Gebhardt, K.N. Watanabe, J.P.T. Valkonen. 2000. Recessive and dominant resistance interfere with the vascular transport of *Potato virus A* in diploid potatoes. Mol Plant-Microbe Interact 13: 400–412.

Hämäläinen, J.H., K.N. Watanabe, J.P.T. Valkonen, A. Arihara, R.L. Plaisted, E. Pehu, L. Miller, and S.A. Slack. 1997. Mapping and marker-assisted selection for a gene for extreme resistance to potato virus Y. Theor Appl Genet 94: 192–197.

Hancock, J.F. 1992. Plant Evolution and the Origin of Crop Species. Prentice-Hall, NJ (USA), 305 pp.

Hawkes, J.G. 1990. The Potato—Evolution, Biodiversity and Genetic Resources. Belhaven Press, London, UK.

Hehl, R., E. Faurie, J. Hesselbach, F. Salamini, S. Whitham, B. Baker, and C. Gebhardt. 1999. TMV resistance gene N homologues are linked to Synchytrium endobioticum resistance in potato. Theor Appl Genet 98: 379–386.

Hosaka, K. 1995. Successive domestication and evolution of the Andean potatoes as revealed by chloroplast DNA restriction endonuclease analysis. Theor. Appl Genet 90: 356–363.

Hosaka, K. and R.E. Hanneman, Jr. 1988. The origin of the cultivated tetraploid potato based on chloroplast DNA. Theor Appl Genet 76: 172–176.

Hosaka, K. and D.M. Spooner. 1992. RFLP analysis of the wild potato species, *Solanum acaule* bitter (*Solanum* sect. Petota). Theor Appl Genet 84: 851–858.

Hosaka, K., Y. Ogihara, M. Matsubayashi, and K. Tsunewaki. 1984. Phlyogenetic relationship between the tuberous *Solanum* species as revealed by restriction endonuclease analysis of chloroplast DNA. Jpn. J Genet 59: 349–369.

Iwanaga, M., R. Ortiz, M.S. Cipar, and S.J. Peloquin. 1991. A restorer gene for genetic-cytoplasmic male sterility in cultivated potatoes. Amer Potato J 68: 19–28.

Jacobs, J.M., H.J. van Eck, P. Arens, B. Verker-Bakker, B. te Lintel Hekkert, H.J.M. Bastiannssen, A. El-Kharbotly, A. Pereira, E. Jacobsen, and W.J. Stiekema. 1995. A genetic map of potato (*Solanum tuberosum*) integrating molecular markers, including transposons, and classical markers. Theor Appl Genet 91: 289–300.

Kanyuka, K., A. Bendahmane, J.N.A.M Rouppe van der Voort, E.A.G. van der Vossen, and D.C. Baulcombe. 1999. Mapping of intra-locus duplications and introgressed DNA: aids to map-based cloning of genes from complex genomes illustrated by physical analysis of the *Rx* locus tetraploid potato. Theor Appl Genet 98: 679–689.

Karp, A. 2001. The new genetic era: will it help us in managing genetic diversity? In: J.M.M. Engels, V. Ramanatha Rao, A.H.D. Brown, and M.T. Jackson (eds.), Managing Plant Genetic Diversity. CABI publishing, CAB International, Wallingford, Oxfordshire, UK. pp. 43–56.

Karp, A. and K.J. Edwards. 1995. Molecular techniques in the analysis of the extent and distribution of genetic diversity. In: W.G. Ayad, T. Hodgkin, A. Jaradat, and V.R. Rao, (eds.), Molecular Genetic Techniques for Plant Genetic Resources. IPGRI Workshop, 9–11 October 1995, Rome, Italy, pp. 11–22.

Karp, A., S. Kresovich, K.V. Bhat, W.G. Ayad, and T. Hodgkin. 1997. Molecular tools in plant genetic resources conservation: a guide to the technologies. IPGR Tech. Bull. 2.

Kasai, K., Y. Morikawa, V.A. Sorri, J.P.T. Valkonen, C. Gebhardt, and K.N. Watanabe. 2000. Development of SCAR markers to the PVY resistance gene *Ryadg* based on a common feature of plant disease resistance genes. Genome 43: 1–8.

Kawagoe, Y. and Y. Kikuta. 1991. Chloroplast DNA evolution in potato (*Solanum tuberosum* L.). Theor Appl Genet 81: 13–20.

Kort, J., H. Ross, H.J. Rumpenhorst, and S.R. Stone. 1977. An international scheme for identifying and classifying pathotypes of potato cyst nematodes *Globodera rostochiensis* and *G. pallida*. Nematologica 23: 333–339.

Kreike, C.M., A.A. Kok-Westeneng, J.H. Vinke, and W.J. Stiekema. 1996. Mapping of QTLs involved in nematode resistance, tuber yield and root development in *Solanum* sp. Theor Appl Genet 92: 463–470.

Kreike, C.M., J.R.A. De Koning, J.H. Vinke, J.W. Van Ooijen, and W.J. Stiekema. 1994. Quantitatively inherited resistance to *Globodera pallida* is dominated by major locus in *Solanum spegazzinii*. Theor Appl Genet 88: 764–769.

Kreike, C.M., J.R.A. De-Koning, J.H. Vinke, J.W. Van Ooijen, C. Gebhardt, and W.J. Stiekema. 1993. Mapping of loci involved in quantitatively inherited resistance to the potato cyst-nematode *Globodera rostochiensis* pathotype R01. Theor Appl Genet 87: 464–470.

Kresovich, S., W.F. Lamboy, R. Li, J. Ren, A.K. Szewc-McFadden, and S.M. Bliek. 1994. Application of molecular methods and statistical analysis for discriminating accessions and clones of vetiver grass. Crop Sci 34: 805–809.

Kuhl, J.C., R.E. Hanneman, Jr., and M.J. Havey. 2001. Characterization and mapping of *Rpi1*, a late-blight resistance locus from diploid (1-EBN) Mexican *Solanum pinnatisectum*. Mol Gen Genet 265: 977–985.

Leister, D., A. Ballvora, F. Salamini, and C. Gebhardt. 1996. A PCR-based approach for isolating pathogen resistance genes from potato with potential for wide application in plants. Nat Genet 14: 421–429.

Leonards-Schippers, C., W. Gieffers, F. Salamini, and C. Gebhardt. 1992. The *R1* gene conferring race-specific resistance to *Phytophthora infestans* in potato is located on potato chromosome V. Mol Gen Genet 223: 278–283.

Leonards-Schippers, C., W. Gieffers, R. Schäfer-Pregl, E. Ritter, E., S.J. Knapp, F. Salamini, and C. Gebhardt. 1994. Quantitative resistance to *Phytophthora infestans* in potato: A case study for QTL mapping in an allogamous plant species. Genetics 137: 67–77.

Li, X., H.J. Van Eck, J. Rouppe van der Voort, D-J. Huigen, P. Stam, and E. Jacobsen. 1998. Autotetraploids and genetic mapping using common AFLP markers: the R2 allele conferring resistance to *Phytophthora infestans* mapped on potato chromosome 4. Theor Appl Genet 96: 1121–1128.

Livingstone, K.D., V.K. Lackney, J.R. Blauth, R. van Wijk, and M.K. Jahn. 1999. Genome mapping in capsicum and the evolution of genome structure in the Solanaceae, Genetics 152: 1183–1202.

Marano, M.R., I. Malcuit, W. De Jong, and D.C. Baulcombe. 2002. High-resolution genetic map of *Nb*, a gene that confers hypersensitive resistance to potato virus X in *Solanum tuberosum*. Theor. Appl. Genet. 105: 192–200.

Marczewski, W., J. Hennig, and C. Gebhardt. 2002. The *Potato virus S* resistance gene *Ns* maps to potato chromosome VIII. Theor Appl Genet. 105: 564–567.

McGregor, C.E., R. van Treuren, R. Hoekstra, and Th.J.L. van Hintum. 2002. Analysis of the wild potato germplasm of the series Acaulia with AFLPs: implications for *ex situ* conservation. Theor Appl Genet 104: 146–156.

Meksem, K., D. Leister, J. Paleman, M. Zebeau, F. Salamini, and C. Gebhardt. 1995. A high-resolution map of chromosome V based on RFLP and AFLP markers in the vicinity of the R1 locus. Mol Gen Genet 249: 74–81.

Meyer, R., F. Salamini, and H. Uhrig. 1993. Isolation and characterization of potato diploid clones generating a high frequency of monoploid or homozygous diploid androgenetic plants. Theor Appl Genet 85: 905–912.

Meyer, R.C., D. Milbourne, C.A. Hackett, J.E. Bradshaw, J.W. McNichol, and R. Waugh. 1998. Linkage analysis in tetraploid potato and association of markers with quantitative resistance to late blight (*Phytophthora infestans*). Mol Gen Genet 259: 150–160.

Milbourne, D., R.C. Meyer, A.J. Collins, L.D. Ramsay, C. Gebhardt, and R. Waugh. 1998. Isolation, characterization and mapping of simple sequence repeat loci in potato. Mol Gen Genet 259: 233–245.

Milbourne, D., R. Meyer, J.E. Bradshaw, E. Baird, N. Bonar, J. Provan, W. Powell, and R. Waugh. 1997. Comparison of PCR-based marker system for the analysis of genetic relationship in cultivated potato. Mol Breed 3: 127–136.

Naess, S.K., J.M. Bradeen, S.M. Wielgus, G.T. Haberlach, J.M. McGrath, and J.P. Helgeson. 2000. Resistance to late blight in *Solanum bulbocastanumis* mapped to chromosome 8. Theor Appl Genet 101: 697–704.

Naess, S.K., J.M. Bradean, S.M. Wielgus, G.T. Haberlech, J.M. McGrath, and J.P. Helgeson. 2001. Analysis of the introgression of *Solanum bulbocastanum* DNA into potato breeding lines. Mol Gen Genet 265: 694–704.

Paterson, A.H., E.S. Lander, J.D. Hewitt, S. Peterson, S.E. Lincoln, and S.D. Tanksley. 1988. Resolution of quantitative traits into Mendelian factors by using a complete linkage map of restriction fragment length polymorphisms. Nature 335: 721–726.

Peloquin, S.J., L. Boiteux, and D. Carputo 1999. Meiotic mutants in potato: valuable variants. Genetics 153: 1493–1499.

Peloquin, S.J., G.L. Yerk, J.E. Werner, and E. Darmo. 1989. Potato breeding with haploids and 2n gametes. Genome 31: 1000–1004.

Perez, F., A. Menedez, P. Dehae, and C.F. Quiros. 1999. Genomic structural differentiation in Solanum comparative mapping of the A-and E-genomes. Theor Appl Genet 98: 1183–1193.

Pineda, O., M.W. Bonierbale, R.L. Plaisted, B. Brodie, and S.D. Tanksley. 1993. Identification of RFLP markers linked to *H1* gene conferring resistance to the potato cyst nematode *Globodera rostochiensis* . Genome 36: 152–156.

Powell, W., M. Morgante, J.J. Doyle, J.W. McNicol, S.V. Tingey, and A.J. Rafalski 1996. Genepool variation in genus *Glycine* subgenus *Soja* revealed by polymorphic nuclear and chloroplast microsatellites. Genetics 144: 793–803.

Prevost, A. and M.J. Wilkinson. 1999. A new system of comparing PCR primers applied to ISSR fingerprinting of potato cultivars. Theor Appl Genet 98: 107–112.

Proval, J., G. Corbett, R. Waugh, J.W. McNicol, M. Morgante, and W. Powell. 1996. DNA fingerprints of rice (*Oryza sativa*) obtained from chloroplast simple-sequence repeats. Proc Roy Soc Lond, Series Sciences 263: 1275–1281.

Ritter, E., T. Debener, A. Barone, F. Salamini, and C. Gebhardt. 1991. RFLP mapping on potato chromosomes of two genes controlling extreme resistance to potato virus X (PVX). Mol Gen Genet 227: 81–85.

Ronning, C.M., J.R. Stommel, S.P. Kowalski, L.L. Sanford, R.S. Kobayashi, and O. Pineda. 1999. Identification of molecular markers associated with leptine production in a population of *Solanum chacoense* Bitter. Theor Appl Genet 98: 39–46.

Ross, H., 1986. Potato Breeding—Problems and Perspectives. Verlag Paul Parey, Berlin, Germany, 132 pp.

Rouppe van der Voort, J., W. Lindeman, R. Folkertsma, R. Hutten, H. Overmars, E. van der Vossen, and J. Bakker. 1998. A QTL for broad-spectrum resistance to cyst nematode species (*Globodera* spp.) maps to a resistance gene cluster in potato. Theor Appl Genet 96: 654–661.

Rouppe van der Voort, J., G.J.W. Janssen, H. Overmars, P.M. van Zandvoort, A. van Norel, O.E. Scholten, R. Janssen, and J. Bakker. 1999a. Development of a PCR based selection assay for root-knot nematode resistance (*Rmc1*) by a comparative analysis of the *Solanum bulbocastanum* and *S. tuberosum* genome. Euphytica. 106: 187–195.

Rouppe van der Voort, J., E. Kanyuka, E. van der Vossen, A. Bendahmane, P. Mooijman, R. Klein-Lankhorst, W. Stiekeman, D. Baulcombe, and J. Bakker. 1999b. Tightly physical linkage of the nematode resistance gene *Gpa2* and the virus resistance gene *Rx* on a single segment introgressed from the wild species *Solanum tuberosum* subsp. *andigena* CPC 1673 into cultivated potato. Mol Plant-Microbe Interact. 12: 197–206.

Rouppe van der Voort, J., P. Wolters, R. Folkertsma, R. Hutten, P. van Zandvoort, H. Vinke, K. Kanyuka, A. Bendahmane, E. Jacobsen, R. Janssen, and J. Bakker. 1997. Mapping of the cyst nematode resistance locus *Gpa2* in potato using a strategy based on comigrating AFLP markers. Theor Appl Genet 95: 874–880.

Rouppe van der Voort, J., E. van der Vossen, E. Bakker, H. Overmars, P.M. van Zandvoort, R. Hutten, R. Klein-Lankhorst, and J. Bakker. 2000. Two additive QTLs conferring broad spectrum resistance in potato *Globodera pallida* are localized on resistance gene clusters. Theor Appl Genet 10: 1122–1130.

Schäfer-Pregl, R., E. Ritter, L. Concilio, J. Hesselbach, L. Lovatti, B. Walkemeier, H. Thelen, F. Salamini, and C. Gebhardt. 1998. Analysis of quantitative trait loci (QTLs) and quantitative trait alleles (QTAs) for potato tuber yield and starch content. Theor Appl Genet 97: 834– 846.

Shiranita, A., K. Kasai, J.H. Hämäläinen, J.P.T. Valkonen, and K.N. Watanabe. 1999a. Selection of resistance to potato Y potyvirus (PVY), using the resistance gene-like fragment ADG2 as an RFLP probe. Plant Biotech 16: 361–369.

Shiranita, A., T. Tsujikawa, K. Kasai, J.P.T. Valkonen, J.A. Watanabe, K.N.Watanabe. 1999b. Use of molecular markers for resistance breeding in potato. VII: Applicability of PCR-based marker for selection for PVY resistance in *S. acaule* introgression lines. Breed Res **1** (Suppl 1): 259.

Sorri, V.A., K.N. Watanabe, and J.P.T. Valkonen. 1999. Predicted kinase 3a motif of a resistance gene-like fragment as a unique marker for PVY resistance. Theor Appl Genet 99: 164–170.

Spooner, D.M. and K.J. Sytsma. 1992. Reexamination of series relationships of Mexican and Central American wild potatoes (*Solanum* sect. Petota): evidence from chloroplast DNA restriction site variation. Syst Bot 17: 432–448.

Spooner, D.M., K.J. Sytsma, and E. Conti. 1991. Chloroplast DNA evidence for genome differentiation in wild potatoes (*Solanum* sect. Petota: Solanaceae). Amer J Bot 78: 1354–1366.

Spooner, D.M., J. Tivang, J. Nienhuis, J.P. Miller, D.S. Douches, and A. Contreras. 1995. Comparison of four molecular markers in measuring relationships among the wild potato relatives of *Solanum* section tuberosum (subgenus *Potatoe*). Theor Appl Genet 91: 1–9.

Tanksley, S.D., M.W. Ganal, J.P. Prince, M.C.de Vincente, M.W. Bonierbale, P. Broun, T.M. Fulton, J.J. Giovannoni, S. Grandillo, and G.B. Martin. 1992. High density molecular linkage maps of the tomato and potato genome. Genetics 132: 1141–1160.

Tommiska, T.J., J.H. Hämäläinen, K.N. Watanabe, and J.P.T. Valkonen. 1998. Mapping of the gene Nx_{phu} that controls hypersensitive resistance to potato virus X in *Solanum phureja* IvP35. Theor Appl Genet 96: 840–843.

Van den Berg, J.H., E.E. Ewing, R. L. Plaisted, S. McMurry, and M.W. Bonierbale. 1996a. QTL analysis of potato tuberization. Theor Appl Genet 93: 307–316.

Van den Berg. J.H., E.E. Ewing, R.L. Plaisted, S. McMurry, and M.W. Bonierbale. 1996b. QTL analysis of potato tuber dormancy. Theor Appl Genet 93 (3): 317–324.

Van Eck, H.J. 1995. Localization of morphological traits on the genetic map of potato using RFLP and isozyme markers. PhD thesis, Univ Wageningen, Netherlands.

Van Eck, H.J., J.M.E. Jacobs, P. Stam, J. Ton, W.J. Stiekema, and E. Jacobsen. 1994. Multiple alleles for tuber shape in diploid potato detected by qualitative and quantitative genetic analysis using RFLPs. Genetics 137: 303–309.

Van Eck, H.J., J. Rouppe van der Voort, J. Draaistra, P. van Zandvoort, E. van Enckevort, B. Segers, J. Peleman, E. Jacoben, J. Helder, and J. Bakker. 1995. The inheritance and chromosomal location of AFLP markers in a non-inbred potato offspring. Mol Breed 1: 397–410.

Verma, S.K., V. Khanna, and N. Singh. 1999. Random amplified polymorphic DNA analysis of Indian scented basmati rice (*Oryza sativa* L.) germplasm for identification of variability and duplicate accessions, if any. Electrophoresis 20: 1786–1789.

Watanabe J.A., M. Orrillo, and K.N. Watanabe. 1999. Resistance to bacterial wilt (*Pseudomonas solanacearum*) of potato evaluated by survival and yield performance at high temperatures. Breed Sci 9: 63–68.

Watanabe, K.N. 1994. Potato molecular genetics. *In:* J.E. Bradshaw and G. Mackey (eds.), Potato Genetics, CAB Intl, Wallingford, UK, pp. 213–235.

Watanabe, K.N., A.M. Golmirzaie, and P. Gregory 1997. Use of biotechnology tools in potato genetic resources management and breeding. In: K.N. Watanabe and E. Pehu (eds.), Plant Biotechnology and Plant Genetic Resources for Sustainability and Productivity. R.G. Landes Company, Austin, Tx. (USA), pp. 143–154.

Watanabe, K, M. Orrillo, S. Vega, R. Masuelli, and K. Ishiki. 1994. Potato germplasm enhancement with disomic tetraploid *Solanum acaule*, II: Assessment to breeding value of tetraploid F_1 hybrids between *S. acaule* and tetrasomic tetraploid potatoes. Theor Appl Genet 88: 135–140.

Watanabe, K.N., M. Orrillo, S. Vega, A. Hurtado, J.P.T. Valkonen, E. Pehu, and S.D. Tanksley. 1995. Overcoming crossing barriers between non-tuber-bearing and tuber-bearing *Solanum* species: Towards potato germplasm enhancement with a broad spectrum of solanaceous genetic resources. Genome 38: 27–35.

Wisman, E. and J. Ohlrogge. 2000. *Arabidopsis* microarray service facilities. Plant Physiol 124: 1468–1471.

Yencho, G.C., S.P. Kowalski, R.S. Kobayashi, S.L. Sinden, and K.L. Deahl. 1998. QTL mapping of *Solanum* steroid alkaloids in interspecific potato crosses: quantitative variation and biosynthetic pathways. Theor Appl Genet 97: 563–574.

Zimnoch-Guzowska, E., W. Marczewski, R. Lebecka, B. Flis, R. Schäfer-Pregl, F. Salamini, and C. Gebhardt. 2000. QTL analysis of new sources of resistance to *Erwinia carotovora* ssp. *atroseptica* in potato done by AFLP, RFLP, and resistance-gene-like markers. Crop Sci 40: 1156–1167.

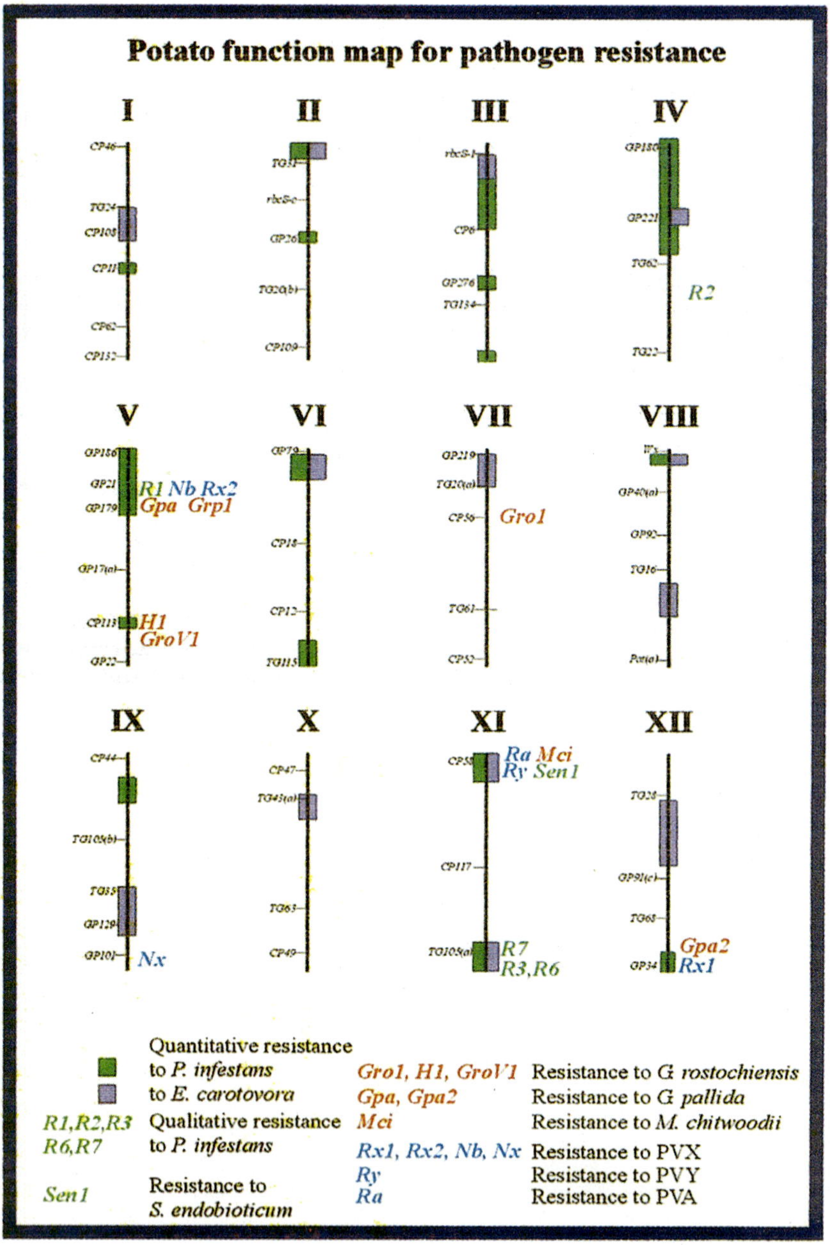

Fig. 6.1 Potato functional genomic map of disease resistance genes (http://www/mips.biochem.mpg.de/proj/gabi/projects gebhardt.htm).

APPENDIX 6.1

Elucidation of Potato Genome

Highly significant advances have recently been made in genetic and genomic research on potatoes by several major research groups and information downloaded on websites:

USA: (http;//www.sgn.cornell.edu/home page.html; (http://potatogenome.org/nsf3/?); (http://www.tigr.org/tdb/potato/). Canada: (http://www.bioatlantech.nb.ca/research-_genome.html),

EU: (http://www.dpw.wau.nl/uhd/links.html), Germany: (http://www/gabi.de/projekte/bereich2/pjgebhardt.html)

The Netherlands: (http://www.stw.nl/projecten/W/wpb5283.html ; http://www.biosystemsgenomics.nl/),

Finland and Japan: (http://www.jstage.jst.go.jp/article/jsbbs/53/2/155/pdf); (http://www.jstage.jst.go.jp/article/jsbbs/53/2/149/ pdf)

Furthermore, efforts are underway among these individual initiatives to integrate the knowledge and information into an international consortium on potato genomics that may take place as sequencing of a specific BAC library.

7

Ploidy Manipulation—Examination of Gene Action and Method of Gene Mapping

GEORGE C.C. TAI AND XINGYAO XIONG
[1]Potato Research Centre, Agriculture & Agri-Food Canada
P.O. Box 20280, Fredericton, N.B. Canada E3B 4Z7
Tel: (506) 459-3260, FAX: (506) 459-3316, e-mail: taig@em.agr.ca

INTRODUCTION

The cultivated potato, *Solanum tuberosum* L. (2n = 4x = 48), is an autotetraploid species. Asexual propagation is the normal mode of reproduction. Almost all genotypes are genetically heterogeneous regardless of cultivars or genetic stocks. The complexity of tetrasomic inheritance and the lack of pure lines add to the difficulty of genetic analysis of even the simplest inherited characteristics. Great progress has been achieved in manipulation of the ploidy level and its genetic consequences in the four decades since the discovery of potato haploids (Hougas and Peloquin 1957; Hougas et al. 1958). This has enabled the transfer of useful genes from related diploid and polyploid species to the cultivated tetraploid genepool. Ploidy manipulation also facilitates the investigation of tetrasomic inheritance and leads to novel strategies of gene mapping. Special mating designs accompanied by ploidy manipulation of chosen parents are used in genetic experiments and breeding for new cultivars. A number of reviews are available on ploidy manipulation and its usefulness in potato genetic and breeding research (e.g. Mendiburu and Peloquin 1976; Hermsen 1984; Iwanaga 1985; Peloquin et al. 1989; Ortiz 1998). This chapter focuses on the consequence and application of ploidy manipulation in association with autotetraploid species. Potential methods are examined for gene action and mapping at the tetrasomic level.

Corresponding author: George C.C. Tai.

MANIPULATION OF PLOIDY LEVELS

Reduction of Ploidy Level from Tetraploids

Maternal Haploids

Haploids can be induced from tetraploid *S. tuberosum* via either male or female organs (Ortiz 1998). These haploids represent diploid *tuberosum* (2n = 2x = 24). Special genotypes from *S. phureja* Juz. et Buk. (2n = 2x = 24) are used as the "pollinator" to hybridize with cultivars (Hougas et al. 1964; Peloquin et al. 1996). Maternal haploids are obtained through parthenogenesis by the union of two chromosome sets of *S. phureja* with the polar nuclei of tetraploid *S. tuberosum* but lack fertilization of the egg. Use of a genetic marker producing an embryo spot (absence of spot on seed in haploids; see Caligari et al. 1988) or electrophoresis (absence of isozyme marker; see Liu and Douches 1993) enables easy detection of haploids after 4x-2x crosses. Both the *Tuberosum* genotype and the pollen source affect the frequency of haploid production (Hougas et al. 1964). Haploids have been generated from a number of *Solanum* species with ploidy and Endosperm Balance Number (EBN) ranging from 2x (2EBN) to 8x (4EBN). The pollinator effect, EBN, and maternal influence are factors affecting the occurrence of pseudogamous parthenogenetic haploid production from these species (Singsit and Hanneman 1991).The parthenogenetic method was also used to produce monoploids (2n = x = 12) from diploid *S. tuberosum* and diploid *Solanum* species (Uijtewaal et al. 1987).

Paternal Haploids

Microspores are haploid cells formed by meiosis. Microspores normally undergo an unequal mitotic division that leads to the formation of a generative and vegetative nucleus. The generative nucleus then divides once more to form pollen grains. Androgenesis is the process whereby an embryo is developed from a microspore. Anther culture induces androgenesis by altering the development process of microspores such that the first mitosis is symmetrical in mode. This leads to the formation of two identical nuclei in the cells that develop into embryos or calli. Plantlets are generated either directly from the embryos or indirectly by inducing roots and shoots from the callus. Diploid *S. tuberosums* have been obtained from paternal haploids, termed dihaploids (Tiainen 1993; Rokka et al. 1996; Uhrig and Salamini 1987). A great deal of research in androgenesis has been directed toward producing monoploids from diploid *S. tuberosum*, wild species, and interspecific hybrids (Cappadocia et al. 1984; Uhrig 1985; Meyer et al. 1993; Shen and Veilleux 1995; Rokka et al. 1995; Aziz et al. 1999). Anther culture in potatoes is genotype dependent. Regenerants from anther culture often vary in ploidy level. Diploid donors yield between 20 to 60% monoploids, the rest comprising diploids and some tetraploids. This may be caused by chromosome doubling of reduced microspores during culture, embryogenesis of a somatic cell, or androgenesis of unreduced microspores.

Wenzel (1994) has discussed various aspects of tissue culture for the production of haploids. Jacobsen and Ramanna (1994) have reviewed production of monoploids.

Increase of Ploidy Level of Haploids

Sexual Polyploidization

Haploids extracted from tetraploid *S. tuberosum* are crossed with other diploid species (2n = 2x = 24) to produce diploid hybrids. These hybrids are found to produce 2n reduced gametes. Tetraploid progenies are produced from either unilateral sexual polyploidization (4x-2x, or 2x-4x) or bilateral sexual polyploidization (2x-2x) mating. The two major modes of 2n gamete formation are the first division restitution (FDR) and second division restitution (SDR) mechanisms (Veilleux 1985; Peloquin et al. 1989). The 2n pollen and 2n egg are caused by different abnormal cytological events during meiosis (see Tai, 1994). Formation of "parallel/fused spindles" during anaphase II of meiosis prevents cell division and consequently two 2n microspores are formed (Mok and Peloquin 1975; Ramanna 1979; Veilleux et al. 1982). This induces FDR gametes. A "premature cytokinesis" following the first division prevents the occurrence of second division during meiosis. Consequently a dyad of two 2n microspores is formed (Mok and Peloquin 1975). This induces SDR gametes. Synaptic mutants are found which cause poor pairing and/or reduced chiasma frequencies in microsporogenesis (Jongedijik and Ramanna 1988; Peloquin et al. 1989). When combined with "parallel spindles" FDR 2n pollen without crossing over, referred to as FDR-NCO gametes, are formed (Okwuagwu and Peloquin 1981; Hermundstad and Peloquin 1987). Parallel spindles, premature cytokinesis, and synaptic mutants have all been identified in microsporogenesis and thus are all mechanisms for the formation of 2n pollen. Omission of the second meiotic division is the predominant mechanism of 2n egg formation (Jongedijik 1985; Stelly and Peloquin 1986)

Asexual Polyploidization—Chromosome Doubling

Colchicine is used for doubling the chromosome set of young (25-30 cm) plants. The procedure used by Ross et al. (1967) is briefly described here. Leaf axils without auxiliary buds or pseudostipules are removed. On average, about six leaf axils per plant are kept. Auxiliary buds in the axils are removed, a pellet of absorbent cotton firmly placed in each axil, and a concentration of 0.25-0.5% colchicine solution dripped on the cotton. The treated plant is then enclosed in a polyethylene bag and placed in a greenhouse for a predetermined length of treatment. The shoot that grows from a treated sub-auxiliary meristem is cut when it reaches 6-8 cm in length, treated with a rooting hormone, placed in moist vermiculite to root, then transplanted into a pot to tuberize. The chromosomes are counted using root tips to confirm the result of doubling. Chromosome doubling of haploids is effective in increasing the homozygosity of the tetraploid parent (De Maine and Jervis, 1989). *S. commersonii* is used as a bridge species to surmount crossability barriers after its 1EBN number is increased to 2EBN by chromosome doubling (Bamberg et al. 1994). Tetraploid progenies produced by diploid hybrid parents and their chromosome doubled counterpart parents have been compared in 2x-2x, 4x-2x and 4x-4x crosses (De Maine, 1994; Tai and De Jong 1997).

Asexual Polyploidization—Somatic Fusion

Somatic fusion uses asexual means to combine two nuclear genomes. Somatic hybridization is especially useful for obtaining hybrids by fusing diploid *S. tuberosum* with sexually incompatible wild species. Protoplast fusion can be induced chemically or via electrofusion (Wenzel 1994). The latter appears to be the preferred method. Leaflets are digested by enzymes to isolate protoplasts. Parental protoplasts are mixed and subjected to electrofusion in a chamber in accordance with a particular fusion protocol (e.g. Thieme et al. 1997) and then cultured in regeneration media (see Wenzel 1994). Shoots are grown from protoplast-derived calli. Verification of successful somatic hybrids is carried out by cytological analysis and species-specific molecular markers (e.g. Novy and Helgeson 1994; Thieme et al. 1997; Barone et al. 2002). A set of chromosome-specific cytogenetic DNA markers was developed for chromosome identification in potato by Dong et al. (2000). Literature on production and inheritance in tetraploid somatic hybrids is extensive.

GENETIC CONSEQUENCES OF 2n GAMETES

Dihaploids from *S. tuberosum* ($2n = 2x = 24$) are crossed with diploid species to obtain diploid hybrids. These diploids are then used as parents in 2x-2x crosses or with tetraploid *S. tuberosum* parents in 4x-2x crosses. Tetraploid progenies are obtained through sexual polyploidization of the diploid parents that form unreduced 2n gametes. Extensive work is available on the genetic consequences of sexual polyploidization through the 2n gametes on both the theoretical and experimental levels (e.g. Mendiburu et al. 1974; Hermsen 1984; Ortiz and Peloquin 1994; Tai 1994; David et al. 1995; Peloquin et al. 1999; and others).

Theoretical Models on Estimation of Genetic Parameters in Tetrasomic Inheritance

The frequencies of gametes from a tetraploid genotype $A_1A_2A_3A_4$ are shown in Table 7.1. The gamete ratios from five types of tetraploid genotypes—monoallelic, balanced diallelic, unbalanced diallelic, triallelic, and tetra-allelic—are extended from the original ratios from a duplex parent $A_1A_1A_2A_2$ given by Fisher and Mather (1943). The parameter α is the double reduction coefficient which measures the probability of sister chromatids going to the same gamete. The range of α is $0 \leq \alpha \leq 1/6$. Chromosomal segregation occurs when $\alpha = 0$ due to complete failure of quadrivalent formation or close linkage of the concerned locus to the centromere. Chromatic segregation, on the other hand, occurs when quadrivalent forms and the concerned locus are located at $\alpha = 1/7$. The proportions of gamete genotypes formed by double- or nondouble reduction events are shown in Table 7.1.

Table 7.1 *Frequencies of 2n gametes of various genotypes in a locus produced by a tetraploid parent with five types of genotypes*

Allelic pattern	*Genotype of 4x parent*	*Genotype*	*2x progenies frequency*		
			DR	*Non-DR*	*Total*
Monoallelic	$A_1A_1A_1A_1$	A_1A_1	—	—	1.0
Unbalanced diallelic	$A_1A_1A_1A_2$	A_1A_1	$3\alpha/4$	$(1-\alpha)/2$	$(2+\alpha)/4$
		A_2A_2	$\alpha/4$	0	$\alpha/4$
		A_1A_2	0	$(1-\alpha)/2$	$(1-\alpha/2)$
Balanced diallelic	$A_1A_1A_2A_2$	A_1A_1	$\alpha/2$	$(1-\alpha)/6$	$(1+2\alpha)/6$
		A_2A_2	$\alpha/2$	$(1-\alpha/6)$	$(1+2\alpha)/6$
		A_1A_2	0	$(2-2\alpha)/3$	$(2-2\alpha)/3$
Triallelic	$A_1A_1A_2A_3$	A_1A_1	$\alpha/2$	$(1-\alpha)/6$	$(1+2\alpha)/6$
		A_2A_2	$\alpha/4$	0	$\alpha/4$
		A_3A_3	$\alpha/4$	0	$\alpha/4$
		A_1A_2	0	$(1-\alpha)/3$	$(1-\alpha)/3$
		A_1A_3	0	$(1-\alpha)/3$	$(1-\alpha)/3$
		A_2A_3	0	$(1-\alpha)/6$	$(1-\alpha)/6$
Tetra-allelic	$A_1A_2A_3A_4$	A_1A_1	$\alpha/4$	0	$\alpha/4$
		A_2A_2	$\alpha/4$	0	$\alpha/4$
		A_3A_3	$\alpha/4$	0	$\alpha/4$
		A_4A_4	$\alpha/4$	0	$\alpha/4$
		A_1A_2	0	$(1-\alpha)/6$	$(1-\alpha)/6$
		A_1A_3	0	$(1-\alpha)/6$	$(1-\alpha)/6$
		A_1A_4	0	$(1-\alpha)/6$	$(1-\alpha)/6$
		A_2A_3	0	$(1-\alpha)/6$	$(1-\alpha)/6$
		A_2A_4	0	$(1-\alpha)/6$	$(1-\alpha)/6$
		A_3A_4	0	$(1-\alpha)/6$	$(1-\alpha)/6$

*DR: produced by double reduction event; Non-DR: produced by nondouble reduction event.

The frequencies of FDR and SDR gametes from a diploid parent are given in Table 7.2. Parameter β is the frequency of single-exchange tetrads resulting from a single crossover between centromere and the concerned locus during meiosis of the diploid parent. The range of β is $0 \leq \beta \leq 1$.

Table 7.2 *Frequencies of FDR and SDR gametes from diploid parents*

Parental genotype	*2n gamete*		
	A_1A_1	A_1A_2	A_2A_2
A_1A_1 [FDR]	1	0	0
A_1A_2 [FDR]	$\beta/4$	$(2 - \beta)/2$	$\beta/4$
A_2A_2 [FDR]	0	0	1
A_1A_1 [SDR]	1	0	0
A_1A_2 [SDR]	$(1 - \beta)/2$	β	$(1 - \beta)/2$
A_1A_2 [SDR]	0	0	1

Estimation of Genetic Parameters by First-Degree Statistics

Tai (1982a,b, 1986, 1994) presented four series of mating schemes involving 2x-2x and/or 4x-2x crosses between two parents and their backcrosses for the estimation of genetic parameters in tetrasomic inheritance. Means of hybrid families in the mating scheme are used for the estimation. The parents are restricted to be either monoallelic ($A_1A_1A_1A_1$ or $A_2A_2A_2A_2$ for tetraploid, and A_1A_1 or A_2A_2 for diploid) or balanced diallelic ($A_1A_1A_2A_2$ or A_1A_2) genotypes. Genetic parameters are estimated based on the expected means of families in the mating scheme according to the additive-dominance model in biometrical genetics (Killick 1971; Mather and Jinks 1982). Monoallelic (d), balanced (h_1) and unbalanced diallelic (h_2 and h_3) genic effects are estimated by a weighted least-squares procedure. Triallelic (e.g. $A_1A_1A_2A_3$) and tetra-allelic ($A_1A_2A_3A_4$) effects are not estimable. The estimation procedure assumes no linkage, no epistasis, and no genotype-environment interactions.

Estimation of Genetic Parameters by Second-Degree Statistics

Second-degree statistics according to the additive-dominance model contain complex genetic parameters involving α/β and cross products, and are thus of limited use in studying tetrasomic inheritance (Killick 1971). A mathematical model on quantitative genetic analysis of tetraploids (see Bradshaw 1994) has been used to investigate 4x-2x hybrids together with the concept of identity of descent (Boudec et al. 1989). The genetic structure of the covariances between 2x parents and 4x offspring in 4x-2x crosses is derived based on the coefficients of coancestry and double coancestry (Haynes 1990). The former approach requires specified α/β values and the latter the assumption of no double reduction and no epistasis. David et al. (1995) established a general model for the genetic value of a 4x-2x cross. Expressions of additive, digenic, trigenic, and tetragenic variances are derived according to their origination from 2x and 4x parents and interaction between them. Covariances between relatives are derived according to the genes from 2x, 4x, or both. A series of 4x-2x families are created using a sample of genotypes from each of a group of full-sib diploid families to separately cross with a 4x parent. These families are treated as "test units" and grouped according to their 2x parents. The variances between and within families provide estimates of additive and dominance variances. Again, the model requires the assumption of no double reduction, linkage, or epistasis.

Experimental Results

Neither of the models described above are presently employed in genetic experiments, as both require complex mating designs. Most experimental works on 2x-2x and 4x-2x crosses use combining-ability analysis based on data obtained from factorial experiments. The contributions of general and specific combining abilities to the genetic variation of traits are found in varying proportions (see Bradshaw and Mackay 1994). General combining ability (GCA) appears to be prevalent over quantitative traits whereas specific combining ability (SCA) between 4x and 2x parents tends to be important for specific traits, such as tuber yield. It has been noted that GCA in tetrasomic inheritance is not entirely due to the additive effects of genes since the 2n gametes of 2x parents transmit, on average, 80% (FDR) and 40% (SDR) of the parental heterozygosity to the 4x offspring (Hermsen 1984). Loci with major effects located between the centromere and proximal crossovers are expected to transmit close to 100% heterozygosity and epistasis to 4x progeny (Peloquin et al. 1999). This leads to the possibility of detecting quantitative trait loci (QTL) proximal to centromeres showing effects on quantitative traits. The proximal hypothesis was proven true for a number of traits (tuber appearance, eye depth, specific gravity, tuber size and yield) as the 4x progenies from 4x-2x FDR-CO (CO, crossover) outperformed their vegetatively doubled counterparts in field experiments (Tai and De Jong 1997), and progenies derived from FDR-CO (transmit 80% heterozygosity) and FDR-NCO (NCO, non-crossover, transmit 100% heterozygosity) showed no significant difference in total tuber yield in field experiments (Buso et al. 1999a,b).

EXAMINATION OF GENE ACTION ON PROGENIES DERIVED FROM ASEXUAL POLYPLOIDIZATION

Asexual means to manipulate ploidy levels, i.e., somatic fusion and chromosomal doubling, can be employed to generate 4x progenies with identifiable genomic structures. The following are the steps used to asexually generate these tetraploids:

1. Targeted parents (4x and 2x) are subjected to ploidy reduction from 4x→2x→1x by parthenogenesis or androgenesis. For example, four genetically distinct monoploids generated from four diploid species are chosen for experiments. The genomes of the four 1x parents are labelled A_I, A_{II}, A_{III}, and A_{IV}.
2. The four 1x parents are subjected to chromosome doubling/somatic fusion to create four diploids with homozygous genomic structures

A_IA_I, $A_{II}A_{II}$, $A_{III}A_{III}$, or $A_{IV}A_{IV}$. A second round of chromosome doubling/somatic fusion of the four diploids creates the first desirable class of four "Monosomic Tetraploids" ($A_IA_IA_IA_I$, $A_{II}A_{II}A_{II}A_{II}$, $A_{III}A_{III}A_{III}A_{III}$, and $A_{IV}A_{IV}A_{IV}A_{IV}$).

3. Somatic fusion between pairs of the monoploids in (1) creates six diploids with heterozygous genomic structures A_IA_{II}, A_IA_{III}, A_IA_{IV}, $A_{II}A_{III}$, $A_{II}A_{IV}$, and $A_{III}A_{IV}$. Chromosome doublilng/somatic fusion of them produces the second desirable class of six "Balanced Diasomic Tetraploids" ($A_IA_IA_{II}A_{II}$, $A_IA_IA_{III}A_{III}$, $A_IA_IA_{IV}A_{IV}$, $A_{II}A_{II}A_{III}A_{III}$, $A_{II}A_{II}A_{IV}A_{IV}$, $A_{III}A_{III}A_{IV}A_{IV}$).
4. Somatic fusion between diploids from (2) and (3) creates the third desirable class of 12 "Unbalanced Diasomic Tetraploids" ($A_IA_IA_IA_{II}$, $A_IA_IA_IA_{III}$, $A_IA_IA_IA_{IV}$, $A_IA_{II}A_{II}A_{II}$, $A_{II}A_{II}A_{II}A_{III}$, $A_{II}A_{II}A_{II}A_{IV}$, $A_IA_{III}A_{III}A_{III}$, $A_{II}A_{III}A_{III}A_{III}$, $A_{III}A_{III}A_{III}A_{IV}$, $A_IA_{IV}A_{IV}A_{IV}$, $A_{II}A_{IV}A_{IV}A_{IV}$, and $A_{III}A_{IV}A_{IV}A_{IV}$.), the fourth desirable class of 12 "Trisomic Tetraploids" ($A_IA_IA_{II}A_{III}$, $A_IA_IA_{II}A_{IV}$, $A_IA_IA_{III}A_{IV}$, $A_IA_{II}A_{II}A_{III}$, $A_IA_{II}A_{II}A_{IV}$, $A_{II}A_{II}A_{III}A_{IV}$, $A_IA_{II}A_{III}A_{III}$, $A_IA_{III}A_{III}A_{IV}$, $A_{II}A_{III}A_{III}A_{IV}$, $A_IA_{II}A_{IV}A_{IV}$, $A_IA_{III}A_{IV}A_{IV}$, $A_{II}A_{III}A_{IV}A_{IV}$.), and the fifth desirable class of one "Tetrasomic Tetraploids" ($A_IA_{II}A_{III}A_{IV}$). A total of 35 4x progenies are obtained. A flow chart is presented in Fig. 7.1 to illustrate the steps of creating the 4x progeny population

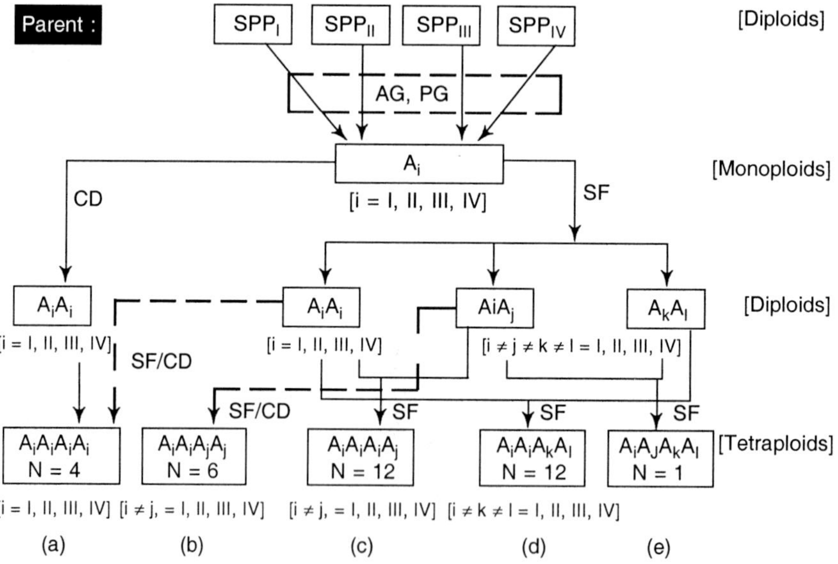

PG = Pathenogenesis, AG = Androgenesis, CD = Chromosome doubling, SF = Somatic fusion (a) Monoallelic, (b) Balanced diallelic, (c) Unbalanced diallelic, (d) Triallelic, (e) Tetra-allelic

Fig. 7.1 Asexual production of tetraploid progenies of all possible genotypes from four diploid species parents.

through asexual polyploidization. Genetic differences would be observed between and within the five classes of progeny populations having multiple alleles of gene(s) with major effects on a quantitative trait distributed over the four genomes of the original monoploids. To estimate the "additive" and "nonadditive" effects of chromosomes originated from the four 1x parents, the following biometric model is considered with respect to the genetic values of the 35 4x progenies with five classes of "reconstituted" genomic structures:

Class	*Genomic structure*	*Genetic value*
Monosomic	$A_IA_IA_IA_I$	m + 4d
Monosomic	$A_{II}A_{II}A_{II}A_{II}$	m + 2d
Monosomic	$A_{III}A_{III}A_{III}A_{III}$	m − 2d
Monosomic	$A_{IV}A_{IV}A_{IV}A_{IV}$	m − 4d
Balanced disomic	$A_IA_IA_{II}A_{II}$	m + 3d + η_{1122}
Balanced disomic	$A_IA_IA_{III}A_{III}$	m + d + η_{1133}
Balanced disomic	$A_IA_IA_{IV}A_{IV}$	m + η_{1144}
Balanced disomic	$A_{II}A_{II}A_{III}A_{III}$	m + η_{2233}
Balanced disomic	$A_{II}A_{II}A_{IV}A_{IV}$	m − d + η_{2244}
Balanced disomic	$A_{III}A_{III}A_{IV}A_{IV}$	m − 3d + η_{3344}
Unbalanced disomic	$A_IA_IA_IA_{II}$	m + 7d/2 + η_{1112}
Unbalanced disomic	$A_IA_IA_IA_{III}$	m + 5d/2 + η_{1113}
Unbalanced disomic	$A_IA_IA_IA_{IV}$	m + 2d + η_{1114}
Unbalanced disomic	$A_IA_{II}A_{II}A_{II}$	m + 5d/2 + η_{1222}
Unbalanced disomic	$A_{II}A_{II}A_{II}A_{III}$	m + d + η_{2223}
Unbalanced disomic	$A_{II}A_{II}A_{II}A_{IV}$	m + d/2 + η_{2224}
Unbalanced disomic	$A_IA_{III}A_{III}A_{III}$	m − d/2 + η_{1333}
Unbalanced disomic	$A_{II}A_{III}A_{III}A_{III}$	m − d + η_{2333}
Unbalanced disomic	$A_{III}A_{III}A_{III}A_{IV}$	m − 5d/2 + η_{3334}
Unbalanced disomic	$A_IA_{IV}A_{IV}A_{IV}$	m − 2d + η_{1444}
Unbalanced disomic	$A_{II}A_{IV}A_{IV}A_{IV}$	m − 5d/2 + η_{2444}
Unbalanced disomic	$A_{III}A_{IV}A_{IV}A_{IV}$	m − 7d/2 + η_{3444}
Trisomic	$A_IA_IA_{II}A_{III}$	m + 2d + η_{1123}
Trisomic	$A_IA_IA_{II}A_{IV}$	m + 3d/2 + η_{1124}
Trisomic	$A_IA_IA_{III}A_{IV}$	m + d/2 + η_{1134}
Trisomic	$A_IA_{II}A_{II}A_{III}$	m + 3d/2 + η_{1223}
Trisomic	$A_IA_{II}A_{II}A_{IV}$	m + d + η_{1224}
Trisomic	$A_{II}A_{II}A_{III}A_{IV}$	m − d/2 + η_{2234}
Trisomic	$A_IA_{II}A_{III}A_{III}$	m + d/2 + η_{1233}
Trisomic	$A_IA_{III}A_{III}A_{IV}$	m − d + η_{1334}
Trisomic	$A_{II}A_{III}A_{III}A_{IV}$	m − 3d/2 + η_{2334}
Trisomic	$A_IA_{II}A_{IV}A_{IV}$	m − d/2 + η_{1244}
Trisomic	$A_IA_{III}A_{IV}A_{IV}$	m − 3d/2 + η_{1344}
Trisomic	$A_{II}A_{III}A_{IV}A_{IV}$	m − 2d + η_{2344}
Tetrasomic	$A_IA_{II}A_{III}A_{IV}$	m + η_{1234}

Here d represents additive effect and η interaction effects between the sources of genomes. The genetic parameters listed above are estimated from observed data on replicated experiments of the 4x progenies generated. For a tetra-allelic locus, the inbreeding coefficient follows the order: monoallelic > unbalanced diallelic > balanced diallelic > triallelic > tetra-allelic (= 0). This is reversely related to the degree and complexity of genic interactions of the genotypes. Each of the 4x progenies generated above carries a unique genetic architecture with a gradual change from homogeneity to heterogeneity on allelic comparisions of genes depending on the source of four chromosome sets from the 1x parents. Comparison of the progenies makes it possible to detect QTL with major effects on quantitative traits and the pattern of genic actions between chromosomes derived from the four genomes of the 1x parents. This could lead to the possibility of "chromosome engineering" for superior progenies by combining chromosome sets from selected 1x parents.

GENE MAPPING THROUGH PLOIDY MANIPULATION

Genetic maps of potatoes based on molecular markers have been constructed based on progenies from crosses of diploid parents (Bonierbale et al. 1988; Gebhardt et al. 1989, 1991; Tanksley et al. 1992). Gene mapping in the polyploid species is much more complex than in the diploid species. Wu et al. (1992) presented a method for the detection and estimation of linkage in polyploids based on the single-dose restriction segments. Meyer et al. (1998), Milbourne et al. (1998), and Hackett et al. (1998) conducted linkage analysis of 4x *S. tuberosum* L. A theoretical method to construct a linkage map in an autotetraploid species from molecular markers was presented by Luo et al. (2000a,b). The method is aimed at constructing linkage maps of codominant or dominant genetic markers in autotetraploid species. An important assumption in their approach is that the pattern of segregation is chromosomal, i.e., the four homologous chromosomes are paired into two bivalents, not a quadrivalent, during meiosis to produce 2n gametes. This avoids complexity of theoretical development due to the phenomenon of double reduction. Manipulation of ploidy levels provides simpler models for gene mapping for the autotetraploid species. Two methods are presented here.

Gene Mapping Based on Monoploids Extracted from a Diploid Hybrid Parent

Use of monoploids or double haploids generated from a diploid parent could represent the simplest way to obtain linkage maps because all segregating genes follow the 1:1 ratio. The approach becomes complicated because a large percentage of molecular markers of many plant species have been reported to yield distorted segregation (see Cloutier and Landry 1994; Chani et al. 2002). The surviving monoploids, however, show unique changes of segregation pattern between pairs of loci when compared with the undistorted situation. Let a marker of the configuration *A-a* be linked with a distortion-causing locus $I_a i_a$. Consider another marker locus *B-b* that is linked with another independent distortion-causing locus $I_b i_b$. *A* and *B* denote presence of unique gel-bands whereas *a* and *b* denote absence of the *A-* or *B-* bands. There are four possible marker configurations: (I_aA/i_aa, I_bB/i_bb), (I_aA/i_aa, I_bb/i_bB), (I_aa/i_aA, I_bB/i_bb), and (I_aa/i_aA, I_bb/i_bB). Gamete cells from these genotypes carrying the alleles I_aI_b are expected to be able to regenerate into shoots or embryos. Gamete cells with the alleles i_a and/or i_b fail to develop in culture. Four genotypes of monoploid are derived from the gamete cells: I_aAI_bB, I_aAI_bb, I_aaI_bB, I_aaI_bb. Let r_{ia} represent the recommendation rate between the first pair of loci and r_{ib} the second pair. Figure 7.2 illustrates the formation of 1x gametes from a diploid parent during meiosis.

The segregation pattern is changed when two loci and linked. Consider two linked loci *A-a* and *B-b* with the *A-a* locus located adjacent to a distortion-causing locus *I-i*. To detect the linkage relationship between them, four marker configurations of the mother plant have to be considered: *IAB/iab, IAb/iaB, IaB/iAb, Iab/iAB*. The recommendation frequencies between *I-i* and *A-a* and between *A-a* and *B-b* are r_{ia} and r_{ab} respectively. Gamete cells with the genotype *i__* are not capable of responding to tissue culture and are therefore unable to undergo morphogenesis. Again, four genotypes of monoploid are derived from the gamete cells: *IAIB, IAIb, IaIB, IaIb*. Figure 7.2 also illustrates the formation of 1x gametes from a diploid parent with the genotype *IAB/iab* or *IAb/iaB*.

The degree and pattern of distortion of the segregration ratio depends on the mother genotype and the intensity of linkage of the markers with the loci. Consider the two corresponding mother genotypes I_aA/i_aa I_bB/i_bb and *IAB/iab*. The segregation ratio of monoploid genotypes from both mother genotypes is *AB:Ab:aB:ab* = 1:1:1:1 when $r_{ia} = r_{ib} = 0.5$ and $r_{ia} = r_{ab} = 0.5$ respectively. The proportions of *AB* and *Ab* are increased whereas those of *aB* and *ab* are decreased when $r_{ia} < 0.5$. Neither *A-a* nor *B-b* show the expected ratio of *A:a* = *B:b* = 1:1 when r_{ia} and $r_{ib} < 0.5$. From the mother genotype *IAB/iab, ab* is expected to have a much larger fraction than that of *aB* when $r_{ia} < 0.5$ and $r_{ab} < 0.5$. The opposite is true for monoploids obtained from the mother genotype I_aA/i_aa I_bB/i_bb, i.e., *ab* has a much smaller fraction than that of *aB*. The foregoing hypothesis can be applied to detection of the distortion pattern of segregation ratios of monoploids obtained from the other three pairs of corresponding mother genotypes in Fig. 7.2. A protocol is presented here to detect and map a series of markers that are located beside a distortion-causing locus (Tai et al. 2000):

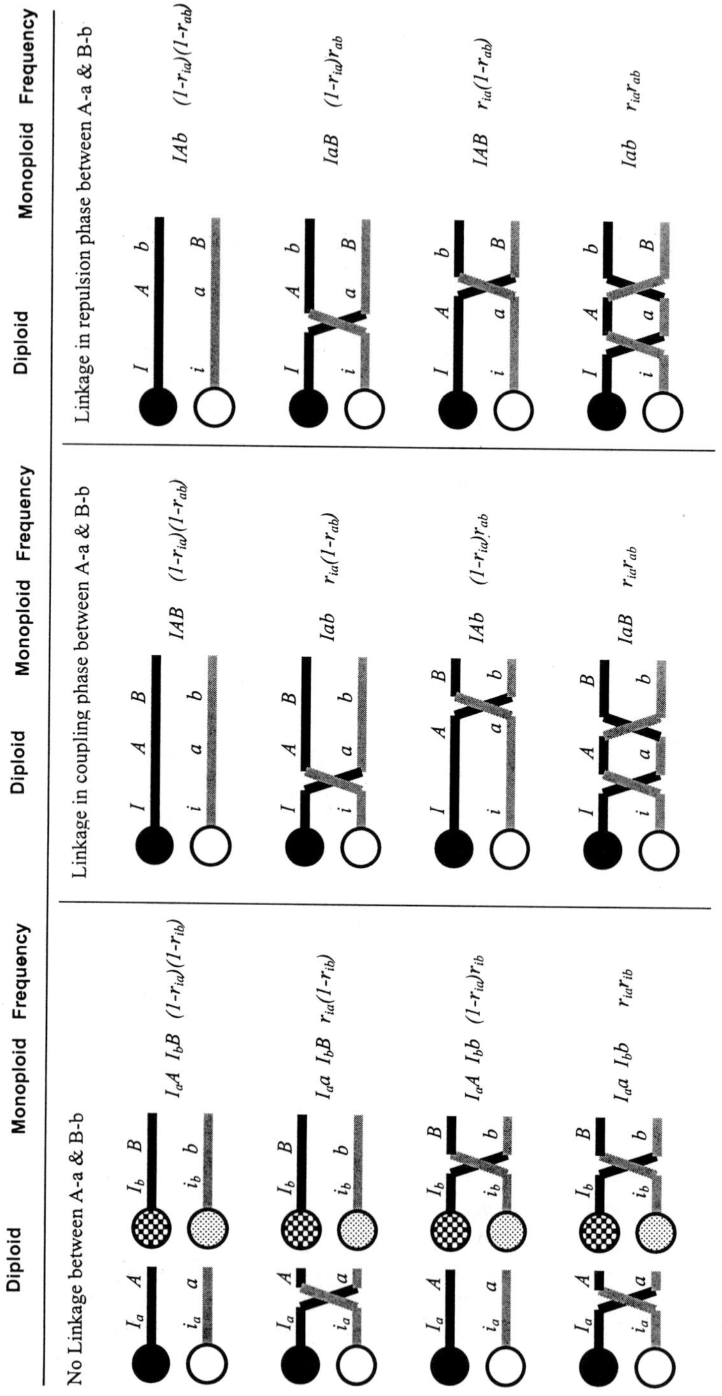

Fig. 7.2 Formation of 1x gametes for two loci *A-a* and *B-b* with each separately linked with a different lethal locus, or both linked with one lethal locus.

1. Segregation data for all markers from the monoploids obtained from a mother genotype are tested for distortion from the expected ratio 1:1. This is achieved by carrying out a statistical test such as the χ^2 goodness-of-fit test for every individual segregating marker.
2. The marker with the highest degree of distortion is identified. A 2 × 2 contingency table is formed between this marker and every other distorted marker. Linkage between two markers is detected when the statistical test for goodness-of-fit is significant because the fraction of the double recombinant monoploid is lower than that of the single recombinant one (e.g. *ab* > *aB* from the mother genotype *IAB*/*iab*). A group of markers linked with a putative distortion-causing locus is identified.
3. A genotype of the group of markers from each of the scored monoploids is used to simulate a hypothetical "mirror image" monoploid which possesses the other alleles of the markers. This seeks to restore those gamete cells that are not able to undergo morphogenesis to form monoploids because of the distortion-causing gene. It restores the segregation ratio for each of the markers in the group to the expected 1:1 ratio in the combined sample of observed and hypothetical monoploids. The genotype of the putative distortion-causing locus is simulated as *I* for all scored monoploids and *i* for all simulated monoploids.
4. Segregation data obtained from (3) are used to map the group of markers and the putative distortion-causing locus. This can be accomplished by means of established software for gene mapping such as MAPMAKER with the F_2 backcross option (Lander et al. 1987). Mapping should be started with the marker that shows the highest degree of distortion, as this marker should be located closest to a distortion-causing locus.
5. The procedure is repeated for another highly distorted marker not included in the above group.

It can be seen that the χ^2-test for the 2 × 2 contingency table of two markers may also show significance because these markers are tightly linked with the distortion-causing locus located between them. This makes, for example, the frequency of *ab* from the mother genotype *AIB*/*aib* lower than that of *aB*, a situation opposite to that from the mother genotype *IAB*/*iab*. But the frequency of *ab* is lower than the expected $r_{ia}r_{ab}$ because of the low frequency of double crossover between the markers, which is different from the expected outcome from the mother genotype *IaA*/*iaa IBb*/*ibb*. Tai et al. (2000) used RAPD marker data from monoploids generated from a diploid hybrid clone 9507-04 (*S. tuberosum* × *S. chacoanse*) for mapping. Twenty-five of 56 markers showed distorted segregation ratios. Seventeen of the distorted markers were mapped into four linkage groups.

Mapping of Gene Markers on a Chromosome Based on 4x-2x Mating

Consider two gene marker loci, $A\text{-}a_i$ and $B\text{-}b_i$. Both A and B markers produce unique marker gel-bands on the electrophoresis map; a_i and b_i represent other marker alleles which yield neither the A nor B marker band. Some mating designs are considered for the purpose of mapping. The mapping model represents an extension of the gene-centromere mapping method developed by Mendiburu and Peloquin (1979). Marker data are generated from progenies of a single 4x-2x cross or a series of 4x-2x crosses.

Marker Data from a Single 4x-2x Cross

Consider at first the use of a single 4x-2x cross. Let a 4x parent of the genotype $a_1a_2a_3a_4\ b_1b_2b_3b_4$ cross with a diploid parent *AaBb* which produces unreduced 2n gametes by the first division restitution (FDR) mechanism. The four alleles of both loci in the tetraploid parent produce neither the A nor B marker band. For simplicity only the tetraploid parent with the genotype *aaaabbbb* will be considered. Three possible segregation models between the two pairs of markers are considered.

Model 1: A-a and B-b are Located on Separate Chromosomes

Let β_a be the rate of single exchange tetrads (SET) between the centromere and the locus *A-a*, and β_b the SET rate between the centromere and the locus *B-b*. The gametes *AA*, *Aa*, and *aa* have the ratio $\beta_a/4$: $(1 - \beta_a/2)$: $\beta_a/4$, $0 \leq \beta_a \leq 1$, whereas those of *BB*, *Bb*, and *bb* have the ratio $\beta_b/4$: $(1 - \beta_b/2)$: $\beta_b/4$, $0 \leq \beta_b \leq 1$. The following segregation ratio is obtained based on the banding patterns from the progenies of the 4x-2x cross:

AAaaBBbb AAaaBbbb AaaaBBbb AaaaBbbb	$= 1 - (\beta_a + \beta_b)/4 + \beta_a\beta_b/16$
AAaabbbb Aaabbbb	$= \beta_a/4 - \beta_a\beta_b/16$
aaaaBBbb aaaaBbbb	$= \beta_a/4 - \beta_a\beta_b/16$
aaaabbbb	$= \beta_a\beta_b/16$

A___B___ genotypes produce both the *A*- and *B*- marker bands, *A___bbbb* the single band for *A* marker, *____B___* the single band for *B* marker, and *aaaabbbb* is absent with both *A*- and *B*- marker bands.

Model 2: A-a and B-b are linked in coupling phase

Assume that *A-a* locus is located closer to the centromere than the *B-b* locus. The situation wherein the two loci are fairly closely linked to each other but located anywhere in the chromosome is considered. This implies that high order crossover events within the segment between *A-a* and *B-b* loci are not likely to occur. No exchange tetrads (NET) are formed when there is no crossover event between the centromere and *B-b* locus. The rate of SET between *A-a* and the centromere is β_a, and that between the *A-a* and *B-b* loci is β_{ab}. Double exchange tetrads (DET) occur when simultaneous single crossover events take place between the centromere and *A-a* locus and between *A-a* and *B-b* loci. The following segregation ratio is obtained from the progenies of the 4x-2x cross:

1. NET—Non crossover:*AaaaBbbb* $= 1 - (\beta_a + \beta_{ab}) + \beta_a\beta_{ab}$
2. SET—Crossover between centromere and *A-a*: 1 (*AAaaBBbb*):2 (*AaaaBbbb*):1(*aaaabbbb*) $= \beta_a + \beta_a\beta_{ab}$
3. SET—Crossover between *A-a* and *B-b*: 1(*AaaaBBbb*):2(*AaaaBbbb*):1 (*Aaaabbbb*) $= \beta_{ab} + \beta_a\beta_{ab}$
4. DET—Crossover between centromere and *A-a* and between *A-a* and *B-b*:2(*AAaaBbbb*):1(*AaaaBBbb*):2(*AaaaBbbb*):1(*Aaaabbbb*):2(*aaaaBbbb*) $= \beta_a\beta_{ab}$

Figure 7.3 illustrates the formation of the 2n gametes from the 2x parent with a single crossover event during meiosis and the frequencies of 4x progenies following the current 4x-2x mating mode. Grouping by the banding patterns, we have

$$A\_\_\_B\_\_\_ = 1 -(\beta_a + \beta_{ab})/4 + \beta_a\beta_{ab}/8$$
$$A\_\_\_bbbb = \beta_{ab}/4 - \beta_a\beta_{ab}/8$$
$$aaaab\_\_\_ = \beta_a\beta_{ab}/4$$
$$aaaabbbb = \beta_a/4 - \beta_a\beta_{ab}/4$$

It can be seen that the term $\beta_a\beta_{ab} = 0$ when there is a complete interference for the crossover event in a chromosome during meiosis. Also, $\beta_a\beta_{ab}$ is negligible when two marker loci are closely linked. The genotype *aaaaB___* is absent from the progenies when only one crossover event per chromosome occurs during meiosis.

Model 3: A-a and B-b are linked in repulsion phase

1. NET—Non crossover: *AaaaBbbb* $= 1- (\beta_a + \beta_{ab}) + \beta_a\beta_{ab}$
2. SET—Crossover between centromere and *A-a*:1(*Aaaabbbb*):2 (*AaaaBa-aa*):1(*aaaaBBbb*) $= \beta_a + \beta_a\beta_{ab}$
3. SET—Crossover between *A-a* and *B-b*:1(*Aaaabbbb*):2(*AaaaBbbb*):1 (*AaaaBBbb*) $= \beta_{ab} + \beta_a\beta_{ab}$
4. DET—Crossover between centromere and *A-a* and between *A-a* and *B-b*:2(*AAaaBbbb*):1(*AaaaBBbb*):2(*AaaaBbbb*):1(*Aaaabbbb*):2(*aaaaBbbb*) $= \beta_a\beta_{ab}$

Grouping by the banding patterns, we have

$$A\_\_\_B\_\_\_ = 1 - \beta_a/2 - \beta_{ab}/4 + 3\beta_a\beta_{ab}/8$$
$$A\_\_\_bbbb = \beta_a/4 + \beta_{ab}/4 - 3\beta_a\beta_{ab}/8$$
$$aaaaB\_\_\_ = \beta_a/4$$

The genotype *aaaabbbb* is absent in the progenies.

In all the three models, β_a and β_a/β_{ab} are estimated by the segregation data of *A-a* and *B-b* marker loci. The estimates are then used to calculate the expected frequencies of the segregation classes in the model. A fitness test (χ^2 test) is applied to each of the three models according to the observed and expected frequencies of the segregation classes. The two marker loci can then be mapped based on estimates of β_a and β_{ab} according

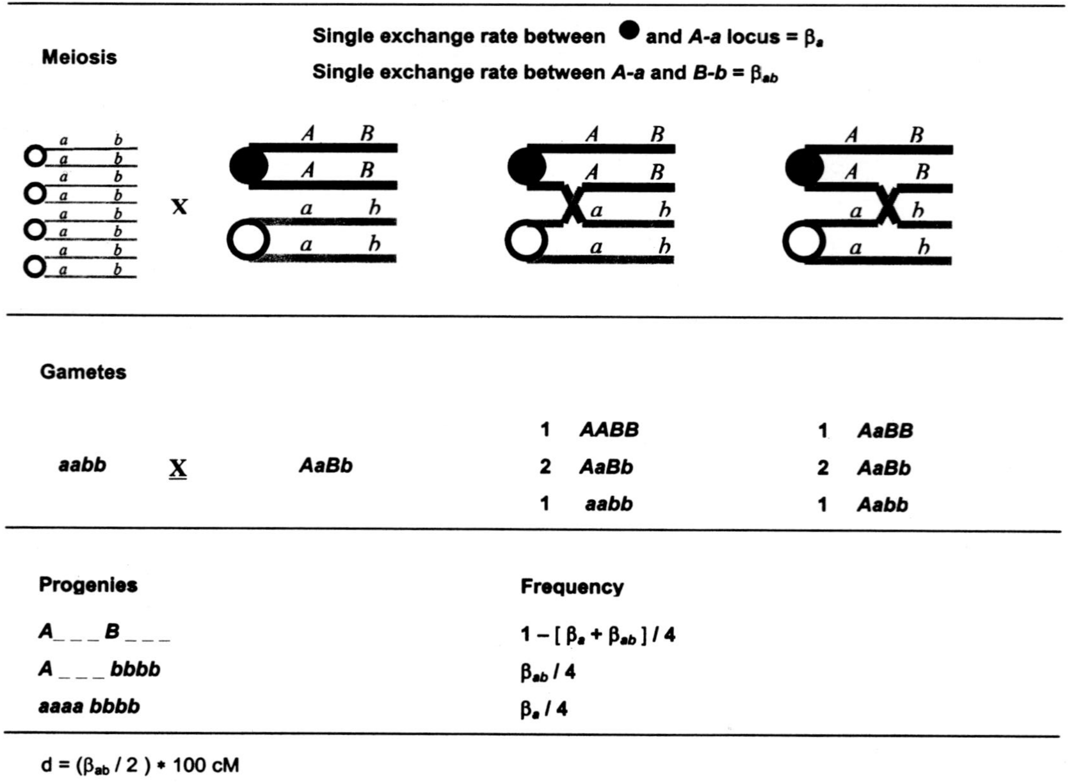

Fig. 7.3 Formation of 2n FDR gametes from the 2x parent with two loci *A-a* and *B-b* linked in coupling phase and frequencies of 4x progenies from 4x-2x mating.

to the fitted model. The map distance between *A-a* and *B-b* is calculated by $d = (\beta_{ab}/2) \times 100$ cM when they are linked.

Marker Data from a Series of 4x-2x Crosses

Let a series of 4x parents cross with a common 2x parent. It is not required to know the genotypes of the 4x parents but they all should produce gametes of *aabb* marker genotypes. The 2x parent, again, has the genotype *AaBb* and produces 2n FDR gametes. Consider the simplest mating design in which two 4x parents T1 and T2 are used to cross with a common 2x parent D. Assume that total of N_1, N_2, and N_3 progenies from the crosses T1 × T2, T1 × D, and T2 × D are used for mapping. The progenies are at first sorted according to the banding phenotypes of the *A-a* locus, then the *B-b* locus, and finally *A-a* and *B-b* loci together. There are n_{a1}, n_{a2}, and n_{a3} progenies of the "nulliplex" *aaaa*____ (i.e., marker genotype without the *A* band), and n_{b1}, n_{b2}, and n_{b3} progenies of the "nulliplex" ____*bbbb* (i.e., marker genotype without the *B* band), and n_{ab1}, n_{ab2} and n_{ab3} progenies of the "nulliplex" *aaaabbbb* (i.e., marker genotype without both bands *A* and *B*) from the three crosses T1 × T2, T1 × D, and T2 × D respectively. In the observed frequencies of *aaaa*____ = $n_{ai}/N_i = w_{ai}$, ____*bbbb* = $n_{bi}/N_i = w_{bi}$, and *aaaabbbb* = $n_{abi}/N_i = w_{abi}$, $i = 1$, 2, and 3 for the three crosses. Their expectations are $E(w_{a1}) = t_{a1} \times t_{a2}$, $E(w_{a2}) = t_{a1} \times f_a$, and $E(w_{a3}) = t_{a2} \times f_a$; $E(w_{b1}) = t_{b1} \times t_{b2}$, $E(w_{b2}) = t_{b1} \times f_b$, and $E(w_{b3}) = t_{b2} \times f_b$; and $E(w_{ab1}) = t_{ab1} \times t_{ab2}$, $E(w_{ab2}) = t_{ab1} \times f_{ab}$, and $E(w_{ab3}) = t_{ab2} \times f_{ab}$. The symbol t represents frequencies of *aa*__, __*bb*, or *aabb* gametes in the 4x parents (T1 or T2), and f for the frequencies of *aa*__, __*bb*, or *aabb* gametes of the 2x parents. Taking the logarithm on the observed frequencies for both loci,

$$\mathbf{w_s} = \begin{bmatrix} \log w_{s1} \\ \log w_{s2} \\ \log w_{s3} \end{bmatrix} \qquad E(\mathbf{w_s}) = \begin{bmatrix} \log t_{s1} + \log t_{s2} \\ \log t_{s1} + \log f_s \\ \log t_{s2} + \log f_s \end{bmatrix}$$

A "design" matrix (**x**) and a "gamete" (**g**) matrix are formed according to the mating design:

$$\mathbf{x} = \begin{bmatrix} 1 & 1 & 0 \\ 1 & 0 & 1 \\ 0 & 1 & 1 \end{bmatrix} \qquad \mathbf{g_s} = \begin{bmatrix} \log t_{s1} \\ \log t_{s2} \\ \log f_s \end{bmatrix}$$

It can be seen that $E(\mathbf{w_s}) = \mathbf{x}\,\mathbf{g_s}$. Let

$$\mathbf{V_s} = \begin{bmatrix} (1-w_{s1})/N_1 w_{s1} & 0 & 0 \\ 0 & (1-w_{s1})/N_2 w_{s2} & 0 \\ 0 & 0 & (1-w_{s1})/N_1 w_{s3} \end{bmatrix}$$

in which the components are the estimates of variances of those in $\mathbf{g_s}$. In the above matrices, s = *a* for the *aa*__ gamete, s = *b* for the __*bb* gamete, and s = *ab* for the *aabb* gamete, $\mathbf{g_s}$; s = *a*,*b*, or *ab* is estimated by $(\mathbf{x}'\mathbf{V_s}^{-1}\mathbf{x})^{-1}$ $\mathbf{x}'\mathbf{V}^{-1}\ \mathbf{w_s}$. The estimates of f_a, f_b, f_{ab} are used to estimate $\beta_a/4 = f_a$, $\beta_b/4 = f_b$, and $\beta_{ab}/4 = f_{ab}$. These estimates are used to determine the map distance of the two marker loci as described in the preceding section. This procedure is expected to work well when the two concerned marker loci are linked in coupling phase. The 2x parent produces no gametes with the genotype *aabb* when the two loci are closely linked in repulsion phase. Thus β_{ab} cannot be estimated. Mapping between *A-a* and *B-b* is only possible when $\beta_b < 1$. In this case, β_a and β_b can be used to determine the distance between the two loci.

Conclusion

There are two major difficulties in genetic analysis of potatoes: nonavailability of pure lines and tetrasomic inheritance. Ploidy manipulation provides a means to bypass these problems. Reduction of ploidy levels by parthenogenesis and androgenesis followed by restoration of ploidy level through chromosome doubling and somatic fusion makes it possible to create tetraploid progenies from complete homozygosity to all levels of heterozygosity. This facilitates the study of gene action in association with tetrasomic inheritance. Monoploids provide the simplest model for gene mapping. Induction of haploids from tetraploid parents, combined with first division restitution (FDR)/second division restitution(SDR) mechanisms of 2n gamete formation, presents special methods for gene mapping. Research works on ploidy manipulation over the past four decades indeed represent unique achievement in potato breeding.

REFERENCES

Aziz, A.N., J.E.A. Seabrook, G.C.C. Tai, and H. De Jong. 1999. Screening diploid *Solanum* genotypes responsive to different anther culture conditions and ploidy assessment of anther-derived roots and plantlets. Amer J Potato Res 76:9–16.

Bamberg. J.B., R.E. Hanneman, Jr., J.P. Palta, and J.F. Harbage. 1994. Using disomic 4x (2EBN) potato species' germplasm via bridge species *Solanum commersonii*. Genome 37:866–870.

Barone, A., J. Li, A. Sebastian, and T. Cardi. 2002. Evidence for tetrasomic inheritance in a tetraploid *Solanum commersonii (+) S. tuberosum* somatic hybrid through the use of molecular markers. Theor Appl Genet 104: 539–546.

Bonierbale, M.W., R.L. Plastid, and S.D. Tansley. 1988. RFLP maps based on a common set of clones reveal modes of chromosomal evolution in potato and tomato. Genetics 120: 1095–1103.

Boudec, P., M. Masson, and Y. Dattee. 1989. A quantitative genetics model for the estimation of variances, covariances between relatives in crosses using 2n gametes in potato. *In:* K.M. Louwes, H.A.J.M. Toussaint, and L.M.W. Dellaert (eds.), Parental Line Breeding and Selection in Potato Breeding. Centre for Agricultural Publishing and Documentation, PUDOC, Wageningen, Netherlands, pp. 43–48.

Bradshaw, J.E. 1994. Quantitative genetics theory for tetrasomic inheritance. *In:* J.E. Bradshaw and G.R. Mackay (eds.). Potato Genetics. CAB Intl, Wallingford, UK, pp. 71–99.

Bradshaw, J.E. and G.R. Mackay. 1994. Breeding strategies for clonally propagated potatoes. *In:* J.E. Bradshaw and G.R. Mackay (eds.), Potato Genetics. CAB Intl., Wallingford, UK, pp. 467–497.

Buso, J.A., L.S. Boiteux, G.C.C. Tai, and S.H. Peloquin. 1999a. Chromosome regions between centromeres and proximal crossovers are the physical size of major effect loci for yield in potato: genetic analysis employing meiotic mutants. Proc Natl Acad Sci USA 96: 1773–1778.

Buso J.A., F.J.B. Reifschneider, L.S. Boiteux, and S.J. Peloquin. 1999b. Effects of 2n-pollen formation by first meiotic division restitution with and without crossover on eight quantitative traits in 4x-2x potato progenies. Theor Appl Genet 98: 1311–1319.

Caligari, P.D.S., W. Powell, K. Liddell, M.J. De Maine, and E.E.L. Swan. 1988. Methods and strategies for detecting *Solanum tuberosum* dihaploids in interspecific crosses with *S. phureja*. Ann Appl Biol 115: 323–328.

Cappadocia, M., D.S.K. Cheng, and R. Ludlum-Simonette. 1984. Plant regeneration from in vitro culture of anthers of *Solanum chacoense* Bitte. and interspecific diploid hybrids *S. tuberosum* L. x *S. chacoense* Bitte. Theor Appl Genet 69: 139–143.

Chani, E., V. Ashkenazi, J. Hillel, and R.E. Veilleux. 2002. Microsatellite marker analysis of an anther-derived potato family: skewed segregation and gene-centromere mapping. Genome 45: 236–242.

Cloutier, S. and B.S. Landry. 1994. Molecular markers applied to plant tissue culture. In Vitro Cell Dev Biol 30P: 32–39.

David, J.L., P. Boudecand, and A. Gallais. 1995. Quantitative genetics of 4x-2x hybrid population with first-division restitution and second-division restitution 2n gametes produced by diploid parents. Genetics 139: 1797–1803.

De Maine, M.J. 1994. Comparison of tetraploid progenies of potato dihaploids, their chromosome-doubling derivatives and second generation dihaploids. Potato Res 37: 173–181.

De Maine, M.J. and L. Jervis. 1989. The use of dihaploids in increasing homozygosity of tetraploid potatoes. Euphytica 44: 37–42.

Dong, F., J. Song, S.K. Naess, J.P. Helgeson, C. Gebhardt, and J. Jiang. 2000. Development and applications of a set of chromosome-specific cytogenetic DNA markers in potato. Theor Appl Genet 101: 1001–1007.

Fisher R.A. and K. Mather. 1943. The inheritance of style length in *Lythrum salicaria*. Ann Eugenics 12: 1–23.

Gebhardt, C., E. Ritter, T. Debener, U. Schachtschabel, B. Walkemeier, H. Uhrig and F. Salamini. 1989. Restriction fragment length polymorphism analysis and linkage map in *Solanum tuberosum*. Theor Appl Genet 78: 65–75.

Gebhardt, C., E. Ritter, A. Barone, T. Debener, B. Walkemeier, U. Schachtschabel, H. Kaufmann, R.D. Thompson, M.W. Bonierbale, M.W. Ganal, S.D. Tanksley, and F. Salamini. 1991.

RFLP maps of potato and their alignment with the homoeologous tomato genome. Theor Appl Genet 83: 49–57.

Hackett, C.A., J.F. Bradshaw, R.C. Meyer, J.W. McNicol, D. Milbourne, and R. Waugh. 1998. Linkage analysis in tetraploid potatoes: a simulation study. Genet Res 71: 143–154.

Haynes, K.G. 1990. Covariances between diploid parent and tetraploid offspring in tetraploid × diploid crosses of *Solanum tuberosum* L.J. Hered 81: 208–210.

Hermsen, J.G.Th. 1984. The potential of meiotic polyploidization in breeding allogamous crops. Iowa State J. Res 58(4): 435–448.

Hermundstad, S.A. and S.J. Peloquin. 1987. Breeding at the 2x level and sexual polyploidization. *In:* G.J. Jellis, and E.E. Richardson (eds.), The Production of New Potato Varieties: Technological Advances. Cambridge Univ Press, Cambridge, UK, pp. 197–210.

Hougas, R.W. and S.J. Peloquin. 1957. A haploid plant of the potato variety Katahdin. Nature 180: 1209–1210.

Hougas, R.W., S.J. Peloquin, and R.W. Ross. 1958. Haploids of the common potato. J Hered 49: 103–107.

Hougas, R.W., S.J. Peloquin, and A.C. Gabert. 1964. Effect of seed parent and pollinator on frequency of haploids in *Solanum tuberosum*. Crop Sci 4: 593–595.

Iwanaga, M. 1985. Haploids, ploidy manipulation, and meiotic mutants in potato breeding. *In:* Biotechnology in International Agricultural Research. IRRI, Manila, Philippines, pp. 139–148.

Jacobsen E. and M.S. Ramanna. 1994. Production of monoploids of *Solanum tuberosum* L. and their use in genetics, molecular biology and breeding. *In:* J.E. Bradshaw and G.R. Mackay (eds.), Potato Genetics. CAB Intl, Wallingford, UK, pp. 155–170.

Jongedijik, E. 1985. The pattern of megasporogenesis and megagametogenesis in diploid *Solanum* species hybrids: its relevance to the origin of 2n eggs and the induction of apomixi. Euphytica 34: 599–611.

Jongedijik, E. and M.S. Ramanna. 1988. Synaptic mutants in potato, *Solanum tuberosum* L. 1: Expression and identity of genes for desynapsis. Genome 30: 664–670.

Killick, R.J. 1971. The biometrical genetics of autotetraploids. I: Generations derived from a cross between two pure lines. Heredity 27: 331–346.

Lander, E.S., P. Green, J. Abrahamson, A. Barlow, M.J. Daly, S.E. Lincoln, and L. Newburg. 1987. MAPMAKER: an interactive computer package for constructing primary genetic linkage maps of experimental and nature populations. Genomics 1: 174–181.

Liu, C.A. and D.S. Douches. 1993. Production of haploids of potato (*Solanum tuberosum* subsp. *tuberosum*) and their identification with electrophoresis analysis. Euphytica 70: 113–126.

Luo, Z.W., C.A. Hackett, J.E. Bradshaw, J.W. McNicol, and D. Milbourne. 2000a. Predicting parental genotypes and gene segregation for tetrasomic inheritance. Theor Appl Genet 100: 1067–1073.

Luo, Z.W., C.A. Hackett, J.E. Bradshaw, J.W. McNicol, and D. Milbourne. 2000b. Construction of a genetic linkage map in tetraploid species using molecular markers. Genetics 157: 1369–1385.

Mather, K. and J.L. Jinks. 1982. Biometrical Genetics (3rd ed.). Chapman Hall, London.

Mendiburu, A.O. and S.J. Peloquin. 1976. Sexual polyploidization and depolyploidization: some terminology and definitions. Theor Appl Genet 48: 137–143.

Mendiburu, A.O. and S.J. Peloquin. 1979. Gene-centromere mapping by 4x × 2x matings in potatoes. Theor Appl Genet 54: 177–180.

Mendiburu, A.O., S.J. Peloquin, and D.W.S. Mok. 1974. Potato breeding with haploids and 2n gametes. *In:* K.J. Kasha (ed.), Haploids in Higher Plants. Guelph Univ Press, Guelph, Ontario (Canada), pp. 249–258.

Meyer, R.C., F. Salamini, and H. Uhrig. 1993. Isolation and characterization of potato diploid clones generating a high frequency of monohaploid or homozygous diploid androgenetic plants. Theor Appl Genet 85: 905–912.

Meyer, R.C., D. Milbourne, C.A. Hackett, J.E. Bradshaw, and J.W. McNicol. 1998. Linkage analysis in tetraploid potato and associations of markers with quantitative resistance to late blight (*Phytophthora infestans*). Mol Gen Genet 259: 233–245.

Milbourne, D., R.C. Meyer, J.E. Bradshaw, E. Baird, N. Bonar, J. Provan, W. Powell, and R. Waugh. 1997. Comparison of PCR-based marker systems for the analysis of genetic relationships in cultivated potato. Mol Breed 3: 127–136.

Mok, D.W.S. and S.J. Peloquin. 1975. Three mechanisms of 2n pollen formation in diploid potatoes. Can J Genet Cytol 17: 217–225.

Novy, R.G. and J.P. Helgeson. 1994. Somatic hybrids between *Solanum tuberosum* and diploid, tuber-bearing *Solanum* clones. Theor Appl Genet 89: 775–782.

Okwuagwu, C.O. and S.J. Peloquin. 1981. A method of transferring the intact parental genotype to the offspring via meiotic mutants. Amer Potato J 58: 512–513.

Ortiz, R. 1998. Potato breeding via ploidy manipulations. Plant Breed Rev 16: 15–86.

Ortiz, R. and S.J. Peloquin. 1994. Use of 24-chromosome potatoes (diploids and dihaploids) for genetical analysis. *In:* J.E. Bradshaw and G.R. Mackay (eds.), Potato Genetics. CAB Intl., Wallingford, UK pp. 133–154.

Peloquin, S.J., G.L. Yerk, and J.E. Werner. 1989. Ploidy manipulation in the potato. *In:* K.W. Adolph (ed.), Chromosomes: Eukaryotic, Prokaryotic, and Viral, Vol. II. CRS Press, Boca Raton, FL (USA), pp. 167–178.

Peloquin, S.J., A.C. Gabert, and R. Ortiz. 1996. Nature of "pollinator" effect in potato haploid production. Ann Bot 77: 539–542.

Peloquin, S.J., L.S. Boiteux, and D. Carputo. 1999. Meiotic mutants in potato: valuable variants. Genetics 153: 1493–1499.

Ramanna, M.S. 1979. A re-examination of the mechanisms of 2n gamete formation in potato and its implications for breeding. Euphytica 28: 537–561.

Rokka, V.M., J.P.T. Valkonen, and E. Pehu. 1995. Production and characterization of haploids derived from somatic hybrids between *Solanum brevidens* and *S. tuberosum* through anther culture. Plant Sci 112: 85–95.

Rokka, V.M., L. Pietila, and E. Pehu. 1996. Enhanced production of dihaploid lines via anther culture of tetraploid potato (*Solanum tuberosum* spp. *tuberosum*) clones. Amer Potato J 73: 1–12.

Ross, R.W., L.A. Dionne, and R.W. Hougas. 1967. Doubling the chromosome number of selected *Solanum* genotypes. Eur Potato J 10: 37–52.

Shen, L.Y. and R.E. Veilleux. 1995. Effect of temperature shock and elevated incubation temperature on androgenic embryo yield of diploid potato. Plant Cell, Tissue, Organ Culture 43: 29–35.

Singsit, C. and R.E. Hanneman, Jr. 1991. Haploid induction in Mexican polyploid species and colchicine-doubled derivatives. Amer Potato J 68: 551–556.

Stelly, D.M. and S.J. Peloquin. 1986. Formation of 2n megagametophytes in diploid tuber-bearing *Solanums*. Amer J Bot 73: 1351–1363.

Tai, G.C.C. 1982a. Estimation of double reduction and genetic parameters in autotetraploids. Heredity 49: 63–70.

Tai G.C.C. 1982b. Estimation of double reduction and genetic parameters in autotetraploids based on 4x-2x and 4x-4x matings. Heredity 49: 331–335.

Tai, G.C.C. 1986. Biometrical genetical analysis of tetrasomic inheritance based on matings of diploid parents which produce 2n gametes. Heredity 57: 315–317.

Tai, G.C.C. 1994. Use of 2n gametes. *In:* J.E. Bradshaw and G.R. Mackay (eds.), Potato Genetics. CAB Intl, Wallingford, UK, pp. 109–132.

Tai, G.C.C. and H. De Jong. 1997. A comparison of performance of tetraploid progenies produced by diploid and their vegetatively doubled (tetraploid) counterpart parents. Theor Appl Genet 94: 303–308.

Tai, G.C.C., J.E.A. Seabrook, and A.N. Aziz. 2000. Linkage analysis of anther-derived monoploids showing distorted segregation of molecular markers. Theor Appl Genet 101: 126–130.

Tanksley, S.D., M.W. Ganal, J.P. Prince, M.C. de Vicente, M.W. Bonierbale, P. Broun, T.M. Fulton, J.J. Giovanonni, S. Grandillo, G.B. Martin, R. Messenguer, J.C. Miller, A.H. Paterson, O. Pineda, M.S. Roder, R.A. Wing, W. Wu, and N.D. Young. 1992. High density molecular linkage maps of the tomato and potato genomes. Genetics 132: 1141–1160.

Thieme, R., U. Darsow, T. Gavrilenko, D. Dorokhov, and H. Tiemann. 1997. Production of somatic hybrids between *S. tuberosum* L. and late blight resistant Mexican wild potato species. Euphytica 97: 189–200.

Tiainen, T. 1993. The influence of hormones on anther culture response of tetraploid potato (*Solanum tuberosum* L.). Plant Sci 88: 83–93.

Uhrig, H. 1985. Genetic selection and liquid medium conditions improve the yield of androgenetic plants from diploid potatoes. Theor Appl Genet 71: 455–460.

Uhrig, H. and F. Salamini. 1987. Dihaploid plant production from 4x-genotype of potato by the use of efficient anther plant producing tetraploid strains (4x EAPP clones)—proposal for a breeding methodology. Plant Breed 98: 228–235.

Uijtewaal, B.A., D.J. Huigen, and J.G. Th. Hermsen. 1987. Production of potato monohaploids (2n = x = 12) through prickle pollination. Theor Appl Genet 73: 751–758.

Veilleux, R.E. 1985. Diploid and polyploid gametes in crop plants: mechanisms of formation and utilization in plant breeding. Plant Breed. Rev. 3: 253–288.

Veilleux, R.E., N.A. McHale, and F.L. Lauer. 1982. 2n gametes in diploid *Solanum*: frequency and types of spindle abnormalities. Can J Genet Cytol. 24: 301–314.

Wenzel, G. 1994. Tissue culture. *In:* J.E. Bradshaw and G.R. Mackay (eds.), Potato Genetics. CAB Intl, Wallingford, UK, pp. 173–195.

Wu, K.K., W. Brunquist, M.E. Sorrells, T.L. Tew, P.H. Moore, and T.D. Tanksley. 1992. The detection and estimation of linkage in polyploids using single-dose restriction fragment. Theor Appl Genet 83: 294–300.

8

Approaches to Gene Isolation in Potato

WALTER DE JONG
Cornell University, Dept. Plant Breeding,
309 Bradfield Hall, Ithaca, NY 14853-1901, USA.
Tel: 1-607-254-5384, Fax: 1-607-255-6683, e-mail:wsd2@cornell.edu

INTRODUCTION

Variation in gene sequence or gene structure underpins the phenotypic variation essential for developing new crop varieties. In a traditional potato (*Solanum tuberosum* L.) breeding program, the creation of a new variety begins when crosses are made between parents with desirable and complementary traits. The progeny are then phenotypically evaluated for 10 to 15 years to determine whether any of them possess sufficient quality and pest resistance characteristics to warrant being named as new varieties and released. Throughout this process the genetic variation responsible for trait variation is, for the most part, not visible and this leads to a wide range of practical difficulties. Some traits, for example, are highly sensitive to environmental variation and hence can only be meaningfully assessed after many years of field evaluation. If the genetic variation responsible for such a trait could be evaluated directly, e.g. with a PCR assay, then judgments about clone performance could be made much earlier. Another practical difficulty arises because many traits are controlled by multiple genes of small or modest effect and so often it is not clear whether two parents share identical or complementary alleles conditioning a genetically complex trait. Were it possible to track all desirable alleles in parental germplasm, crosses could be made in a more informed manner. A third issue concerns gene dosage: the clone of an autotetraploid potato may have zero, one, two, three, or four copies of a given allele. A higher dosage is preferable should a breeder desire most or all progeny to receive the particular allele. While it is possible to evaluate dosage for dominant genes by making test crosses and counting the number of progeny expressing the trait, it would be quicker and simpler to assess dosage directly by amplifying and characterizing all the alleles present in a given clone, e.g. by pyrosequencing (Rickert et al. 2002).

Isolating genes and characterizing allelic variation provide tremendous flexibility and potential for manipulating phenotypic variation. Naturally, cloned genes also provide opportunities for improving potato by transgenic approaches. The past decade has seen many improvements in both the tools and approaches to isolating economically important genes in both potato and related solanaceous plants. Recent advancements in gene isolation methods as they have been or could reasonably be applied to potato are reviewed here.

POTATO MOLECULAR GENETICS: BACKGROUND AND AVAILABLE RESOURCES

Prior to initiating any gene isolation project, it is necessary to have some idea of the magnitude of the task, e.g. the size of the potato genome, the number of genes it encodes and how they are distributed. Precise knowledge of these aspects will not be available until the potato genome is completely sequenced, however, which is not likely to be accomplished anytime soon. Nevertheless, what is known or can be reasonably estimated is briefly outlined below.

The potato genome is distributed across twelve chromosomes. Using flow cytometric analysis Arumuganathan and Earle (1991) estimated the haploid genome content to be 800 to 930 million base pairs. A reasonable estimate for the number of potato genes would be between 25,000 and 55,000, which represent the number of genes in the two plant species that have been completely sequenced to date, *Arabidopsis* and rice (The Arabidopsis Genome Initiative, 2000; Yu et al. 2002; Goff et al. 2002). Consistent with this is a recent analysis of cDNA sequences from tomato—closely related to potato—concluding that tomato likely codes for 35,000 genes (Van der Hoeven et al. 2002). In the past few years almost 80,000 randomly selected potato cDNA clones have been partially or completely sequenced, largely through the efforts of a National Science Foundation-funded collaborative project in the USA. These sequenced cDNA clones represent expressed genes and are commonly referred to as expressed sequence tags or ESTs. In any large collection of ESTs there is inevitably sequence redundancy, particularly among highly expressed genes. In release 6.0 of their potato gene index, The Institute for Genomic Research (www.tigr.org) reported that the 79,657 potato ESTs publicly available on May 3, 2002 represented a maximum of 21,941 unique genes.

Sequence data on which to base an estimate of potato gene spacing is scant. Only one large contiguous region of potato has been sequenced to date, a 187 kb region surrounding the *Gpa2* resistance gene (sequence accession AF265664), but it has not yet been annotated for predicted gene number and location. In *Arabidopsis*, genes are spaced on average 4.5 kb apart (The Arabidopsis Genome Initiative 2000). Six large contiguous regions, ranging in size from 70 to 128 kb, have been sequenced in tomato (Van der Hoeven et al. 2002). Gene density ranged from one gene every five kb to one gene every 17 kb. Gene density is likely to be similar in potato since potato and tomato have similar sized genomes. The lowest gene densities in tomato were observed in regions within or adjacent to heterochromatin.

Both diploid and tetraploid potatoes are highly heterozygous. The frequency of single nucleotide polymorphism between any two homologous chromosomes has not yet been systematically evaluated. Based on the author's limited experience (unpublished), at several dispersed loci the frequency appears to be of the order of one polymorphism every one hundred bases. This frequency is considerably higher than in *Arabidopsis*, where comparison of the ecotypes Landsberg *erecta* and Columbia revealed an average of one single nucleotide polymorphism every 3,300 bases (The Arabidopsis Genome Initiative 2000). A significant benefit of high levels of potato sequence polymorphism is the relative ease in mapping a given gene. Chen et al. (2001), for example, amplified 50 potato loci with gene specific primers. Four loci were polymorphic after amplification. Forty-three required digestion with just one of three restriction enzymes (*Taq*I, *Alu*I or *Rsa*I) to reveal polymorphism. Two others required digestion with other enzymes (*Hae*III or *Dde*I) to detect polymorphism. Only one gene could not be mapped by digesting a PCR product but was successfully mapped, following hybridization to digested genomic DNA, as a conventional RFLP.

Many reference genetic maps have been described for wild and cultivated potato species (including Bonierbale et al. 1988; Gebhardt et al. 1989; Gebhardt et al. 1991; Tanksley et al. 1992; Jacobs et al. 1995; Rivard et al. 1996; Perez et al. 1999). The total length of each map varies considerably, ranging from 206 cM (Rivard et al. 1996) to 1120 cM (Jacobs et al. 1995). All these maps are based on roughly one to two hundred RFLP markers. Unfortunately, only a few markers are shared between many of these maps, which hinders high-resolution cross-referencing. A map delineating 89 microsatellite loci defined by 65 primer pairs has also been published (Milbourne et al. 1998), as has a map describing the position of 264 AFLP marker alleles (Van Eck et al. 1995). A highly ambitious genetic map of potato, which will be denser than such maps of any other crop, is currently being developed by a European consortium. This ultrahigh density (UHD) map (www.dpw.wageningen-ur.nl/uhd) aims to localize well over 8,000 AFLP markers across the potato genome.

Libraries of large-insert DNA clones are indispensable for many gene isolation projects. Bacterial artificial chromosome (BAC) vectors (Shizuya et al. 1992) have made it possible to routinely propagate large fragments of potato genomic DNA, up to 300 kb in length. Many potato BAC libraries have been constructed in recent years, from both wild and cultivated potato species (e.g. Kanyuka et al. 1999; Song et al. 2000; Ballvora et al. 2002). A typical potato BAC library consists of approximately 20,000 to 100,000 independent clones, and each clone carries an average of 75 to 150 kb of genomic DNA. Poisson distribution predicts that when a genomic library is sufficiently large to represent the genome on average five times, there will be a 99% chance of any given gene being present in the library.

MAP-BASED CLONING: GENERAL CONSIDERATIONS

The identity of genes controlling expression of many economically important potato traits often cannot be predicted in advance. This is especially true for tuber-related traits, because there is almost never a clear precedent of a cloned gene controlling a similar trait in another tuber-bearing plant. In cases where gene identity is a complete mystery, map-based (or position-based) cloning is usually the method of choice for gene isolation. As detailed below, map-based cloning involves carefully analyzing the segregation of molecular marker alleles and the trait of interest in a large segregating population to progressively narrow down the region where the target gene must reside. The only prerequisites for map-based cloning are a phenotype that can be unambiguously scored in each segregant and a sufficiently large population, typically a thousand plants or more. The most attractive feature of map-based cloning is that, if performed with due care, it will always succeed; map-based cloning thus represents a technical "gold-standard". The least attractive feature is the amount of time required; with current tools about five person-years are needed to isolate a new gene in potato. With luck, the time may be lessened, but the time may also be considerably longer. The labor requirements contrast significantly with (completely sequenced) *Arabidopsis*, where only a few person-months are needed to isolate a gene using similar methods. Because map-based cloning requires so much effort, in most gene isolation projects other approaches such as the candidate gene approach (see below) are attempted, or at least considered first. Nevertheless, if shortcuts fail, and a gene is important enough, it is reassuring to know that map-based cloing always provides a means to achieve its isolation.

Two technical advances of the past decade are primarily responsible for making map-based cloning feasible in potato. The first was the development of BAC technology, described above. A second important technical advance was the development of AFLP methodology (Vos et al. 1995).

AFLP markers are useful for map-based cloning primarily because they allow very dense genetic maps to be rapidly constructed around genes of interest. To generate an AFLP marker profile genomic DNA is first digested into several million small fragments with a pair of restriction enzymes. Enzyme combinations commonly used in potato are *Eco*RI + *Mse*I or *Pst*I + *Mse*I. Next, a few additional nucleotides (adaptors) are ligated onto the fragment ends. PCR is then performed with primers that are complementary to the adaptors and also have a few extra bases at their 3' ends; these extra bases serve to ensure that only a subset of the fragments are amplified. Normally 50-150 fragments are amplified in any one reaction. Following amplification the fragments are visualized on high-resolution acrylamide gels. In diploid potato crosses as many as 73% of the amplified fragments have been observed to be polymorphic (Rouppe van der Voort et al. 1997a). Thus, when comparing two pools of DNA from a segregating population, wherein one pool is constructed from plants that carry the allele of interest, and the other pool from plants without that allele (bulked segregant analysis; Michelmore et al. 1991), it is possible in just one to two months time for a single investigator to assess thousands of polymorphic loci for linkage to a gene of interest. Another convenient feature of AFLP markers is that, in most cases, identical-size AFLP marker alleles in unrelated clones map to identical locations in the potato genome (Van Eck et al. 1995; Rouppe van der Voort et al. 1997a).

One of the limitations of AFLP technology is the requirement for extensive sample preparation before a marker profile can be generated. This makes these markers less attractive for screening thousands of samples, a scale routinely encountered in map-based cloning projects. AFLP markers (as well as RFLP markers) can nevertheless be converted into other types of markers more suitable for high-throughput screening. A common approach is to excise an AFLP marker allele from an acrylamide gel for cloning and sequencing (De Jong et al. 1997). New primers based on this sequence can be used to amplify genomic DNA directly without the need for extensive template preparation. It is rarely evident from the sequence of an AFLP marker, however, what sequence polymorphism differentiates it from other alleles. In practice, therefore, the new primers generally amplify more than one allele. This necessitates a further search for useful polymorphism, e.g. by comparing the sequence of all alleles amplified by the new primer pair. If sequence differences result in a restriction site polymorphism, alleles can be distinguished by digesting the PCR products with a restriction enzyme. This type of marker is commonly referred to as a CAPS marker (Konieczny and Ausubel, 1993). An alternative approach is to design additional primers to specifically amplify just one allele. One important practical distinction between CAPS markers and allele specific markers concerns data interpretation following PCR. DNA prepared quickly and cheaply for high-throughput screening is rarely pure and this occasionally causes PCR to fail. Lack of an amplification product following PCR with CAPS primers is interpreted to indicate that the PCR reaction has failed. However, lack of product following PCR with allele-specific primers yields an ambiguous result: it might indicate that the PCR has failed, or it might mean that the target allele is not present in the template DNA.

The ideal outcome from high-density mapping is to identify two markers which flank the gene of interest and are separated by no more than 75-150 kb, a typical insert size for a BAC library. This result is called "chromosome landing" (Tanksley et al. 1995) since if two flanking markers are present in a single large-insert clone, then the gene of interest must lie somewhere in that clone as well. If the two closest flanking markers are separated by a larger distance, then "chromosome walking" must be employed to obtain cloned DNA corresponding to the entire interval. To "walk" along a chromosome, end fragments from large-insert clones that contain individual flanking markers are used to identify overlapping large-insert clones from the BAC library, and the process repeated until the walks from flanking markers converge. Once a single BAC or collection of overlapping BACs is identified that spans the entire marker interval, attention shifts to sifting through the interval to identify the gene of interest.

"Sifting" normally involves a combination of sequencing and functional testing. Just a few years ago functional testing always preceded sequencing, but this order is commonly reversed now. "Functional testing" here means genetically transforming a plant with the allele of interest and observing whether the engineered plants display the predicted phenotype. Because methods for potato transformation are well known (e.g. Barrell et al. 2002) they are not detailed here. One advance in transformation technology with particular relevance for map-based cloning is noteworthy, nevertheless, namely the development of BIBAC vectors (Hamilton et al. 1996). These vectors make it possible to transform a plant with the entire insert of a BAC clone and thus determine whether a given BAC contains the gene of interest. This approach is especially useful after a lengthy chromosome walk since many BAC clones can be identified between the closest flanking markers. Once a BAC is known to contain a gene, subsets of the insert can be transformed individually to further narrow down gene location. Since DNA sequencing costs have fallen, this process is now within the reach of many research groups for sequencing the entire region between the closest flanking markers before functional testing. With sequence information it is often possible to identify one or a few genes in the interval likely to correspond to the gene of interest, thereby allowing priority to be given to their evaluation.

While molecular geneticists would prefer to perform map-based cloning at the diploid (2x) level since genetics is simpler here, the vast majority of segregating potato populations generated and maintained by breeders are tetraploid (4x). Populations are bred to segregate for economically important traits and so it is only natural to ask whether existing 4x progenies can be used to isolate genes by map-based means. From a purely technical standpoint the answer is undoubtedly "yes"; the potato virus X resistance gene *Rx1* was isolated in this manner. However, there are several practical constraints to be considered before undertaking such a project. First of all, much more effort is required to saturate a region with markers at the 4x level (Meyer et al. 1998). While performing bulked segregant analysis (BSA) with a tetraploid F_1 population, the only marker alleles detected are those linked in coupling to the allele for the dominant trait, i.e., on the same physical strand of DNA. This contrasts with BSA at the 2x level, where markers linked in repulsion can also be detected. Moreover, twice as many homologous chromosomes segregate in a 4x cross than in a 2x cross, which increases the chances that an arbitrary marker allele linked to the trait allele of interest is also present on a homologous chromosome not carrying the trait allele. Such marker alleles cannot be detected by BSA. Another issue is that the best 4x parents for mapping a trait are not the best sources of DNA from which to construct a BAC library. The best type of 4x cross for map-based cloning results from crossing simplex (Aaaa) and nulliplex (aaaa) parents. Here the progeny are expected to segregate 1:1 for the trait of interest. Any other type of cross will produce unequal numbers of progeny with and without the trait, requiring larger populations to be maintained to obtain a meaningful number of individuals of the rarer phenotypic class for BSA and high-resolution linkage analyses. A simplex parent is not ideal for constructing a BAC library, however, since only one clone in four that harbors the desired gene will be expected to carry the allele of interest. Planning ahead by developing lines that are duplex, triplex or quadruplex prior to constructing a BAC library makes it easier to find the desired allele. Lastly, it should be noted that, whatever the ploidy level, chromosome walking is more challenging when a library is constructed from a heterozygous individual. This is because a positive hybridization result alone does not suffice to show that two BAC clones are derived from the same chromosome. To prove that two BAC clones overlap, it must also be shown that they share a common marker allele.

MAP-BASED CLONING IN PRACTICE

Three genes in potato have recently been isolated using variants of the map-based cloning strategy described above. It is no accident that all three—*Rx1, R1,* and *Gpa2*—encode disease resistance genes. *Rx1* confers immunity to most strains of PVX, *R1* confers race-specific resistance to *Phytophthora infestans,* while *Gpa2* confers race-specific resistance to *Globodera pallida.* Until recently, very little was known about resistance genes, so that when these gene isolation projects began it wasn't possible to predict what each gene would look like. Like many other resistance genes, these genes also confer unambiguous phenotypes, a necessary prerequisite for map-based cloning. Two of these genes, *Rx1* and *R1,* were isolated completely from scratch (Bendahmane et al. 1999; Ballvora et al. 2002). In contrast, isolation of *Gpa2,* which is located very close to *Rx1,* was much simpler, as it utilized the high-resolution genetic and large-insert clone maps constructed for purposes of cloning *Rx1* (van der Vossen et al. 2000).

A map location for *Rx1* was first described by Ritter et al. (1991) who analyzed a diploid population and reported that *Rx1* was located in a region near the end of one arm of chromosome 12 with RFLP marker GP34 positioned slightly centromeric to the gene. Several years later an independent research group began the process of cloning *Rx1* by map-based means in an unrelated tetraploid population. Somewhat surprisingly, Bendahmane et al. (1997) found that *Rx1* mapped to a slightly different interval, with GP34 positioned on the telomeric side of the gene. The reason for the difference in map locations between these studies is not yet known but it nevertheless highlights the need to carefully re-evaluate any map location before committing extensive resources to map-based cloning.

To saturate the region around *Rx1* with additional markers, pools of resistant and susceptible progeny were evaluated with 728 AFLP primer combinations (Bendahmane et al. 1997). This led to the identification of 57 additional tightly linked markers. The only plants of value for constructing a high-resolution map around a locus are those with a recombination event close to the gene of interest since these can be used to determine the map order of linked markers relative to each other and to the gene. Thus, flanking RFLP markers CT99 and CT129 were converted to CAPS markers in order to screen 1350 F_1 plants for individual recombinant events (Bendahmane et al. 1997). By analyzing marker and trait segregation in these recombinants it was possible to restrict the location of *Rx1* to a 0.23 centiMorgan (cM) interval, between AFLP markers PM3 and PM4. To estimate the physical distance between PM3 and PM4, each of these markers was hybridized individually to size-fractionated high molecular weight fragments of genomic DNA. The smallest fragment that both markers hybridized to was 80 kb in length (Bendahmane et al. 1997), suggesting—incorrectly, it turned out—that *Rx1* had been delimited to an acceptably small interval for chromosome landing.

A BAC library was then constructed from an individual plant known to be duplex for *Rx1*. This library consisted of over 160,000 clones with an average insert size of 100 kb (Kanyuka et al. 1999). Screening the library with CAPS markers derived from PM3 and PM4 revealed no BAC clones that contained both markers, i.e., chromosome landing had failed. The process of chromosome walking was then initiated. The conventional strategy of using the ends of newly isolated BAC clones to identify additional BAC clones covered some of the distance between PM3 and PM4. However, the two walks failed to converge since both eventually encountered the problem that no new clones could be isolated using the ends of previously isolated clones. Given the large size of the BAC library, predicted to cover the genome about 18 times, this result was rather unexpected. Kanyuka et al. (1999) reported that their library was constructed with a single restriction enzyme which may have resulted in uneven representation across the genome. They therefore recommended that future gene isolation projects avoid this problem by screening multiple libraries constructed with a variety of restriction enzymes.

Having exhausted the standard approaches, a creative approach was required to isolate *Rx1*. In the process of chromosome walking, Kanyuka et al. (1999) observed that PM4 and several BAC end sequences shared sequence similarity with known resistance genes. Because resistance genes are often clustered in plant genomes they reasoned that it might be possible to identify additional markers linked to *Rx1* by using low stringency PCR with existing primers based on PM4 sequence. This strategy was successful and led to the identification of an additional BAC clone between PM3 and PM4, which had not been identified by earlier chromosome walking (Kanyuka et al. 1999). Using existing recombinants it was shown that one end of this new BAC cosegregated with *Rx1*, while the other end was separated by a single recombination event. A novel approach was then employed to search for additional recombinants that could be used to determine precisely where the cosegregating BAC end marker mapped relative to *Rx1* (Kanyuka et al. 1999). Potato cultivars known to contain *Rx1* were screened for crossovers near *Rx1*, and one variety was found with a recombination event between the previously cosegregating BAC end marker and *Rx1*. Thus, it was established that *Rx1* must be contained within the new BAC (Kanyuka et al. 1999). An elegant transient infection assay was used to test this BAC, followed by subsets of the BAC, for ability to confer resistance to PVX (Bendahmane et al. 1999). Sequencing ultimately revealed that *Rx1* codes for a protein of 107.5 kD similar to other resistance genes of the nucleotide binding site/leucine rich repeat (NBS-LRR) class (Bendahmane et al. 1999).

The *R1* gene conferring resistance to late blight was also isolated by position-based means. *R1* was first localized to chromosome 5 in a population of 97 segregating diploid plants between RFLP markers GP21 and GP179 (Leonards-Schippers et al. 1992). These RFLP markers were converted into PCR-based markers and used to screen an additional 952 plants for recombination events (Meksem et al. 1995; Ballvora et al. 2002). Precisely 31 crossovers were found in the 1,069 plants (Ballvora et al. 2002). One hundred and eight AFLP primer combinations were used to screen bulks of DNA, identifying six AFLP markers linked in coupling and 23 markers linked in repulsion (Meksem et al. 1995). Marker AFLP1 was found to be located 0.1 cM to one side of *R1*. As luck would have it, however, the most useful marker came from a completely unrelated study. CAPS marker SPUD237, developed in the course of mapping hypersensitive resistance to PVX in unrelated germplasm, was fortuitously found to map 0.1 cM to the other side of *R1* (Ballvora et al. 2002). SPUD237 was used to start a chromosome walk, with the *R1*-oriented ends of each new BAC clone being used to identify additional clones (Ballvora et al. 2002). Several steps into the walk, and before reaching the AFLP1 marker known to be on the other side of *R1*, it was observed that an end fragment of one newly isolated BAC clone shared sequence similarity with known resistance genes (Ballvora et al. 2002). This sequence was used to isolate additional BAC clones, all of which were linked without recombination to *R1*. One of these cosegregating BAC clones was derived from the *R1*-bearing chromosome and contained an entire resistance-like gene. The insert of this BAC was found to confer resistance to LB when transformed into a susceptible potato line (Ballvora et al. 2002). The successful result of this short-cut meant that no further effort was required to complete the chromosome walk across the SPUD237-AFLP1 marker interval. After further subcloning, functional testing, and sequencing, *R1* was ultimately revealed to encode a protein of 149.4 kD, with leucine zipper, NBS, and LRR motifs common to many other resistance genes (Ballvora et al. 2002).

Undoubtedly the best situation for a map-based cloning approach is when high-resolution genetic and physical maps already exist around the gene of interest. Through careful observation Rouppe van der Voort et al. (1997b) noticed that the nematode resistance gene *GPa2* mapped close to *Rx1*, and because of a common wild ancestor, possibly on the same introgressed fragment. High-resolution genetic mapping with existing CAPS markers of *Rx1* origin confirmed a very tight linkage of 0.02 cM between these two genes (Rouppe van der Voort et al. 1999). Ultimately a BAC clone identified in the *Rx1* project during the attempted chromosome walk from PM4 to PM3 was shown to contain *Gpa2* (van der Vossen et al. 2000). While there are currently very few regions of the potato genome for which similar high-resolution maps have been constructed to facilitate cloning of other genes, this situation will gradually improve over time as more map-based cloning (or whole genome sequencing) projects are initiated *de novo*.

CANDIDATE GENE APPROACH

When evidence from other systems suggests that a particular gene might control a trait of interest, this gene is commonly referred to as a "candidate gene". Similarly, testing whether such genes actually control a trait of interest has come to be known as the "candidate gene approach" (reviewed in Pflieger et al. 2001). Although few potato genes have been isolated in this manner to date (Appendix 8.1), in the coming decade it is almost certain that many more will be isolated by the candidate gene approach than by map-based cloning.

Intensive efforts of plant geneticists over the last decade have generated vast data on gene sequence, gene function and genome organization from a wide array of plants, especially rice and *Arabidopsis*. It is thus increasingly possible to make an educated guess about the identity of a gene that controls a particular trait by searching for genes known to control related traits in other plant species. In some cases it is not even necessary to look at other species. The *Rx2* locus in potato maps to chromosome 5 and, like *Rx1*, confers immunity to PVX. Although it required much effort to isolate *Rx1*, *Rx2* was isolated quite easily by employing low-stringency PCR with primers based on the *Rx1* sequence (Bendahmane et al. 2000). With the large number of potato ESTs currently available, the chances of finding a potato gene with unambiguous sequence similarity to a known gene of another organism are already quite high. As additional ESTs are generated in the future, the chances of finding potato homologues can only improve.

Map location of a gene can also be useful in deciding whether or not it makes a good candidate. To date several hundred genes have been mapped in potato. One paper of broad interest reported recently the positions of 85 loci, defined by 69 genes known to be involved in carbohydrate metabolism and transport (Chen et al. 2001). Given the obvious importance of starch to potatoes it is likely that economically imporatnt loci will ultimately be found to colocalize with some of these mapped genes. Thirty-two regions defined by 14 probes with sequence similarity to known resistance genes have also been mapped in potato (Leister et al. 1996). At least some of these will correspond to genes which control resistance to potato pathogens and pests, the relationship becoming apparent when genes controlling resistance traits are found to map to the same regions. To begin understanding the genetic basis of quantitative resistance to LB, Trognitz et al. (2002) recently evaluated over 300 loci related to 27 defense genes and compared their segregation with known QTLs. Fifteen loci, corresponding to genes such as phenylalanine ammonia lyase and a cytochrome P450 oxidase, mapped to the same regions as known QTLs (Trognitz et al. 2002), thus making them obvious candidates for further testing.

It is worth mentioning that positional relationships that hint at gene identity can often be found by examining map locations of known genes in other solanaceous plants. Genetic maps of tomato, pepper, and eggplant provide excellent sources of gene map data and can be useful in identifying or eliminating candidate genes from consideration. Potato and tomato in particular show extensive conservation of gene order (Bonierbale et al. 1988; Gebhardt et al. 1991; Tanksley et al. 1992). Indeed, because more genes have been mapped in tomato than in potato (Ganal et al. 1998; Fulton et al. 2002), tomato is often the best source of positional candidates in potato. Potato also shares marker order with pepper (Livingstone et al. 1999) although a larger number of chromosomal inversions and translocations sometimes obscure the similarities. It was nevertheless recently possible for some pepper researchers to make a prediction about the identity of the potato gene *Y*, which conditions the yellow flesh trait. Thorup et al. (2000) mapped many carotenoid biosynthetic genes in pepper and observed that beta-carotene hydroxylase maps to the same region as *Y*. A publication that summarizes all gene and marker mapping in tomato, pepper, and potato in a simple and single genetic map similar to the elegant circular map of cereal genomes (Moore et al. 1995) does not yet exist, although it is hoped that such a map will eventually be developed. In any case, given the rapid pace of progress, the most up-to-date maps are usually available on the internet. One highly recommended site for solanaceous map data is the Solanaceae Genomics Network (www.sgn.cornell. edu).

Currently the densest plant gene map is that of the almost completely sequenced *Arabidopsis* genome (The Arabidopsis Genome Initiative 2000). An issue of considerable importance is the extent and scale to which gene order has been conserved in potato relative to this model species. If gene order is ultimately found to be highly conserved, then it should be possible to jump back and forth between species maps to identify candidate genes for potato traits. For example, in an ideal scenario of extensive one-to-one similarity between potato and *Arabidopsis*, if a genetic marker were tightly linked to a trait in potato, then identifying candidate genes for the potato trait could be performed quite easily by sifting through *Arabidopsis* genes that surround the same marker. In practice, only genetic markers based on conserved coding regions are likely to be able to detect polymorphism in both potato and *Arabidopsis*, since non-coding regions are poorly conserved between these species.

While there is little direct evidence to date to show that potato shares gene order with *Arabidopsis*, there is good evidence that tomato and *Arabidopsis* do. Support for conservation of gene order has come primarily from the sequencing of large contiguous stretches of various plant genomes and comparing these with existing *Arabidopsis* sequence. Meyer et al. (1996) sequenced a 4.3 kb potato clone containing a protein kinase and found that the clone also contained part of an adjacent gene encoding a sucrolytic enzyme. These two genes are also adjacent in *Arabidopsis* (Meyer et al. 1996). Ku et al. (2000) determined the sequence of 105 kb of DNA surrounding the *ovate* locus in tomato and identified 17 putative genes. Comparison with *Arabidopsis* revealed a "network of synteny", with four regions of the *Arabidopsis* genome sharing sequence similarity with 12 of these tomato genes (Ku et al. 2000). No single *Arabidopsis* region matched all 12 genes; rather each region matched between two and seven tomato genes. Nevertheless, the order of the genes in each *Arabidopsis* region was the same as the order of the tomato genes. This and other evidence (Vision et al. 2000) has been interpreted to indicate that the *Arabidopsis* genome has undergone at least two large-scale duplications in its evolutionary history, with frequent gene loss and occasional gene relocalization occurring after one or both duplications. If the pattern of correspondence for other regions of the tomato or potato genomes to *Arabidopsis* is predominantly one region to many regions, or many regions to many regions, the process of identifying candidate genes by examining genes known to be present around a shared marker would require considerably more effort than implied in the ideal scenario above. In particular, it would be necessary to integrate information from all related regions in *Arabidopsis* to identify a more comprehensive set of candidate genes. Moreover, even with this conceptual integration, given the occasional relocalization of genes in *Arabidopsis* relative to tomato or potato, there would always be a chance that the *Arabidopsis* gene corresponding to the potato gene of interest would not be present anywhere near the shared marker. Whether or not the patterns of conservation of gene order observed on the scale of 100 kb extend to much larger regions, or other parts of the tomato or potato genomes, remains to be seen.

If conservation of gene order between plant species is common then the conserved ortholog set (COS) markers developed by Fulton et al. (2002) may prove highly useful. About 1,025 genes that are highly conserved between tomato and *Arabidopsis* have been identified through comparisons of tomato and *Aradbidopsis* sequence databases. Over 500 of these have already been mapped as RFLPs in tomato (Fulton et al. 2002, www.sgn.cornell.edu). A very attractive feature of these COS markers is that they come with built-in knowledge of the genes which flank a locus in *Arabidopsis*. Thus, if COS markers were used to map a trait in potato, it would be possible to make predictions about potato gene content around each and every trait-linked marker.

In addition to providing a rich source of positional candidate genes, conserved gene order between solanaceous plants and *Arabidopsis* might also be exploited to identify new markers for map-based cloning. While working to isolate the *ovate* locus in tomato, Ku et al. (2001) observed several markers closely linked to this gene that matched genes from a region of *Arabidopsis* chromosome 4. It was found that several other *Arabidopsis* genes from this region could be mapped close to *ovate* (Ku et al. 2001). Several tomato ESTs similar to *Arabidopsis* genes from this region were also successfully mapped close to *ovate*. If this approach to increasing marker density proves reproducible for other regions of solanaceous genomes, it should prove valuable in accelerating many map-based cloning projects.

ADDITIONAL METHODS OF ISOLATING GENES

While map-based cloning and the candidate gene approach are currently the most favored approaches for isolating genes, several other approaches may also be employed, depending on how much is already known about the trait in question. Perhaps the most obvious is when a gene is thought to code for an enzyme in a biochemical pathway. In such cases it is sometimes possible to purify the enzyme in sufficient quantities so that it can be partially or wholly sequenced. An oligonucleotide probe based on the protein sequence can then be used as a hybridization probe to search for cDNA or genomic DNA clones that code for the enzyme.

A method that has found widespread use in maize (Federoff et al. 1984) and occasional use in tomato (Bishop et al. 1996), is to employ transposons to insertionally inactivate, and simultaneously tag, genes of interest. No potato genes have yet been isolated with this approach, although a system based on the *Ac-Ds* system of maize is being developed for this purpose (van Enckevort et al. 2001). *Ac* is a complete transposon, which can transpose autonomously, while *Ds* elements require functions provided by *Ac* in order to move. *Ac* and *Ds* are known to preferentially transpose to closely linked sites, both in maize and in heterologous systems. A typical gene isolation strategy with *Ac-Ds* involves several steps. First, a plant line is identified in which a *Ds* element is tightly linked to the gene of interest. Second, *Ds* is then induced to transpose, usually by crossing with another line that expresses *Ac* transposase. Third, progeny plants in which *Ds* has been activated are screened to determine whether any plant possesses a phenotype consistent with the gene having been inactivated. Once *Ds* is inserted into a gene, it becomes a relatively straightforward process to isolate it using inverse PCR or plasmid rescue.

Several hundred diploid potato lines have been developed that contain one or more *Ds* elements on separate T-DNAs (El-Kharbotly et al.

1996). The T-DNA construct in these lines was designed so that hygromycin phosphotransferase would be expressed following excision of *Ds*, providing a means to select for plants in which transposition has occurred (El-Kharbotly et al. 1996). Many of these T-DNAs have been mapped in the potato genome (Jacobs et al. 1995; El-Kharbotly et al. 1996). Thus, if a gene has been mapped, it is easy to identify a plant with a *Ds* element close to the gene. The *Ds* elements in these lines become active after crossing to other lines expressing the *Ac* transposase (El-Kharbotly et al. 1996; van Enckevort et al. 2001), thereby substantiating their potential value for regional mutagenesis. Nevertheless, diploid potatoes being self-incompatible (SI), the current *Ac/Ds* system is not as easy to implement as similar systems are in other plants. SI makes it difficult to obtain homozygous insertion events so that this system is essentially limited to tagging genes where a heterozygous mutant yields a detectable phenotype. The heterozygosity associated with SI also makes it difficult to ensure that a trait of interest doesn't segregate in the progeny. Such segregation masks variation resulting from gene inactivation. To circumvent this impediment it is possible to isolate protoplasts from the somatic tissue of a plant containing an active *Ds* element (van Enckevort et al. 2000). Plants regenerated from such protoplasts and selected on media containing hygromycin must harbor transposed *Ds* elements and can be produced in large numbers to screen for a phenotype of interest (Van Enckevort et al. 2000). One remaining point that should be considered with transposon-tagging is that many potato genes are likely to be duplicated in the genome. One-sixth of genes in *Arabidopsis* are present in tandem duplications or higher-order arrays (The Arabidopsis Genome Initiative 2000). For genes present in tandem arrays, insertional inactivation of just a single member will not necessarily result in a mutant phenotype.

Lastly, a powerful and refined approach to gene isolation is possible when the gene of interest is predicted to provide a detectable phenotype in yeast or *Escherichia coli.* It is technically straightforward to express large collections of potato cDNA clones in either of these two hosts. Sorting through colonies of *E. coli* or yeast with appropriate screens or selection processes can then be used to identify those few that express the potato gene of interest. The gene encoding SGT, an enzyme which catalyzes the addition of the first sugar moiety to an alkaloid backbone in the synthesis of some potato glycoalkaloids, was isolated using such an approach. Moehs et al. (1997) observed that yeast is very sensitive to the presence of the alkaloid solasodine in culture media but could grow when the alkaloid was glycosylated. This differential sensitivity was exploited to isolate the SGT gene; following transformation with a potato cDNA library, yeast cells expressing SGT were identified by their ability to survive in the presence of solasodine (Moehs et al. 1997).

POTATO GENE ISOLATION IN FUTURE

Given the rapid pace of progress in the entire field of genetics, it is virtually certain that methods for gene isolation will continue to be developed in the coming years, and that existing methods will become both more standardized and routine. The number of mapped and/or sequenced genes in potato will continue to increase, further facilitating candidate gene approaches. Indeed, as genome sequencing costs continue to fall it is not unreasonable to think that potato, or its close relative tomato, might be completely sequenced in the next 5-10 years. In lieu of a complete genome sequence, the development of a physical map of potato, consisting of overlapping large-insert clones, would also be a huge boost for map-based cloning efforts.

Whether it is a step on the road to systematically testing a gene's function, or a means for tracking and selecting desirable natural variation in an applied potato breeding program, for most researchers successful gene isolation just constitutes a beginning. Ten years ago it was almost impossible to isolate potato genes that didn't control simple biochemical traits. Many interesting genes have been isolated since, and many more will be described in the decade ahead. For those whose research depends on cloned and sequenced genes, the future has never been brighter.

REFERENCES

Arumuganathan, K. and E.D. Earle. 1991. Nuclear DNA content of some important plant species. Plant Mol Biol Reporter 9: 208–219.

Ballvora, A., M.R. Ercollano, J. Weiss, K. Meksem, C.A. Bormann, P. Oberhagemann, F. Salamini, and C. Gebhardt. 2002. The *R1* gene for potato resistance to late blight (*Phytophthora infestans*) belongs to the leucine zipper/NBS/LRR class of plant resistance genes. Plant J 30: 361–371.

Barrell, P.J., S. Yongquin, P.A. Copper, and A.J. Conner. 2002. Alternative selectable markers for potato transformation using minimal T-DNA vectors. Plant Cell, Tissue Organ Culture 70: 61–68.

Bendahmane, A., K. Kanyuka, and D.C. Baulcombe. 1997. High-resolution genetical and physical mapping of the *Rx* gene for extreme resistance to potato virus X in tetraploid potato. Theor Appl Genet 95: 153–162.

Bendahmane, A., K. Kanyuka, and D.C. Baulcombe. 1999. The *Rx* gene from potato controls separate virus resistance and cell death responses. Plant Cell 11: 781–791.

Bendahmane, A., M. Querci, K. Kanyuka, and D.C. Baulcombe. 2000. *Agrobacterium* transient expression system as a tool for the isolation of disease resistance genes: application to the *Rx2* locus in potato. Plant J 21: 73–81.

Bishop, G., K. Harrison, and J.D.G. Jones. 1996. The tomato *dwarf* gene isolated by heterologous transposon tagging encodes the first member of a new cytochrome P450 family. Plant Cell 8: 959–969.

Bonierbale, M.W., R.L. Plaisted, S.D. Tanksley. 1988. RFLP maps based on a common set of clones reveal modes of chromosome evolution in potato and tomato. Genetics 120: 1095–1103.

Chen, X., F. Salamini, and C. Gebhardt. 2001. A potato molecular-function map for carbohydrate metabolism and transport. Theor Appl Genet 102: 284–295.

De Jong, W., A. Forsyth, D. Leister, C. Gebhardt, and D. Baulcombe. 1997. A potato hypersensitive gene against potato virus X maps to a resistance gene cluster on chromosome 5. Theor Appl Genet 95: 246–252.

El-Kharbotly, A., J.M.E. Jacobs, B. te Linkel Hekkert, E. Jacobsen, M.S. Ramanna, W.J. Stiekema, and A. Pereira. 1996. Localization of Ds-transposon containing T-DNA inserts in the diploid transgenic potato: linkage to the *R1* resistance gene against *Phytophthora infestans* (Mont.) de Bary. Genome 39: 249–257.

Federoff, N.V., D.B. Furtek, O.E. Nelson. 1984. Cloning of the *bronze* locus in maize by a simple and generalizable procedure using the transposable element *Activator* (*Ac*). Proc Natl Acad Sci USA 81: 3825–3829.

Fulton, T.M., R. Van der Hoeven, N.T. Eannetta, and S.D. Tanksley. 2000. Identification, analysis, and utilization of conserved ortholog set markers for comparative genomics in higher plants. Plant Cell 14: 1457–1467.

Ganal, M.W., R. Czihal, U. Hannapel, D-U. Kloos, A. Polley, and H-Q. Ling. 1998. Sequencing of cDNA clones from the genetic map of tomato (*Lycopersicon esculentum*). Genome Res 8: 842–847.

Gebhardt, C., E. Ritter, T. Debener, U. Scachtschabel, B. Walkemeier, H. Uhrig, and F. Salamini. 1989. RFLP analysis and linkage mapping in *Solanum tuberosum*. Theor Appl Genet 78: 65–75.

Gebhardt, C., E. Ritter, A. Barone, T. Debener, B. Walkemeier, U. Schactschabel, H. Kaufmann, R.D. Thompson, M.W. Bonierbale, M.W. Ganal, S.D. Tanksley, and F. Salamini. 1991. RFLP maps of potato and their alignment with the homoeologous tomato genome. Theor Appl Genet 83: 49–57.

Goff, S.A. et al. 2002. A draft sequence of the rice genome (*Oryza sativa* L. ssp. *japonica*). Science 296: 92–100.

Hamilton, C.M., A. Frary, C. Lewis, and S.D. Tanksley. 1996. Stable transfer of intact high molecular weight DNA into plant chromosomes. Proc Natl Acad Sci USA 93: 9975–9979.

Jacobs, J.M.E., H.J. Van Eck, P. Arens, B. Verkerk-Bakker, B. te Lintel Hekkert, H.J.M. Bastiaanssen, A. El-Kharbotly, A. Pereira, E. Jacobsen, and W.J. Stiekema. 1995. A genetic map of potato (*Solanum tuberosum*) integrating molecular markers, including transposons, and classical markers. Theor Appl Genet 91: 289–300.

Kanyuka, K., A. Bendahmane, J.N.A.M. Rouppe van der Voort, E.A.G. van der Vossen, and D.C. Baulcombe. 1999. Mapping of intra-locus duplications and introgressed DNA: aids to map-based cloning of genes from complex genomes illustrated by physical analysis of the *Rx* locus in tetraploid potato. Theor Appl Genet 98: 679–689.

Konieczny, A. and F.M. Ausubel. 1993. A procedure for mapping *Arabidopsis* mutations using co-dominant ecotype-specific PCR-based markers. Plant J 4: 403–410.

Ku, H.K., T. Vision, J. Liu, and S.D. Tanksley. 2000. Comparing sequenced segments of the tomato and *Arabidopsis* genomes: large scale duplication followed by selective gene loss creates a network of synteny. Proc Natl Acad Sci USA 97: 9121–9126.

Ku, H-M., J. Liu, S. Doganlar, and S.D. Tanksley. 2001. Exploitation of *Arabidopsis*-tomato synteny to construct a high-resolution map of the *ovule*-containing region in tomato chromosome 2. Genome 44: 470–475.

Leister, D., A. Ballvora, F. Salamini, and C. Gebhardt. 1996. A PCR-based approach for isolating pathogen resistance genes from potato with potential for wide application in plants. Nature Genet 14: 421–429.

Leonards-Schippers, C., W. Gieffers, F. Salamini, and C. Gebhardt. 1992. The *R1* gene conferring race specific resistance to *Phytophthora infestans* in potato is located on potato chromosome. V. Mol Gen Genet 233: 278–283.

Livingstone, K.D., V.K. Lackney, J.R. Blauth, R. van Wijk, and M.K. Jahn. 1999. Genome mapping a *Capsicum* and the evolution of genome structure in the *Solanaceae*. Genetics 152: 1183–1202.

Meksem, K., D. Leister, J. Peleman, M. Zabeau, F. Salamini, and C. Gebhardt. 1995. A high-resolution map of the vicinity of the *R1* locus on chromosome V of potato based on RFLP and AFLP markers. Mol Gen Genet 249: 74–81.

Meyer, R.C., P.E. Hedley, R. Waugh, and G.C. Mackay. 1996. Organisation and expression of a potato (*Solanum tuberosum*) protein kinase gene. Plant Sci 118: 71–80.

Meyer, R.C., D. Milbourne, C.A. Hackett, J.E. Bradshaw, J.W. McNichol, and R. Waugh. 1998. Linkage analysis in tetraploid potato and association of markers with quantitative resistance to late blight (*Phytophthora infestans*). Mol Gen Genet 259: 150–160.

Michelmore, R.W., I. Paran, and R.V. Kesseli. 1991. Identification of markers linked to disease resistance genes by bulked segregant analysis: a rapid method to detect markers in specific genomic regions by using segregating populations. Proc Natl Acad Sci USA 88: 9828–9832.

Milbourne, D., R.C. Meyer, A.J. Collins, L.D. Ramsay, C. Gebhardt, and R. Waugh. 1998. Isolation, characterisation and mapping of simple sequence repeat loci in potato. Mol Gen Genet 259: 233–245.

Moehs, C.P., P.V. Allen, M. Friedman, and W.R. Belknap. 1997. Cloning and expression of solanidine UDP-glucosyltransferase from potato. Plant J 11: 227–236.

Moore, G., K.M. Devos, Z. Wang, and M.D. Gale. 1995. Grasses, line up and form a circle. Curr Biol 5: 737–739.

Perez, F., A. Menendez, P. Dehal, C.F. Quiros. 1999. Genomic structural differentiation in *Solanum*: comparative mapping of the A- and E-genomes. Theor Appl Genet 98: 1183–1193.

Pflieger, S., V. Lefebvre, and M. Causse. 2001. The candidate gene approach in plant genetics: a review. Mol Breed 7: 275–291.

Rickert, A.M., A. Premstaller, C. Gebhardt, and P.J. Oefner. 2002. Genotyping of SNPs in a polyploid genome by pyrosequencing (TM). Biotech 32: 592–

Ritter, E., T. Debener, A. Barone, F. Salamini, and C. Gebhardt. 1991. RFLP mapping on potato chromosomes of two genes controlling resistance to potato virus X (PVX). Mol Gen Genet 227: 81–85.

Rivard, S.R., M. Cappadocia, and B.S. Landry. 1996. A comparison of RFLP maps based on anther culture derived, selfed, and hybrid progenies of *Solanum chacoense*. Genome 39: 611–621.

Rouppe van der Voort, J.N.A.M., P. van Zandvoort, H.J. van Eck, R.T. Folkertsma, R.C.B. Hutten, J. Draaistra, F.J. Gommers, E. Jacobsen, J. Helder, and J. Bakker. 1997a. Use of allele specificity of comigrating AFLP markers to align genetic maps from different potato genotypes. Mol Gen Genet 255: 438–447.

Rouppe van der Voort, J., P. Wolters, R. Folkertsma, R. Hutten, P. van Zandvoort, H. Vinke, K. Kanyuka, A. Bendahmane, E. Jacobsen, R. Janssen, and J. Bakker. 1997b. Mapping of the cyst nematode resistance locus *Gpa2* in potato using a strategy based on comigrating AFLP markers. Theor Appl Genet 95: 874–880.

Rouppe van der Voort, J., K. Kanyuka, E. van der Vossen, A. Bendahmane, P. Mooijman, R. Klein-Lankhorst, W. Stiekema, D. Baulcombe, and J. Bakker. 1999. Tight physical linkage of the nematode resistance gene Gpa2 and the virus resistance gene *Rx* on a single segment introgressed from the wild species *Solanum tuberosum* subsp. andigena CPC 1673 into cultivated potato. Mol Plant-Microbe interact 12: 197-206.

Shizuya, H., B. Birren, U-J. Kim, V. Mancino, T. Slepak, Y. Tachiiri, and M. Simon. 1992. Cloning and stable maintenance of 300-kilobase-pair fragments of human DNA in *Escherichia coli* using an F-factor-based vector. Proc Natl Acad Sci USA 89: 8794–8797.

Song, J.Q., F.G. Dong, and J.M. Jiang. 2000. Construction of a bacterial artificial chromosome (BAC) library for potato molecular cytogenetics research. Genome 43: 199–204.

Tanksley, S.D., M.W. Ganal, and G.B. Martin. 1995. Chromosome landing: a paradigm for map-based gene cloning in plants with large genomes. Trends Genet 11: 63–68.

Tanksley, S.D., M.W. Ganal, J.P. Prince, M.C. de Vicente, M.W. Bonierbale, P. Broun, T.M. Fulton, J.J. Giovannoni, S. Grandillo, G.B. Martin, R. Messeguer, J.C. Miller, L. Miller, A.H. Paterson, O. Pineda, M.S. Roder, R.W. Wing, W. Wu, and N.D. Young. 1992. High density molecular linkage maps of the tomato and potato genomes. Genetics 132: 1141–1160.

The Arabidopsis Genome Initiative. 2000. Analysis of the genome sequence of the flowering plant *Arabidopsis thaliana*. Nature 408: 796–815.

Thorup, T.A., B. Tanyolac, K.D. Livingstone, S. Popovsky, I. Paran, and M. Jahn. 2000. Candidate gene analysis of organ pigmentation loci in the *Solanaceae*. Proc Natl Acad Sci USA 97: 11192–11197.

Trognitz, F.T., P. Manosalva, R. Gysin, D. Nino-Liu, R. Simon, M. del Rosario Herrera, B. Trognitz, M. Ghislain, and R. Nelson. 2002. Plant defense genes associated with quantitative resistance to potato late blight in *Solanum phureja* × dihaploid *S. tuberosum* hybrids. Mol Plant Microbe Interact 15: 587–597.

van der Hoeven, R., C. Ronning, J. Giovannoni, G. Martin, and S.D. Tanksley. 2002. Deductions about the number, organization, and evolution of genes in the tomato genome based on analysis of a large expressed sequence tag collection and selective genomic sequencing. Plant Cell 12: 1441–1456.

van der Vossen, E.A.G., J.N.A.M. Rouppe van der Voort, K. Kanyuka, A. Bendahmane, H. Sandbrink, D.C. Baulcombe, W.J. Stiekema, and R.M. Klein-Lankhorst, 2000. Homologues of a single resistance gene cluster in potato confer resistance to distinct pathogens: a virus and a nematode. Plant J 23: 567–576.

Van Eck, H.J., J. Rouppe van der Voort, J. Draaistra, P. van Zandvoort, E. van Enckevort, B. Segers, J. Peleman, E. Jacobsen, J. Helder, and J. Bakker. 1995. The inheritance and chromosomal localisation of AFLP markers in a non-inbred potato offspring. Mol Breed 1: 397–410.

Van Enckevort, L.J.G., J.E.M. Bergervoet, W.J. Stiekema, A. Pereira, and E. Jacobsen. 2000. Selection of independent Ds transposon insertions in somatic tissue of potato by protoplast regeneration. Theor Appl Genet 101: 503–510.

Van Enckevort, L.J.G., J. Lasschuit, W.J. Stiekema, E. Jacobsen, and A. Pereira. 2001. Development of *Ac* and *Ds* transposon tagging lines for gene isolation in diploid potato. Mol Breed 7: 117–129.

Vision, T.J., D.G. Brown, and S.D. Tanksley. 2000. The origins of genomic duplications in *Arabidopsis*. Science 290: 2114–2117.

Vos, P., R. Hogers, M. Bleeker, M. Rijans, T. Van de Lee, M. Hornes, A. Frijters, J. Pot, J. Peleman, M. Kuiper, and M. Zabeau. 1995. AFLP: a new technique for DNA fingerprinting. Nucl Acids Res 23: 4407–4414.

Yu, J. et al. 2002. A draft sequence of the rice genome (*Oryza sativa* L. ssp. *indica*). Science 296: 79–92.

APPENDIX 8.1

Candidate Gene Approach

Considering what is known in other, better characterized plant/organisms (e.g. tomato, Arabidopsis, or yeast): make an educated guess about what gene might control the trait

↓

Test the hypothesis by one or more of the following:

- Does an allele of the gene cosegregate with the trait?
- Is the gene/allele expressed in the correct tissue?
- Is the gene/allele expressed at the correct time?

↓

Once sufficiently confident in the hypothesis, validate it by constructing a transgenic potato plant

- Is expected phenotype observed?

9

Cell and Tissue Culture of Potato (Solanaceae)

RICHARD E. VEILLEUX
Virginia Polytechnic Institute & State University,
Dept. of Horticulture, Blacksburg, VA (USA) 24061-0327, Tel: (540) 231-5584,
Fax: (540) 231-3083, e-mail: potato@vt.edu

INTRODUCTION

Potato (*Solanum tuberosum* L.) is one of the most amenable of major crop plants to tissue and cell culture manipulations. Explants used successfully for *in vitro* regeneration and micropropagation of potato plants include various types of meristem, protoplasts, leaf, tuber, and internodes. Perhaps more than any other crop save tobacco (*Nicotiana tabacum* L.), cell and tissue culture methods have greatly contributed to improvement of potatoes. Meristem culture and micropropagation have become routine in potato seed certification schemes (Struik and Wiersema 1999). Anther-derived haploids of potato have been used in mapping studies and breeding strategies. Potato somatic hybrids have been used to introgress traits from sexually incompatible germplasm into the cultivated genepool and somaclones derived from regenerated protoplasts released as improved cultivars. Many of the tissue culture techniques applied to potato have been reviewed from time to time over the years. This review discusses the various tissue culture techniques applied to potato and follows the format of a college course or textbook on plant tissue culture, i.e., it commences with disorganized cultures (callus and suspension), then takes up various routes of regeneration, and finishes with breeding applications such as anther or microspore culture and protoplast fusion.

CALLUS CULTURES

Callus can be initiated from almost any part of the potato plant using media containing a wide array of growth regulator combinations

(Table 9.1). Tubers, leaf sections, or internodes are the usual explants. Usually both an auxin and a cytokinin are used; however, cytokinin is occasionally omitted, depending on the purpose for which use of the callus is necessary. Callus cultures have been used for physiological studies, especially with regard to starch biosynthesis (Hagen and Muneta 1993; Hagen et al. 1991, 1993). The starch content of potato callus derived from tubers has generally been lower than that of the tubers although it can be increased by optimizing auxin concentration and sugar source (Hagen et al. 1993). Hagen and Muneta (1993) studied various enzyme activities in cultured callus cells and concluded that the system was an adequate model for studying carbohydrate metabolism. Other researchers have studied potato callus cells from a cytological perspective. Anjum (2001) compared the cytology of callus derived from frost-tolerant and control lines of potato and concluded that a frost-tolerant line had greater starch grains in the plastids and more microbodies containing protein crystals than the control line. Svetek et al. (1999) observed membrane domains of potato callus and concluded that callus had a greater proportion of membranes with disordered domains compared to root cells, indicating higher membrane fluidity. Dhingra et al. (1991) found that PVX infection of callus cells varied with cytokinin treatment.

Table 9.1 *Explant sources and media used to initiate callus in potato*

Explant	*Medium*[1]	*Auxin*	*Cytokinin*	*Reference*
Leaf	MS	10.7 μM NAA	0.46 μM kin	Ochatt et al. (1998)
Leaf	MS	13.6 μM, 2, 4-D	1.4 μM kin	Anjum (2001)
Tubers	MS	10 μM picloram	—	Hagan et al. (1991)
Internodes	SH	9.9 μM NAA	2.2 μM BA	Pavingerova et al. (2001)

[1]MS: Murashige and Skoog (1962); SH: Schenk and Hildebrandt (1972)

Studies on potato callus growth have provided an opportunity to select for cells that resist various stresses applied *in vitro*. Plants regenerated from a frost-tolerant callus line exhibited greater frost tolerance in the leaves than the susceptible control (Anjum 1998). Likewise, Ochatt et al. (1998) regenerated plants from a potato cell line that had been selected for its ability to grow rapidly in the presence of high levels of salt in the medium. The regenerated plants exhibited higher fresh and dry weights and greater tuber production under salt stress in greenhouse conditions compared to unselected controls. The salt-tolerant lines also demonstrated a genotypic change, evidenced by loss of the RAPD band present in the controls.

SUSPENSION CULTURES

A change of culture medium from solid to liquid will readily convert friable potato callus cultures to suspension cultures that have been used for a variety of physiological and metabolic studies. Growth regulator composition of the culture medium has been simple, generally requiring only an auxin but occasionally supplemented with a cytokinin (Table 9.2). The response of potato suspensions to stress has been quite well studied. Keller et al. (1996) documented changes in accumulation patterns of soluble and cell wall-bound phenolics of potato suspension cultures in response to exposure to a culture filtrate of *Phytophthora infestans* (Mont.) de Bary. The phenolic defense compounds were thought to reinforce in cell walls a sort of active defense response system similar to that found in *Nicotiana* (Schmidt et al. 1998). Polkowska-Kowalczyk and Maciejewska (2001) reported that culture filtrates of *P. infestans* induced similar oxidative processes in cell lines derived from both susceptible and resistant lines, but that the kinetics and intensity of the response differed. Increased oxygen uptake was documented after potato suspension cultures were exposed to heat-killed bacteria (*Pseudomonas syringae* cv. *syringae*), a pathogen that causes hypersensitive response in intact potato plants (Baker et al. 2001). Dörnenburg and Knorr (1997) examined the response of potato suspension cultures to the application of two types of stress-chitosan (an elicitor-active polysaccharide) and hydrostatic pressure. Chitosan-induced stress resulted in induction of the defense-related enzyme, phenylalanine ammonia lyase (PAL), whereas increasing hydrostatic pressure accelerated the activity of polyphenol oxidase (PPO). The possibility of selecting potato cultivars with increased activity of stress-related enzymes in suspension cultures was looked into for identification of potato processing characteristics. The response of potato suspension cultures to extension, a hydroxyproline-rich glycoprotein, was studied by Dey et al. (1997) to investigate its role in pathogenesis-related response. Selection for salinity tolerance in suspension culture has been applied in the hope of regenerating plants that better withstand salt stress. Naik and Widholm (1993)

Table 9.2 *Explant sources and media used to initiate suspension cultures of potato*

Explant	*Medium*	*Auxin*	*Cytokinin*	*Reference*
Leaf	MS	9 μM 2, 4-D		Keller et al. (1996)
Tubers	MS	4.5 μM 2, 4-D		Dörnenburg and Knorr (1997)
Leaf rachis	MS	10.7 μM NAA	2.2 μM BA	Naik and Widholm (1993)
		9.9 μM NAA	2.2 μM BA	

found, however, that the relative salt sensitivities of six potato cultivars did not correlate well with that of cells in suspension.

Potato suspensions have been used to study nutrient absorption from liquid culture medium; nicotinic acid, pyridoxine, picloram, and thiamine were rapidly utilized by actively growing cells of cv. Lemhi Russet (Hagen et al. 1991). Collings and Emons (1999) studied the cytology of cell division of potato suspension cultures using fluorescence microscopy. Kosegarten et al. (1995) used protoplasts derived from a heterotrophic cell suspension of potato to isolate and study amyloplasts.

PLANT REGENERATION

Because of the regenerative properties of potato and many other solanaceous species, many aspects of tissue culture applications could be studied in them. There was some speculation that glycoalkaloids present in most solanaceous species mimic cytokinins and predispose solanaceous plants to regeneration in plant tissue culture. However, even though regeneration from callus, anther culture, and protoplasts has been relatively easy with potato, considerable variation is found among genotypes for disposition toward regeneration. Dale and Hampson (1995) found that only half of the 34 cultivars tested for regeneration from tuber disc explants were able to regenerate; of those that regenerated, all but one could be transformed. Such variation has led to considerable research on tailoring plant growth regulator regimes to optimize regeneration for particular genotypes and also to better understand the inheritance of regenerability in various systems.

Hulme et al. (1992) defined what is probably the most widely used media combination to obtain regeneration from potato leaves and stem internodes. It is a three-step process consisting of a growth regulator pulse, followed by a callus induction phase, then a shoot regeneration phase. The overnight pulse occurs on liquid medium [MS (Murashige and Skoog 1962) + 10g l^{-1} sucrose + 80mg l^{-1} NH_4NO_3 + 147mg l^{-1} $CaCl_2$ + 54 μM NAA + 44 μM BA]. The explants are transferred to semisolid callus medium (MS + 1g l^{-1} sucrose + 4g l^{-1} mannitol + 0.1 μM IAA + 10 μM BA + 0.8% agar). After 7 days, explants are transferred to regeneration medium (MS + 15g l^{-1} sucrose + 10 μM BA + 14μM GA + 0.8% agar). This process resulted in less variation among genotypes for frequency of regeneration than had been observed using other protocols. Leaf explants have generally been found to be more regenerative than internodal segments (Carputo et al. 1995; Paz and Veilleux 1999) although Zel and Medved (1999) reported that callus derived from internode tissue was the most regenerative in their system using a particular potato cultivar.

Hansen et al. (1999) found that the frequency of regeneration from leaf explants decreased with increase in leaf age; even among the leaflets of a compound leaf, regeneration increased from the apical to the basal pair of leaflets. Regardless of growth regulator treatments, regeneration from leaf explants of a poorly responding cultivar could not be improved. Yee et al. (2001) described a petiole-with-intact-leaflet explant that exhibited high regeneration rates across seven cultivars. Opatrna et al. (1997) reported that the growth retardant paclobutrazol, when applied to plants from which explants were derived, increased the frequency of buds regenerated from internodal segments. Lozoya-Saldana (1992) described a procedure for inducing flowering *in vitro* from apical explants of potato.

The ability of a potato genotype to regenerate in a particular tissue culture system has occasionally been studied as a predictor of regeneration in other tissue culture systems. Carputo et al. (1995) found a high positive correlation between regeneration from leaf and stem internodes and regeneration from protoplasts in a study using 11 genotypes representing five *Solanum* spp. Nadolska-Orczyk et al. (1995) found a strong correlation between regeneration ability and transformation efficiency among 12 potato cultivars. Hoogkamp et al. (2000) attempted to develop vigorous monoploid lines with altered starch composition with the ability to tuberize *in vitro* and regenerate shoots from leaf explants. From 26 sibling monoploids, only two were found to possess all the characteristics desired.

A few analyses of genetic control of regeneration have been undertaken, primarily with diploid populations. Through an analysis of dihaploids extracted from the same teraploid cultivar, Coleman et al. (1990) proposed that tissue culture responses in potato were controlled by blocks of genes. Véronneau et al. (1992) studied F_1 hybrids between selected clones of the diploid potato species, *S. chacoense* Bitt. and reported a positive significant correlation between anther culture response and leaf disk regeneration, suggesting common genetic control. Birhman et al. (1994) proposed that three genes controlled shoot regeneration from leaf explants in *S. chacoense.* The three-gene hypothesis was supported in a subsequent study of 51 progeny of two unresponsive *S. chacoense* clones (Jan et al. 1996).

SOMATIC EMBRYOGENESIS

Although carrot somatic embryos have been programmable since the 1950s, and the phenomenon has been induced in hundreds of species, it is only recently that some progress has been made in developing a system to induce somatic embryos of potato (Table 9.3). One of the possibilities for a reliable somatic embryo system for potato, in which somaclonal variation can be kept to a minimum, is the development of synthetic

Table 9.3 *Growth regulator concentrations used in series of media for initiation of somatic embryo from explants of potato*

Callus initiation	*Maintenance*	*Regeneration*	*Comments*	*Reference*
18 μM 2, 4-D	9 μM 2, 4-D	0.3 μM GA or 4.4 μM BA	Unsynchronized; plantlets developed	Degarcia and Martinez (1995)
16 μM NAA + 1.1 μM BA	16 μM NAA + 4.4 μM BA	0.3 μM GA + 0.6 μM zeatin	Plants to greenhouse	Fiegert et al. (2000)
19 μM IAA + 0.15 μM TDZ + 0.15 μM BA		12 μM zeatin + 50 nM IAA + 550 nM GA	Plants to greenhouse	Seabrook and Douglass (2001)
0.9 μM 2,4-D (*or* 1.1 μM NAA) + 10 μM BA		14.4 μM GA+ 10 μM BA *or* 14.4 μM GA + 4.6 μM zeatin *or* 10 μM BA + 14.6-22.8 μM zeatin	Embryos obtained in culture	JayaSree et al. (2001)

seed. Propagation of potato from true potato seed (TPS) has been an objective of scientists working at CIP for many years. The benefits of TPS include the fact that most viruses and other pathogens of potato are not transmitted through the seed, botanical seed is light and easily transported compared to bulky "seed" pieces (certified seed tubers), and special storage conditions are not required for botanical seed. However, uniformity of a TPS crop depends upon development of sufficiently homozygous parents that are able to generate an F_1 that conforms to industry standards. Such parents have not been developed as yet. An alternative to TPS that would circumvent the need of developing parents that generate uniform hybrids is synthetic seed. If somatic embryos could be reliably produced and stored from a standard cultivar, the resultant synthetic seed would be expected to yield a uniform crop similar to that expected from planting certified seed tubers of that cultivar, but with all of the advantages of TPS.

Degarcia and Martinez (1995) reported that somatic embryos of *S. tuberosum* cv. Desiree could be obtained from callus initiated from stem nodal sections after 90 days in culture, with appropriate changes of medium, lowering 2, 4-D levels and finally adding gibberellic acid (GA) or benzyladenine (BA) at the last transfer. Fiegert et al. (2000) regenerated plants from somatic embryos of potato cv. Tomensa. Callus had been initiated from shoot tip meristems on medium with naphthalene acetic acid (NAA) and BA. A recent breakthrough in somatic embryogenesis from a wide range of cultivars and species of potato was made wherein the maintenance step was omitted and embryos obtained after only 14-28 days (Seabrook and Douglass 2001). Several different explants from freshly micropropagated plant material were placed on medium with indole acetic acid (IAA), thidiazuron (TDZ), and BA. After 7-14 days, the explants were transferred to medium with zeatin, IAA, and GA; somatic embryos appeared directly within 2-3 weeks. The broad range of germplasm tested and the ease of obtaining somatic embryos in this system afford the opportunity of exploiting synthetic potato seed.

The ability of potato stem internodal explants to form somatic embryos was found to be genetically controlled (Seabrook et al. 2001). By examining segregation for somatic embryo production in three seedling populations of tetraploid potato, Seabrook et al. (2001) demonstrated significant differences for crosses and seedlings within crosses. High heritability estimates were obtained, suggesting simple genetic control. Somatic embryogenesis has also been obtained from leaf explants of potato cv. Jyothi using a more complex protocol (JayaSree et al. 2001). Patel et al. (2000) described improvements to the encapsulation process for somatic embryos of carrot that was also successful on callus cells and shoot tips of potato.

MICROPROPAGATION/MICROTUBERIZATION

Micropropagation and microtuberization of potato have become indispensable techniques for production of certified seed tubers and germplasm exchange. However, the topics are so extensive that they deserve a separate review article. For recent coverage of the subjects, the reader is referred to Ranalli (1997), Boxus (1999), Coleman et al. (2001), and Struik and Wiersema (1999).

CRYOPRESERVATION

Cryopreservation of potato germplasm has become an important component of maintaining germplasm collections. Not only it is essential for reduction of labor involved in routine subculture of *in-vitro* stocks, it also facilitates germplasm exchange across international borders. Research on cryopreservation of potato has concentrated both on the methods for improving survival of plantlets or explants and analysis of plants revived from cryopreserved cultures for genetic fidelity. Although cryopreservation can be defined strictly as storage of germplasm under ultralow temperatures, for the purpose of this review, I shall define it more liberally as any form of cold storage of potato germplasm. Mix-Wagner (1999) described three such maintenance protocols for storage of a German potato collection: (1) slow growth conservation where 4-week-old (3-4 cm) microplants on basal medium are maintained at 10°C, 2 klux, 16-h days for 2-3 years; (2) microtuber conservation wherein microplants are grown in sucrose-enriched medium (6-10%), short days (8 h) at 10°C for 2-4 months until microtubers form; the microtubers are then stored in darkness for 12-15 months; (3) cryoconservation wherein nodal cuttings are grown on a medium enriched with growth regulators (2.3 μM zeatin riboside, 0.6 μM GA, and 2.9 μM IAA); shoot tips are then transferred to cryoprotectant (similar medium with 10% dimethylsulfoxide), placed on aluminum foil and stored in vials filled with liquid nitrogen. Microtubers proved to be more convenient for survival during shipping. Of 245 cultivars for which 10-12 shoot tips were cryopreserved, approx. 80% survival was reported but at least one representative of each cultivar revived.

Minimal growth conservation (equivalent to slow growth conservation) has been the subject of many media and growth condition experiments. Sarkar and Naik (1998a) found that MS medium with 40g l^{-1} sucrose, 20 gL^{-1} mannitol, with plants grown under 16 h photoperiod, was optimal. Addition of 6-9 μg ml^{-1} silver thiosulfate (STS) improved the growth of microplants by reducing ethylene-induced abnormalities (Sarkar et al. 1999). The STS could be encapsulated in alginate capsules that were placed on the surface of the slow growth medium for a similar effect (Sarkar et al. 2002). Lopez-Delgado et al. (1998) substituted 100 μM acetylsalicylic acid for mannitol in slow growth medium and found similar efficiencies of recovery. Sarkar et al. (2001) reported that ancymidol had a beneficial effect on culture viability after prolonged maintenance. For distribution of germplasm, Sarkar and Naik (1998b) recommended encapsulation of nodal cuttings in MS medium containing 2 or 3% sodium alginate.

Procedures for encapsulation of vitrified meristems prior to cryopreservation were successfully applied to 14 potato cultivars (Hirai and Sakai 1999) and, in a separate study, five potato cultivars (Sarkar and Naik 1998c). Recovery was much faster than with encapsulation drying methods. Plants recovered from cryopreservation were studied with various molecular techniques to determine whether genetic changes accompany the process. Using rDNA probes, Harding (1991) reported that two of 16 plant samples recovered from slow growth exhibited RFLP variation. RAPD fingerprints (six primers generating 68 bands) revealed no detectable changes due to ancymidol treatments after 16 months of cold storage of potato microplants (Sarkar et al. 2001). Likewise, Hirai and Sakai (1999) found no difference in RAPD patterns (17 primers) between cryopreserved (encapsulation/vitrification) and nontreated plants. Harding (1994) reported that slow growth of potato cultivars in a mannitol-containing medium resulted in DNA methylation changes as detected by analysis of banding patterns following restriction with methylation-sensitive or insensitive isoschizomers. Identical cp DNA fragments were observed in recovered plants following the encapsulation/dehydration method of cryopreservation compared to control (Harding and Benson 2000). Of 161 plants recovered following cryopreservation, one likely polyploid and five with poor growth were observed (Schäfer-Menuhr et al. 1996). Flowering of plants recovered after cryopreservation was reported to be inhibited compared to control plants, although tuberization was normal (Harding and Benson 1994).

ANTHER/MICROSPORE CULTURE

Derivation of plants of reduced ploidy (dihaploid and monoploid) by anther culture of potato has been possible since Irikura and Sakaguchi first reported success with the technique using various tuber-bearing solanaceous species in 1972. Veilleux reviewed progress with haploid extraction of potato through 1996. The following is a review of subsequent literature. Genotype specificity of the anther culture response in potato has been well documented with the frequent finding that agronomically desirable cultivars have been marginally responsive or totally recalcitrant whereas primitive cultivars or selected wild species have responded with an abundance of embryos and subsequent plantlets. Some refinements to the media have facilitated anther culture response but the selection of medium tends to be genotype specific. Aziz et al. (1999) screened 23 diploid potato clones for their response to anther culture on seven different media; they found that only seven of the 23 clones yielded anther-derived tissues and that no one medium elicited the maximum response from all seven clones. Only two of the seven responsive clones regenerated plantlets and the yield of monoploid regenerants varied with the clone. In a study of 48 potato cultivars and seven tetraploid breeding lines, Rokka et al. (1996) reported that 33 produced embryos and 23 regenerated shoots from embryos. Říhová and Tupý (1996) found that the improvement in embryogenic response in anther culture of two genotypes by substituting lactose for sucrose in the medium was only evident in medium that also contained 2,4-D. Shen and Veilleux (1995a) found that a high temperature shock (35°C for 12 h) followed by elevated incubation temperature (30°C for 16 h/20°C for 8 h) yielded 11 times as many anther-derived embryos of diploid potato clones as the control treatment (20°C). Likewise, Rokka et al. (1996) were able to increase embryo yield in anther culture by using an elevated incubation temperature (28°C compared to 20°C or 24°C); reducing incubation temperature for the regeneration step increased shoot production. Chani et al. (2000) reconfirmed the beneficial effect of temperature shock to anthers of a group of diploid hybrids between *S. chacoense* and *S. phureja* Juz. et Buk.; they also found that plants grown at an elevated greenhouse temperature (30°C day/

20°C night) and short days (12 h) yielded more embryos in anther culture. Teparkum and Veilleux (1998) did not find that colchicine improved the anther culture response of potato, even though it had been shown to be a beneficial additive to anther culture medium of other crops.

Since Uhrig (1985) first demonstrated the beneficial effect of using liquid rather than solid media for potato anther culture, liquid shake culture has become routine. However, a disadvantage is that all of the embryos produced within a flask of many cultured anthers are mostly released into the medium. Shen and Veilleux (1995b) continued to obtain embryos from cultured anthers if they were replaced into culture medium after the first embryo harvest at 6 weeks postinoculation: however the regeneration potential of the sequentially harvested embryos was reduced. Reusing the same medium generated more sequential embryos than replacing the anthers in fresh medium. This suggested that there may have been some minute embryos released into the medium that continued to develop in the prolonged culture. Such embryos may actually be secondary embryos derived by secondary embryogenesis of primary embryos. Teparkum and Veilleux (1998) examined the genetic fingerprints obtained by RAPD analysis of anther-derived plants from common flasks of anthers from a highly heterozygous potato clone; they found evidence that many of these plants fell into genetically distinct groups of genetically indistinguishable plants. Without the precaution of identifying and removing genetically similar plants, researchers using an anther-derived population for genetic studies will not have a random population due to presentation of the same genotype many times through secondary embryogenesis. Lough et al. (2001) circumvented this problem by using micro-well plates for culturing individual anthers and then selecting only a single regenerated plantlet per cultured anther. This allowed the benefit of liquid culture medium while avoiding the necessity of fingerprinting every regenerant. Skewed segregation of molecular markers in anther-derived populations of potato has been common (Tai et al. 2000; Chani et al. 2002).

Although the response to anther culture of potato is somewhat unpredictable, sufficient regenerants of both tetraploid and diploid clones have been obtained for analysis of traits in anther-derived populations. Rokka et al. (1998) obtained more than 250 dihaploid lines from a selection of the wild tetraploid, (*Solanum acaule* Bitt.). However, there was little variation for traits of interest and little fertility among the dihaploid lines. By analysis of the reaction of a population of 58 dihaploid lines extracted from *S. tuberosum* cv. Pito by anther culture, Valkonen et al. (1998) concluded that control of the hypersensitive response to infection with PVY by the dominant resistance gene *Ny* was influenced by another temperature sensitive gene. Grammatikaki et al. (1999) screened gametoclones of three anther donor genotypes for resistance to four species of RKN and found segregation for resistance among the anther-derived lines. Paz and Veilleux (1997, 1999) doubled the chromosome number of anther-derived monoploids of the diploid primitive cultivated species, *S. phureja,* to generate homozygous lines. These lines were all male sterile but varied for female fertility. After crossing them to heterozygous diploid pollinators, Paz and Veilleux (1997) used the subsequent generation in a field study. Valkonen et al. (1999) observed many dwarf mutants among anther-derived dihaploids of *S. tuberosum* cv. Pito; these mutants were deficient in GA synthesis but normal growth could be restored by exogenous application of GA.

Anther-derived dihaploids and monoploids of potato have been components of various somatic hybridization schemes. Rokka et al. (1998) fused anther-derived dihaploids of *S. tuberosum* cv. White Lady to *S. acaule* in an attempt to introgress frost tolerance and other traits of *S. acaule* into cultivated germplasm. Rokka et al. (1996) also fused dihaploids with other dihaploids of potato and regenerated mostly tetraploid somatic hybrids; reduction of the potato genome to the dihaploid level is thought to be a partial screen against lethal and deleterious genes, such that reconstructed tetraploids from dihaploid-dihaploid fusions have the potential of a reduced genetic load and hence the possibility of greater vigor. Monoploids, on the other hand, have no tolerance to lethal alleles and greatly reduced tolerance to deleterious genes, as most genotypes carrying them are expected to perish through the "monoploid sieve" (Wenzel et al. 1979). Johnson et al. (2001) constructed intermonoploid somatic hybrids from selected *S. phureja* clones. However, most of the regenerated fusions were tetraploid or hexaploid due to endoreduplicated source tissue in the monoploids or doubling during the lengthy regeneration phase of protoplast-derived calli. One of the limitations of protoplast fusion is that somatic hybrids are generally of a high ploidy status due to combined somatic chromosomes of both parents and/or chromosome doubling in culture. By synthesizing somatic hybrids upon fusion of protoplasts of androgenic dihaploids, Rokka et al. (1998a) reapplied anther culture to regenerate what were called somatohaploids, which were later studied for genomic constitution using fluorescence *in situ* hybridization (FISH). Somatohaploids were subsequently used in protoplast fusions to derive second generation somatic hybrids (Rokka et al. 2000). Gavrilenko et al. (2001) derived somatohaploids from an intergeneric somatic hybrid between *Lycopersicon esculentum* Mill. and the wild nontuberous potato species, *S. etuberosum* Lindl. Each of four anther-derived plants had a unique chromosome composition revealed by genomic *in situ* hybridization (GISH). Reduction of the chromosome number of an allotetraploid somatic hybrid to the dihaploid level may facilitate inter-genomic chromosome recombination.

Even with the genotypic limitations of anther culture response, plants derived by anther culture are finding many applications in genetic research of potato. Boluarte-Medina and Veilleux (2002) studied the response to anther culture in segregating populations derived from an interspecific hybrid between *S. chacoense* and *S. phureja*. By characterizing individual plants in backcross populations for their androgenic response, they were able to use bulk segregant analysis to identify RAPD markers associated both in coupling and repulsion to the androgenic response. This is the first step in approaching genes required for response to anther culture. Once identified, such genes may be used to transform recalcitrant genotypes into responsive ones.

PROTOPLAST CULTURE

Regeneration of potato plants from protoplasts has been possible for more than two decades and one of the earliest and most extensive studies of somaclonal variation in plants was conducted on plants regenerated from protoplasts of *S. tuberosum* cv. Russet Burbank. The original procedures described by Shepard et al. (1980) have been revised and improved over the years such that many genotypes have now been regenerated. One difficulty in protoplast isolation of potato has been the explosive growth of latent bacteria that are not obvious in routine micropropagation. Gilbert et al. (1991) described antibiotic cocktails that allowed regeneration of infected cultures if applied during plasmolysis and enzyme digestion stages of the process of protoplast isolation. Regeneration from protoplasts, however, remains genotype dependent (Cheng and Veilleux 1991) and not very predictable from other regenerative phenomena in potato (Taylor and Veilleux 1992). Coleman et al. (1991) described intraclonal variation for regenerability of protoplasts isolated from eight different lines of *S. tuberosum* cv. Record. The ability of callus to undergo xylogenesis correlated positively with shoot-forming ability of protoplast-derived callus (Barr et al. 1996). The source tissue from which protoplasts have been isolated is important. Anjum (1998) found that calluses derived from leaf mesophyll protoplasts regenerated earlier than those derived from protoplasts extracted from suspension cultures. Likewise, Szczerbakowa et al. (2000, 2001) found that protoplasts derived from leaf mesophyll were more regenerative than those derived from suspension cultures. Langille et al. (1993) attempted to direct somaclonal variation by culturing protoplasts of *S. tuberosum* cv. Russet Burbank in the presence of 5-methyltryptophan (5-MT) to improve nutritional qualities of tubers of regenerated plants. Although the plating efficiency of protoplasts was only marginally affected by the treatment, high levels of 5-MT prevented regeneration and there was no beneficial effect of selecting somaclones from the moderate treatments. Through protoplast regeneration of a diploid potato genotype that was heterozygous for major gene resistance (*R1*) to *Phytophthora infestans* and constructed to contain transposable elements *Ac* and *Ds*, van Enckevort et al. (2000) derived a population useful for searching for transposon-tagged *R1* mutants.

SOMACLONAL VARIATION

Potato was one of the first crop plants for which somaclonal variation was well-documented in the extensive descriptions of protoclones of cv. Russet Burbank (Shepard et al. 1980). Although many of these protoclones were described as having one or the other agronomic trait exceeding that of the parent cultivar, most displayed too many accompanying undesirable changes to merit continued breeding effort. Over a 5-year period, Thieme and Griess (1996) evaluated the agronomic traits of approximately 13,000 somaclones from 14 potato cultivars regenerated from stem and leaf explants. The frequency of observed variants varied with the cultivar and most of the variants correlated negatively with agronomic performance. However, 0.2-2.1% of the variants represented plants with superior performance and the variation persisted over three field generations. From 33 protoclones regenerated from *S. tuberosum* cv. Crystal, Taylor et al. (1993) selected 12 that were more resistant to *Erwinia* soft rot, 20 with improved resistance to bruising and six with improved processing characteristics compared to the parent clone. Langille et al. (1998) directed somaclonal variation toward improved nutritional quality (increased methionine) by selecting protoplast-derived calli in the presence of the amino acid analogue, ethionine; of the 48 protoclones selected, free methionine significantly increased in the tubers of six. Sebastiani et al. (1994) regenerated plants from callus derived from stem explants of cv. Desiree in the presence of culture filtrates of *Verticillium dahliae* and identified one somaclone that showed resistance similar to the resistant control. However, selection of cells in the presence of culture filtrates of *Phytophthora infestans* did not result in a higher frequency of resistant somaclones than those regenerated without the filtrates (Cerato et al. 1993). Grammatikaki et al. (1999) evaluated 46 anther-derived gametoclones from three potato genotypes for resistance to four species of RKN. Although the resistance trait showed segregation, it was retained through ploidy reduction in several gametoclones.

Attempts to associate molecular variation with somaclonal variation have met with mixed results. Binsfield et al. (1996) found isozyme variation in some somaclones regenerated from stem internodes of three potato cultivars. Using four ISSR primers, Albani and Wilkinson (1998) observed altered band profiles in only two of 40 somaclones of cv. Skirma.

Cytological abnormalities have also been observed among somaclones (Benzine-Tizroutine et al. 1993; Jelenic et al. 2001). Several studies on genetic transformation have shown somaclonal variation inadvertently introduced through transgenesis due to reliance on *Agrobacterium*-mediated transformation in leaf or tuber discs. Dale and McPartlan (1992) compared β-glucoronidase (GUS) transformed and untransformed regenerants transplanted in the field and found that both groups of plants exhibited generally negative characteristics (reduced height and tuber weight) in contrast to the micropropagated controls. However, the field performance of transgenic plants was significantly worse than for plants simply regenerated from callus, suggesting that insertional mutagenesis augments somaclonal variation. Valkonen et al. (1995) found that somaclonal variation enhanced accumulation of PVY and also increased resistance to TMV in transgenic *S. brevidens* carrying a 30 kDa gene of TMV.

SOMATIC HYBRIDS

Somatic hybridization of potato has been accomplished intraspecifically, interspecifically and even intergenerically. The somatic hybrids described exhibit genetic elements of *Lycopersicon esculentum, Solanum nigrum* L., and a wide range of both cross-compatible and cross-incompatible tuber-bearing *Solanum* spp. (Table 9.4). Considerably more research, which is briefly summarized below, has been published on somatic hybridization of potato since the recent review of the subject by Johnson and Veilleux (2001).

Table 9.4 *Potato somatic hybrids in recent literature*

Parent 1	*Parent 2*	*Purpose of fusion*	*Reference*
2x *S. tuberosum*	2x *S. tuberosum*	Genetic complementation	Gavrilenko et al. (1999)
1x *S. phureja*	1x *S. phureja*	Genetic complementation	Johnson et al. (2001)
2x *S. tuberosum*	2x *S. tuberosum*	Genetic complementation	Rokka et al.\ (1996)
2x *S. tuberosum*	2x *S. stenotomum*	BW resistance	Fock et al. (2001)
2x *S. tuberosum*	2x *S. phureja*	BW resistance	Fock et al. (2000)
2x *S. tuberosum*	2x *S. commersonii*	*Verticillium* wilt resistance, frost resistance	Cardi et al. (1999), Bastia et al. (2000)
2x *S. tuberosum*	2x *S. megistacrolobum*, 2x *S. sanctaerosae*, 2x *S. sparspilum*		Harding and Millam (1999, 2000); Mathews et al. (1999)

(Contd.)

(Contd.)

2x *S. tuberosum*	2x *S. circaeifolium*	*Phytophthora infestans* resistance	Oberwalder et al. (2000)
2x *S. tuberosum*	4x *S. acaule*	Disease and stress resistance	Rokka et al. (1998)
2x *S. tuberosum*	3x interspecific somatohaploid (4x *S. tuberosum* and 2x *S. brevidens*)	PLRV and *Erwinia* resistance	Rokka et al. (2000)
2x *S. tuberosum*	*S. nigrum, S. bulbocastanum*	LB resistance	Szczerbakowa et al. (2001)
4x *S. tuberosum*	2x *S. brevidens*	Virus resistance	Valkonen and Rokka (1998)
4x *S. tuberosum* + *Lycopersicon esculentum* fusion	2x *Lycopersicon pennellii*	Trigenomic hybrid for cytogenetic analysis	Ali et al. (2001)

The main purpose of intraspecific hybridization in a highly heterozygous species, such as potato, has been to reduce the genetic load of lethal and deleterious genes prior to reconstructing more vigorous hybrids. Tetraploid cultigens can be reduced to the dihaploid level either through androgenesis or gynogenesis with some loss of deleterious alleles. Diploid cultigens can also be reduced to the monoploid level, a process that should completely eliminate lethal and severely harmful alleles (monoploid sieve). Partial elimination of lethals in dihaploids can be followed by somatic hybridization in order to resynthesize tetraploids, as an alternative to crossing of tetraploids, a process generally expected to circumvent allelic discrimination. Gavrilenko et al. (1999) reported that 18 of 73 tetraploid intraspecific hybrids constructed from fusion combination of ten *S. tuberosum* dihaploids exceeded the performance of control cultivars in field trials. Sterility of dihaploids, a common effect of inbreeding imposed by chromosome reduction, was reversed in the somatic hybrids. Rokka et al. (1996) developed similar tetraploid somatic hybrids by fusion of anther-derived dihaploid potato lines but did not discuss field performance. Monoploid extraction of *S. tuberosum* has been accompanied by severe inbreeding depression; however, the primitive cultivated species *S. phureja* proved more amenable, yielding an array of monoploids, some of which have been field tested (Johnson et al. 2001; Lough et al. 2001). Protoplast fusion of anther-derived monoploids of *S. phureja* resulted primarily in plants with elevated ploidy (4x, 6x). In addition to elevated ploidy, the primitive traits of *S. phureja* (many small tubers per plant, elongated tubers, secondary growth, late onset of tuberization) have interfered with the ability to test the hypothesis of beneficial effects of passage of genomes through the monoploid sieve (Johnson et al. 2001; Veilleux, unpubl. data).

Both the wild diploid species *S. stenotomum* Juz. et Buk. and the primitive cultivated diploid species *S. phureja* are cross compatible with *S. tuberosum*. They can be hybridized at the diploid level with dihaploids of *S. tuberosum* or through 4x-2x hybridization by selection of unreduced male or female gametes in the diploid. However, it is not always possible to obtain desired crosses due to problems such as sterility, incompatibility, or absence of 2n gamete formation. Fock et al. (2000) developed somatic hybrids between dihaploid *S. tuberosum* and a diploid clone of *S. phureja*. Two of the ten somatic hybrids obtained expressed tolerance to two races of *Ralstonia solanacearum* (bacterial wilt) equivalent to the parent *S. phureja*. In a similar study, Fock et al. (2001) identified six tetraploid somatic hybrids between a dihaploid of *S. tuberosum* and a diploid clone of *S. stenotomum*; the hybrids expressed resistance to one race of *R. solanacearum* and tolerance to the other equivalent to that of *S. stenotomum*. The authors concluded that somatic fusion was a valid means of introducing traits from wild to cultivated species to complement conventional breeding.

Despite the possibilities of chromosomal reduction and sexual polyploidization through interspecific hybridization between potato species, several species still remain sexually isolated due to differences in endosperm balance number (EBN) or other genomic incompatibilities. Somatic hybridization is uniquely applicable for introgression of such germplasm into the cultivated potato. Somatic hybrids between dihaploid *S. tuberosum* and sexually incongruous *S. commersonii* Dun. segregated for male fertility and organellar composition, with male fertile somatic hybrids having predominantly *S. commersonii* mtDNA fragments (Cardi et al. 1999). The desirable traits of *S. commersonii* include high specific gravity, resistance to *Verticillium dahliae*, and frost tolerance. Bastia et al. (2000) found that somatic hybrids between dihaploid *S. tuberosum* and *S. commersonii* were more resistant to *Verticillium* wilt and frost than the cultivated parent. However, they preferentially inherited mtDNA fragments of *S. tuberosum* and were uniformly male sterile. By analyzing molecular marker segregation among the progeny of a single fertile tetraploid somatic hybrid between dihaploid *S. tuberosum* and *S. commersonii*, Barone et al. (2002) determined that genetic recombination had occurred between the homoeologous chromosomes.

S. sanctae-rosae Hawkes has been used as a source of nematode resistance for somatic hybridization with potato cv. Brodick (Harding and Millam 1999). Mathews et al. (1999) identified the somatic hybrids by ISSR polymorphism using anchored di-and trinucleotide primers. By digesting DNA of parents and somatic hybrids with pairs of isoschizomers, Harding and Millam (1999) demonstrated that DNA methylation patterns had changed through the somatic hybridization process. Resistance to deoxyribonuclease I activity within the rDNA nucleosomal array of some somatic hybrids suggested that sequences within the ribosomal RNA gene tandem array were not actively expressed (Harding and Millam 2000). Oberwalder et al. (2000) compared symmetric and asymmetric somatic hybrids of dihaploid *S. tuberosum* and the wild diploid Bolivian species, *S. circaeifolium* ssp. *quimense* (Bitt.). The irregular chromosome number and lack of fertility of the asymmetric hybrids limited their utility in a breeding program.

Solanum acaule, a tetraploid species that is cross incompatible with potato, is desirable for its frost tolerance and immunity to bacterial ring rot. Rokka et al. (1998a) developed hexaploid somatic hybrids between *S. acaule* 4x and *S. tuberosum* 2x. The somatic hybrids tuberized more readily in the greenhouse than the *S. acaule* parent and appeared to be both male and female fertile. Valkonen and Rokka (1998) combined different virus resistance mechanisms, the hypersensitive response of *S. tuberosum* cv. Pentland Dell and the slow virus movement of *S. brevidens* Phil., into hexaploid somatic hybrids. Resistance levels varied among six somatic hybrids even though they all had derived from a single callus. The chromosome number of the hexaploid somatic hybrid was later reduced by anther culture to regenerate triploid somatohaploids (Rokka et al. 2000). Subsequent fusion of the somatohaploids with *S. tuberosum* (2x) resulted in "second generation" pentaploid somatic hybrids. FISH analysis of the genomes of hexaploid (6x) somatic hybrids and anther-derived triploid (3x) somatohaploids revealed they were composed primarily of *S. brevidens* (Rokka et al. 1998b). The second generation somatic hybrids with a greater proportion of *S. tuberosum* germplasm retained resistance to PLRV, but PVY resistance was partially lost compared to the somatohaploid (Rokka et al. 2000). Although tuberization was greater in the second generation hybrids, compared to the somatohaploid, many still did not produce tubers under field conditions.

Intergeneric hybrids between tomato (*Lycopersicon esculentum* Mill.) and potato were one of the early successes of somatic hybridization. The sterility of such hybrids, however, has limited gene flow between the two cultivated species. By backcrossing a tetraploid potato + tomato fusion to *L. pennellii* and using ovule culture to rescue immature embryos, Ali et al. (2001) developed and studied a "trigenomic" hybrid. GISH analysis revealed homoeologous pairing among the chromosomes of all three genomes comprising the hybrid. However, it was completely sterile even after crossing the female parent with many different pollinators. Hexaploids derived by chromosome doubling of the trigenomic hybrid were also sterile. Tetraploid (amphidiploid) somatic hybrids between *L. esculentum* and *S. etuberosum* (2x), a nontuberous species, formed a low frequency of multivalents as determined by GISH analysis (Gavrilenko et al. 2001). Tomato chromosomes were preferentially eliminated during meiosis.

CONCLUSION

Plant cell and tissue culture techniques have become commonplace in production practices and genetic research on potato. Even traditional potato breeding involving crossing, evaluation, and selection of seedlings, followed by years of evaluation prior to release of a new cultivar rely on tissue culture propagation and preservation of disease-free stock. Genetic transformation depends upon cell culture regeneration systems. Reduction of tetraploid potato to the dihaploid and monoploid levels can simplify genetic analyses. And the introgression of desirable traits through somatic hybridization will eventually find its way into cultivar improvement. Perhaps more than with any other crop, researchers concentrating on potato need a thorough understanding of cell and tissue culture techniques for continued improvement in production methodology and cultivar development.

REFERENCES

Albani, M.C. and M.J. Wilkinson. 1998. Intersimple sequence repeat polymerase chain reaction for the detection of somaclonal variation. Plant Breed 117: 573-575.

Ali, S.N.H., D.J. Huigen, M.S. Ramanna, E. Jacobsen, and R.G.F. Visser. 2001. Genomic *in situ* hybridization analysis of a trigenomic hybrid involving *Solanum* and *Lycopersicon* species. Genome 44: 299-304.

Anjum, M.A. 1998a. Effect of protoplast source and media on growth and regenerability of protoplast-derived calluses of *Solanum tuberosum* L. Acta Physiol Plant 20: 129-133.

Anjum, M.A. 1998b. Selection of hydroxyproline-resistant cell lines from *Solanum tuberosum* L. callus, II: Plant regeneration and frost tolerance of regenerated plants. Acta Biotechnol 18: 361-366.

Anjum, M.A. 2001. Cytology of potato callus cells in relation to their frost hardiness. Biol Plant 44: 325-331.

Aziz, A.N., J.E.A. Seabrook, G.C.C. Tai, and H. De Jong. 1999. Screening diploid *Solanum* genotypes responsive to different anther culture conditions and ploidy assessment of anther-derived roots and plantlets. Amer J Potato Res 76: 9-16.

Baker, C.J., E.W. Orlandi, and K.L. Deahl. 2001. Oxidative metabolism in plant/bacteria interactions: characterization of a unique oxygen uptake response of potato suspension cells. Physiol Mol Plant Pathol 59: 25-32.

Barone, A., J. Li, A. Sebastiano, T. Cardi, and L. Frusciante. 2002. Evidence for tetrasomic inheritance in a tetraploid *Solanum commersonii* (+) *S. tuberosum* somatic hybrid through the use of molecular markers. Theor Appl Genet 104: 539-546.

Barr, S.N.R., L.A. Payne, M.F.B. Dale, and M.J. Wilkinson. 1996. Predictive correlates of shoot regeneration from potato protoplast culture. Plant Cell Rep 15: 350-354.

Bastia, T., N. Carotenuto, B. Basile, A. Zoina, and T. Cardi. 2000. Induction of novel organelle DNA variation and transfer of resistance to frost and *Verticillium* wilt in *Solanum tuberosum* through somatic hybridization with 1EBN *S. commersonii*. Euphytica 116: 1-10.

Benzine-Tizroutine, S., L. Rossignol, M. Rossignol, A. Ambroise, and C. Gaisne. 1993. Somaclonal variation in potato —phenomenons correlated with flowering. Acta Bot Gall 140: 5-16.

Binsfeld, P.C., J.A. Peters, and E. Augustin. 1996. Isoenzymatic variation in potato somaclones (*Solanum tuberosum* L). Braz J Genet 19: 117-121.

Birhman, R.K., G. Laublin, and M. Cappadocia. 1994. Genetic control of *in vitro* shoot regeneration from leaf explants in *Solanum chacoense* Bitt. Theor Appl Genet 88: 535-540.

Boluarte-Medina, T. and R.E. Veilleux. 2002. Phenotypic characterization and bulk segregant analysis of anther culture response in two backcross families of diploid potato—RAPD markers for androgenesis in potato. Plant Cell, Tissue Organ Cult 68: 277-286.

Boxus, P. 1999. Proc Conf on Potato Seed Production by Tissue Culture, held at Brussels, Belgium, 25th-28th February, 1998 within the framework of COST 822. Development of integrated systems for large-scale production of elite plants using *in vitro* techniques, granted by the European Commission, Brussels, Belgium, Preface. Potato Res 42: 409-409.

Cardi, T., T. Bastia, L. Monti, and E.D. Earle. 1999. Organelle DNA and male fertility variation in *Solanum* spp. and interspecific somatic hybrids. Theor Appl Genet 99: 819-828.

Carputo, D., T. Cardi, T. Chiari, G. Ferraiolo, and L. Frusciante. 1995. Tissue culture response in various wild and cultivated *Solanum* germplasm accessions for exploitation in potato breeding. Plant Cell, Tissue Organ Cult 41: 151-158.

Cerato, C., L.M. Manici, S. Borgatti, R. Alicchio, R. Ghedini, and A. Ghinelli. 1993. Resistance to late blight (*Phytophthora infestans* (Mont.) Debary) of potato plants regenerated from *in vitro* selected calli. Potato Res 36: 341-351.

Chani, E., R.E. Veilleux, and T. Boluarte-Medina. 2000. Improved androgenesis of interspecific potato and efficiency of SSR markers to identify homozygous regenerants. Plant Cell, Tissue Organ Cult 60: 101-112.

Chani, E., V. Ashkenazi, J. Hillel, and R.E. Veilleux. 2002. Microsatellite marker analysis of an author-derived potato family; skewed segregation and gene-centromere mapping. Genome 45: 236-242.

Cheng, J. and R.E. Veilleux. 1991. Genetic analysis of protoplast culturability in *Solanum phureja*. Plant Sci 75: 257-265.

Coleman, M., R. Waugh, and W. Powell. 1990. Genetic analysis of *in vitro* cell and tissue culture response in potato. Plant Cell, Tissue Organ Cult 23: 181-186.

Coleman, W.K., D.J. Donnelly, and S.E. Coleman. 2001. Potato microtubers as research tools: A review. Amer J Potato Res 78: 47-55.

Coleman, M., P. Davie, J. Vessey, and W. Powell. 1991. Intraclonal genetic variation for protoplast regenerative ability within *Solanum tuberosum* cv. Record Ann Bot 67: 459-461.

Collings, D.A. and M.C. Emons. 1999. Microtubule and actin filament organization during acentral divisions in potato suspension culture cells. Protoplasma 207: 158-168.

Dale, P.J. and H.C. McPartlan. 1992. Field performance of transgenic potato plants compared with controls regenerated from tuber discs and shoot cuttings. Theor Appl Genet 84: 585-591.

Dale, P.J. and K.K. Hampson. 1995. An assessment of morphogenic and transformation efficiency in a range of varieties of potato (*Solanum tuberosum* L.). Euphytica 85: 101-108.

Degarcia, E. and G. Martinez. 1995. Somatic embryogenesis in *Solanum tuberosum* L. cv. Desiree from stem nodal sections. J Plant Physiol 145: 526-530.

Dey, P.M., M.D. Brownleader, A.T. Pantelides, M. Trevan, J.J. Smith, and G. Saddler. 1997. Extension from suspension-cultured potato cells: A hydroxyproline-rich glycoprotein, devoid of agglutinin activity. Planta 202: 179-187.

Dhingra, M.K., S.M.P. Khurana, T.N. Lakhanpal, and R. Chandra. 1991. Effect of cytokinins and light on the growth and virus content of potato leaf callus. Natl Acad Sci Lett, India 14: 117-120.

Dörnenburg, H. and D. Knorr. 1997. Evaluation of elicitor-and high-pressure-induced enzymatic browning utilizing potato (*Solanum tuberosum*) suspension cultures as a model system for plant tissues. J Agric Food Chem 45: 4173-4177.

Fiegert, A.K., G. Mix-Wagner, and K.D. Vorlop. 2000. Regeneration of *Solanum tuberosum* L. cv. Tomensa: induction of somatic embryogenesis in liquid culture for the production of "artificial seed". Landbauforsch Volk 50: 199-202.

Fock, I., C. Collonnier, A. Purwito, J. Luisetti, V. Souvannavong, F. Vedel, A. Servaes, A. Ambroise, H. Kodja, G. Ducreux, and D. Sihachakr. 2000. Resistance to bacterial wilt in somatic hybrids between *Solanum tuberosum* and *Solanum phureja*. Plant Sci 160: 165-176.

Fock, I., C. Collonnier, J. Luisetti, A. Purwito, V. Souvannavong, F. Vedel, A. Servaes, A. Ambroise, H. Kodja, G. Ducreux, and D. Sihachakr. 2001. Use of *Solanum stenotomum* for introduction of resistance to bacterial wilt in somatic hybrids of potato. Plant Physiol Biochem 39: 899-908.

Gavrilenko, T., R. Thieme, and H. Tiemann. 1999. Assessment of genetic and phenotypic variation among intraspecific somatic hybrids of potato, *Solanum tuberosum* L. Plant Breed 118: 205-213.

Gavrilenko, T., R. Thieme, and V.M. Rokka. 2001. Cytogenetic analysis of *Lycopersicon esculentum* + *Solanum etuberosum* somatic hybrids and their androgenetic regenerants. Theor Appl Genet 103: 231-239.

Gilbert, J.E., S. Shohet, and P.D.S. Caligari. 1991. The use of antibiotics to eliminate latent bacterial contamination in potato tissue cultures. Ann Appl Biol 119: 113-120.

Gopal, J. and J.L. Minocha. 1998. Effectiveness of *in vitro* selection for agronomic characters in potato. Euphytica 103: 67-74.

Grammatikaki, G., N. Vovlas, P.J. Kaltsikes, and A. Sonnino. 1999. Response of potato gametoclones to infection of four root-knot nematode (*Meloidogyne*) species. Russ J Nematol 7: 155-159.

Hagen, S.R., and P. Muneta. 1993. Effect of temperature on carbohydrate content, ADP glucose pyrophosphorylase, and ATP-dependent and PPI-dependent phosphofructokinase activity of potato-tuber callus-tissue. Plant Cell, Tissue Organ Cult 32: 115-121.

Hagen, S.R., P. Muneta, J. Augustin, and D. Letourneau. 1991. Stability and utilization of picloram, vitamins, and sucrose in a tissue culture medium. Plant Cell, Tissue Organ Cult 25: 45-48.

Hagen, S.R., S. Harrison, P. Muneta, and D. Letourneau. 1993. Methods to increase the starch content of potato tuber callus tissue. Potato Res 36: 293-299.

Hansen, J., B. Nielsen, and S.V.S. Nielsen. 1999. *In vitro* shoot regeneration of *Solanum tuberosum* cultivars: interactions of medium composition and leaf, leaflet, and explant position. Potato Res 42: 141-151.

Harding, K. 1991. Molecular stability of the ribosomal RNA genes in *Solanum tuberosum* plants recovered from slow growth and cryopreservation. Euphytica 55: 141-146.

Harding, K. 1994. The methylation status of DNA derived from potato plants recovered from slow growth. Plant Cell, Tissue Organ Cult 37: 31-38.

Harding, K. and E.E. Benson. 1994. A study of growth, flowering, and tuberization in plants derived from cryopreserved potato shoot tips—implications for *in vitro* germplasm collections. Cryo-Lett 15: 59-66.

Harding, K. and S. Millam. 1999. Analysis of ribosomal RNA genes in somatic hybrids between wild and cultivated *Solanum* species. Mol Breed 5: 11-20.

Harding, K. and E.E. Benson. 2000. Analysis of nuclear and chloroplast DNA in plants regenerated from cryopreserved shoot tips of potato. Cryo-Lett 21: 279-288.

Harding, K. and S. Millam. 2000. Analysis of chromatin, nuclear DNA and organelle composition in somatic hybrids between *Solanum tuberosum* and *Solanum sanctae-rosae*. Theor Appl Genet 101: 939-947.

Hirai, D. and A. Sakai. 1999. Cryopreservation of *in vitro*-grown meristems of potato (*Solanum tuberosum* L.) by encapsulation-vitrification. Potato Res 42: 153-160.

Hoogkamp, T.J.H., R.G.T. Van den Ende, E. Jacobsen, and R.G.F. Visser. 2000. Development of amylose-free (amf) monoploid potatoes as new basic material for mutation breeding *in vitro*. Potato Res 43: 179-189.

Hulme, J.S., E.S. Higgins, and R. Shields. 1992. An efficient genotype-independent method for regeneration of potato plants from leaf tissue. Plant Cell, Tissue Organ Cult 31: 161-167.

Irikura, Y. and S. Sakaguchi. 1972. Induction of 12-chromosome plants from anther culture in a tuberous *Solanum*. Potato Res 15: 170-173.

Jan, V.V., G. Laublin, R.K. Birhman, and M. Cappadocia. 1996. Genetic analysis of leaf explant regenerability in *Solanum chacoense*. Plant Cell, Tissue Organ Cult 47: 9-13.

JayaSree, T., U. Pavan, M. Ramesh, A.V. Rao, K.J.M. Reddy, and A. Sadanandam. 2001. Somatic embryogenesis from leaf cultures of potato. Plant Cell, Tissue Organ Cult 64: 13-17.

Jelenic, S., J. Berljak, D. Papes, and S. Jelaska. 2001. Mixoploidy and chimeric structures in somaclones of potato (*Solanum tuberosum* L.) cv. Bintje. Food Technol Biotechnol 39: 13-17.

Johnson, A.A.T. and R.E. Veilleux. 2001. Somatic hybridization and applications in plant breeding. Plant Breed Rev 20: 167-225.

Johnson, A.A.T., S.M. Piovano, V. Ravichandran, and R.E. Veilleux. 2001. Selection of monoploids for protoplast fusion and generation of intermonoploid somatic hybrids of potato. Amer J Potato Res 78: 19-29.

Keller, H., H. Hohlfeld, V. Wray, K. Hahlbrock, D. Scheel, and D. Strack. 1996. Changes in the accumulation of soluble and cell wall-bound phenolics in elicitor-treated cell suspension cultures and fungus-infected leaves of *Solanum tuberosum*. Phytochemistry 42: 389-396.

Kosegarten, H., K. Zetsche, and K. Mengel. 1995. Isolation of intact storage tissue amyloplasts from suspension-cultured potato cells (*Solanum tuberosum*) and determination of their intermembrane and stroma volumes. J Appl Bot-Angew Bot 69: 211-214.

Langille, A.R., K.L. Prouty, and W.A. Halteman. 1993. Effects of the amino acid analog, 5-methyltryptophan on protoplast survival, plating efficiency and free typtophan levels in tubers of regenerated potato plants. Amer Potato J 70: 735-741.

Langille, A.R., Y. Lan, and D.L. Gustine. 1998. Seeking improved nutritional properties for the potato: Ethionine-resistant protoclones. Amer J Potato Res 75: 201-205.

Lopez-Delgado, H., M. Jimenez-Casas, and I.M. Scott. 1998. Storage of potato microplants *in vitro* in the presence of acetylsalicylic acid. Plant Cell, Tissue Organ Cult 54: 145-152.

Lough, R.C., J.M. Varrieur, and R.E. Veilleux. 2001. Selection inherent in monoploid derivation mechanisms for potato. Theor Appl Genet 103: 178-184.

Lozoya-Saldana, H. 1992. Photoperiod, gibberellic acid, and kinetin, in potato flower differentiation *in vitro*. Amer Potato J 69: 265-274.

Mathews, D., J. McNicoll, K. Harding, and S. Millam. 1999. 5'-Anchored simple-sequence repeat primers are useful for analysing potato somatic hybrids. Plant Cell Rep 19: 210-212.

Mix Wagner, G. 1999. The conservation of potato cultivars. Potato Res 42: 427-436.

Murashige, T. and F. Skoog. 1962. A revised medium for rapid growth and bioassays with tobacco tissue cultures. Physiol Plant 15: 473-497.

Nadolska-Orczyk, A., L. Milkowska, A. Palucha, P. Czembor, and W. Orczyk. 1995. Regeneration and transformation of Polish cultivars of potato. Acta Soc Bot Pol 64: 335-340.

Naik, P.S. and J.M. Widholm. 1993. Comparison of tissue culture and whole plant responses to salinity in potato. Plant Cell, Tissue Organ Cult 33: 273-280.

Oberwalder, B., L. Schilde-Rentschler, B. Loffelhardt-Ruoss, and H. Ninnemann. 2000. Differences between hybrids of *Solanum tuberosum* L. and *Solanum circaeifolium* Bitt. obtained from symmetric and asymmetric fusion experiments. Potato Res 43: 71-82.

Ochatt, S.J., P.L. Marconi, S. Radice, P.A. Arnozis, and O.H. Caso. 1998. *In vitro* recurrent selection of potato: production and characterization of salt tolerant cell lines and plants. Plant Cell, Tissue Organ Cult 55: 1-8.

Opatrna, J., P. Novak, and Z. Opatrny. 1997. Paclobutrazol stimulates bud regeneration in *Solanum tuberosum* L. primary explant cultures. Biol Plant 39: 151-158.

Patel, A.V., I Pusch, G. Mix-Wagner, and K.D. Vorlop. 2000. A novel encapsulation technique for the production of artificial seeds. Plant Cell Rep 19: 868-874.

Pavingerová, D., J. Briza, and H. Niedermeierová. 2001. Timing of transportation of *Ac* mobile element in potato. Biol Plant 44: 347-353.

Paz, M.M. and R.E. Veilleux. 1997. Genetic diversity based on randomly amplified polymorphic DNA (RAPD) and its relationship with the performance of diploid potato hybrids. J Amer Soc Hortic Sci 122: 740-747.

Paz, M.M. and R.E. Veilleux. 1999. Influence of culture medium and *in vitro* conditions on shoot regeneration in *Solanum phureja* monoploids and fertility of regenerated doubled monoploids. Plant Breed 118: 53-57.

Polkowska-Kowalczyk, L. and U. Maciejewska. 2001. The oxidative processes induced in cell suspensions of *Solanum* species by culture filtrate of *Phytophthora infestans*. Z Naturforsch (C) 56: 235-244.

Ranalli, P. 1997. Innovative propagation methods in seed tuber multiplication programmes. Potato Res 40: 439-453.

Říhová, L. and J. Tupý. 1996. Influence of 2,4-D and lactose on pollen embryogenesis in anther culture of potato. Plant Cell, Tissue Organ Cult 45: 269-272.

Rokka, V.M., L. Pietila, and E. Pehu. 1996. Enhanced production of dihaploid lines via anther culture of tetraploid potato (*Solanum tuberosum* L. ssp. *tuberosum*) clones. Amer Potato J 73: 1-12.

Rokka, V.M., C.A. Ishimaru, N.L.V. Lapitan, and E. Pehu. 1998a. Production of androgenic dihaploid lines of the disomic tetraploid potato species *Solanum acaule* ssp. *acaule*. Plant Cell Rep 18: 89-93.

Rokka, V.M., N.L.V. Lapitan, D.L. Knudson, and E. Pehu. 1998b. Fluorescence *in situ* hybridization of potato somatohaploids and their somatic hybrid donors using two *Solanum brevidens* specific sequences. Agric Food Sci Finland 7: 31-38.

Rokka, V.M., A. Tauriainen, L. Pietila, E. Pehu. 1998c. Interspecific somatic hybrids between wild potato *Solanum acaule* Bitt. and anther-derived dihaploid potato (*Solanum tuberosum* L.). Plant Cell Rep 18: 82-88.

Rokka, V.M., Y.S. Xu, P. Tanhuanpaa, L. Pietila, and E. Pehu. 1996. Electrofusion of protoplasts of anther-derived dihaploid lines of commercial potato cultivars. Agric Food Sci Finland 5: 449-460.

Rokka, V.M., J.P.T. Valkonen, A. Tauriainen, L. Pietila, R. Lebecka, E. Zimnoch-Guzowska, and E. Pehu. 2000. Production and characterization of "second generation" somatic hybrids derived from protoplast fusion between interspecific somatohaploid and dihaploid *Solanum tuberosum* L. Amer J Potato Res 77: 149-159.

Sarkar, D. and P.S. Naik. 1998a. Cryopreservation of shoot tips of tetraploid potato (*Solanum tuberosum* L.) clones by vitrification. Ann Bot 82: 455-461.

Sarkar, D. and P.S. Naik. 1998b. Factors affecting minimal growth conservation of potato microplants *in vitro*. Euphytica 102: 275-280.

Sarkar, D. and P.S. Naik. 1998c. Nutrient-encapsulation of potato nodal segments for germplasm exchange and distribution. Biol Plant 40: 285-290.

Sarkar, D. and P.S. Naik. 1998d. Synseeds in potato: an investigation using nutrient-encapsulated *in vitro* nodal segments. Sci Hortic 73: 179-184.

Sarkar, D., S.K. Kaushik, and P.S. Naik. 1999. Minimal growth conservation of potato microplants: silver thiosulfate reduces ethylene-induced growth abnormalities during prolonged storage *in vitro*. Plant Cell Rep 18: 897-903.

Sarkar, D., S.K. Chakrabarti, and P.S.Naik. 2001. Slow-growth conservation of potato microplants: efficacy of ancymidol for long-term storage *in vitro*. Euphytica 117: 133-142.

Sarkar, D., K.C. Sud, S.K. Chakrabarti, and P.S. Naik. 2002. Growing of potato microplants in the presence of alginate-silver thiosulfate capsules reduces ethylene-induced culture abnormalities during minimal growth conservation *in vitro*. Plant Cell, Tissue Organ Cult 68: 79-89.

Schäfer-Menuhr, A., E. Muller, and G. Mix-Wagner. 1996. Cryopreservation: an alternative for the long-term storage of old potato varieties. Potato Res 39: 507-513.

Schenk, R.V. and A.C. Hildebrandt. 1972. Medium and techniques for induction and growth of monocotyledonous and dicotyledonous plant cell cultures. Can J Bot 50: 199-204.

Schmidt, A.,D. Scheel, and D. Strack. 1998. Elicitor-stimulated biosynthesis of hydroxycinnamoyltyramines in cell suspension cultures of *Solanum tuberosum*. Planta 205: 51-55.

Seabrook, J.E.A and L.K. Douglass. 2001. Somatic embryogenesis on various potato tissues from a range of genotypes and ploidy levels. Plant Cell Rep 20: 175-182.

Seabrook, J.E.A., L.K. Douglass, and G.C.C. Tai. 2001. Segregation for somatic embryogenesis on stem-internode explants from potato seedlings. Plant Cell, Tissue Organ Cult 65: 69-73.

Sebastiani, L., A. Lenzi, C. Pugliesi, and M. Fambrini. 1994. Somaclonal variation for resistance to *Verticillium dahliae* in potato (*Solanum tuberosum* L.) plants regenerated from callus. Euphytica 80: 5-11.

Shen, L.Y. and R.E. Veilleux. 1995a. Effect of temperature shock and elevated incubation temperature on androgenic embryo yield of diploid potato. Plant Cell, Tissue Organ Cult 43: 29-35.

Shen, L.Y. and R.E. Veilleux. 1995b. Yield of embryos from sequential anther culture. Amer Potato J 72: 689-700.

Shepard, J.F., D. Bidney, and E. Shahin. 1980. Potato protoplasts in crop improvement. Science 208: 17-24.

Struik, P.C. and S.G. Wiersema. 1999. Seed Potato Technology. Wageningen Press, Wageningen, Netherlands.

Svetek, J., B. Kirn, B. Vihar, and M. Schara. 1999. Lateral domain diversity in membranes of callus and root cells of potato as revealed by EPR spectroscopy. Physiol Plant 105: 499-505.

Szczerbakowa, A., M. Borkowska, and B. Wielgat. 2000. Plant regeneration from the protoplasts of *Solanum tuberosum*, *S. nigrum* and *S. bulbocastanum*. Acta Physiol Plant 22: 3-10.

Szczerbakowa, A., U. Maciejewska, P. Pawlowski, J.S. Skierski, and B. Wielgat. 2001. Electrofusion of protoplasts from *Solanum tuberosum*, *S. nigrum* and *S-bulbocastanum*. Acta Physiol Plant 23: 169-179.

Tai, G.C.C., J.E.A. Seabrook, and A.N. Aziz. 2000. Linkage analysis of anther-derived monoploids showing distorted segregation of molecular markers. Theor Appl Genet 101: 126-130.

Taylor, R.J., G.A. Secor, C.L. Ruby, and P.H. Orr. 1993. Tuber yield, soft rot resistance, bruising resistance and processing quality in a population of potato (cv. Crystal) somaclones. Amer Potato J 70: 117-130.

Taylor, T.E. and R.E. Veilleux. 1992. Inheritance of competences for leaf disk regeneration, anther culture, and protoplast culture in *Solanum phureja* and correlations among them. Plant Cell, Tissue Organ Cult 31: 95-103.

Teparkum, S. and R.E. Veilleux. 1998. Indifference of potato anther culture to colchicine and genetic similarity among anther-derived monoploid regenerants determined by RAPD analysis. Plant Cell, Tissue Organ Cult 53: 49-58.

Thieme, R. and H. Griess. 1996. Somaclonal variation of haulm growth, earliness, and yield in potato. Potato Res 39: 355-365.

Uhrig, H. 1985. Genetic selection and liquid medium conditions improve the yield of androgenetic plants from diploid potatoes. Theor Appl Genet 71: 455-460.

Valkonen, J.P.T. and V.M. Rokka. 1998. Combination and expression of two virus resistance mechanisms in interspecific somatic hybrids of potato. Plant Sci 131: 85-94.

Valkonen, J.P.T., V.M. Rokka, and K.N. Watanabe. 1998. Examination of the leaf-drop symptom of virus-infected potato using anther culture-derived haploids. Phytopathology 88: 1073-1077.

Valkonen, J.P.T., K. Koivu, S.A. Slack, and E. Pehu. 1995. Modified resistance of *Solanum brevidens* to potato Y potyvirus and tobacco mosaic tobamovirus following genetic transformation and explant regeneration. Plant Sci 106: 71-79.

Valkonen, J.P.T., T. Moritz, K.N. Watanabe, and V.M. Rokka. 1999. Dwarf (di) haploid Pito mutants obtained from a tetraploid potato cultivar (*Solanum tuberosum* subsp. *tuberosum*) via anther culture are defective in gibberellin biosynthesis. Plant Sci 149: 51-57.

van Enckevort, L.J.G., J.E.M. Bergervoet, W.J. Stiekema, A. Pereira, and E. Jacobsen. 2000. Selection of independent *Ds* transposon insertions in somatic tissue of potato by protoplast regeneration. Theor Appl Genet 101: 503-510.

Veilleux, R.E. 1996. Haploidy in important crop plants—potato. *In:* S.M. Jain, S.K. Sopory and R.E. Veilleux (eds.), *In Vitro* Haploid Production in Higher Plants, vol. 3. Kluwer Acad Publ, Dordrecht, Netherlands, pp. 37-49.

Veilleux R.E. and A.A.T. Johnson. 1998. Somaclonal variation: molecular evidence and utilization in plant breeding. Plant Breed Rev 16: 229-268.

Véronneau, H., G. Lavoie, and M. Cappadocia. 1992. Genetic analysis of anther and leaf disk culture in 2 clones of *Solanum chacoense* Bitt. and their reciprocal hybrids. Plant Cell, Tissue Organ Cult 30: 199-209.

Wenzel, G.,O. Schieder, T. Przewozny, S.K. Sopory, and G. Melchers. 1979. Comparison of single cell culture derived *Solanum tuberosum* L. plants and a model for their application in breeding programs. Theor Appl Genet 55: 49-55.

Yee, S., B Stevens, S. Coleman, J.E.A. Seabrook, and X.Q. Li. 2001. High-efficiency regeneration *in vitro* from potato petioles with intact leaflets. Amer J Potato Res 78: 151-157.

Zel, J. and M.M. Medved. 1999. The efficient regeneration of the potato (*Solanum tuberosum* L.) cv. Igor *in vitro*. Phyton-Ann REI Bot 39: 277-282.

10

Starch-Sugar Metabolism in Potato (*Solanum tuberosum* L.) Tubers in Response to Temperature Variations

THEOPHANES SOLOMOS[1] AND AUTAR K. MATTOO[2]

[1]*Department of Natural Resource Sciences and Landscape Architecture, University of Maryland, College Park, MD 20742, USA*

[2]*Vegetable Laboratory, Building 010A, U.S. Department of Agriculture, Beltsville Agricultural Research Center, Beltsville, MD 20705-2350, USA*

INTRODUCTION

Potato (*Solanum tuberosum* L.) is of paramount importance worldwide. The planet's potato crop ranks fourth after rice, wheat, and maize (Horton and Sawyer 1985). About 50 percent of the US crop is mainly processed into French fries and chips (USDA statistics). Potatoes are stored for considerable lengths of time, usually at 8-15°C, both to prevent seasonal gluts and to ensure their availability throughout the year (Jadhav et al. 1991). During storage at these temperatures, however, potatoes sprout. In addition, their processing quality decreases because of the increase in reducing sugars due to senescence (Burton 1969). Sprouting greatly diminishes the storage life of potatoes. The breaking of tuber dormancy varies with the cultivar and season. It usually occurs between 6 and 12 weeks after harvest (Hemberg 1985). At present, sprouting is controlled chemically with a mixture of propham (isopropyl-N-phenylcarbamate-IPC) and chloropham (chloroisopropyl-N-phenyl carbamate-CIPC) (Corsini et al. 1978; Mondy et al. 1978). These chemicals are considered a health hazard and in some European countries their use is restricted. In view of the enormous economic and nutritional significance of potatoes, it is clear that nonchemical control of sprouting is of great significance to the potato industry.

Corresponding author: Autar K. Mattoo. Tel: 01-301-504-7380, fax: 01-301-504-5555. *e-mail: mattooa@ba.ars.usda.gov.*

Storage of potatoes at cooler temperatures (2-4°C) inhibits sprouting, decreases losses due to microbial spoilage, and prevents weight-loss caused by dehydration. At these temperatures, however, reducing sugars—namely, glucose and fructose—accumulate (Isherwood 1973; Coffin et al. 1987; Sowokinos et al. 1987; Barichello et al. 1990; Sowokinos 1990; Hill et al. 1996; Zhou and Solomos 1998). Reducing sugars also accumulate at 18-20°C in response to the plant hormone ethylene (C_2H_4) (Solomos and Laties 1975). Color is one of the most important quality factors of these products (Smith 1987). Tubers containing high levels of reducing sugars produce an unacceptable dark color when fried in oil, due to the chemical reaction of reducing sugars with the amino groups of amino acids (Maillard reaction) (Burton 1969; Coffin et al. 1987). Methods used to decrease the level of reducing sugars after cold storage have had only partial success (Burton 1978). Extensive breeding research has not yet produced a cultivar that can be stored at low temperatures without the accompanying rise in reducing sugars.

This chapter is a synthesis of studies on the regulation of enzymes involved in the starch-sugar interconversion during tuber storage at low temperatures and subsequent reconditioning at higher temperatures.

RESPIRATION

Dizengremel (1985) described the characteristics of potato tuber respiration. The rate of CO_2 output from tubers kept at 10°C remains constant for long periods of time (Barker 1968; Isherwood 1973; ap Rees et al. 1981; Zhou and Solomos 1998). In contrast, the respiratory rate of tubers transferred to lower temperatures exhibits a distinctive pattern: it decreases initially, then increases rapidly, reaching a peak two- to threefold higher than at 10°C. Thereafter, it gradually decreases to low levels which, depending on cultivar and time, could be lower than at 10°C (Barker 1968; Isherwood 1973; ap Rees et al. 1981; Zhou and Solomos 1998).

The rate of O_2 uptake in uncoupled mitochondria isolated from tubers kept at 10°C remained constant for long periods. However, the rate in mitochondria isolated from tubers transferred to 1°C decreased initially, then increased within 5-6 days to a rate similar to that of tubers kept at 10°C. This pattern is associated with a decrease in cytochrome oxidase capacity and a concomitant increase in alternative oxidase, resulting in a total specific mitochondrial activity similar to that in tubers kept in air at 10°C (Zhou and Solomos 1998). The increase in alternative oxidase is due to *de novo* synthesis (unpubl. observations).

An increase in respiration, sugar accumulation, and induction of the alternative oxidase was observed in tubers treated with either ethylene (C_2H_4) or cyanide at 20°C (Solomos and Laties 1975). The rise in respiration at low temperatures and in response to C_2H_4 is associated with a decrease in metabolites phosphoglyceric acid (3-PGA) and phosphoenol pyruvate (PEP) (Barker 1968; Pollock and ap Rees 1975; Solomos and Laties 1975; Isherwood 1976; Zhou and Solomos 1998). A similar decrease in these metabolites in growing tubers exposed to high temperatures and water stress was accompanied by an inhibition of starch biosynthesis (Geigenberger et al. 1997, 1998). The sharp decline in PEP and 3-PGA is usually linked with an increase in the glycolytic flux (Paxton 1996).

The pattern of CO_2 output is closely related to sucrose biosynthesis (Isherwood 1973; Zhou and Solomos 1998). In fact, the changes effected by lowering the storage temperature from 10°C to 2°C, or by raising it from 2°C to 10°C, in the "King Edward" cultivar are quantitatively and respectively related to either the synthesis of sucrose (10°C to 2°C) or to its conversion to starch (2°C to 10°C). In addition, the rate of respiration in low-sugar cultivars is lower than in those with high sweetening capacity.

KINETICS OF SUGAR ACCUMULATION

Transfer from 10°C to 1-4°C

Starch is the sole source of sugars and respiratory CO_2 during sweetening at low temperatures (Isherwood 1973). Further, the first sugar seen to accumulate is sucrose, followed, after an appreciable delay, depending on cultivar and temperature, by a rise in glucose and fructose (Isherwood 1973; Pollock and ap Rees 1975; Hill et al. 1996; Zhou and Solomos 1998). For instance, in "Russet Burbank" tubers, sucrose content increased after four days at 1°C, whereas the rise in glucose and fructose occurred after 12 days. Moreover, the ratio of the increment of glucose/fructose was 1.05, indicating that sucrose was their immediate precursor (Zhou and Solomos 1998). Similar observations concerning the glucose/fructose ratio have been reported with cultivars other than "Russet Burbank" (Sowokinos 1990; Sowokinos et al. 1987). Changes in sucrose concentration are biphasic, i.e., it increases rapidly, then peaks, after which it gradually declines and, depending on cultivar, temperature and storage duration, tendes to approach the levels found in tubers kept at 10°C (Isherwood 1973; Zhou and Solomos 1998). The content of reducing sugars also increases rapidly, then decreases with time, although the level of reducing sugars remains high and does not decrease during storage at low temperatures (Isherwood 1973; Zhou and Solomos 1998).

ENZYMATIC PATTERNS DURING SWEETENING

The enzymes involved in starch-sugar conversion are well known and have been extensively studied (Isherwood 1976; Preiss 1988; Sowokinos 1990; Smith et al. 1997; Geigenberger 2003). Figure 10.1 presents the reactions they catalyze and the cellular localization where they function.

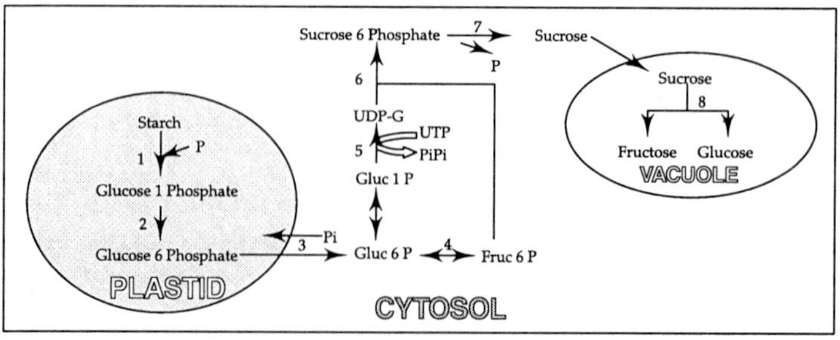

Fig. 10.1 1: Starch phosphorylase, 2: Phosphoglucomutase, 3: Hexose P/Pi translocator, 4: Phosphohexose isomerase, 5: UDP-glucose pyrophosphorylase, 6: Sucrose phosphate synthase, 7: Sucrose phosphate phosphatase, 8: Invertase

Starch Phosphorylase

The kinetics of sugar accumulation indicate that starch is converted to sugars via starch phosphorylase (α-1, 4-glucan phosphorylase (EC 2.4.1.1)). Specifically, the kinetics show that: (a) sucrose is the first sugar to accumulate; (b) increase in reducing sugars—glucose and fructose—begins after 8-12 days, depending on cultivar and storage temperature; (c) glucose and fructose accumulate together and their ratio is close to unity, indicating that sucrose is their immediate precursor; and (d) there is no increase in the usual products of amylolytic starch degradation, i.e., maltose or glucose polymers higher than maltose (Preiss 1982; Dennis and Blakeley 2000).

In higher plants, there are two types of phosphorylases, designated Pho 1 and Pho 2 respectively. These were previously designated as L and H types (Sonnewald et al. 1995). Pho 1 is located in the plastids, and Pho 2 in the cytosol (Duwenig et al. 1997). In potato, two isoforms-Pho 1a and Pho 1b are expressed in the plastids (Steup 1988; Sonnewald et al. 1995). It is not yet clear whether cold temperatures change the expression of phosphorylases in tubers. However, neither the extractable activity nor the two isoforms present in tubers kept at 10°C change during storage at low temperature (Kennedy and Isherwood 1975; Zhou and Solomos 1998). Plant starch phosphorylases do not appear to be subject to metabolite fine-tuning (Lee 1960; Duwenig et al. 1997). Spinach starch phosphorylase does not degrade intact starch granules (Steup et al. 1983). Moreover, the addition of amylases from poplar wood enhances the phosphorolytic activity of type II poplar wood starch phosphorylase (Witt and Sauter 1996). It has been suggested that starch needs to be modified by the debranching enzyme before the modified starch is used as a substrate by starch phosphorylase (Kennedy and Isherwood 1975; Beck and Ziegler 1989). Likewise, the size and crystalline nature of starch granules change during sweetening of potato tubers at low temperatures (Barichello et al. 1990).

It is noteworthy that keeping tubers under 1.52 kPa O_2 at 1°C, which strongly inhibits sucrose accumulation, has no effect either on total extractable activity or on any of the starch phosphorylase isoforms (Zhou and Solomos 1998).

Amylases

Amylase activity increases during the course of sweetening of potato tubers at low temperatures (Cottrell et al. 1993; Hill et al. 1996; Deiting et al. 1998; Zhou and Solomos 1998). Nielsen et al. (1997) observed that low temperature induces the synthesis of β-amylase. In "Russet Burbank" tubers kept at 1°C, amylase activity began to increase after 4 days and reached a peak (10-fold increase) after about 10 days. IEF zymograms showed that "Russet Burbank" tubers contain four amylase isoforms, with pI values of 5.05, 5.10, 5.23, and 5.27. However, only two isoforms, those with pI values of 5.23 and 5.57, increased during storage at low temperature. Two additional amylase isoforms were detected at low temperatures (Deiting et al. 1998; Zhou and Solomos 1998). It is not yet known whether the isoforms correspond to different genes or are the result of posttranslational modifications.

As described in the previous section, several observations indicate that starch is converted to sugars via starch phosphorylase. That amylolytic activity might contribute to glucose accumulation is not favored by the following observations: (a) there is no accumulation of maltose or of glucose polymers higher than maltose; (b) sucrose is the first sugar to accumulate (Isherwood 1973; Sowokinos 1990; Hill et al. 1996; Zhou and Solomos 1998); and (c) amylase activity may not be sufficient to support the rate of sugar biosynthesis. One could hypothesize that the activity of the enzymes involved in the phosphorylation of glucose, mainly hexokinase, is much higher than that of the enzymes producing free glucose, amylases and maltase, thereby rapidly metabolizing glucose. No data exist, however, to support this hypothesis. Further, maltose could not be detected in "Russet Burbank" tubers even by GC-mass spectrometry (Zhou and Solomos 1998). Nevertheless, amylases could play a role in sweetening by modifying the structure of the starch granule, particularly if a cold-induced amylase is shown to have a debranching activity. Modification of the starch granule, in turn, could enhance starch degradation (Steup et al. 1983; Beck and Ziegler 1989; Witt and Sauter 1996).

Phosphoglucomutase (PGM)

PGM catalyzes reversible interconversion of glucose phosphates (Glc-1-P and Glc-6-P). It is widely distributed throughout plant tissues. There are two types of PGM: one expressed in the cytosol and the other in plastids (Mühlbach and Schnarrenberger 1978). Both isoforms require glucose diphosphate (Glc-1,6-P_2) as a cofactor (Salvucci et al. 1990) and are inhibited by Fru-1,6-P. The reaction of PGM with Glc-1,6-P_2 as a cofactor is thought to occur near equilibrium (Turner and Turner 1980). Yet the Glc-1-P/Glc-6-P ratio changes during sweetening of potato tubers at low temperatures (Barker 1968). PGM may not always be near equilibrium, especially under conditions where the production of Glc-1-P exceeds the capacity of PGM (Hattenbach and Heineke 1999). If Glc-6-P is the form of glucose transported through the amyloplast membranes, then cytosolic PGM would be required for the synthesis of sucrose from phosphorolytic starch degradation as well as for the synthesis of starch from sucrose. In *rug3* pea mutants deficient in plastidic PGM, starch synthesis in both embryos and chloroplasts was strongly inhibited (Harrisson et al. 1998). Similar results were observed with *Arabidopsis* (Caspar et al. 1985; Periappuram et al. 2000). In growing potato tubers, a reduction of plastidic PGM using antisense RNA technology resulted in the inhibition of starch synthesis (Tauberger et al. 2000; Fernie et al. 2001). Thus, it seems that potato tuber plastidial PGM is one of the enzymes necessary for starch biosynthesis, and that Glc-6-P is the main form of glucose translocated into the amyloplast from the cytosol. Changes in plastidial PGM activity during sweetening at low temperatures are not yet well established. However, PGM is unlikely to be a key regulatory step in the conversion of starch to sucrose. It is a very important enzyme in plant metabolism, catalyzing an easily reversible reaction and is highly active than several other enzymes involved in starch synthesis (Tauberger et al. 2000; Fernie et al. 2001).

UDP-Glucose Pyrophosphorylase (UGPase)

Synthesis of sucrose from the phosphorolytic breakdown of starch requires conversion in the cytosol of Glc-1-P to UDP-glucose, the limiting substrate for sucrose phosphate synthase (Sowokinos 1990). Also, for the synthesis of starch from sucrose, UDP-G must be converted to Glc-1-P, a reversible reaction catalyzed by UGPase (UTP: α-D-glucose-1-phosphate uridylyl transferase EC 2.7.7.9):

$$\text{UTP} + \text{Glc-1-P} \leftrightarrow \text{UDP-G} + P_iP_i$$

In addition, UDP-G is the glycosyl donor for the synthesis of cell wall polysaccharides and can be derivatized to other nucleotide diphosphate sugars utilized in the synthesis of polysaccharides (Delmer and Amor 1995; Carpita and McCann 2000). A biological role for UGPase was demonstrated when a mutant with diminished ability to produce UDP-G via UGPase was found to be deficient in cellulose (Valla et al. 1989).

Two cDNA sequences of potato tuber UGPase are known (Katsube et al. 1990; Spychalla et al. 1994). Although the length of the reading frame is the same for both sequences, there are differences at 28 bases and in five amino acids, two of which lead to a change in charge. These differences result in two restriction patterns (Sowokinos et al. 1997). It has been suggested that these two cDNAs are present in two different alleles (Sowokinos et al. 1997), based on the evidence that a single gene encodes UGPase (Borovkov et al. 1996). Restriction mapping of the UGPase cDNAs showed that, in most tested cultivars, UGPase is represented by at least two kinds of alleles differentiated by the presence or absence of a BamHI restriction site. Significantly, the cultivars that lack a BamHI site are associated with those resistant to cold sweetening (Sowokinos et al. 1997).

UGPase exhibits negative cooperativity with respect to substrates Glc-1-P and UTP (Sowokinos et al. 1993). For instance, when the concentration of Glc-1-P is higher than 0.2 mM, the K_m for the substrate increases from 0.08 to 0.68 mM (Sowokinos et al. 1993). The latter study suggested that under environmental stress requiring mobilization of starch to sugars, UGPase limits sucro-neogenesis. Support for this suggestion is given by experiments in which silencing of this gene resulted in decreased sweetening at low temperatures (Spychalla et al. 1994; Borovkov et al. 1996). However, Zrenner et al. (1993) reported that a 95-96% reduction in UGPase activity in antisense RNA transgenic potato plants had no effect on either tuber growth or carbohydrate metabolism.

Arguments against UGPase as the single limiting regulatory step in the conversion of starch to sugars at low temperatures are: (a) UGPase is not subject to fine-tuning by cellular metabolites; (b) it catalyzes an easily reversible reaction; and (c) its activity is high compared with the other enzymatic steps involved in the starch-to-sucrose conversion during sweetening at low temperatures. It is likely that *in vivo* the availability of substrates may determine its activity. However, Gupta and Sowokinos (2003) identified a new UGPase, which is mainly expressed in cultivars resistant to cold-induced sweetening.

Hexose Phosphates Transported across Amyloplast Membranes

The amyloplasts of heterotrophic and storage plant tissues are not capable of generating hexose phosphates from imported triose phosphates because they lack fructose-1, 6-bisphosphatase (Entwistle and ap Rees 1990; Kossmann et al. 1992). Instead, they import cytosolic hexose phosphates as substrates for starch biosynthesis and for other pathways. During the import of hexose phosphates into the amyloplast they are exchanged with inorganic phosphate in a variety of plant tissues (Hill and Smith 1991; Neuhaus et al. 1993; Schott et al. 1995; Kammerer et al. 1998). The form in which hexose phosphate may exist differs with the tissue. Usually Glc-6-P is the preferred substrate of the transporter protein. In certain Graminaceous species, however, Glc-1-P rather than Glc-6-P is the precursor of starch synthesis (Tyson and ap Rees 1988; Tetlow et al.

1994). In isolated potato tuber amyloplasts, only Glc-1-P supports *in vitro* starch synthesis (Naeem et al. 1997). However, experiments in which the isolated amyloplasts were separated into two sizes—small and large—showed that the contribution to starch synthesis by either Glc-6-P or Glc-1-P is dependent on their size and integrity (Wischmann et al. 1999). Kammerer et al. (1998) isolated from maize endosperm, pea root tissue, and potato tubers the cDNA of the glucose-6-phosphate/phosphate translocator (GPT). Sequence analysis of the cDNA showed that GPT proteins have a high degree of identity with each other and their expression appears to be restricted to heterotrophic and storage tissues. In addition, Schott et al. (1995) reported that only Glc-6-P is able to provide redox equivalents to support the activity of glutamine synthase in isolated potato tuber amyloplasts, indicating that this is the translocated form. In short, the existing experimental evidence indicates that Glc-6-P is the translocated hexose-P across the inner potato tuber amyloplast membrane.

It is obvious that the transport of Glc-6-P from the amyloplast to the cytosol is a prerequisite for the synthesis of sucrose during sweetening. Likewise its transport from the cytosol to the amyloplast is a prerequisite for the synthesis of starch during reconditioning (Figs. 10.1 and 10.2, see later discussion). There is no detailed information concerning the contribution of GPT to the regulation of low temperature sweetening.

Sucrose Phosphate Synthase (SPS)

Sucrose is vital for plant growth and development because it is not only a source of carbon in heterotrophic tissues, but also a signaling molecule (Huber and Huber 1996; Smeekens 2000; Rolland et al., 2002). Sucrose is synthesized by the following reactions involving sucrose phosphate synthase (SPS) in conjunction with sucrose-6-phosphatase:

$$\text{UDP-G} + \text{Frc-6-P} \leftrightarrow \text{UDP} + \text{sucrose-6-P} + H^+$$

$$\text{sucrose-6-P} + H_2O \rightarrow \text{sucrose} + P_i$$

The combined action of the two enzymes renders the synthesis of sucrose irreversible (Huber and Huber 1996; Dennis and Blakeley 2000). SPS is considered a key enzyme in sucrose biosynthesis (Leloir and Cardini 1955; Stitt et al. 1987; Huber and Huber 1996; Geigenberger 2003). It is generally accepted that saturation kinetics for both substrates, UDP-G and Frc-6-P, are hyperbolic rather than sigmoidal (Huber and Huber 1996). SPS is regulated at several levels: expression of its transcripts, covalent modification of the protein through reversible phosphorylation, and allosteric modification of the enzyme by glucose-6-phosphate (activator) and P_i (inhibitor) (see Huber and Huber 1996 for review). The latter modulators affect the K_m for both substrates (fructose-6-phosphate and UDP-G) (Reimholz et al., 1994; Huber and Huber, 1996). SPS becomes phosphorylated at two serine residues, serine 158 and serine 424, and the phosphorylation state correlates with modification in the activity of spinach SPS (McMichael et al. 1993; Toroser and Huber 1997). Phosphorylation of ser 158 inactivates the enzyme and is considered a control step in the light/dark modulation transitions of SPS (McMichael et al. 1993), whereas phosphorylation of ser 424 results in enzyme activation; ser 424 phosphorylation is likely responsible for the activation of SPS in response to osmotic stress (Toroser and Huber 1997; Geigenberger et al. 1999).

SPS has been cloned from a variety of plants, including potatoes (for review, see Huber and Huber 1996). Potato contains multiple forms of SPS: SPS-1a, SPS-1b, SPS-2 and SPS-3 (Reimholz et al. 1997). SPS-1a is constitutively expressed in tubers, whereas SPS-1b is induced during cold storage. The onset of sucrose accumulation in tubers stored at low temperatures is accompanied by an increase in SPS activity (Sowokinos 1990; Reimholz et al. 1994; Hill et al. 1996; Illeperuma et al. 1998). The mechanism of SPS activation is not fully understood. Since immunoblot analysis indicated no change at the protein level, it was suggested that changes in the kinetic properties of SPS are the cause of increased activity (Hill et al. 1996; Illeperuma et al. 1998). This suggestion is supported by the findings that an increase in SPS activity in response to water stress in intact potato tubers and tuber slices is accompanied by changes in its kinetic properties, and the increased SPS activity is blocked by inhibitors of protein phosphatases (Geigenberger et al. 1997). Studies have revealed that induction of a new form of SPS is a prerequisite for an increase in its activity as well as in sweetening of potato tubers at cooler temperatures (Deiting et al. 1998). Further, upon transfer of the tubers from cold to 20°C, this new form of SPS decreases in abundance as does the SPS activity. A decrease in SPS activity during reconditioning is also observed in "Russet Burbank" tubers (Illeperuma et al. 1998). Interestingly, a 70-80% reduction in SPS gene expression via antisense and cosuppression resulted in only a moderate decrease in sucrose synthesis (Krause et al. 1998). The authors concluded that SPS expression *per se* does not regulate sucrose biosynthesis. Rather, the changes in its kinetic properties, such as those induced by low temperatures, provide a more effective stimulus for sucrose synthesis.

Invertase

Plant invertases (β-fructofuranosidase EC 3.2.1.26) are a group of enzymes that irreversibly catalyze the conversion of sucrose to glucose and fructose. Two forms of invertases are known, each characterized by its optimum pH: acid invertase has a pH optimum of 4 to 5, and alkaline invertase is most active under alkaline pH (Dennis and Blakeley 2000). Acid invertases are further subdivided into two forms: soluble (vacuolar) and extracellular (Unger et al. 1992; Dennis and Blakeley 2000). In potato tubers, acid invertase is localized in the vacuoles and in cell walls (Pressey 1969; Bracho and Whitaker 1990; Isla et al. 1991, 1998, 1999). Acid invertase is regulated by product inhibition. Inhibition by fructose is competitive but that by glucose noncompetitive (Isla et al. 1991; Burch et al. 1992). Potato tubers contain a proteinaceous inhibitor(s) of acid invertase.

The role of invertase in the synthesis of reducing sugars during sweetening at low temperatures is not unequivocally established. Reports have been published that argue both for and against invertase playing a crucial role in the regulation of the synthesis of hexoses (Pressey 1969; ap Rees et al. 1981; Richardson et al. 1990; Zrenner et al. 1996). The discrepancies could be related to the age of the tubers investigated, as well as to the time in the storage period at which the temporal correlations between invertase activity and hexose content were assessed. In young tubers, the accumulation of sucrose and hexoses occurs almost simultaneously, whereas in dormant mature tubers the rise in the former precedes that of hexoses (Isherwood 1973). Further, upon longer storage at low temperatures, the concentration of sucrose begins to decrease, whereas the hexoses continue to increase, though at a diminished rate (see the section on Kinetics of Sugar Accumulation). It is thus possible that late in storage the correlations between invertase activity and hexose content may not show a close temporal relationship. Experimental results with "Russet Burbank" potatoes stored at 1°C showed that acid invertase activity correlates with hexose synthesis since: (a) a rise in glucose and fructose followed the increase in sucrose, with a lag of 12 days; (b) the ratio of increment in glucose/fructose was 1.05, indicating that sucrose is their immediate precursor; (c) the accumulation of acid invertase mRNA followed the increase in sucrose but preceded the increases in both invertase activity and hexose accumulation; (d) acid invertase mRNA was not detected in tubers stored at 10°C in air or at 1°C under 1.52 kPa O_2, conditions that inhibit by about 82% the rise in hexoses; and (e) a very close temporal correlation existed between invertase activity and rise in hexose content during the initial phase of rapid increase in sugar accumulation (Zhou and Solomos 1998; Zhou et al. 1999). Suppression of invertase activity by hypoxia has also been observed in other tissues (Zeng et al. 1999). Moreover, Zrenner et al. (1996) reported that the glucose-fructose ratio is determined by acid invertase.

Sucrose synthase (SuSy) activity is low in potatoes during storage at 1°C (Illeperuma et al. 1998). This, together with the equimolar increase in glucose and fructose, indicates that it does not have a substantial role in the synthesis of hexoses. Further, during reconditioning of tubers at 10°C, the decline in hexose content was accompanied by a decrease in acid invertase activity (Illeperuma et al. 1998). In addition, in transgenic potato plants that express the acid invertase gene in the antisense orientation, a 22% suppression of activity resulted in a 34% decrease in hexose content without affecting sucrose synthesis (Zrenner et al. 1996). Moreover, in transgenic potato tubers that express a putative homologue of a tobacco cell wall invertase inhibitor, the cold-induced increase in hexoses was reduced by 75%, with no effect on potato tuber yield (Greiner et al. 1999). Also, as expected, the color of the chips produced from these potatoes is acceptable. The pattern of changes in sucrose and acid invertase indicates that sucrose may be the signal for the induction of soluble acid invertase, especially in view of the ability of sucrose to modulate gene expression (Stitt and Sonnewald 1995; Koch 1996; Smeekens 2000; Rolland et al. 2002; Hajirezael et al. 2003). It should also be noted that in transgenic potato tubers with diminished SuSy activity, the resultant increase in sucrose was associated with a 40-fold increase in soluble invertase activity and a concomitant increase in glucose and fructose (Zrenner et al. 1995).

Since even constitutive suppression of vacuolar acid invertase produces no detrimental effects on potato tuber yield, and since it is the final step in the production of reducing sugars, it is a potential target for genetic engineering to eliminate the increase in reducing sugars during cold storage, particularly if the genetic suppression is placed under the control of a cold-inducible promoter.

RECONDITIONING

Reconditioning is defined as the return of tubers from low temperature storage to higher or ambient temperatures.

Respiration

Isherwood (1973) showed that the return of tubers held at low temperature for periods of 30-60 days to 10°C resulted in the conversion of accumulated sugars to starch. This conversion is a sink of ATP utilization in dormant potato tubers since conversion of 1 mole of either glucose or fructose requires 2 moles of ATP. The combined molar concentration of hexoses at low temperatures is much larger than that of sucrose. Therefore, it is not surprising that reconditioning is associated with a sharp increase in respiration (Barker 1968; Isherwood 1973; Sowokinos 1990; Illeperuma et al. 1998). The rate of CO_2 output initially increased sharply and after reaching a peak, gradually declined to the rate observed with tubers held continuously at 10°C (Isherwood 1973; Illeperuma et al. 1998). Hypoxia suppressed both the decrease in sugars and the rise in respiration, and the latter remained slightly above that of tubers held continuously at 10°C (Illeperuma et al. 1998).

Sugars

The rate of sugar decrease differs between sucrose and hexoses (Isherwood 1976; Illeperuma et al. 1998). For instance, in "Russet Burbank," sucrose decreased within 7 days to levels found in tubers held continuously at 10°C, whereas it took about 30 days for the hexoses to reach their steady-state level (Illeperuma et al. 1998). Again, the ratio of glucose/fructose remained close to one. Hypoxia inhibited by 2.5-fold the initial decrease in sucrose; thereafter its level remained constant at values threefold higher than those at 10°C. On the other hand, hypoxia completely inhibited the decrease in hexoses (Illeperuma et al. 1998). The pattern of changes in sucrose content and SuSy activity showed that the rapid decrease in sucrose coincides with a parallel increase in SuSy, which eventually decreases to its initial level once sucrose has reached its steady-state concentration. Somewhat unexpectedly, hypoxia prevented the initial increase in SuSy and only at the end of the experiment (30 days) did it increase slightly.

SPS and invertase activities declined during reconditioning of tubers in both air and hypoxia, indicating that they may not play a role in the conversion of hexoses to starch (Illeperuma et al. 1998). The sucrose-to-starch conversion mechanism has been studied extensively in growing potato tubers, the individual enzymes involved in this process have been characterized at both the biochemical and molecular levels, and their activities have been altered in transgenic plants (Morrell and ap Rees 1986; Preiss et al. 1991; Martin and Smith 1995; Smith et al. 1997; Pozueta-Romero et al. 1999; Kossmann and Lloyd 2000; Geigenberger 2003). This corpus of studies indicates that regulation of the overall process involves transcription, covalent modification, allosteric metabolite regulation, and substrate concentration. In growing tubers, SuSy and ADP-glucose pyrophosphorylase are key enzymes in the sucrose-starch conversion (Geigenberger 2003). However, certain differences exist among young, growing tubers and dormant tubers that have been transferred from 1°C to 10°C. In particular, in dormant tubers during 30 days of storage at 1°C, the combined concentration of glucose and fructose was about twofold higher than that of sucrose (Illeperuma et al. 1998), whereas in growing tubers sucrose was the predominant sugar (see Geigenberger 2003 for review). Exogenous addition of glucose or sucrose to growing tubers resulted in differences in starch synthesis (Sweetlove et al. 2002). Although studies on the regulation of the sugar-to-starch conversion during reconditioning are not extensive, there is no reason to believe that the overall biochemical pathway differs from that occurring in growing tubers (Fig. 10.2).

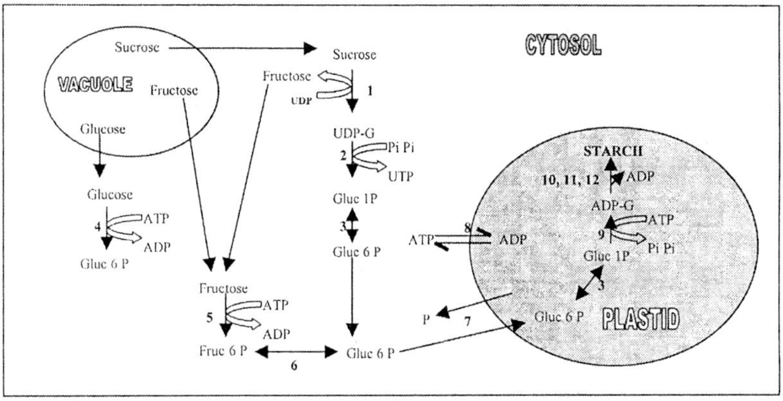

Fig. 10.2 1: Sucrose synthase, 2: UDP-G glucose pyrophosphorylase, 3: Phosphoglucoisomerase, 4: Hexokinase, 5: Fructose kinase, 6: Phosphohexose isomerase, 7: Hexose P/Pi translocator, 8: ATP/ADP translocator, 9: ADP-glucose pyrophosphorylase, 10: Soluble starch synthase, 11: Granule-bound starch synthase, 12: Branching enzyme

Sucrose Synthase

SuSy (EC 2.4.1.13) is a ubiquitous enzyme in higher plants and catalyzes the conversion of sucrose and UDP to UDP-G and fructose, a readily reversible reaction *in vivo* (Avigad 1982; Geigenberger and Stitt 1993). Salanoubat and Belliard (1987) characterized potato SuSy cDNA and showed that its transcript level is highest in growing tubers. They also detected SuSy transcripts in leaves incubated with sucrose, indicating that sucrose may regulate its expression. SuSy transcripts are upregulated by wounding and anoxia (Salanoubat and Belliard 1989). Interestingly, maize SuSy is also differentially regulated by hypoxia and sucrose (Zeng et al. 1998). Potatoes contain two differentially expressed SuSy genes: Sus3 and Sus4. Sus4 is expressed primarily in tubers but is inducible in leaves treated with 60 mM sucrose (Fu and Park 1995). Data concerning the changes in sucrose during sweetening at low temperatures and subsequent conversion of sugars to starch during reconditioning indicate that additional factors may be involved in the regulation of SuSy expression. For instance, contrary to expectations, the increase in sucrose at low temperatures is not associated with an increase in SuSy (Salanoubat and Belliard 1989; Fu and Park 1995). Neither did hypoxia enhance its activity during desweetening at 10°C (Illeperuma et al. 1998). However, sucrose content may still be involved in the initial increase in SuSy activity during reconditioning, since its activity decreases with decrease in sucrose. Studies with maize SuSy, Sus 1, and Sh1 showed that they respond to hypoxia and prolonged anoxia respectively (Zeng et al. 1998). Further, enhancement by hypoxia decreased with time. Extending these findings to potato genes, it is possible that potato tuber Sus4 may indeed be induced by hypoxia but that the response period is early and transient (24-48 h). If this is in fact the case, then changes in its activity will be undetectable if measurements are taken after several days (Illeperuma et al. 1998).

In view of the ability of SuSy to form ADP-G in the cytosol, from where it is translocated into the amyloplasts, it has been proposed that SuSy plays a crucial role in starch formation by bypassing AGPase (Pozueta-Romero et al. 1999; Baroja-Fernandez et al. 2003). The relative contributions of SuSy and AGPase to starch formation in potato tubers are still debatable. However, the fact that the inhibition of phophoglucomutase by reverse genetics results in decreased starch synthesis (Tauberger et al. 2000; Fernie et al. 2001) indicates that in potato tubers AGPase is the principal enzyme in starch biosynthesis. In addition, as pointed out earlier, during reconditioning sucrose content is smaller than that of glucose and fructose, and decreases quickly during the first 7-9 days to its steady-state level, whereas it takes about 30 days for the conversion of hexoses to starch. It is thus unlikely that SuSy contributes to the conversion of sugars to starch during this process. Furthermore, the rise in its activity is restricted to a short period after the transfer of the tubers to higher temperature.

ADP-GLUCOSE PYROPHOSPHORYLASE (AGPase)

AGPase catalyzes the first committed step in starch biosynthesis, converting ATP and G-1-P to ADP-glucose and P_iP_i (Preiss 1988; Preiss et al. 1991; Martin and Smith 1995; Smith et al. 1997; Fig. 10.2). It is generally accepted that AGPase is the sole enzyme that catalyzes the synthesis of ADP-G. However, as mentioned above, SuSy has also been shown to catalyze this synthesis (Baroja-Fernandez et al. 2003). The enzyme from higher plants is a heterotetramer comprised of two small (catalytic) and two large (regulatory) subunits (Copeland and Preiss 1981; Okita et al. 1990; Smith-White and Preiss 1992). A comparison of AGPase proteins from different sources revealed that the amino acid sequences of the small subunit are highly conserved, whereas those of the large subunit are divergent (Smith-White and Preiss 1992). The molecular weights of potato tuber subunits are 50,000 and 51,000 (Okita et al. 1990), and genes encoding both subunits have been cloned (Nakata et al. 1991). Expression of cDNAs in *Escherichia coli* produced a heterotetrameric protein with catalytic properties similar to the native potato tuber proteins (Iglesias et al. 1993). Subsequent studies showed that the small subunit carries out the catalysis, whereas the larger subunit modulates the properties of the small subunit (Smith-White and Preiss 1992; Ballicora et al. 1995). Plant AGPases, including the potato tuber enzyme, are subjected to allosteric regulation, activation by 3-PGA and inhibition by P_i (Ghosh and Preiss

1966; Sowokinos and Preiss 1982; Preiss 1988). The 3-PGA-mediated activation is enhanced when cysteine[12] bisulfide linkage is reduced by the addition of DTT (Fu et al. 1998). This led to the finding that the enzyme is redox regulated. Likewise, spinach leaf thioredoxins f and m both activate or reduce the AGPase activity, depending upon whether they are reduced (activator) or oxidized (inhibitor) (Ballicora et al. 2000).

The localization of AGPase differs with the tissue. In potato tubers it is exclusively plastidial, whereas in graminaceous endosperm it is cytosolic (Beckles et al. 2001; Farré et al. 2001).

In a number of experiments, changes in starch synthesis could not be attributed either to the changes in AGPase activity, concentrations of substrates, or 3-PGA/P_i ratio (Geigenberger 2003 and references therein). However, it was subsequently demonstrated in elegant studies that the decrease in starch synthesis upon detachment of growing potato tubers from the parent plant was due to a novel posttranslational inactivation of AGPase via a reversible mechanism that involves a redox-dependent dimerization of the AGPB subunits (Tiessen et al. 2002).

AGPase expression is enhanced by sucrose and decreased by nitrate and phosphate (Müller-Röber et al. 1990; Scheible et al. 1997; Nielsen et al. 1998; Sokolov et al. 1998). Therefore, the supply of sucrose to growing tubers plays a major role in the transcriptional regulation of AGPase. For instance, diurnal changes in sucrose in the tubers were reflected in the accumulation of AGPase transcripts, which were high at the end of the day and low overnight (Geigenberger et al. 1994; Geigenberger and Stitt 2000). Nevertheless, diurnal changes in AGPase transcripts were not reflected in changes in maximal enzymatic activity (Geigenberger and Stitt 2000). In detached growing tubers, the AGPase enzymatic activity remained high for several days but the transcripts did not accumulate (Geigenberger et al. 1994). In addition, in transgenic potato plants expressing antisense AGPase, reduction in AGPase activity required a large decrease in AGPase transcripts (Müller-Röber et al. 1990). It was observed that the contribution of AGPase to the regulation of starch biosynthesis is decreased by such environmental factors as water stress (Geigenberger et al. 1999). In short, AGPase is subject to transcriptional, posttranslational, and allosteric regulation.

The regulatory role played by AGPase in the conversion of sugars to starch at high temperatures during reconditioning has not been assessed. However, in view of its pivotal role in starch biosynthesis, it is expected to participate in the overall control of the sugar-starch conversion.

EFFECTS OF NUCLEOTIDE LEVELS

Conversion of sucrose to starch via SuSy requires uridine diphosphate (UDP) whose concentration in potato tubers is lower than the K_m of potato tuber UDPase (Loef et al. 1999). Indeed, an increase in uridine nucleotides resulted in enhanced degradation of sucrose and an increase in starch synthesis (Loef et al. 1999). As already mentioned, participation of SuSy in the conversion of sugars to starch during reconditioning is probably limited, since both the sucrose levels and SuSy activity return to their steady-state levels within seven days (Illeperuma et al. 1998). Thereafter, conversion of glucose and fructose to starch does not require UDP (Fig. 10.2). The fact that the ratio of glucose to fructose remains close to unity throughout 30 days of reconditioning indicates that they are converted to G-6-P, which is then transported to the amyloplast (Fig. 10.2). It is obvious that this procedure requires ATP in both cytosol and plastids. The ATP required for the synthesis of ADPG in the amyloplasts is carried via an ATP/ADP transporter localized in the inner membrane of the plastid (Schünemann et al. 1993; Kampfenkel et al. 1995). It has been demonstrated that alterations in the activity of potato tuber ATP/ADP transporter alter the syntheis of starch (Tjaden et al. 1998; Geigenberger et al. 2001). Further, analytical data indicate that the plastidial concentration of ATP during potato tuber growth is within the range of the K_m for AGPase (Sowokinos and Preiss 1982; Tiessen et al. 2002). Also, the addition of adenine to growing tuber discs increased the level of ATP, an increase which was accompanied by an enhancement of starch synthesis (Loef et al. 2001). Very little is known about the changes in uridine nucleotides or in the activity of the ATP/ADP transporter. However, the levels of ATP increased during sweetening at 1°C (Barker 1968). Since a rise in respiration occurred during reconditioning (Barker 1968; Isherwood 1973; Illeperuma et al. 1998), it is reasonable to assume that the levels of ATP may also increase.

EFFECTS OF HYPOXIA

For purposes of clarity, the term "hypoxia" is defined as the range of O_2 concentrations that are below their level in air, affecting plant metabolism, but not inducing partial anaerobiosis. The diffusivity of O_2 through the skin and flesh of "Russet Burbank" tubers was determined (Abdul-Baki and Solomos 1994). Using these values, in conjunction with the rate of respiration at the equivalent storage temperatures, the O_2 level at the center of the tuber was estimated to be in a range sufficient to saturate cytochrome oxidase (Illeperuma et al. 1998). No increase in ethanol concentration occurred in the tubers under hypoxic conditions (Zhou and Solomos, unpubl. data). It is well known that hypoxic pretreatments enhance the ability of plant tissues to tolerate anoxia (Saglio et al. 1988; Drew 1997, for review). It is important to note that hypoxia suppresses the expression of developmentally regulated genes associated with fruit ripening, while at the same time inducing the expression of anoxic proteins (Kanellis et al. 1991).

Retardation by low O_2 of senescence in detached plant organs in general, and fruit ripening in particular, is saturable in that only below a certain level of O_2 is fruit ripening delayed. This level also induces the anoxic isoforms of alcohol dehydrogenase (ADH) and initiates a diminution in respiration that is attended by a decrease in the activity of pyruvate kinase (Mattoo et al. 1988; Kanellis et al. 1991; Chen and Solomos, 1996; Solomos 1999). The fact that in apples stored under 1.5% O_2, a 50% decrease in respiration resulted in a delay of 150-200 days in the onset of ripening indicates that the turnover of ATP must be greatly reduced, as is the case with animals (Storey and Storey 1990). Thus, it appears that hypoxia apparently engenders a metabolic depression (for review, see Storey and Storey 1990).

Storage of fruits under low O_2 has been in extensive commercial use for a long time (Kidd and West 1927). The levels of O_2 that initiate retardation of apple fruit ripening are about 5.15 kPa (Solomos 1999). Their effectiveness increases with decreasing O_2 concentration until its level begins to restrict cytochrome oxidase, thereby inducing anaerobic fermentation, which is injurious to the tissue. This is known as "low oxygen injury."

In potatoes, low O_2 (1.52 kPa) totally suppressed the rise in CO_2 output, the rate of which remained at the same levels as that attained by tubers that had been held in air at 1°C for two days. Low O_2 did not prevent the decrease in cytochrome oxidase observed in tubers held in air at 1°C. Nevertheless, it suppressed the induction of alternative oxidase, resulting in a diminution of the total mitochondrial O_2 uptake (Zhou and Solomos 1998). Hypoxia strongly inhibited the accumulation of sugars during sweetening at 1°C. This was accompanied by suppression of rise in the activities of SPS, invertase, and amylases. In addition, it suppressed induction of acid invertase transcripts and new amylase isoforms. On the other hand, it had no effect on the activity of starch phosphorylase (Zhou and Solomos 1998). Hypoxia (2.03 kPa O_2) inhibited the decrease in sugars during reconditioning of tubers held at 1°C for 28 days and then transferred to 10°C. Furthermore, it prevented a rise in respiration. Yet low oxygen did not prevent the decline in the activities of SPS and invertase (Illeperuma et al. 1998). It is therefore probable that hypoxic treatment may alter the pattern of changes during the conversion of starch to sugars at low temperatures, as well as during the reverse conversion at 10°C. Thus, hypoxia inhibits the induction and/or enhancement of enzymatic activities in response to temperature treatments but has no effect on enzymes that are either constitutively expressed, such as starch phosphorylase, or that decline in response to treatments, such as cytochrome oxidase, SPS and invertase. In this respect, hypoxic treatments may be employed to discern the regulatory steps in the sucrose-starch interconversion during storage of potato tubers. Studies with growing intact potato tubers and slices showed that a decrease in O_2 concentration to levels that do not restrain mitochondrial terminal oxidases, resulted in a decrease in starch synthesis (Tiessen et al. 2002; Geigenberger, 2003).

SUMMARY

Regulation of the starch-sugar conversion during storage of potato tubers at low temperatures has not yet been completely characterized at the molecular level. Available data concerning the kinetics of sugar accumulation indicate that: (a) the conversion of starch to sucrose is mediated via starch phosphorylase; (b) sucrose synthesis is carried out by SPS; and (c) invertase regulates the levels of glucose and fructose. Results with enzyme activity measurements show: (a) no change in either total extractable phosphorylase or in its two isoforms; (b) an increase in sucrose phosphatase (the mechanism of this enhancement has been attributed in turn to the induction of a new gene and to changes in its kinetic properties due to covalent modification); and (c) an increase in soluble acid invertase induced by cold storage. Furthermore, the pattern of changes in sucrose, invertase mRNA accumulation, enzyme activity and hexoses suggests that the rise in sucrose signals the induction of acid invertase.

Low temperatures enhance the activity of amylases and the mitochondrial alternative oxidase. Since the initial temporal relationship between the rise in amylase activity and glucose content is very weak, and since the ratio of glucose/fructose is close to unity, it is unlikely that amylases contribute to the synthesis of hexoses. Nevertheless, they may make an important contribution to the sweetening process by inducing alterations in starch granule structure. The role of alternative oxidase in cold-induced sweetening is not yet clear. Regulation of the starch-sugar conversion should also be affected by the enzymes participating in the overall process, viz., GPT transporters and cytosolic UGPase, as well as the levels of ATP, UTP, and probably 3-PGA. The mechanism of the reverse conversion of sugars to starch remains to be studied in as much detail as that of starch to sugars. Analytical and enzymatic data point to the probability that enzymes contributing tc the regulation of this process are cytosolic hexose kinase(s), GPT and ATP/ADP transporters, and AGPase. SuSy may play a limited role initially.

The effects of hypoxia are interesting because it inhibits the temperature-enhanced changes but not the enzymes expressed constitutively nor those that decline as a result of temperature treatment. To date, the mechanism by which temperature stress induces changes in the starch-sucrose interconversion has not been identified. Starch biosynthesis in growing potato tubers does not appear to be regulated by an individual enzymatic step, however. Only soluble acid invertase regulates the synthesis of reducing sugars and since even its constitutive expression does not affect potato yield, it is a good target for genetic engineering of potato tubers that can be stored at low temperature with no detrimental effects on processed products, especially if the antisense invertase gene is placed under the control of a promoter specifically induced in tubers by cold temperatures.

REFERENCES

Abdul-Baki, A.A. and T. Solomos. 1994. Diffusivity of carbon dioxide through the skin and flesh of "Russet Burbank" potato tubers. J Amer Soc Hort Sci 119: 742–746.

ap Rees, T., W.L. Dixon, C.J. Pollock, and F. Franks. 1981. Low temperature sweetening of higher plants. *In*: J. Friend and M.J.C. Rhodes (eds.), Recent Advances in the Biochemistry of Fruits and Vegetables. Acad. Press, London, UK, pp. 41–61.

Avigad, G. 1982. Sucrose and other disaccharides. *In*: T.A. Loewus and W. Tanner (eds.), Encyclopedia of Plant Physiology. Springer-Verlag, Heidelberg, vol. 18A, pp. 217–347.

Ballicora, M.A., J.B. Frueauf, Y. Fu, P. Schürrmann, and J. Preiss. 2000. Activation of the potato tuber ADP glucose pyrophosphorylase by thioredoxin. J Biol Chem 275: 1315–1320.

Ballicora, M.A., M.J. Laughlin, Y. Fu, T.W. Okita, G.F. Barry, and J. Preiss. 1995. Adenosine 5-diphosphate-glucose pyrophosphorylase from potato tuber: Significance of the N terminus of the small subunit for catalytic properties and heat stability. Plant Physiol 109: 245–251.

Barichello, V., R.Y. Yada, R.H. Coffin, and D.W. Stanley. 1990. Low temperature sweetening in susceptible and resistant potatoes: Starch structure and composition. J Food Sci 55: 1054–1059.

Barker, J. 1968. Studies in the respiratory and carbohydrate metabolism of plant tissues. XXIV. The influence of a decrease in temperature on the contents of certain phosphate esters in plant tissues. New Phytol 67: 487–493.

Baroja-Fernandez, E., F.J. Munoz, T. Saikusa, M. Rodriguez-Lopez, T. Akazawa, and J. Pozueta-Romero. 2003. Sucrose synthase catalyzes the de novo production of ADPglucose linked to starch biosynthesis in heterotrophic tissues of plants. Plant Cell Physiol 44: 500–509.

Beck, E. and P. Ziegler. 1989. Biosynthesis and degradation of starch in higher plants. Ann Rev Plant Physiol. 40: 95–117.

Beckles, D.M., A.M. Smith, and T. ap Rees. 2001. A cytosolic ADP-glucose pyrophosphorylase is a feature of graminaceous endosperms, but not of other starch-storing organs. Plant Physiol 125: 818–827.

Borovkov, A.Y., P.E. McClean, J.R. Sowokinos, S. Ruud, and G. Secor. 1996. Effect of expression of UDP-glucose pyrophosphorylase ribozyme and antisense RNAs on the enzyme activity and carbohydrate composition of field-grown transgenic potato plants. J Plant Physiol 147: 644–652.

Bracho, G.E. and J.R. Whitaker. 1990. Purification and partial characterization of potato (*Solanum tuberosum*) invertase and its endogenous proteinaceous inhibitor. Plant Physiol 92: 386–394.

Burch, L.R., H.V. Davies, E.M. Cuthbert, G.C. Machray, P. Hedley, and R. Waugh. 1992. Purification of soluble invertase from potato. Phytochem 31: 1901–1904.

Burton, W.G. 1969. The sugar balance in some British potato varieties during storage: 2. Effect of tuber age, previous storage temperature and intermittent refrigeration upon low temperature sweetening. Eur Potato J 12: 81–95.

Burton, W.G. 1978. The physics and physiology of storage. *In*: P.M. Harris (ed.). The Potato Crop. Chapman and Hall Publ., London, UK.

Carpita, N. and M. McCann. 2000. The cell wall. *In:* B.B. Buchanan, W. Gruissem, and R.L. Jones (eds.), Biochemistry and Molecular Biology of Plants. Amer Soc Plant Biol, Washington, DC, pp. 52–108.

Caspar, T., S.C. Huber, and C. Somerville. 1985. Alterations in growth, photosynthesis, and respiration in a starchless mutant of *Arabidopsis thaliana* (L.) deficient in chloroplast phosphoglucomutase activity. Plant Physiol 79: 11–17.

Chen, X. and T. Solomos. 1996. Effects of hypoxia on cut carnation flowers (*Dianthus caryophyllus* L.): Longevity, ability to survive under anoxia, and activities of alcohol dehydrogenase and pyruvate kinase. *Postharvest Biol. Tech. 7*: 317–329.

Coffin, R.H., R.Y. Yada, B. Parkin, B. Grodzinski, and D.W. Stanley. 1987. Effect of low temperature on sugar concentration and chip color of certain processing potato cultivars and selections. J Food Sci 52: 639–645.

Copeland, L. and J. Preiss. 1981. Purification of spinach leaf ADP-glucose pyrophosphorylase. Plant Physiol 68: 996–1001.

Cottrell, J.E., C.M. Duffus, L. Paterson, G.R. Mackay, M.J. Allison, and H. Bain. 1993. The effect of storage temperature on the reducing sugar concentration and the activities of three amylolytic enzymes in tubers of the cultivated potato, *Solanum tuberosum* L. Potato Res 36: 107–117.

Corsini, D., G.F. Stallknecht, and W.C. Sparks. 1978. A simplified method for determining sprout-inhibiting levels of chloroisopropham (CICP). Potato J Agric Food Chem 26: 990–991.

Deiting, U., R. Zrenner, and M. Stitt. 1998. Similar temperature requirement for sugar accumulation and for the induction of new forms of sucrose phosphate synthase and amylase in cold-stored potato tubers. Plant, Cell Environ 21: 127–138.

Delmer, D.P. and Y. Amor. 1995. Cellulose biosynthesis. Plant Cell 7: 987–1000.

Dennis, D.T. and S.D. Blakeley. 2000. Carbohydrate metabolism. *In*: B.B. Buchanan, W. Gruissem, and R.L. Jones (eds.). Biochemistry and Molecular Biology of Plants. Amer Soc Plant Physiol., Rockville, MD, pp. 630–675.

Dizengremel, P. 1985. Potato respiration: Electron transport pathways. *In:* P.H. Li (ed.). Potato Physiology. Acad. Press, London, UK, pp. 59–121.

Drew, M.C. 1997. Oxygen deficiency and root metabolism: Injury and acclimation under hypoxia and anoxia. Ann Rev Plant Physiol Plant Mol Biol 48: 223–250.

Duwenig, E., M. Steup, and J. Kossmann. 1997. Induction of genes encoding plastidic phosphorylase from spinach (*Spinacia oleracea* L.) and potato (*Solanum tuberosum* L.) by exogenously supplied carbohydrates in excised leaf discs. Planta 203: 111–120.

Entwistle, G. and T. ap Rees. 1990. Lack of fructose-1,6-bisphosphatase in a range of higher plants that store starch. Biochem J 271: 467–472.

Farré, E.M., A. Tiessen, U. Roessner, P. Geigenberger, R.N. Trethewey, and L. Willmitzer. 2001. Analysis and compartmentation of glycolytic intermediates, nucleotides, sugars, amino acids, and sugar alcohols in potato tubers by a non-aqueous fractionation method. Plant Physiol 127: 685–700.

Fernie, A.R., U.Rössner, R.N. Trethewey, and L. Willmitzer. 2001. The contribution of plastidial phosphoglucomutase to the control of starch synthesis within the potato tuber. Planta 213: 418–426.

Fu, H. and W.D. Park. 1995. Sink- and vascular-associated sucrose synthase functions are encoded by different gene classes in potato. Plant Cell 7: 1369–1385.

Fu, Y., M.A. Ballicora, J.F. Leykam, and J. Preiss. 1998. Mechanism of reductive activation of potato tuber ADP-glucose pyrophosphorylation. J Biol Chem 273: 25045–25052.

Geigenberger, P. 2003. Regulation of sucrose to starch conversion in growing potato tubers. J Experim Bot 54: 457–465.

Geigenberger, P. and M. Stitt. 1993. Sucrose synthase catalyzes a readily reversible reaction *in vivo* in developing potato tubers and other plant tissues. Planta 189: 329–339.

Geigenberger, P. and M. Stitt. 2000. Diurnal changes in sucrose, nucleotides, starch synthesis, and AGPS transcript in growing potato tubers that are suppressed by decreased expression of sucrose phosphate synthase. Plant J 23: 795–806.

Geigenberger, P., M. Geiger, and M. Stitt. 1998. High-temperature perturbation of starch synthesis in growing potato tubers is attributable to inhibition of ADP-glucose pyrophosphorylase by decreased levels of glycerate-3-phosphate as a result of high rates or respiration. Plant Physiol 117: 1307–1316.

Geigenberger, P., L. Merlo, R. Reimholz, and M. Stitt. 1994. When growing potato tubers are detached from their mother plant there is a rapid inhibition of starch synthesis, involving inhibition of ADP-glucose pyrophosphorylase. Planta 193: 486–493.

Geigenberger, P., R. Reimholz, U. Deiting, U. Sonnewald, and M. Stitt. 1999. Decreased expression of sucrose phosphate synthase strongly inhibits water stress-induced synthesis of sucrose in growing potato tubers. Plant J 19: 119–129.

Geigenberger, P., A.R. Fernie, Y. Gibon, M. Christ, and M. Stitt. 2000. Metabolic activity decreases as an adaptive response to low internal oxygen in growing potato tubers. Biol Chem 381: 723–740.

Geigenberger, P., R. Reimholz, M. Geiger, L. Merlo, V. Canale, and M. Stitt. 1997. Regulation of sucrose and starch metabolism in potato tubers in response to short-term water deficit. Planta 201: 502–518.

Geigenberger, P., C. Stamme, J. Tjaden, A. Schulz, P.W. Quick, T. Betsche, H.J. Kersting, and H.E. Neuhaus. 2001. Tuber physiology and properties of starch from tubers of transgenic potato plants with altered plastidic adenylate transporter activity. Plant Physiol 125: 1667–1678.

Ghosh, H.P. and J. Preiss. 1966. Adenosine diphosphate glucose pyrophosphorylase: A regulatory enzyme in the biosynthesis of starch in spinach leaf chloroplasts. J Biol Chem 241: 4491–4504.

Greiner, S., T. Rausch, U. Sonnewald, and K. Herbers. 1999. Ectopic expression of a tobacco invertase inhibitor homolog prevents cold-induced sweetening of potato tubers. Nature Biotech 17: 708–711.

Gupta, S.K. and J.R., Sowokinos. 2003. Physicochemical and kinetic properties of unique isoenzymes of UDP-Glc pyrophosphorylase that are associated with resistance to sweetening in cold-stored potato tubers. J Plant Physiol. 160: 589–600.

Hajirezael, M.-R., F. Börnke, M. Peisker, Y. Takahata, J. Lerchl, A. Kirakosyan, and U. Sonnewald. 2003. Decreased sucrose content triggers starch breakdown and respiration in stored potato tubers (*Solanum tuberosum*). J Exp Bot 54: 477–488.

Harrison, C.J., C.L. Hedley, and T.L. Wang. 1998. Evidence that the rug3 locus of pea (*Pisum sativum* L.) encodes plastidial phosphoglucomutase confirms that the imported substrate for starch synthesis in pea amyloplasts is glucose-6-phosphate. Plant J 13: 753–762.

Hattenbach, A. and D. Heineke. 1999. On the role of chloroplastic phosphoglucomutase in the regulation of starch turnover. Planta 207: 527–532.

Hemberg, T. 1985. Potato rest. *In*: P.H. Li (ed.). Potato Physiology. Acad. Press, London, UK, pp. 353–388.

Hill, L.M. and A.M. Smith. 1991. Evidence that glucose-6-phosphate is imported as the substrate for starch synthesis by the plastids of developing pea embryos. Planta 185: 91–96.

Hill, L.M., R. Reimholz, R. Schroder, T.H. Nielsen, and M. Stitt. 1996. The onset of sucrose accumulation in cold-stored potato tubers is caused by an increased rate of sucrose synthesis and coincides with low levels of hexose-phosphates, an activation of sucrose phosphate synthase and the appearance of a new form of amylase. Plant, Cell Environ 19: 1223–1237.

Horton, D. and R.L. Sawyer. 1985. The potato as a world crop, with special reference to developing areas. *In*: P.H. Li, (ed.), Potato Physiology. Acad. Press, New York, NY, pp. 1–34.

Huber, S.C. and J.L. Huber. 1996. Role and regulation of sucrose-phosphate synthase in higher plants. Ann Rev Plant Physiol Plant Molec Biol 47: 431–444.

Iglesias, A.A., G.F. Barry, C. Meyer, L. Bloksberg, P.A. Nakata, T. Greene, M.J. Laughling, T.W. Okita, G.M. Kishore, and J. Preiss. 1993. Expression of the potato tuber ADP-glucose pyrophosphorylase in *Escherichia coli*. J Biol Chem 268: 1061–1086.

Illeperuma, C., D. Schlimme, and T. Solomos. 1998. Changes in sugars and activities of sucrose phosphate synthase, sucrose synthase, and invertase during potato tuber (Russet Burbank) reconditioning at 10°C in air and 2.53 kPa oxygen after storage for 28 days at 1°C. J Amer Soc Hort Sci 123: 311–316.

Isherwood, F.A. 1973. Starch-sugar interconversion in *Solanum tuberosum*. Phytochem 12: 2579–2591.

Isherwood, F.A. 1976. Mechanism of starch-sugar interconversion in *Solanum tuberosum*. Phytochemistry 15: 33–41.

Isla, M.I., M.A. Vattuono, and A.R. Sampietro. 1991. Modulation of potato invertase activity by fructose. Phytochem 30: 423–426.

Isla, M.I., M.A. Vattuone, and A.R. Sampietro. 1998. Hydrolysis of sucrose within isolated vacuoles from *Solanum tuberosum* L. tubers. Planta 205: 601–605.

Isla, M.I., M.A. Vattuone, R.M. Ordonez, and A.R. Sampietro. 1999. Invertase activity associated with the walls of *Solanum tuberosum* tubers. Phytochem 50: 525–534.

Jadhav, S.J., G. Mazza, and W. T. Desai. 1991. Postharvest handling and storage. *In*: D.K. Salunkhe, S.S. Kadam, and S.J. Jadhav (eds.), Potato: Production, Processing and Products. RC Press, Boca Raton, FL, pp. 69–109.

Kammerer, B., K. Fischer, B. Hilpert, S. Schubert, M. Gutensohn, A. Weber, and U.I. Flügge. 1998. Molecular characterization of a carbon transporter in plastids from heterotrophic tissues: The glucose-6-phosphate/phosphate antiporter. Plant Cell 10: 105–117.

Kampfenkel, K., T. Möhlmann, O. Batz, M. van Montagu, D. Inze, and H.E. Neuhaus. 1995. Molecular cloning of an *Arabdiposis thaliana* cDNA encoding a novel putative adenylate translocator of higher plants. FEBS Lett 374: 351–355.

Kanellis, A.K., T. Solomos, and K.A. Roubelakis-Angelakis. 1991. Suppression of cellulase and polygalacturonase and induction of alcohol dehydrogenase isoenzymes in avocado fruit mesocarp subjected to low oxygen stress. Plant Physiol 96: 269–274.

Katsube, T., Y. Kazuta, H. Mori, K. Nakano, K. Tanizawa, and T. Fukui. 1990. UDP-glucose pyrophosphorylase from potato tuber: cDNA cloning and sequencing. J Biochem 108: 321–326.

Kennedy, M.G.H. and F. Isherwood. 1975. Activity of phosphorylase in *Solanum tuberosum* during low temperature storage. Phytochem. 14: 667–670.

Kidd, F. and C. West. 1927. Gas storage of fruits. Spec. Report, Fd. Invest., DSIRO.

Koch, K.E. 1996. Carbohydrate-modulated gene expression in plants. Ann Rev Plant Physiol Plant Molec Biol 47: 509–540.

Kossmann, J. and J. Lloyd. 2000. Understanding and influencing starch biochemistry. Crit Revs Plant Sci 19: 171–226.

Kossmann, J., B. Müller-Röber, T.A. Dyer, C.A. Raines, U. Sonnewald, and L. Willmitzer. 1992. Cloning and expression analysis of the plastidic fructose-1,6-bisphosphatase coding sequence from potato: Circumstantial evidence for the import of hexoses into chloroplasts. Planta 188: 7–12.

Krause, K.P., L. Hill, R. Reimholz, T.H. Nielsen, U. Sonnewald, and M. Stitt. 1998. Sucrose metabolism in cold-stored potato tubers with decreased expression of sucrose phosphate synthase. Plant, Cell Environ 21: 285–299.

Lee, Y.P. 1960. Potato phosphorylase. I. Purification, physicochemical properties and catalytic activity. Biochim Biophys Acta 43: 18–24.

Leloir, L.F. and C.E. Cardini. 1955. The biosynthesis of sucrose phosphate. J Biol Chem 214: 212–218.

Loef, I., M. Stitt, and P. Geigenberger. 1999. Orotate leads to a specific increase in uridine nucleotide levels and a stimulation of sucrose degradation and starch synthesis in disks from growing potato plants. Planta 209: 314–323.

Loef, I., M. Stitt, and P. Geigenberger. 2001. Increased levels of adenine nucleotide modify the interaction between starch synthesis and respiration when adenine is supplied to disks from growing potato tubers. Planta 212: 782–791.

Martin, C. and A. Smith. 1995. Starch biosynthesis. Plant Cell 7: 971–985.

Mattoo, A., A.K. Kanellis, T. Solomos, and A.M. Mehta. 1988. Low oxygen atmospheres and gene expression in relation to ethylene action. *In:* R.Jona (ed.), Physiology of Fruit Drop: Ripening, Storage and Postharvest Processing of Fruits. Univ. Torino, Italy, pp. 1–10.

McMichael, Jr., R.W., R.R. Klein, M.E. Salvucci, and S.C. Huber. 1993. Identification of major regulatory phosphorylation site in sucrose phosphate synthase. Arch Biochem Biophys 307: 248–252.

Mondy, N.I., A. Tymiak, and S. Chandra. 1978. Inhibition of glycoalkaloid formation in potatoes by the sprout inhibitor maleic hadrazine. J Food Sci 43: 1033–1035.

Morrell, S. and T. ap Rees. 1986. Sugar metabolism in developing tubers of *Solanum tuberosum*. Phytochem 25: 1579–1585.

Mühlbach, H. and C. Schnarrenberger. 1978. Properties and intracellular distribution of two phosphoglucomutases from spinach leaves. Planta 141: 65–70.

Müller-Röber, B.T., J. Kossmann, L.C. Hannah, L. Willmitzer, and U. Sonnewald. 1990. One of two different ADP-glucose phosphorylase genes from potato responds strongly to elevated levels of sucrose. Molec Gen Genet 224: 136–146.

Naeem, M., I.J. Tetlow, and M.J. Emes. 1997. Starch synthesis in amyloplasts purified from developing potato tubers. Plant J 11: 1095–1103.

Nakata, P.A., T.W. Green, J.M. Anderson, B.J. Smith-White, T.W. Okita, and J. Preiss. 1991. Comparison of the primary sequences of two potato tuber ADP-glucose phosphorylase subunits. Plant Molec Biol 17: 1089–1093.

Neuhaus, H.E., O. Batz, E. Thom, and R. Scheibe. 1993. Purification of highly intact plastids from various heterotrophic plant tissues: Analysis of enzymatic equipment and precursor dependency for starch biosynthesis. Biochem J 296: 395–401.

Nielsen, T.H., U. Deiting, and M. Stitt. 1997. A beta-amylase in potato tubers is induced by storage at low temperature. Plant Physiol 113: 503–510.

Nielsen, T.H., A. Krapp, U. Roeper-Schwarz, and M.T. Stitt 1998. The sugar-mediated regulation of genes encoding the small subunit of Rubisco and the regulatory subunit of ADP glucose pyrophosphorylase is modified by nitrogen and phosphate. Plant, Cell Environ 21: 443–454.

Okita, T.W., P.A. Nakata, J.M. Anderson, J.R. Sowokinos, M. Morrell, and J. Preiss. 1990. The subunit structure of potato tuber ADP glucose pyrophosphorylase. Plant Physiol 93: 785–790.

Paxton, W.C. 1996. The organization and regulation of plant glycolysis. Ann Rev Plant Physiol Plant Molec Biol 47: 185–214.

Periappuram, C., L. Steinhauer, D.L. Barton, D.C. Taylor, B. Chatson, and J. Zou. 2000. The plastidic phosphoglucomutase from *Arabidopsis*. A reversible enzyme reaction with an important role in metabolic control. Plant Physiol 122: 1193–1199.

Pollock, C.J. and T. ap Rees. 1975. Activities of enzymes of sugar metabolism in cold stored tubers of *Solanum tuberosum*. Phytochem 14: 613–617.

Pozueta-Romero, J., P. Perata, and T. Akazawa. 1999. Sucrose-starch conversion in heterotrophic tissues of plants. Crit Revs Plant Sci 18: 489–525.

Preiss, J. 1982. Regulation of the biosynthesis and degradation of starch. Ann Rev Plant Physiol 33: 419–458.

Preiss, J. 1988. Biosynthesis of starch and its regulation. *In:* J. Preiss (ed.), The Biochemistry of Plants. Acad. Press, San Francisco, CA, pp. 181–254.

Preiss, J., K. Bali, B. Smith-White, A. Iglesias, G. Kakefuda, and L. Li. 1991. Starch biosynthesis and its regulation. Biochem Soc Trans 19: 539–546.

Pressey, R. 1969. Role of invertase in accumulation of sugars in cold-stored potatoes. Amer Potato J 46: 291–297.

Reimholz, R., P. Geigenberger, and M. Stitt. 1994. Sucrose-phosphate synthase is regulated via metabolites and protein phosphorylation in potato tubers, in a manner analogous to the enzyme in leaves. Planta 192: 480–488.

Reimholz, R., M. Geiger, V. Haake, U. Deiting, K.P. Krause, U. Sonnewald, and M. Stitt. 1997. Potato plants contain multiple forms of sucrose phosphate synthase, which differ in their tissue distributions, their levels during development, and their responses to low temperature. Plant, Cell Environ 20: 291–305.

Richardson, D.L., H.V. Davies, H.A. Ross, and G.R. Mackay. 1990. Invertase activity and its relation to hexose accumulation of potato tubers. J Exp Bot 41: 95–99.

Rolland, F., B. Moore, and J. Sheen. 2002. Sugar sensing and signaling in plants. Plant Cell (suppl), 14: 185–205.

Saglio, P.H., M.C. Drew, and A. Pradet. 1988. Metabolic acclimation to anoxia induced by low (2-4 kPa partial pressure) oxygen treatments (hypoxia) in root tips of *Zea mays*. Plant Physiol 86: 61–66.

Salanoubat, M. and G. Belliard. 1987. Molecular cloning and sequencing of sucrose synthase cDNA from potato (*Solanum tuberosum*): Preliminary characterization of sucrose synthase mRNA distribution. Gene 60: 47–56.

Salanoubat, M. and G. Belliard. 1989. The steady-state level of potato sucrose synthase mRNA is dependent on wounding, anaerobiosis and sucrose concentration. Gene 84: 181–185.

Salvucci, M.E., R.R. Drake, K.P. Broadbent, B.E. Haley, K.R. Hanson, and N.A. McHale. 1990. Identification of the 64-kilodalton chloroplast stromal phosphoprotein as phosphoglucomutase. Plant Physiol 93: 105–109.

Scheible, W.R., A. González-Fontes, M. Lauerer, B. Müller-Röber, M. Caboche, and M. Stitt. 1997. Nitrate acts as a signal to induce organic acid metabolism and repress starch metabolism in tobacco. Plant Cell 9: 783–798.

Schott, K., S. Borchert, B. Mueller-Roeber, and H.W. Heldt. 1995. Transport of inorganic phosphate and C_3- and C_6-sugar phosphates across the envelope membranes of potato tuber amyloplasts. Planta 196: 647–652.

Schünemann, D., S. Borchert, U.-I. Flügge, and H.W. Heldt. 1993. ATP/ADP translocator from pea root plastids: Comparison with translocators from spinach chloroplasts and pea leaf mitochondria. Plant Physiol 103: 131–137.

Smeekens, S. 2000. Sugar-induced signal transduction in plants. Ann Rev Plant Physiol Plant Molec Biol 51: 49–81.

Smith, A.M., K. Denyer, and C. Martin. 1997. The synthesis of the starch granule. Ann Rev Plant Physiol Plant Molec Biol 48: 67–87.

Smith, D. 1987. Effect of cultural and environmental conditions on potatoes for processing. *In:* W.F. Talburt and O. Smith (eds.), Potato Processing. AVI Publ. Co., Westport, CT, pp. 73–147.

Smith-White, B.J. and J. Preiss. 1992. Comparison of proteins of ADP-glucose pyrophosphorylase from diverse sources. J Molec Evol 34: 449–464.

Sokolov, L.N., A. Dejardin, and L.A. Kleczkowski. 1998. Sugars and light/dark exposure trigger differential regulation of ADP-glucose pyrophosphorylase genes in *Arabidopsis thaliana* (thale cress). Biochem J 336: 681–687.

Solomos, T. 1999. Interactions between oxygen concentration and climactieric onset of ethylene evolution. *In:* A.K. Kanellis, C. Chang, H. Klee, A.B. Bleecker, J.C. Pech, and D. Grierson (eds.), Biology and Biotechnology of the Plant Hormone Ethylene. Kluwer Acad. Publ., Amsterdam, vol. II, pp. 313–319.

Solomos, T. and G.G. Laties. 1975. The mechanism of ethylene and cyanide action in triggering the rise in respiration in potato tubers. Plant Physiol 55: 73–78.

Sonnewald, U., A. Basner, B. Greve, and M. Steup. 1995. A second L-type isozyme of potato glucan phosphorylase: Cloning, antisense inhibition and expression analysis. Plant Molec Biol 27: 567–576.

Sowokinos, J. 1990. Effect of stress and senescence on carbon partitioning on stored potatoes. Amer Potato J 67: 849–857.

Sowokinos, J.R. and J. Preiss. 1982. Phosphorylases in *Solanum tuberosum*: III. Purification, physical and catalytic properties of ADP-glucose pyrophosphorylase in potatoes. Plant Physiol 69: 1459–1466.

Sowokinos, J.R., J.P. Spychalla, and S.L. Desborough. 1993. Pyrophosphorylases in *Solanum tuberosum*. IV. Purification, tissue localization, and physicochemical properties of UDP-glucose pyrophosphorylase. Plant Physiol 101: 1073–1080.

Sowokinos, J.R., C. Thomas, and M.M. Burrell. 1997. Pyrophosphorylases in potato. V. Allelic polymorphism in UDP-glucose pyrophosphorylase in potato cultivars and its association with tuber resistance to sweetening in the cold. Plant Physiol 113: 511–517.

Sowokinos, J.R., P.H. Orr, and J.A. Knoper, and J.L. Varns, 1987. Influence of potato storage and handling stress on sugars, chip quality and integrity of the starch (amyloplast) membrane. Amer Potato J 64: 213–226.

Spychalla, J.P., B.E. Scheffler, J.R. Sowokinos, and M.W. Bevan. 1994. Cloning, antisense RNA inhibition, and the coordinated expression of UDP-glucose pyrophosphorylase with starch biosynthetic genes in potato tubers. J Plant Physiol 144: 444–453.

Steup, M. 1988. Starch degradation. *In:* J. Preiss (ed.), The Biochemistry of Plants. Acad. Press, London, vol. 14, 255–296.

Steup, M., H. Robenek, and M. Melkonian. 1983. *In-vitro* degradation of starch granules isolated from spinach chloroplasts. Planta 158: 428–436.

Stitt, M. and U. Sonnewald. 1995. Regulation of metabolism in transgenic plants. Ann Rev Plant Physiol Plant Molec Biol 46: 341–368.

Stitt, M., S.C. Huber, and P. Kerr. 1987. Control of photosynthetic sucrose synthesis. *In:* M.D. Hatch and M.K. Boardmann (eds.), Biochemistry of Plants. Acad Press, New York, NY, vol. 10, pp. 327–409.

Storey, K.B. and J.M. Storey. 1990. Metabolic rate depression and biochemical adaptation in anaerobiosis, hibernation, and estivation. Quart Rev Biol 65: 145–174.

Sweetlove, LJ., K.L. Tomlinson, and S.A. Hill. 2002. The effect of exogenous sugars on the control of flux by adenosine 5′-diphosphoglucose pyrophosphorylase in potato tuber discs. Planta 214: 741–750.

Tauberger, E., A.R. Fernie, M. Emmermann, A. Renz, J. Kossmann, L. Willmitzer, and R.N. Trethewey. 2000. Antisense inhibition of plastidial phosphoglucomutase provides compelling evidence that potato tuber amyloplasts import carbon from the cytosol in the form of glucose-6-phosphate. Plant J 23: 43–53.

Tetlow, I.J., K.J. Blissett, and M.J. Emes. 1994. Starch synthesis and carbohydrate oxidation in amyloplasts from developing wheat endosperm. Planta 194: 454–460.

Tiessen, A., J.H.M. Hendriks, M. Stitt, A. Branscheid, Y. Gibon, E.M. Farré and P. Geigenberger. 2002. Starch synthesis in potato tubers is regulated by post-translational redox modification of ADP-glucose pyrophosphorylase: A novel regulatory mechanism linking starch synthesis to the sucrose supply. Plant Cell 14: 2191–2213.

Tjaden, J., T. Möhlmann, K. Kampfenkel, G. Henrichs, and H.E. Neuhaus. 1998. Altered plastidic ATP/ADP-transporter activity influences potato (*Solanum tuberosum* L.) tuber morphology, yield and composition of starch. Plant J 16: 531–540.

Toroser, D. and S.C. Huber. 1997. Protein phosphorylation as a mechanism for osmotic-stress activation of sucrose phosphate synthase in spinach leaves. Plant Physiol 114: 947–955.

Turner, J.F. and D. H. Turner. 1980. The regulation of glycolysis and the pentose phosphate pathway. *In:* D.P. Davies (ed.), Biochemistry of Plants. Acad Press, New York, NY, vol. 2, pp. 279–316.

Tyson, R.H. and T. ap Rees. 1988. Starch synthesis by isolated amyloplasts from wheat endosperm. Planta 175: 33–38.

Unger, C., J. Hofsteenge, and A. Sturm. 1992. Purification and characterization of a soluble β-fructofuranosidase from *Daucus carota*. Eur J Biochem 204: 915–921.

Valla, S., D.H. Coucheron, E. Fjaervik, J. Kjosbakken, H. Welnhouse, P. Ross, D. Amidan, and M. Benzleman. 1989. Cloning of a gene involved in cellulose biosynthesis in *Acetobacter xylinum*: Complementation of cellulose-negative mutants by the UDPG pyrophosphorylase structural gene. Molec Gen Genet 217: 26–30.

Wischmann, B., T.H. Nielson, and B.L. Moller. 1999. *In vitro* biosynthesis of phosphorylated starch in intact potato amyloplasts. Plant Physiol 119: 455–462.

Witt, W. and J.J. Sauter. 1996. Purification and properties of a starch granule-degrading alpha-amylase from potato tubers. J Exp Bot 47: 1789–1795.

Zeng, Y., Y. Yu, W.T. Avigne, and K.E. Koch. 1998. Differential regulation of sugar-sensitive sucrose synthases by hypoxia and anoxia indicates complementary transcriptional and posttranscriptional responses. Plant Physiol 116: 1573–1580.

Zeng, Y., Y. Yu, W.T. Avigne, and K.E. Koch. 1999. Rapid repression of maize invertases by low oxygen. Invertase/Sucrose synthase balance, sugar signaling potential, and seedling survival. Plant Physiol 121: 599–608.

Zhou, D. and T. Solomos. 1998. Effect of hypoxia on sugar accumulation, respiration, activities of amylase and starch phosphorylase, and induction of alternative oxidase and acid invertase during storage of potato tubers (*Solanum tuberosum*, cv. Russet Burbank) at 1°C. Physiologia Plantarum 104: 255–265.

Zhou, D., T. Solomos, H. Imaseki, and A.K. Mattoo. 1999. Low temperature storage induces acid invertase in potato tubers (*Solanum tuberosum*). J Plant Physiol 154: 346–350.

Zrenner, R., L. Willmitzer, and U. Sonnewald. 1993. Analysis of the expression of potato uridinediphosphate-glucose pyrophosphorylase and its inhibition by antisense RNA. Planta 190: 247–252.

Zrenner, R., K. Schuler, and U. Sonnewald, 1996. Soluble acid invertase determines the hexose-to-sucrose ratio in cold-stored potatoes. Planta 198: 246–252.

Zrenner, R., M. Salanoubat, L. Willmitzer, and U. Sonnewald. 1995. Evidence of the crucial role of sucrose synthase for sink strength using transgenic potato plants (Solanum tuberosum). Plant J 7: 97–107.

11

Transformation for Insect Resistance

David S. Douches and Edward J. Grafius
Michigan State University, Depts. Crop & Soil Science and Entomology, East Lansing, MI 48824-1325, USA

INTRODUCTION

The cultivated potato, *Solanum tuberosum* L. (2n = 4x = 48), is the most important vegetable crop, constituting the fourth most important food crop in the world; it was grown on more than 19 million ha in 2001 (FAO 2002). A potato crop produces, on average, more food energy and protein than cereals, and the lysine content of potato complements cereal- or maize-based diets deficient in this essential amino acid. Not only is the potato an important food for the fresh market, it is also the raw material for French fry, chip, and starch-processing industries. It is highly productive on a per hectare basis and, because of its adaptability, can be grown commercially in all of the 50 states of the USA.

In fact, the United States produces 22 million metric tons of potatoes annually on approximately 567,000 ha, with a farmgate value of greater than $2.7 billion.

IMPACT OF INSECT PESTS ON POTATO

The potato crop is attacked by a diversity of insect pests representing all the major insect orders (except Hymenoptera) including aphids, leafhoppers, plant bugs, whiteflies, symphylans, thrips, psyllids, crickets, a number of caterpillars, several beetles, and dipterous leafminers. Leaves, stems, and tubers are prone to infestation and injury may be so severe in some cases that an entire crop is lost. Probably the first wide-scale use of insecticides was undertaken in 1864 when they were applied to control the Colorado potato beetle (Gauthier et al. 1981), and remain the primary means for controlling this insect (Casagrande 1987). In part because of

Corresponding author: David S. Douches, Tel: 01-517-353-3145; Fax: 01-517-353-5174; *e-mail: douchesd@msu.edu* our reliance on insecticides, severe problems with insecticide resistance have arisen with regard to the Colorado potato beetle and green peach aphid. Costs for control led to losses exceeding \$250 ha^{-1} for Colorado potato beetle in Michigan in the early 1990s, when insecticide resistance problems were severe (Grafius 1997).

The Colorado potato beetle, *Leptinotarsa decemlineata* Say, is one of the most economically significant pests of potato in northern latitudes (US, Canada, Europe, and Russia) (Fig. 11.1). As little as 12.5% to 25% defoliation can significantly decrease potato yields (Mailloux and Bostanian 1989) and complete defoliation of a crop can reduce potato yields by as much as two-thirds (Hare 1980). Currently, about 5,90,000 kg active ingredients of insecticides are applied annually to potato crops to control the Colorado potato beetle in the top eight potato-producing states in the US. (Wiese et al. 1998). However, the Colorado potato beetle has shown a remarkable ability to develop resistance to every insecticide used for its control (Bishop and Grafius 1996) and has done so at an increasingly rapid rate (Forgash 1985; Ioannidis et al. 1991; Heim et al. 1990). At present, only four insecticides are registered as effective for Colorado potato beetle control in much of northeastern US and Canada. Industry relies heavily on one of these, namely imidacloprid, for most of its control projects. Although high level of resistance to imidacloprid were observed on Long Island, NY within two years of its first use (Zhao et al. 2000), it still remains effective in other parts of the US and Canada.

Nonchemical controls for Colorado potato beetle have been unsuccessful in commercial potato production. Several biological control agents have proven ineffective unless released in very high numbers in the potato crop (Groden and Casagrande 1986; Hazzard and Ferro 1991; Hough-Goldstein and Keil 1991). Crop rotation is an important part of Colorado potato beetle management (Wright 1984) but does not suffice for maintaining the Colorado potato beetle below economically significant levels in most commercial situations. Likewise, propane flamer treatment of small potato plants can be used to control Colorado potato beetle adults in the spring (Moyer 1992; Grafius et al. 1993), but is expensive and not economically viable. On the other hand, host plant resistance may be compatible with biological control agents and crop rotation in a synergistic manner.

The potato tuber moth *Phthorimaea operculella* Zeller is the most serious insect pest of potatoes worldwide (Fig. 11.2). It is generally recognized as having originated in South America where the potato also originated and is well adapted to plants with high glycoalkaloid levels (Goldson and Emberson 1985). Other species related to the potato tuber moth (e.g. *Symmetrischema tangolias*) are found in South and Central America and have similar life cycles and habits. *P. operculella* is of greatest importance

Fig. 11.1 A. Adult Colorado potato beetle (*Leptinotarsa decemlineata* Say). B. Mature larvae of Colorado potato beetle. C. Field experiment at the Montcalm Research Farm, Entrican, Michigan, comparing *Bt-cry3A* (left) and susceptible (right) potato lines under natural beetle pressure (approximately 75 days after planting) D. Defoliation of susceptible potato line by insecticide-resistant Colorado potato beetles at Long Island Horticultural Research and Extension Center, Riverhead, New York (approximately 50 days after planting). E. Predation of Colorado potato beetle larvae in an unsprayed field study comparing various host plant resistant lines of potato at the Montcalm Research Farm, Entrican, Michigan.

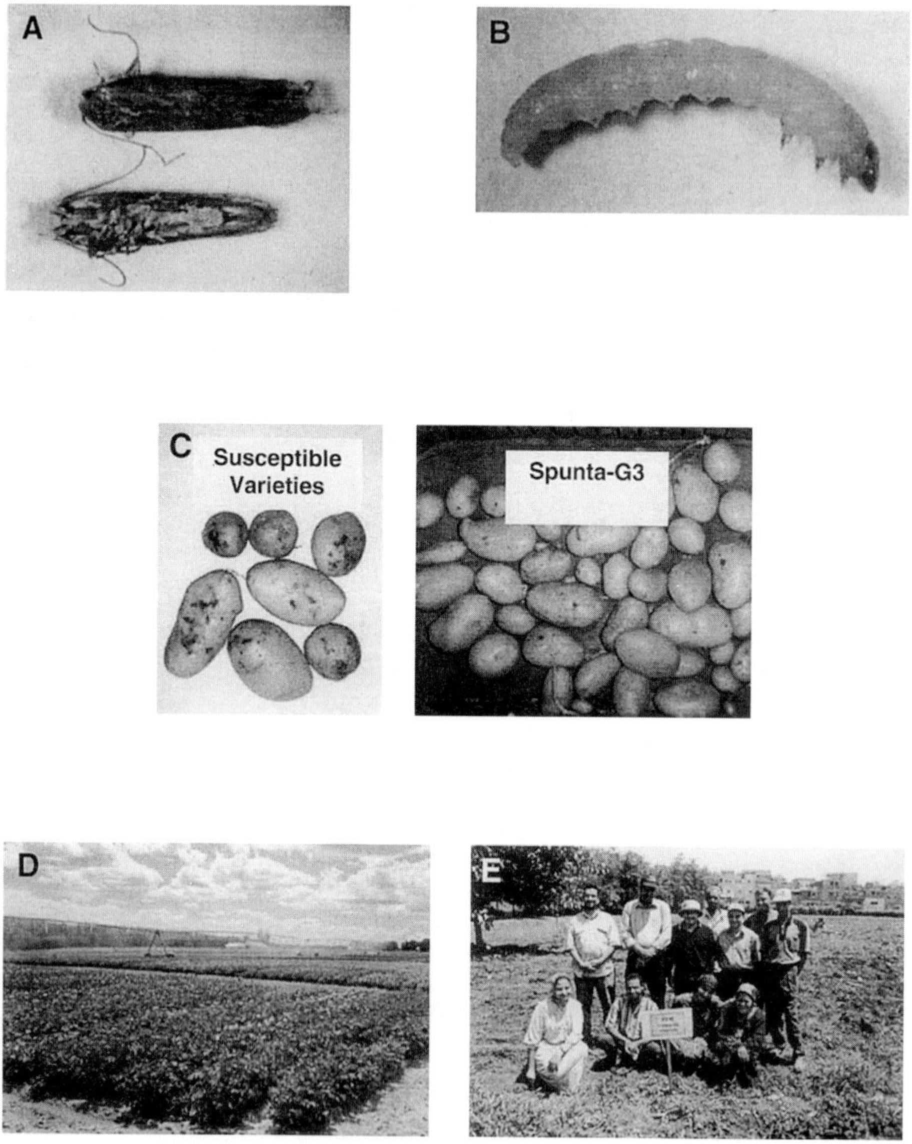

Fig. 11.2 A. Adult potato tuber moth (*Phthorimaea operculella* Zeller). B. Potato tuber moth larva. C. Tubers from a field trial in Egypt showing susceptible tubers with potato tuber moth infestation (left) and transgenic *Bt-cry1la1* resistant tubers (right). D. Agronomic field trials of *Bt-cry1la1* transgenic potato lines. E. Cooperators of a potato tuber moth trial of *Bt-cry1la1* transgenic potato lines under natural infestion at Kafyr-el-Zyat, Egypt.

in subtropical and tropical latitudes, including southern and southwestern US. The larvae mine the foliage, stems, and tubers in the field and also in storage. Tuber infection causes dramatic losses as damaged tubers are attacked by various secondary pests and diseases. Damage is caused only by the larval stage (Raman 1980; Trivedi and Rajagopal 1992; Goldson and Emberson 1985). Average oviposition rate at peak production is about 60 eggs female^{-1} day^{-1} and depends on the physiological state of the female. Larvae penetrate the leaves and feed by tunneling within the leaf veins and stems of the plant. This damage causes loss of leaf tissue, death of growing points and weakening or breakage of stems (Raman 1980; Bald and Helson 1944). The tubers are infested by eggs deposited on the surface of the soil near the stem or those laid near the tuber eyes when in storage. The larvae mine into the tuber causing irregular tunnels both near the surface and deep inside the tuber, rendering it unfit for human consumption. Duration of pupation varies with temperature, but generally takes 5 days (Goldson and Emberson 1985). The life cycle of the potato tuber moth may be completed within 20-30 days at 28°C, with as few as two generations and as many as twelve generations per year (Raman 1980). The potato tuber moth is not strictly confined to potatoes. When necessary, its feeding habits may extend to numerous Solanaceous species including tobacco (*Nicotiana tabacum* L.), tomato (*Lycospersicon esculentum* Mill.) and eggplant (*solanum melongena* L.) (Goldson and Emberson 1985).

Annual losses in storage alone due to the potato tuber moth range from 30 to 70% in India, and similar losses are encountered in the Middle East, Northern Africa, and South America (Raman and Palacios 1982). Effective control of the potato tuber moth is a challenge. Traditionally, pesticides have been used to control potato tuber moth in both field and storage with 12-20 applications during the growing season and 3-4 treatments during storage (Madkour et al. 1999). The utility of insecticides is, however, limited by their high costs, persistence of residues in tubers and by development of insecticide resistance (Collantes et al. 1986).

PRODUCTION OF TRANSGENIC PLANTS

Since the late 1980s potato has been targeted for insertion of genes to produce insect-resistant plants. *Agrobacterium tumefaciens*-mediated transformation has been the method of choice for transforming potato. Transformation protocols based on a binary Ti vector system have been used (An et al. 1986). Simple chimeric gene constructs have also been used to achieve gene expression. A typical T-DNA region of a vector construct consists of the gene with promoter and a selectable marker (Fig. 11.3). To identify transformed plant cells, neomycin phosphotransferase II (NPT II) has been the selectable marker of choice. One variable in these constructs

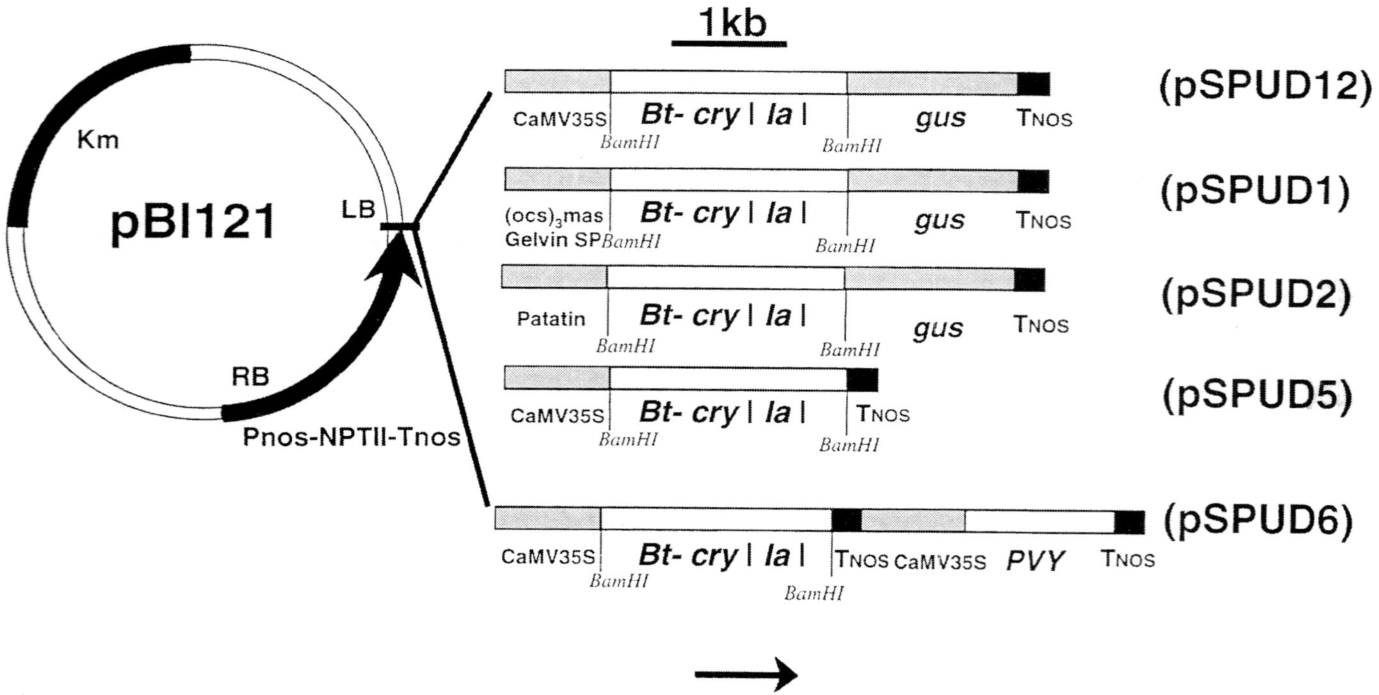

Fig. 11.3 Example of different *Bt-cry1Ia1* vector constructs used in potato transformation for engineered host plant resistance against the potato tuber moth.

has been the promoter for gene expression. The CaMV 35S promoter has frequently been used for constitutive gene expression; however, some alternative promoters include the Rubisco small subunit promoter from *Arabidopsis thaliana* for leaf expression (De Almeida et al. 1989), a tuber-specific promoter (class-I patatin element) (Li et al. 1999), the enhanced 35S promoter (Kay et al. 1987), the *(ocs)$_3$mas* super promoter (Ni et al. 1995; Li et al. 1999; Coombs et al. 2002, 2003), and the ubiquitin 3 promoter (Garbarino and Belknap 1994) for high constitutive expression.

Cocultivation of *Agrobacterium tumefaciens* cultures with leaf segments from tissue culture-derived plants has led to successful transformation of potato (An et al. 1986; De Block 1988; Douches et al. 1998; Sheerman and Bevan 1988; Tavassa et al. 1988; Visser et al. 1989; Wenzler et al. 1989). Despite these numerous transformation/regeneration protocols proving successful in generating transgenic potato lines, several protocols lack genotype independence. Yadav and Sticklen (1995) developed a highly efficient two-step regeneration protocol for leaf explants. Douches et al. (1998) combined this adventitious shoot regeneration protocol with *Agrobacterium tumefaciens*-mediated transformation procedures to obtain transgenic plants. This regeneration protocol has broad genotypic adaptability (Li et al. 1999; Coombs et al. 2002, 2003; Felcher et al. 2003) and a frequency of transgenic plant recovery up to 50%. Contrarily, Sheerman and Bevan (1988) experienced highly genotype-dependent responses to their regeneration protocol and recovered no transgenic lines for some potato genotypes. The De Block (1988) transformation/shoot regeneration protocal gave mixed results in our laboratory in previous studies (data not shown), producing many regenerants and transgenic plants of some cultivars, and few to none for others. Stiekema et al. (1988) used tuber discs as the explant source for *Agrobacterium*-mediated transformation and recovered only 1% transgenic shoots. Newell et al. (1991) proposed a transformation technique using potato stem sections; transgenic plant recovery was 2-5%.

Growth and performance of transgenic plants can be affected by three distinct factors: somaclonal variation, insertional mutagenesis, and pleiotropy (Dale and McPartlan 1992). Somaclonal variation can be observed among the regenerated plants and can lead to distinct phenotypic differences or yield performance. Somaclonal variation from the regeneration process may also affect the naturally bred resistance factors, leptines and glandular trichomes. Dale and McPartlan (1992) and Belknap et al. (1994) reported that the *gus* gene had a negative effect upon the physiology of agronomic traits in transgenic potato. Li et al. (1999) and Coombs et al. (2002, 2003) used vector constructs that did not contain the *gus* gene. The deletion of *gus* eliminated the possibility of its negative action on *Bt-cry1Ia1* gene expression. Douches et al. (2002) reported that preselected *Bt-cry1Ia1* transgenic lines of Atlantic and Spunta were comparable to the nontransgenic cultivars. Contrarily, phenotypic differences between transgenic lines and nontransgenic parent lines have been reported (Felcher et al. 2002; Dale and McPartlan 1992; Belknap et al. 1994; Conner et al. 1994). Insertional mutagenesis can occur when DNA, via transformation, is inserted into or close to a plant's gene and disrupts its function. Changes in plant phenotype may be obvious or subtle depending on the gene(s) affected. Dale and McPartlan (1992) considered T-DNA insertional mutagenesis unlikely to have a significant influence in potatoes because the majority of mutations are known to be recessive and further, a high proportion of the DNA of potatoes is not actively coding. Pleiotropy occurs when a gene has multiple effects on the plant phenotype. Jefferson (1990) reported that constructs that differed in regulation of tuber to shoot expression of a *gus* gene paralleled field performance. The construct with greater shoot expression of GUS was more disruptive on plant growth.

Bacillus Thuringiensis

The majority of efforts to engineer insect resistant transgenic potatoes have focused on genes derived from *Bacillus thuringiensis* (Bt). Bt is an aerobic, gram-positive, soil bacterium that accumulates high levels of insecticidal crystal (Cry) proteins during sporulation (McGaughey and Whalon 1992; Barton and Miller 1993). These Cry proteins, also known as δ-endotoxins, are the principal active ingredients in Bt formulations currently in use (McGaughey and Whalon 1992). For over 30 years, Cry proteins have been used as biological insecticides to control agricultural pests. The spores dissolve in the midgut of the insect, generally requiring alkaline pH conditions, releasing one or more Cry proteins that are activated by gut proteases. The activated Cry proteins bind specifically to receptors in the midgut of the insect and form pores or lesions, causing lysis of cells in the gut wall and septicemia (Lambert and Peferoen 1992). Affected insects stop feeding and die from starvation and/or septicemia (Hilder and Boulter 1999). Small larvae are often more affected than large larvae or adults because of their high feeding rates in relation to body size. Cry proteins are highly host specific. For example, Cry1A is toxic to Lepidoptera, Cry3A is toxic to Colorado potato beetle and cottonwood leaf beetle, and Cry2 is toxic to Diptera. Host specificity is due to conditions in the gut required for solubilization and activation and, most importantly, due to specific receptors on the midgut binding sites of susceptible insects. This narrow host specificity is an advantage of the Bt toxin over conventional chemical insecticides, providing safety to nontarget organisms including beneficial insects and humans.

Use of Bt Toxin Genes for Insect Control

The δ-endotoxins are encoded by single genes (*Cry* genes) and transgenic plants expressing Bt genes are the most effective means of delivering Bt-based protein toxins. Barton et al. (1987), Fischhoff et al. (1987) and Vaeck et al. (1987) produced the first Bt-expressing transgenic plants (tobacco and tomato). These initial reports indicated poor transcription of *cry* genes in transgenic plants compared to many other heterologous genes (Barton et al. 1987; Fischhoff et al. 1987; Vaeck et al. 1987). Insufficient production of toxic protein *in situ* led to poor host plant resistance. A-T content, higher than observed in plant DNA sequences, was found in native δ-endotoxin coding regions, reducing the stability of mRNA (Fujimoto et al. 1993). In the following years, Bt gene modifications focused on the truncation of A-T rich regions and codon-modification in native *cry* protein genes (such as *cry1* and *cry3A*), resulting in a distinct increase in Bt gene expression in certain crops (Perlak et al. 1991; Murray et al. 1991; Adang et al. 1993; Koziel et al. 1993; Wunn et al. 1996). These engineered changes in the native Bt genes in conjunction with high expressing promoters led to very high levels of Bt expression in potato foliage (Sutton et al. 1992; Perlak et al. 1993). Adang et al. (1993) reported zero survival of 250,000 Colorado potato beetle larvae fed on high expressing potato Bt lines. Recently Kota et al. (1999) achieved extremely high Bt expression in tobacco by targeting the Bt transgene to the chloroplast genome.

Development and Testing of Bt Potatoes for Insect Resistance

Use of *cry* genes to provide insect control in potato has been attempted by numerous research groups since the initial efforts in the late 1980s. Sticklen et al. (1993) transformed the potato with a wild type *Bt-cry1A(c)* HD 73 gene. The resultant transgenic plants demonstrated control on *Manduca sexta*. Jansens et al. (1995) transformed cvs. Bintje and Kennebec with a native *cry1A(c)* gene and used a modified truncated *cry1A(c)* gene for transformation into Kennebec and cv. Yesmina. The transgenic lines expressing the codon-modified Bt gene were significantly better than the noncodon-modified transgenic plants in controlling potato tuber moth in leaf and tuber bioassays. In addition, in storage bioassays infested with neonate larvae, almost 100% tuber moth control was observed for approximately two months. Beuning et al. (2001) modified the truncated version of a *cry1Ac9* nucleotide sequence to achieve a significant increase in expression and insecticidal activity. This gene was placed under the control of the CaMV 35S promoter and transformed into tobacco. Control of potato tuber moth was achieved. Davidson et al. (2002) transformed the potato with this codon-modified *cry1Ac9*. Phenotypically normal transgenic plants of cvs. Ilam Hardy and Iwa were field tested and many significantly inhibited larval growth of potato tuber moth when fed field-grown foliage. Mortality was not reported.

In an effort to expand the array of *cry* genes available for expression in plants, Gleave et al. (1998) developed *Bt-cry9Aa2* tobacco plants with increased expression of the gene by increasing the stability of the *cry* mRNA transcript via minor modification of the nucleotide sequence. This level of gene expression conferred resistance to the potato tuber moth. Kuvshinov et al. (2001) placed a synthetic (24% nucleotide change) *cry9Aa* gene under the control of the double 35S promoter and transformed potato cv. Pito. Mortality of potato tuber moth larvae ranged from 50-100% in 9-day feeding tests.

Douches et al. (1998) transformed potato with a codon-modified *Bt-cry1la1* (*Bt-cry1Ia1*) gene to combine natural and engineered mechanisms to control infestation of potato tuber moth. The purpose of this research was to augment natural resistance. The susceptible cv. Lemhi Russet and two clones with different host plant resistance mechanisms, USDA8380-1 (foliar leptines) and NYL235-4 (glandular trichomes), were transformed with the *Bt-cry1la1* gene. Detached leaf bioassays showed that high levels of Bt gene expression occurred in the *Bt-cry1la1* transgenic lines, providing up to 96% control in growth of potato tuber moth larvae, compared with 3 and 54% control in NYL235-4 (glandular trichome line) and USDA8380-1 (high foliar leptine line) respectively. Li et al. (1999) incorporated the *Bt-cry1la1* gene into binary vectors with different promoters and the presence or absence of the *gus* reporter gene. These constructs were integrated into cv. Spunta by *Agrobacterium tumefaciens* mediated-transformation. Highest expression of *Bt-cry1la1* gene, determined by mRNA levels and insect mortality, was obtained using CaMV 35S promoter without the *gus* gene configuration. Detached-leaf and tuber bioassays showed a mortality rate of up to 100% for potato tuber moth in these transgenic lines. These results demonstrated that transcription of the *gus* gene negatively affects the expression level of *Bt-cry1la1*. Bt gene expression was also facilitated by using the $(ocs)_3mas$ superpromoter, whereas Bt gene expression regulated by the patatin promoter (tuber specific) was too low to exert any effect on mortality of the potato tuber moth. Mohammed et al. (2000) evaluated potato tuber moth resistance in these tubers and other *Bt-cry1la1* transgenic potato lines. Potatoes, stored 11-12 months and newly harvested tubers of cv. Lemhi Russet and cv. Atlantic potato lines showed high larval mortality (up to 100%) except in the newly harvested tubers of *Bt-cry1la1* Atlantic lines, which showed a mortality range from 47.5% to 67.5%. The *Bt*-Şpunta lines tested were the products of three different *Bt-cry1la1* constructs. Mortality was high (100%) in the lines transformed with CaMV 35S promoter and two of three lines transformed with $(ocs)_3$ *mas* promoter. Lines with the patatin promoter showed the lowest mortality of the three constructs (31.25 and 32.5%). Lagnaoui et al. (2001) evaluated the *Bt-cry1la1* potatoes against both *P. operculella* and *S. tangolias*

(Lepidoptera: Gelechiidae) using detached-leaf bioassays. All *Bt-cry1la1* Spunta lines showed high levels of mortality with both species (80-98%). The *Bt-cry1la1* Spunta lines were also effective against European corn borer (*Ostrina nubilalis*), but not against cabbage looper (*Trichoplusia ni*) in detached-leaf bioassays (Santos, Grafius and Douches, data not shown).

Douches et al. (2004) reported on field and storage studies to evaluate these *Bt-cry1la1* potato lines for resistance to potato tuber moth in Egypt under natural infestations and their agronomic performance in both Egypt and Michigan. A total of 27 *Bt* transgenic potato lines from six different *Bt* constructs were evaluated over a five-year period. The 1997 field trial was the first field test of genetically engineered crops in Egypt. The *Bt-cry1la1*-Spunta lines (Spunta-G2, Spunta-G3, and Spunta-6a3) were the most resistant lines in the field with 99-100% of tubers free of damage. In Nawhalla storage studies, these three lines were also over 90% free of tuber moth damage after 3 months. NYL235-4.13, which combines glandular trichomes with the *Bt-cry1la1/gus* fusion construct, also had a high percentage of clean tubers in the field studies. In agronomic field trials in Michigan from 1997-2001, in most instances the Bt transgenic lines performed similar to the non transgenic line in agronomic trials; however, in Egypt (1998-1999) yields were less than half of those in Michigan. Expression of the *Bt-cry1la1* gene in both the potato tuber and foliage provides the seed producer and grower a tool with which to reduce potato tuber moth damage to the tuber crop in the field and in storage.

Douches et al. (2001) studied the effectiveness of natural and engineered host plant resistance in potato to the Colorado potato beetle using 72-h detached leaf bioassays from greenhouse grown foliage. Insecticide-resistant, first instar Colorado potato beetles were used for this study. Potato lines tested included nontransgenic cultivars (Russet Burbank, Lemhi Russet, and Spunta), the glandular trichome line (NYL235-4), the high foliar leptine line (USDA8380-1), and transgenic lines expressing either codon-modified *Bt-cry3A* or *Bt-cry1la1*. *Bt-cry3A* transgenic lines, foliar leptine line, and foliar leptine lines with *Bt-cry1la1* reduced feeding compared to nontransgenic cultivars. Glandular trichome lines and glandular trichome lines with *Bt-cry1la1* did not reduce feeding in this no-choice feeding study. Some *Bt-cry1la1* transgenic lines, using either the constitutive promoters CaMV 35S or $(ocs)_3mas$, were moderately effective in reducing larval feeding. Feeding on *Bt-cry1la1* transgenic lines with the tuber-specific patatin promoter did not differ significantly from feeding on susceptible cultivars. Mortality of first instars was highest when fed on the *Bt-cry3A* lines (68-70%) and intermediate (38%) on the *Bt-cry1la1* line Spunta-G3. From this work it was concluded that host plant resistance from foliar leptines is a candidate mechanism to pyramid with either *Bt-cry3A* or *Bt-cry1la1* expression in potato foliage against Colorado potato beetle.

Combining natural host plant resistance mechanisms in potato with Bt should enhance the efficacy and sustainability of control of Colorado potato beetle in commercial potato production. Coombs et al. (2002) combined the resistance mechanisms of leptine glycoalkaloids and glandular trichomes with the synthetic *Bt-cry3A* gene. *Bt-cry3A* transgenic plants were developed for three different host plant resistant potato lines: cv. Yukon Gold (susceptible), USDA8380-1, and NYL235-4. Detached-leaf bioassays of the *Bt-cry3A* engineered transgenic lines demonstrated that resistance effectively controlled feeding by first instar Colorado potato beetles. The susceptible Yukon Gold control had 32.3% defoliation, USDA8380-1 had 3.0% defoliation, and NYL235-4 had 32.9% defoliation. Mean percent defoliation for all transgenic lines ranged between 0.1 and 1.9%. Mean mortality ranged from 0.0 to 98.9% among the *Bt-cry3A* transgenic lines, compared to 20% for the susceptible Yukon Gold control, 32.2% for USDA8380-1 and 16.4% for NYL 235-4.

Coombs et al. (2003) conducted a field evaluation of these natural, engineered, and combined resistance mechanisms in potato for control of Colorado potato beetle. Nine different potato clones representing five different host plant resistance mechanisms were evaluated under natural Colorado potato beetle infestations in Michigan. The *Bt-cry3A* transgenic lines, the high leptine line (USDA8380-1), and the high foliar glycoalkaloid line (ND5873-15) were most effective for controlling defoliation by Colorado potato beetle adults and larvae. The *Bt-cry1la1* line (Spunta-G2) was not as effective as the *Bt-cry3A* transgenic lines. The glandular trichome (NYL235-4) and *Bt-cry1la1* + glandular trichome lines proved ineffective. Based on these results in this choice field study, the *Bt-cry3A* transgenic lines, the high leptine line, and the high total glycoalkaloid line proved effective host plant resistance mechanisms for control of Colorado potato beetle.

Coombs et al. (2001) also conducted field studies to compare natural (glandular trichomes and glycoalkaloid-based), engineered (*Bt-cry3A)*, and combined (glandular trichomes + *Bt-cry3A* and glycoalkaloids + *Bt-cry3A* transgenic potato lines) host plant resistance mechanisms of potato for control of Colorado potato beetle. Twelve different potato lines were evaluated in a choice situation under natural Colorado potato beetle pressure in Michigan and Long Island, New York. The high glycoalkaloid line, all *Bt-cry3A* transgenic, and the combined resistance lines were effective in controlling feeding by Colorado potato beetle adults and larvae. Virtually no feeding was observed in the glycoalkaloid + *Bt-cry3A* transgenic line while markedly less feeding was observed in the glandular trichome line than the susceptible control. Based on these results, the *Bt-cry3A* transgenic and glycoalkaloid-based host plant resistance mechanisms proved effective tools for control of Colorado potato beetle.

ALTERNATIVE GENES FOR INSECT CONTROL

The strategy of expressing Bt genes in crop plants has been well exploited as a means of protecting the plants from insect attack. Hilder and Boulter (1999) define the ideal transgenic technology as possessing the following characteristics:

- — relatively economical to produce,
- — environmentally benign,
- — easily deployed in the field,
- — have a wide spectrum of activity against pests, but consumer friendly and beneficial to useful insects,
- — flexible in technology to target vulnerable site of a pest or a particular pest species,
- — adaptable to allow development of alternatives when resistance develops, and
- — producing acute rather than chronic effects on the pest.

Bt-transgenics meet the first three characteristics of the ideal transgenic technology. Development and testing of other genes to complement, enhance or replace Bt genes has and continues to be explored. Other transgenic strategies may possess more of the ideal characteristics described above. Plant-based proteins may provide some level of crop protection. One class of alternative insect control proteins is protease inhibitors. A common trait of these compounds is that the effect is usually chronic rather than acute. The result may be an increase in mortality and retardation of insect growth and development. For example, serine protease inhibitors have been targeted against Lepidoptera since they depend on serine protease as their primary protein digestive enzyme, leading to a decrease in amino acid availability (Hilder and Boulter 1999). Gatehouse et al. (1997) produced transgenic potatoes (cv. Desiree) that expressed the cowpea trypsin inhibitor gene. Insect bioassays with tomato moth (*Lacanobia oleracea*) reduced total insect biomass and survival, but did not protect the plants from insect damage. Cloutier et al. (2000) examined the effects of feeding Colorado potato beetle with foliage from the cv. Kennebec priorly transformed with the oryzacystatin I gene. Oryzacystatin I is a cysteine proteinase inhibitor against newly emerged adult female Colorado potato beetles. Interestingly, it was determined that the nutritional stress to the females led to increased foliage consumed per egg laid. In contrast, Lecardonnel et al. (1999) reported 43 and 53% mortality of potato beetle larvae in a 13-day bioassay with oryzacystatin-transgenic lines of cv. Bintje and BF15.

Gatehouse et al. (1997) also transformed the potato with three other plant-derived genes: snowdrop lectin (GNA), bean chitinase (BCH) and wheat α-amylase (WAI). In addition, transgenic potatoes were produced that expressed the genes in pairwise combinations. Plant lectins effect development and survival of coleopteran and homopteran insects fed artificial diets. GNA has also been shown to be toxic to aphids (Sauvion et al. 1996). It is also believed that inhibitors of α-amylase could interfere with nutrient utilization such as protease inhibitors (Gatehouse et al. 1997). Chitinases could disrupt the insect gut lining and possibly interfere with normal digestive physiology. The GNA and BCH transgenic lines accumulated the proteins at levels up to 2% of the total soluble proteins. The WAI-transgenic potatoes did not readily express a detectable level of mRNA transcript. When tested against the tomato moth, GNA-expressing plants showed increased resistance (50% reduction in leaf damage), while BCH-expressing potatoes were ineffective. Potatoes containing the double gene constructs (BCH + GNA and WAI + GNA) performed similar to the GNA-transgenic lines in tomato moth bioassays. Gatehouse et al. (1999) also produced concanavalin A (Con A) transgenic potato lines that expressed the glucose/mannose-specific lectin from jackbean. Despite low expression levels of ConA, bioassays with *L. oleracea* showed that larval development was retarded and also reduced tissue consumption by the larvae. Even though these plant-encoding genes are not as effective as Bt genes, their use has potential for instilling resistance to a wide range of chewing and sucking insect pests. These transgenic strategies should complement other crop protection management strategies.

Marwick et al. (2001) tested the insecticidal activity of two biotin-binding proteins, avidin and streptavidin, against four lepidopteran pests. Biotin deficiency leads to stunted growth and mortality among several insects. Potato tuber moth larvae were the most susceptible of four pests in dose mortality tests. Kramer et al. (2000) transformed maize and expressed the avidin glycoprotein in maize grains. Avidin maize gave excellent control of three grain storage pests when the kernels contained >100 ppm avidin, suggesting that this transgenic strategy can be used against a spectrum of storage pests. Avidin expressing transgenic potatoes were developed with up to 100% mortality of potato tuber moth in four-day leaf bioassays (A. Conner, pers. comm.).

The insecticidal properties of δ-endotoxins from *B. thuringiensis* have been the primary focus of transgenic strategies. However, *B. thuringiensis* offers alternative insecticidal proteins to the δ-endotoxins. Estruch et al. (1996) cloned and characterized genes (*vip3A(a)* and *vip3A(b)*). These genes encode vegetative insecticidal proteins from *B. thuringiensis* cultures produced prior to sporulation. The gene products show activity against some Lepidoptera and represent a novel class of insecticidal proteins that can be engineered for transgenic-based host plant resistance. Cyt proteins, produced in sporulating *B. thuringiensis*, have similar intoxication pathways as Cry proteins in susceptible insects. Federici and Bauer (1998) have shown that Cyt1Aa, originally thought to be toxic only to dipterans, is highly toxic to *Chrysomela scripta* (cottonwood leaf beetle). In addition, they showed that the Cyt1Aa protein suppresses high levels of resistance in *C. scripta* selected for resistance to Cry3Aa. These results suggest that δ-endotoxins, used in combination or rotation, may provide a more effective resistance management strategy due to a lack of cross-resistance. Cloning and expression of the Cyt proteins in plants have not been reported to date.

Photorhabdus and *Xenorhabdus* species are potential sources of genes for engineered Colorado potato beetle resistance in potato. They are facultative aerobic Gram-negative bacteria that produce novel toxins that are orally infectious to insects. *Photorhabdus* and *Xenorhabdus* species are symbionts in the guts of the infective juvenile stage of entomopathogenic nematodes belonging to families Heterorhabditidae and Steinernematidae respectively (Forst et al. 1997). When nematodes enter the insect the bacteria are released into the hemocoel and, in concert with the nematode, kill the insect and help break down its tissues to provide nutrition for both the nematode and bacteria. The bacteria have been isolated from nematodes (as non free-living microbes) and can be cultured as free-living organisms using standard laboratory conditions. Under these conditions they secrete a number of products into the culture medium. The secreted products include antibiotics, toxins, proteinases, and lipases that presumably aid the nematode in killing and utilizing insect prey by suppressing the insect's immune system and growth of competing microbes as well as digestion of insect tissues. A number of toxin complexes have been isolated from the supernatants of cultured *P. luminescens* (Bowen et al. 1998). Two of these, Tca and Tcd, are orally infectious to lepidopteran larvae. Tca exerts a specific action on the midgut whether it is ingested or injected into the hemocoel, although the mechanism is not currently known. Similar toxins have been isolated from *X. nematophilus*. (french-Constant and Bowen, 2000). A heterorhabditid sp., *Heterorhabditis marelatus,* isolated in Seaside, OR (Liu et al. 1997) was especially effective against the Colorado potato beetle, resulting in nearly 100% mortality in laboratory tests (Berry et al. 1997). Based on 16S rRNA sequences the symbiotic bacteria isolated from this nematode differed significantly from previously characterized *Photorhabdus* species (Liu et al. 1997), suggesting that it may produce novel toxins that would be particularly effective for engineering resistance to the Colorado potato beetle.

COMMERCIAL GROWTH OF BT POTATOES

The debate about use of biotechnology should focus on comparing transgenic technology with existing technologies in the areas of food safety, human health, environmental compatibility, beneficial as well as risk factors to the producer and consumer, effects on general food systems, and issues of social justice (Shelton et al. 2002). Moreover, the risks and benefits of Bt potatoes are neither certain nor universal. Temporal and spatial effects make assessment of risks or benefits a case-by-case analysis. Emphasis on the safety assessment of transgenic potatoes should focus on whether transgenic potatoes pose any new risks over existing potato cultivars, or on new cultivars released from traditional breeding efforts, in the context of environmental or food safety factors (Conner 1993). Scientific evaluation to predict ecological impacts of the transgenic potato is imprecise and data collected for assessing potential ecological impacts have limitations (e.g. long-term and higher-order interactions). Conner et al. (1994) stated that most field evaluations of transgenic potatoes are based on small-scale trials. It is hard to extrapolate performance to full-scale commercial production. Furthermore, unidentified benefits and risks may exist that published data do not currently address. The track record of conventional potato breeding for the safe introduction of new cultivars with new gene combinations should set a positive foundation for assessing risks in releasing transgenic potatoes.

Bt potatoes meet key criteria as a new pest control product: technical feasibility, need, efficacy and safety. Microbial Bt-based products have been used safely for almost 40 years. Betz et al. (2000) reviewed the human health implications of Bt micorbial pesticides and Cry proteins in Bt plants. Bt proteins are not generally toxic to humans and animals. Cry proteins exhibit a high degree of specificity to the target insect and have no contact activity. They also conclude that Cry and marker proteins are not toxic to humans and pose no significant concern for allergenicity. Cry3 and Cry1 proteins are rapidly degraded *in vitro* using gastric fluids. This rapid degradation minimizes the potential for the protein to induce an allergic reaction. They also reasoned that the modified Cry proteins pose no unique concerns since modified Cry proteins have already been generated in nature. Lavrik et al. (1995) showed that NewLeaf potatoes are substantially equivalent to non transgenic parental lines for key nutrients and antinutrients. Therefore, Bt potatoes are safe for consumption.

Birch et al. (1999) reported negative effects of 2-spot ladybird beetle (*Adalia bipunctata* L.) fecundity, egg viability, and adult longevity when feeding on aphids colonizing potatoes expressing snowdrop lectin. In contrast, convergent lady beetles (*Hippodamia convergens* L.) feeding on aphids colonizing Bt potatoes had no reported effect on growth, fecundity, or female offspring longevity (Dogan et al. 1996). Betz et al. (2000) reported that Cry proteins were practically nontoxic to nontarget organisms such as ladybird beetles, Collembola, honeybees, earthworms, parasitic wasps, green lacewings and Bobwhite quail.

Kanamycin-based selection during transformation has been a common feature in the development of transgenic plants. Concerns revolving around the gene and its protein expression products include horizontal gene transfer to microbe populations, immunologic and allergenic response to the antibiotic, and loss of an antibiotic for oral therapeutic use (Wolfenbarger and Phifer 2000). Despite positive food toxicology and safety assessments for the NPTII protein (Fuchs et al. 1993a,b), negative public perception of antibiotic selection exists. Herbicide resistance can be used as an alternative to antibiotic selection in other Bt crops (NatureMark 2002). A transformation system that excises the selectable marker from the T-DNA after transformation has been developed for tobacco and poplar (Ebinuma et al. 1997). This MAT vector system is currently being attempted in potato (Zarka, pers. comm.).

Considering the outcrossing potential of Bt potatoes, there is little possibility of gene capture and expression of Bt endotoxins by wild and weedy relatives in the United States. This statement would probably hold for most countries except those in Central and South America. Crawley et al. (2001) studied whether Bt potatoes could potentially become weeds in agricultural fields or invasive to natural habitats. They concluded that the ability of Bt potatoes to become a dominant invasive plant in the wild is limited. Conner (1993) detected only minimal transgenic pollen dispersal.

In potato production, broad-spectrum insecticides have traditionally controlled the Colorado potato beetle. NewLeaf potatoes require no insecticide applications to control the Colorado potato beetle. Therefore, these Bt potatoes reduce the grower's need to handle pesticides. This reduces the risk of misuse and misapplication of pesticides and worker exposure to pesticides. With broad-spectrum insecticide use, secondary pests can become a problem when predators of these pests are reduced. Bt potatoes, without additional insecticide treatments, may provide supplemental pest control by preserving populations of beneficial organisms. For example, spider mite infestation in NewLeaf plus Russet Burbank potatoes was lower than in Russet fields treated with insecticides and miticide (Betz et al. 2000). Howerver, indirect effects can occur; decreased numbers of a predatory specialist in a *Bt-cry3A* potato field may be explained by effective control of the Colorado potato beetle (Riddick and Barbosa 1998).

NatureMark, a Monsanto-affiliated company, commenced commercial release of NewLeaf potato cultivars (*Bt-cry3A* potatoes) in 1995 in the USA (Russet Burbank, Atlantic, Superior) NewLeaf Plus Russet Burbank was released in 1998 and New Leaf Y Russet Burbank and Shepody in 1999. Additional commercial releases have occurred in Romania (1999) and Canada (1995 and 1999). Limited production of Bt potatoes has occurred because of public perception of market concerns for transgenic food crops. Introduction of imidacloprid, a systemic insecticide, is an effective alternative to Bt potatoes. Bt potato acreage peaked at 50,000 acres (4% of the market). Based upon EPA estimates, Bt potatoes resulted in 89,000 fewer acre treatments of insecticides on potato (Shelton et al. 2002). In 2001, NatureMark stopped marketing Bt potatoes.

RESISTANCE MANAGEMENT

The effectiveness or sustainability of insect-resistant potatoes is intertwined with the evolution of resistance. There are many similarities between insecticide resistance and host plant resistance. Both insecticide and host plant resistances will be widely and frequently applied to agricultural fields in future and both select strongly against pests adapted to survive. Modes of action and detoxification mechanisms are also similar between many insecticides and plant allelochemicals. For example, host plant alkaloids, glycoalkaloids, carbamate and organophosphate insecticides inhibit acetylcholinesterase in the Colorado potato beetle (Bushway et al. 1987). In fact, alterations in acetylcholinesterase result in resistance to organophosphate and carbamate insecticides. They also affect acetyl cholinesterase inhibition caused by the potato glycoalkaloids α-chaconine and α-solanine (Wierenga and Hollingworth 1992, 1993; Zhu and Clark 1995). Detoxification enzymes in the Colorado potato beetle also act on both insecticides and plant allelochemicals (Mahdavi et al. 1991).

Theories of insecticide resistance management have potentially broad application to managing host plant resistance factors (McGaughey and Whalon 1992). Most insecticide resistance management depends on alternation of chemical use, allowing refuges for susceptible insects, and use of nonchemical mortality factors (Georgiou and Taylor 1986). Computer models describing insect adaptation to transgenic plants are made using inheritance characteristics of the resistant gene, seed mixtures, and refuges to predict occurrence of insect resistance. The validity of many resistance management systems is only as good as the validity of the assumptions that make up the model since empirical data are scarce or lacking (Tabashnik et al. 1992; Tabashnik 1994; Kennedy and Whalon 1995; Alstad and Andow 1995). We hypothesize that these same approaches can be used to maintain the effectiveness of host plant resistance factors.

Chemical insecticide resistance management techniques have been emphasized in commerical potato production which include alternations of insecticides, use of insecticide mixes, and high does strategies. These have been largely ineffective for the Colorado potato beetle because insecticide resistance often involves a stable, dominantly inherited factor, conferring a high level of resistance, with little fitness cost (Bishop and Grafius 1996). To maintain effectiveness of host plant resistance factors, mixes (gene pyramids) have proven effective for managing adaptation to disease resistant cultivars (Lamberti et al. 1981) and may also be effective for managing adaptation of the Colorado potato beetle.

Targeting a single specific site has not offered much of an evolutionary barrier in the development of resistance. The single toxin approach often leads to low mortality of heterozygous insects. In an ideal situation, only homozygous resistant insects will survive treatment and mate primarily with the susceptible insects from a nearby refuge, which produces mostly susceptible and heterozygous offspring. If both homozygous resistant and heterozygous insects survive, a larger number in the field are more likely to mate among themselves producing homozygous resistant and heterozygous offspring. By pyramiding effects of two or more distinctly different toxins, it was found that survivorship of the heterozygous insects decreased significantly (Roush 1999).

When insecticides with very different modes of action are combined, resistance development is delayed by orders of magnitude suggesting that pyramiding may be as effective as the host plant resistance (Mani 1985). With a 30% refuge, Gould (1986) found resistance to a single toxin evolved after 14 generations, but with two toxins resistance evolved only after 123 generations. The benefits are apparent. Computer simulated models of pyramiding host plant resistance genes also suggest that pyramiding will extend the effective life of the toxin. Roush (1999) modeled cultivars showing both single and two toxin resistance with a 10% refuge. Insects developed resistance to the single toxin in only 5 generations, whereas in two-toxin cultivars resistance arose only after 50 generations. In the model, the refuges for two-toxin cultivars could be greatly reduced, from 30-40% to 10% but still express resistance successfully (Roush 1999). Colorado potato beetles are not highly mobile, hence refuges may be largely ineffective (Alyokhin et al. 1999). Therefore, the two-toxin cultivar is of even greater benefit to the Colorado potato beetle-potato system because the refuge is less effective.

Bt Resistance Management

The Colorado potato beetle is widely known for its adaptability to resist any insecticide used for its control (Bishop and Grafius 1996). The potato tuber moth has also been able to adapt to resist insecticides (Collantes et al. 1986). Thus there is a high likelihood that the two will adapt to the Bt toxin if it is widely deployed. In fact, a strain of Colorado potato beetle with high levels of Bt resistance was selected in the laboratory (Whalon et al. 1993) and this resistance proved to be polygenic and unstable in the absence of selection. However, it is not known whether resistance selected for field deployment would be similar in nature. If Bt resistance appears in the field, its inheritance and stability will be critical in determining our ability to manage this resistance; resistance inherited recessively and unstable in the absence of selection is much easier to manage than resistance inherited in a dominant or semidominant manner and stable in the absence of selection. Although we won't know the status of resistance to Bt in the field until it appears, resistance to many insecticides in Colorado potato beetle is inherited as a dominant trait and is stable for long periods of time even without selection (Bishop and Grafius 1996).

Many strategies for managing Bt crops have been implemented, including: (1) high level of a single toxin; (2) mixture of nonresistant and resistant plants in the field; (3) use of low level toxins and biocontrol agents; (4) toxins deployed sequentially; and (5) pyramiding multiple toxins (Gould 1986). Programs to manage pest adaptation to Bt transgenic crops rely on a high dose of Bt toxin and structured refuges (US EPA 2001). The refuges in these systems provide sources of large numbers of susceptible individuals that readily mix in time and space with resistant individuals appearing in the transgenic crop. The basic assumptions are: (1) Bt resistance will be inherited recessively in relation to the Bt dose present in the crop; (2) the number of Bt resistant individuals will initially be very low; (3) the refuges will produce a large number of susceptible individuals; and (4) these susceptible individuals will mix with resistant individuals arising in the crop before mating has occurred. The result will be heterozygous offspring, susceptible to Bt at the levels expressed in the transgenic crop. Grower compliance on a regional scale is also critical for managing resistance to Bt in transgenic crops using any system relying heavily on refuges. Even with strict government regulation and industry self-monitoring, one or a few growers ignorant of the regulations or not following them would suffice to cause failure (Grafius 2002). Thus, pyramiding of Bt toxin genes with other resistance genes may be a much more effective resistance management strategy than single factor resistance, relying on high toxin doses and structured refuges for resistance management (Shelton et al. 2002).

Foliar expression of the Bt gene may be critical to minimizing foliar damage by the potato tuber moth and reducing potato tuber moth population levels in the field. Moreover, Bt expression in leaves may reduce tuber damage because newly hatched larvae typically feed on the foliage before dropping to the ground and feeding on the tuber. As an alternate host resistance management strategy, it might be advisable to deploy Bt protein only in the tuber. According to this principle, we introduced a tuber-specific promoter (class-I patatin element) into the *Bt-cry1Ia1* construct (Li et al. 1999).

Natural Host Plant Resistance

If Bt expression is to be used as a tool in crop protection, then the next step is to determine how the lines with combined resistances can be used in an insect management strategy to build a more durable host plant resistance. Currently, there are two defined host plant resistance factors available in the *Solanum* gene pool that contribute to plant defenses against insects: glandular trichomes and leptine glycoalkaloids. Small insects exhibit modified behavior in the presence of trichomes, such as host avoidance and restlessness, reduced feeding, delayed development, and diminished longevity (Tingey 1991). Three wild *Solanum* species, *S. berthaultii*, *S. polyadenium*, and *S. tarijense*, have high densities of glandular trichomes (Tingey et al. 1984) that have been bred into cultivated potato. Breeding line NYL235-4 has glandular trichomes derived from *S. berthaultii* and is available for further research and breeding (Plaisted et al. 1992). This line had a positive effect on both the potato tuber moth and the Colorado potato beetle.

Glycoalkaloids are the most common form of antibiosis in potato (Sinden et al. 1986) and have been shown to inhibit acetyl cholinesterase (Bushway et al. 1987). They have also demonstrated membrane disruption by lysis of sterol-containing liposomes (reviewed in Lawson et al. 1993). Steroid glycoalkaloids (solanine and chaconine) are present in all plant tissues including potato tubers and processed products (Sinden 1987). Glycoalkaloids below 20 mg/100 g fresh weight (mg %) in tubers are considered safe for human consumption. *S. tuberosum* generally contains only 2-10 mg% glycoalkaloid whereas wild potato species such as *S. chacoense* and *S. commersonii*, can have concentrations of 230 mg% and 500 mg% glycoalkaloids respectively (Sinden 1987). Acetylated glycoalkaloids are the most active form of glycoalkaloids present in potato. Leptines are acetylated analogues of the common potato glycoalkaloids, solanine and chaconine. Leptines as found in USDA8380-1 and other acetylated glycoalkaloids are only reported to be synthesized by some accessions of *S. chacoense* and synthesized only in leaves and not the tubers (Sanford et al. 1996). The strong antifeedant properties of high leptine glycoalkaloid levels found in a few accessions of *Solanum chacoense* confer natural host plant resistance against the Colorado potato beetle (Stürckow and Löw 1961; Sinden et al. 1980) and the potato tuber moth (Douches et al. 2001).

Both the glandular trichomes and leptines may affect the insect pests (e.g. potato tuber moth and Colorado potato beetle) at the host acceptance stage (Sinden et al. 1986; Yencho and Tingey 1994). Under field conditions, these natural host resistance mechanisms may reduce the number of insects accepting the plant as a host and ovipositing. This could reduce selection for adaptation to Bt gene in lines with combined resistance mechanisms.

As of now high leptine expression has been combined with high *Bt-cry3A* expression in a single genotype (Coombs et al. 2001). This line, along with lines exhibiting either *Bt-cry3A* or leptine expression alone, allowed us to test the effectiveness of combined resistance mechanisms versus individual mechanisms. Field studies were conducted in Michigan and Long Island, NY in 2000, comparing combined versus individual mechanisms (glandular trichomes, high total glycoalkaloids, and *Bt-cry3A*) of resistance (Fig. 11.4). Based on field and lab results in Michigan and Long Island, NY, both leptine and *Bt-cry3A*-based resistance give strong control of the Colorado potato beetle.

The ability to integrate conventional breeding and transgenic strategies may provide the best opportunity to build durable host plant resistance to insects in potato. Breeding programs could focus breeding efforts on the development of cultivars with natural host plant resistance mechanisms. The superior selections of this breeding effort would be candidates for combining the *Bt-cry3A* gene via *Agrobacterium*-mediated transformation (Douches et al. 1998). If there are problems with accumulation of glycoalkaloids above acceptable levels in the tubers of these superior selections, then antisense TGA technology could be applied (Stapleton et al. 1991; Moehs et al. 1997). This same strategy of combining natural and engineered traits could be applied to the glandular trichome-mediated resistance. The combination of *Bt-cry3A* gene and glandular trichomes may provide a broader based insect resistance giving control of small-bodied insects such as potato leafhoppers and aphids along with the Colorado potato beetle. Moreover, transgenic strategies that constitutively express plant-based proteins (Gatehouse et al. 1997; Gutierrez-Campos et al. 1999) could be utilized and provide new opportunities to pyramid multiple resistance factors in potato. For example, snowdrop lectin or bean chitinase could be combined with a lepidopteran or coleopteran-targeted Bt gene to develop double-gene vector constructs for transformation. These double-gene constructs could then be targeted for transformation with leptine-based resistance. As gene discovery progresses, other transgenes will become available for pyramiding in potato. Moreover, to support the strategy of integrated pest management, it may be necessary to develop transgenic potatoes with several Bt genes that bind at different receptor sites in the target pest to inhibit the formation of resistant insect populations. In future, transgenic approaches may focus on the exploitation of tissue-specific promoters and co-integration of potato cultivars with multiple Bt genes, such as *cryIA(b), cry3A, and cry1Ia1*. A combination of natural and genetically engineered host plant resistance factors should delay or eliminate the insect adaptation to resistant crop varieties.

CONCLUSIONS

Over the past 15 years the feasibility of expressing Bt genes in potato to provide protection against the Colorado potato beetle and potato tuber moth has been demonstrated. Other transgenic strategies have not been as dramatic in effect, but offer the possibility of enhanced resistance in combination with other resistance factors. Incorporation of host plant resistance into crop management offers the grower flexibility in design-

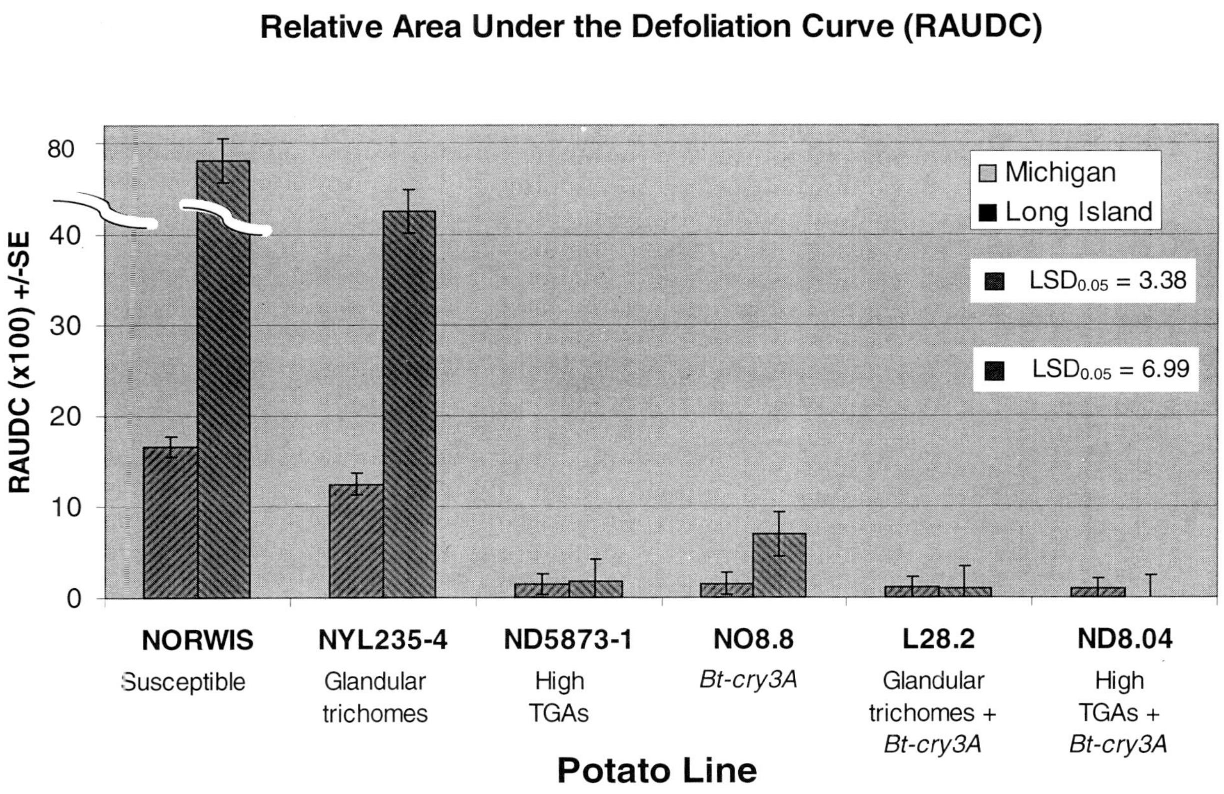

Fig. 11.4 Comparison of defoliation by Colorado potato beetles for five different host plant resistant potato lines in Montcalm County, Michigan and Long Island, New York. TGA: total glycoalkaloids. GT: glandular trichomes.

ing integrated pest management strategies that employ nonchemical management components. If pesticides are not used, harmful insects (parasites, predators) or adverse environmental conditions for insect survival could contribute to insect control.

Without multiple sources of host plant resistance, long-term sustainability is dubious for a highly adaptable insect such as the Colorado potato beetle. Each host plant resistance approach merits attention as potatoes expressing insect resistance promise to provide a sustainable potato production system that minimizes the environmental and human health costs associated with excessive use of pesticides. Further, it is anticipated that genes mediating resistance to major insect pests such as the Colorado potato beetle and potato tuber moth, which possess different modes of activity and pest control spectra, will eventually be pyramided into a single background, thereby producing potatoes with durable resistance to a broad range of insect pests. If durable host plant resistance can be established, then the use of pesticides and pest control costs could be reduced, resulting in simplification of production management strategies and reduced production cost.

A balanced breeding approach involving the biotechnologist, breeder and entomologist is necessary, since cultivars with superior disease and pest resistance, but lacking marketability traits would be of little commercial value. Likewise, potatoes with excellent marketability traits but highly susceptible to insect or disease pest would cause management problems in the field. The traits required by potato cultivars destined for both fresh market and processing include high yield, yield stability across locations, and appropriate tuber traits such as shape, size distribution, dry matter content, and storability. Fresh market potatoes need superior culinary quality for multiple consumer uses such as microwave cooking, boiling, and baking.

Acknowledgments

This research was supported by the office of USAID/CAIRO/AGR/A under cooperative agreement no. 263-0152-A-00-3036-00. It was also supported by the Michigan Agriculture Experimental Station.

REFERENCES

Adang, M.J., M.S. Brody, G. Cardineau, N. Eagen, R Roush, C.K., Shewmaker, A. Jones, J.V. Oakes, and K. McBride. 1993. The reconstruction and expression of *Bacillus thuringiensis cyr* IIIA gene in protoplasts and potato plants. Plant Molec Biol, 21: 1131–1145.

Alstad, D.N. and D.A. Andow. 1995. Managing the evolution of insect resistance to transgenic plants: Science 268: 1894–1896.

Alyohkhin, A., D. N. Ferro, C.W. Hoy, and G. Head. 1999. Laboratory assessment of flight activity displayed by Colorado potato beetles (Coleoptera: Chrysomelidae) fed on transgenic and Cry3a toxin-treated potato foliage. J Econ Entomol 92: 115–120.

An, G., B.D. Warson, and C.C. Chiang. 1986. Transformation of tobacco (*Nicotiana* spp.), tomato (*Lycopersicon esculentum*), potato (*Solanum tuberosum*) and *Arabidopsis thaliana* using a binary *Ti* vector system. *Plant Physiol* 81: 301-305.

Bald, J.G. and G.A.H. Helson. 1944. Estimation of damage to potato foliage by potato moth, *Gnorimoschema operculella* (Zell.). J Council Sci Indust Res 17(1): 30–48.

Barton, K.A. and M.J. Miller. 1993. Production of *Bacillus thuringiensis* insecticidal proteins in plants. *In:* S. Kung and R. Wu (eds), Transgenic Plants, vol. 1, Engineering and Utilization. Academic Press, San Diego, CA, pp. 297–315.

Barton, K.A., H.R. Whiteley, and N. Yang. 1987. *Bacillus thuringiensis* δ-Endotoxin expressed in transgenic *Nicotiana tabaccum* provides resistance to Lepidopteran insects. Plant Physiol 85: 1103–1109.

Belknap W.R., D. Corsini, J.J. Pavek, G.W. Snyder, D. R. Rockhold, and M.E. Vayda. 1994. Field performance of transgenic Russet Burbank and Lemhi Russet potatoes. Amer Potato J 71: 285–297.

Berry, R.E., J. Liu, and G. Reed. 1997. Comparison of endemic and exotic entomopathogenic nematode species for the control of Colorado potato beetle (Coleoptera: Chrysomelidae). J Econ Entomol 90: 1528-1533.

Betz F. S., B.G. Hammond, and R.L. Fuchs. 2000. Safety and advantages of*Bacillus thuringiensis*-protected plants to control insect pests. Reg Toxicol Pharmacol. 32: 156–173.

Beuning, L., D.S. Mitra, N.P. Markwick, and A.P. Gleave. 2001. Minor modifications to *cry* 1Ac9 nucleotide sequence are sufficient to generate transgenic plant resistant to *Phthorimaea operculella*. Ann Appl Biol. (in press).

Birch, A.N.E., I.E. Geoghegan, M.E.N. Majerus, J.W. McNicol, C.A. Hackett, A.M.R. Gatehouse, and J.A. Gatehouse. 1999. Tri-trophic interactions involving pest aphids, predatory 2-spot ladybugs and transgenic potatoes expressing snowdrop lectin for aphid resistance. Mol Breeding 5: 75–83.

Bishop, B.A. and E. Grafius. 1996. Insecticide resistance in the Colorado potato beetle. *In:* Joliet and Hsaio (eds.), *Chrysomelidae Biology*. SBP Acad. Publ. Amsterdam, vol. 1, pp. 355–377.

Bowen D., T.A. Rocheleau., M. Blackburn, O. Andreev, E. Golubeva, R. Bhartia, and R.H. french-Constant. 1998. Insecticidal toxins from the bacterium *Photorhabdus luminescens*. Science 280: 2129–2132.

Bushway R.J., S.A. Savage, and B.S. Ferguson. 1987. Inhibition of acetylcholinesterase by solanaceous glycoalkaloids and alkaloids. Amer Potato J 64: 409–413.

Casagrande, R.A. 1987. The Colorado potato beetle: 125 years of mismanagement. Bull Entomol Soc Amer 33: 142–150.

Cloutier, C., C. Jean, M. Fournier, S. Yelle, and D. Michaud. 2000. Adult Colorado potato beetles (*Lepinotarsa decemlineata*) compensate for nutritional stress on oryzacystatin I-transgenic potatoes by hypertrophic behavior and over-production of insensitive protease. Arch Insect Biochem Physiol. 44: 69–81.

Collantes, L.G., K.V. Raman, and F.H Cisneros. 1986. Effect of six synthetic pyrethroids on two populations of potato tuber moth, *Phthorimaea operculella* (Zeller) (Lepidoptera: Gelechiidae) in Peru. Crop Protection 5: 355–357.

Conner, A. J. 1993. Monitoring 'escapes' from field trials of transgenic potatoes: a basis for assessing environmental risks. *In:* Seminar of Scientific Approaches for the Assessment of Research Trials with Genetically Modified Plants, Jouy-en-Josas, France, April 1992. Organization for Economic Co-operation and Development, Paris, pp. 34–40.

Conner, A.J., M.K. Williams, D.J. Abernathy, P.J., Fletcher, and R.A. Genet. 1994. Field performance of transgenic potatoes. New Zealand J Crop Hort Sci 22: 361–371.

Coombs, J., D.S. Douches, E. Grafius, W. Pett, and D. Moyer. 2001. Field evaluation of natural and engineered potato (*Solanum tuberosum* L.) resistance mechanisms for control of Colorado potato beetle (*Leptinotarsa decemlineata* Say). Amer J Potato Res 78–448.

Coombs, J.J., D.S Douches, W. Li, E.J. Grafius, and W.L. Pett. 2002. Combining engineered (*Bt-cry3A*) and natural resistance mechanisms in potato (*Solanum tuberosum* L.) for control of the Colorado potato beetle. J Amer Soc Hort Sci. 127: 62–68.

Coombs, J.J., D.S Douches, W. Li, E.J. Grafius, and W. L. Pett. 2003. Field evaluation of natural, engineered, and combined resistance mechanisms is potato (*Solanum tuberosum* L.) for control of Colorado potato betle (*Leptinotarsa decemlineata* Say). J Amer Soc Hort Sci 128: 219-224.

Crawley, M.J., S.L. Brown, R.S. Hails, D.D. Kohn, and M. Rees. 2001. Biotechnology: transgenic crops in natural habitats. Nature 409: 682–683.

Dale, P.J. and H.C. McPartlan. 1992. Field performance of transgenic potato plants compared with controls regenerated from tuber discs and shoot cuttings. Theor Appl Genet 84: 585–591.

Davidson, M.M., J.M.E. Jacobs, J.K. Reader, R.C. Butler, C.M. Frater, N.P. Markwick, S.D. Wratten, and A.J. Conner. 2002. Development and evaluation of potatoes transgenic for a *cry*1Ac9 gene conferring resistance to potato tuber moth. J Amer Soc Hort Sci 127(4): 590–596.

De Almeida, V. Gossele, C.G. Muller, J. Dockx, A. Reynaerts, J. Botterman, E. Krebbers, and M.P. Timko 1989. Transgenic expression of two marker genes under the control of an *Arabidopsis rbcS* promoter: Sequences encoding the Rubisco transit peptide increase expression levels. Mol Gen Genet 218: 78–86.

De Block, M. 1988. Genotype-independent leaf disc transformation of potato (*Solanum tuberosum*) using *Agrobacterium tumefaciens*. Theor Appl Genet 76: 767–774.

Dogan, E.B., R.E. Berry, G.L. Reed, and P.A. Rossignol. 1996. Biological parameters of convergent lady beetle (Coleoptera: Coccinellidae) feeding on aphids (Homoptera: Aphididae) on transgenic potato. J Econ Entomol 89: 1105–1108.

Douches, D.S., A.L. Westedt, K. Zarka, B. Schroeter, and E.J. Grafius. 1998. Transformation of *Cry V-Bt* transgene combined with natural resistance mechanisms for resistance to tuber moth in potato (*Solanum tubersoum* L.). HortSci 33(6): 1063–1056.

Douches, D.S., T.J. Kisha, W. Li, W.L. Pett, and E.J. Grafius. 2001. Effectiveness of natural and engineered host plant resistance in potato to the Colorado potato beetle (*Leptinotarsa decemlineata* (Say)). Hort Sci 36: 967–970.

Douches, D.S., W. Pett, F. Santos, J. Coombs, E. Grafius, W. Li, E.A. Metry, T. Nasr El-Din, and M. Madkour. 2004. Field and storage testing *Bt*-potatoes for resistance to potato tuber moth (Lepidoptera: Gelichiidae). J Econ Entomol. (in press)

Ebinuma, H.K. Sugita, E. Matsunaga and M. Yamakado. 1997. Selection of marker-free transgenic plants using the isopentyl transferase gene. Proc Natl Acad Sci 94: 2117-2121.

Estruch, J.J., G.W. Warren, M.A. Mullins, G.J. Nye, J.A. Craig, and M.G. Koziel. 1996. Vip3A, a novel *Bacillus thuringiensis* vegetative insecticidal protein with a wide spectrum of activities against lepidopteran insects. Proc Natl Acad Sci 93: 5389–5394.

Federici, B.A. and L.S. Bauer. 1998. Cyt1A protein of *Bacillus thuringiensis* is toxic to the cottonwood leaf beetle, *Chrysomela scripta*, and suppresses high levels of resistance to Cry3Aa. Appl Environ Microbiol 64(11): 4368–4371.

Felcher, K.J., D.S. Douches, W.W. Kirk, R. Hammerschmidt, and W. Li. 2003. Expression of a fungal glucose oxidase gene in three potato (*Solanum tuberosum* L.) cultivars with different susceptibility to late blight (*Phytophthora infestans* Mont. de Bary). J Amer Soc Hort Sci 128: 238–245.

French-Constant, R.H. and D.J. Bowen. 2000. Novel insecticidal toxins from nematode-symbiotic bacteria. Cell Mol Life Sci 57: 828–833.

Fischhoff, D.A., S.B. Katherine, and J.P. Frederick. 1987. Insect tolerant transgenic tomato plants. Bio/Tech 5: 807–813.

Food and Agriculture Organization, UN (FAO) 2002. Agricultural data. http:// apps.fao.org/ page/collections.

Forgash, A.J. 1985. Insecticide resistance in the Colorado potato beetle. *In:* D.N. Ferro (ed.), Proc XVII Intl Cong Entomol. Res Bull 704. Univ Massachusetts, Amherst, MA.

Forst, S., B. Dowds, N. Boemare, and E. Stackebrandt. 1997. *Xenorhabdus* and *Photorhabdus* spp.: Bugs that kill bugs. 51: Ann Rev Microbiol 51: 47–72.

Fuchs, R.L., R.A. Heeren, M.E. Gustafson, G.J. Rogan, D.E. Bartnicki, R.M. Leimgruber, R.F. Finn, A. Hershman and S.A. Berberich. 1993a. Purification and characterization of microbially expressed neomycin phosphotransferase II (NPTII) protein and its equivalence to the plant expressed protein. Bio/Techn 11: 1537–1542.

Fuchs, R.L., J.E. Ream, B.G. Hammond, M.W. Naylor, M.W. Leimgruber, and S.A. Berberich. 1993b. Safety assessment of neomycin phosphotransferase II (NPTII) protein. Bio/Tech 11: 1543–1547.

Fujimoto, H., I, Kimiko, and Y. Mikihiro. 1993. Insect resistant rice generated by introduction of a modified-δ-endotoxin gene of *Bacillus thuringiensis*. Bio/Tech 11: 1151–1155.

Garbarino, J.E. and W.R. Belknap. 1994. The use of ubiquitin promoters for transgene expression in potato. *In:* W.R. Belknap, M.E. Vayda, and W.D. Park (eds.). Molecular and Cellular Biology of the Potato. CAB Intl, Wallingford, UK, pp 173–185.

Gatehouse, A.M.R., G.M. Davison, C.A. Newell, A. Merryweather, W.D.O. Hamilton, E.P.J. Burgess, R.J.C. Gilbert, and J.A. Gatehouse. 1997. Transgenic potato plants with enhanced resistance to the tomato moth, *Lacanobia oleracea*—growth room trials. Mol Breeding 3: 49–63.

Gatehouse, A.M.R., G.M. Davison, J.N. Stewart, L.N. Gatehouse, A. Kumar, I.E. Geoghegan, A.N.E. Birch, and J.A. Gatehouse. 1999. Concanavalin A inhibits development of tomato moth (*Lacanobia oleracea*) and peach-potato aphid (*Mysus persicae*) when expressed in transgenic potato plants. Mol Breeding 5: 153–165.

Gauthier, N.L., R.N. Hofmaster, and M. Semel. 1981. History of Colorado potato beetle control. *In:* J.H. Lashomb and R. Casagrande (eds.), Advances in Potato Pest Management. Hutchinson Ross Publ. Co., Stouchsburg, PA, pp. 13–33.

Georgiou, G.P. and C.E. Taylor. 1986. Factors influencing the evolution of resistance. *In:* Pesticide Resistance: Strategies and Tactics for Management. Natl Acad Press, Washington, DC.

Gleave, A.P., D.S. Mitra, N.P. Markwick, B.A.M. Morris, and L. Beuning. 1998. Enhanced expression of *Bacillus thuringiensis cry9*Aa2 gene in transgenic plants by nucleotide sequence modification confer resistance to potato tuber moth. Mol Breeding 4: 459–472.

Goldson, S.L. and R.M. Emberson. 1985. The potato moth *Phthorimaea operculella* (Zeller)—its habits, damage potential and management. Special publ., Agron Soc New Zealand, Christchurch, NZ, pp. 61–66.

Gould, F. 1986. Simulation models for predicting durability of insect-resistant germ plasm: a deterministic diploid, two-locus model. Environ Entomol 15(1): 1–10.

Grafius, E. 1997. Economic impact of insecticide resistance in the Colorado potato beetle (Coleoptera: Chrysomelidae) on the Michigan potato industry. J Econ Entomol 90: 1144–1151.

Grafius, E. 2002. Insecticide resistance management. *In:* D. Pimentel (ed.), Encyclopedia of Pest Management. Marcel Dekker, Inc., NewYork, NY. Published on line at: http://www.dekker.com/servlet/product/productid/E-EPM.

Grafius, E., B. Bishop, W. Boylan-Pett, J. Sirota, and J. Altre. 1993. Colorado potato beetle management. 1992. Michigan Potato Research Report 24: 85–102.

Groden, E. and R.A. Casagrande. 1986. Population dynamics of the Colorado potato beetle, *Leptinotarsa decemlineata* (Coleoptera: Chrysomelidae), on *Solanum berthaultii*. J Econ Entomol 79: 91–97.

Gutierrez-Campos, R., J.A. Torres-Acosta, L.J. Saucedo-Arias, and M.A. Gomez-Lim. 1999. The use of cysteine proteinase inhibitors to engineer resistance against potyviruses in transgenic tobacco plants. Nat Biotech 17: 1223–1226.

Hare, J.D. 1980. Impact of defoliation by the Colorado potato beetles and potato yields. J Econ Entomol 73: 369–373.

Hazzard, R.V. and D.N. Ferro. 1991. Feeding responses of adult *Coleomegilla maculata* (Coleoptera: Coccinellidae) to eggs of Colorado potato beetle (Coleoptera: Chrysomelidae) and green peach aphids (Homoptera: Aphididae). Environ Entomol 20(2): 644–651.

Heim, D.C., G.G. Kennedy, and J.W. Vanduyn. 1990. Survey of insecticide resistance among North Carolina Colorado potato beetle (Coleoptera, Chrysomelidae) populations. J Econ Entomol 83: 1229–1235.

Hilder, V.A. and D. Boulter. 1999. Genetic engineering of crop plants for insect resistance—a critical review. Crop Protection 18: 177–191.

Hough-Goldstein, J. and C.B. Keil. 1991. Prospects for integrated control of the Colorado potato beetle (Coleoptera: Chrysomelidae) using *Perillus bioculatus* (Hemiptera: Pentatomidae) and various pesticides. J Econ Entomol 84: 1645–1651.

Ioannidis, P.M., E. Grafius, and M.E. Whalon. 1991. Patterns of insecticide resistance to Azinphosmethyl carbofuran and Permethrin in Colorado potato beetle (Coleoptera: Chrysomelidae). J Econ Entomol 84(5): 1417–1423.

Jansens, S., M. Cornlissen, R. De Clercq, A. Reynaerts, and M. Peferoen. 1995. *Phthorimaea operculella* (Lepidoptera: Gelechiidae) resistance in potato by expression of the *Bacillus thuringiensis* CryIA(b) insecticidal protein. J Econ Entomol 88(5): 1469–1476.

Jefferson, R.A. 1990. New approaches for agricultural molecular biology: from single cells to field analysis. *In:* J.P. Gustafson (ed.), Gene Manipulation in Plant Improvement. Plenum Press, New York, NY, vol. II, pp. 365–400.

Kay, R.A. Chan, M. Daly, and J. McPherson. 1987. Duplication of CaMV 35S promoter sequences creates a strong enhancer for plant genes. Science 236: 1299–1302.

Kennedy, G.G. and M.E. Whalon. 1995. Managing pest resistance to *Bacillus thuringiensis* endotoxins: Constraints and incentives to implementation. J Econ Entomol 88: 454–460.

Kota, M., H. Daniell, S. Varma, S.F. Garczynski, F. Gould, and W.J. Moar. 1999. Overexpression of the *Bacillus thuringiensis* (Bt) Cry2Aa2 protein in chloroplasts confers resistance to plants against susceptible and Bt-resistant insects. Proc Natl Acad Sci 96: 1840–1845.

Koziel, N.G., G.L. Beland, and C. Bowman. 1993. Field performance of Elite transgenic maize plants expressing an insecticidal protein derived from *Bacillus thuringiensis.* Bio/Tech 11: 194–200.

Kramer, K.J., T.D. Morgan, J.E. Throne, F.E. Dowell, M. Bailey, and J.A. Howard. 2000. Transgenic avidin maize is resistant to storage insect pests. Nature Biotech 18: 670–674.

Kuvshinov, V., K. Koivu, A . Kanerva, and E. Pehu. 2001. Transgenic crop plants expressing synthetic *cry9*Aa gene are protected against insect damage. Plant Sci 160: 341–353.

Lagnaoui, A., V. Canedo, and D.S. Douches. 2001. Evaluation of *Bt-cry1Ia1*(*cryV*) transgenic potatoes on two species of potato tuber moth, *Phthorimaea operculella* and *Symmetrischema tangolias* (Lepidoptera: Gelechiidae) in Peru. CIP Program Report 1999–2000, pp. 117–121.

Lambert, B. and M. Peferoen. 1992. Insecticidal promise of *Bacillus thuringiensis.* Bio Sci 42(2): 112–121.

Lamberti, F., J.M. Waller, N.A. Van der Graaff (eds.). 1981. Durable Resistance in Crops. NATO Adv. Sci. Inst. Ser. 55. Plenum Publ. Corp., New York, NY, 454 pp.

Lavrik, P.B., D.E. Bartnicki, J. Feldman, B.G. Hammond, P.J. Keck, S.L. Love, M.W. Naylor, G.J. Rogan, S.R. Sims, and R.L. Fuchs. 1995. Safety assessment of potatoes resistant to Colorado potato beetle. *In:* K.H. Engel, G.R. Takeoka and R. Teranishi (eds.), Genetically Modified Foods: Safety Issues. ACS, Washington, DC, pp. 148–158.

Lawson, D.R., R.E. Veilleux, and A.R. Miller. 1993. Biochemistry and genetics of *Solanum chacoense* steroidal alkaloids: Natural resistance factors to the Colorado potato beetle. Curr Topics Bot Res 1: 335–352.

Lecardonnel A., G. Prevost, A. Beaujean, R.S. Sangwan, and B.S. Sangwan-Norreel. 1999. Genetic transformation of potato with *nptII-gus* marker genes enhances foliage consumption by Colorado potato beetle larvae. Mol Breeding. 5: 441–451.

Li, W., K. Zarka, D.S. Douches, J.J. Coombs, W.L. Pett, and E.J. Grafius. 1999. Co-expression of potato PVY^{O} coat protein gene and *cry V-Bt* genes in potato (*Solanum tuberosum* L.). J Amer Soc Hort Sci 124(3): 218–223.

Liu, J.R. Berry, G. Poinar, and A. Moldenke. 1997. Phylogeny of *Photorhabdus* and *Xenorhabdus* species and strains as determined by comparison of partial 16S rRNA gene sequences. Intl J Syst Bacteriol 47: 948–951.

Madkour, M. 1999. Addressing agricultural development in Egypt through genetic engineering. Adv Agric Res Egypt 2(2): 115–135.

Mahadavi, A., K.R. Solomon, and J.J. Hubert. 1991. Effect of Solanaceous hosts on toxicity and synergism of permethrin and fenvalerate in Colorado potato beetle (Coleoptera: Chrysomelidae) larvae. Environ Entomol 20: 427–432.

Mailloux, G. and N.J. Bostanian. 1989. Effect of manual defoliation on potato yield at maximum abundance of different stages of Colorado potato beetle, *Leptinotarsa decemlineata* (Say), in the field. J Agric Entomol 6(4): 217–226.

Mani. G.S. 1985. Evolution of resistance in the presence of two insecticides. Genetics 109: 761–783.

Marwick, N.P., J.T. Christeller, L.C. Docherty, and C.M. Lilley. 2001. Insecticidal activity of avidin and streptavidin against four species of pest Lepidoptera. Entomologia. Experimentalis et Applicata 98: 59–66.

McGaughey, W.H. and M.E. Whalon. 1992. Managing insect resistance to *Bacillus thuringiensis* toxins, Science 285: 1451–1454.

Moehs, C.P., P.V. Allen, M. Friedman and W.R. Belknap. 1997. Cloning and expression of solanidine UDP-glucose glucosyltransferase from potato. Plant J 11(2): 227–236.

Mohammed, A., D.S. Douches, W. Pett, E. Grafius, J. Coombs, Liswidowati, W. Li, and M.A. Madkour. 2000. Evaluation of potato tuber moth (Lepidoptera: Gelechiidae) resistance in tubers of *Bt-cry1la1* transgenic potato lines. J Econ Entomol 93(2): 472–476.

Moyer, D.D. 1992. Fabrication and Operation of a Propane Flamer for Colorado Potato Beetle Control. Cornell Coop. Extension, Suffolk Co; New York, NY.

Murray, E.E., T. Rocheleau, M. Eberle, C. Stock, V. Sekar, and M. Adang. 1991. Analysis of unstable RNA transcripts of insecticidal crystal protein genes of *Bacillus thuringiensis* in transgenic plants and electroporated protoplasts. Plant Molec Biol 16: 1035–1050.

NatureMark. 2002. http://www.naturemark.com/pages/Home.html

Newell, C.A., R. Rozman, M.A. Hinchee, E.C. Lawson, L. Haley, P. Sanders, W. Kaniewski, T.E. Tumer, R.B. Horsch, and R.T. Fraley. 1991. *Agrobacterium*-mediated transformation of *Solanum tuberosum* L. cv. 'Russet Burbank'. Plant Cell Rep 10: 30–34.

Ni, M., D. Cui, J. Einstein, S. Narasimhulu, C.E. Vergara, and S.B. Gelvin. 1995. Strength and tissue specificity of chimeric promoters derived from the octopine and mannopine synthase genes. Plant J 7(4): 661–676.

Perlak, R.J., R.L. Fuchs, D.A. Dean, S.L. McPherson, and D.A. Fischhoff. 1991. Modification of the coding sequence enhances plant expression of insect control protein genes. Proc Natl Acad Sci 88: 3324–3328.

Perlak, R.J., T.B. Stone, Y.M. Muskopf, L.J. Petersen, G.B. Parker, S.A. McPherson, J. Wyman, S. Love, G. Reed, D. Biever, and D.A. Fischhoff. 1993. Genetically improved potatoes: protection from damage by Colorado potato beetles. Plant Molec Biol 22: 313–321.

Plaisted, R.L., W.M. Tingey, and J.C. Steffens. 1992. The germplasm release of NYL 235-4, clone with resistance to the Colorado potato beetle. Amer Potato J 60: 843–847.

Raman, K.V. 1980. Potato tuber moth. Tech Inform Bull 3. International Potato Center, Lima, Peru, 14 pp.

Raman, K.V. and M. Palacios. 1982. Screening potato for resistance to potato tuberworm. J Econ Entomol 75: 47–48.

Riddick, E.W. and P. Barbosa. 1998. Impact of Cry3A-intoxicated *Leptinotarsa decemlineata* (Coleoptera: Chrysomelidae) and pollen on consumption, development, and fecundity of *Coleomegilla maculata* (Coleoptera: Coccinellidae). Soc Amer 91(3): 303–307.

Roush, R.T. 1999. Two-toxin strategies for management of insecticidal transgenic crops: can pyramiding succeed where insecticide mixtures have not? Phil Trans Roy Soc Lond B 353: 1777–1786.

Sanford, L.L., R.S. Kobayshi, K.L. Deahl, and S.L. Sinden. 1996. Segregation of leptines and other glycoalkaloids in *Solanum tuberosum* (4x) × *S. chacoense* (4x) crosses. Amer Potato J 73: 21–31.

Sauvion, N., Y. Rahbe, W.J. Peumans, E. van Damme, J.A. Gatehouse, and A.M.R. Gatehouse. 1996. Effects of GNA and other mannose-binding lectins on development and fecundity of the peach potato aphid. Entomol Exp Appl 79: 285–293.

Sheerman, S. and M.W. Bevan. 1998. A rapid transformation method for *Solanum tuberosum* using binary *Agrobacterium tumefaciens* vector. Plant Cell Rep 7: 13–16.

Shelton, A.M., J.Z. Zhao, and R.T. Roush. 2002. Economic, ecological, food safety, and social consequences of the deployment of Bt transgenic plants. Ann Rev Entomol 47: 845–881.

Sinden, S.L. 1987. Potato glycoalkaloids. Acta Hortic 207: 41–47.

Sinden, S.L., L.L. Sanford, and S.F. Osman. 1980. Glycoalkaloids and resistance to the Colorado potato beetle in *Solanum chacoense* Bitter. Amer Potato J 57: 331–343.

Sinden, S.L., L. Sanford, L. Cantelo, and K.L. Deahl. 1986. Leptine glycoalkaloids and resistance to the Colorado potato beetle (Coleoptera: Chrysomelidae) in *Solanum chacoense*. Environ Entom 155: 1057–1062.

Stapleton, A., P.V. Allen, M. Friedman, and W.R. Belknap. 1991. Purification and characterization of solanidine glycosyltransferase from potato (*Solanum tuberosum*). J Agric Food Chem 39 (6): 1187–1193.

Sticklen, M.H., R.V. Ebora, J. Cheng, M.M. Ebora, M.G. Bolyard, R.C. Saxena, and D.L. Miller. 1993. Genetic transformation of potato with *Bacillus thuringiensis* HD 73 Cry1A(c) gene and development of insect resistant plants. *In:* C.B. You et al. (eds.), Biotechnology in Agriculture. Kluwer Acad. Publ. Dardrecht, Netherlands, pp. 233–236.

Stiekema, W.J., F. Heidekamp, J. Louwerse, H.A. Verhoeven, and P. Dijkhuis. 1988. Introduction of foreign genes into potato cultivars Bintje and Desirée using an *Agrobacterium tumefaciens* binary vector. Plant Cell Rep 7: 47–50.

Stürckow, B. and I. Löw. 1961. Die wirkung einger *Solanum*-alkaloidglykoside auf den kartoffelkafer, *Leptinotarsa decemlineata* Say. Entomol Exp Appl 4: 133–142.

Sutton, D.W., P.K. Havstad, and J.D. Kemp. 1992. Synthetic *cryIIIA* gene from *Bacillus thuringiensis* improved for high expression in plants. Transgenic Res 1: 228–236.

Tabashnik, B.E. 1994. Evolution of resistance to *Bacillus thuringiensis*. Ann Rev Entomol 39: 47–79.

Tabashnik, B.E., J.A. Rosenheim, and M.A. Caprio. 1992. What do we really know about management of insecticide resistance? *In:* I. Denholm, Devonshire, A.L. and Hollomon, D.W. (eds.), Resistance 91. Achievements and Developments in Combating Insecticide Resistance. Elsevier Appl. Science, London, UK, pp. 124–135.

Tailor, R., J. Tippett, G. Gibb, S. Rells, D. Pike, L. Jordan, and S. Ely. 1992. Identification and characterization of a novel *Bacillus thuringiensis* endotoxin entomocidal to coleopteran and lepidopteran larvae. Molec Microbiol 6: 1211–1217.

Tavassa, R., M. Tavassa, R.J. Ordas, G. Ancora, and E. Benvenuto. 1988. Genetic transformation of potato (*Solanum tuberosum*): An efficient method to obtain transgenic plants. Plant Sci 59: 175–181.

Tingey, W.M., P. Gregory, R.L. Plaisted, and M.J. Tauber. 1984. Research Progress: Potato glandular trichomes and steroid glycoalkaloids. Rept XXII Planning Conf. Integrated Pest Management. International Potato Center, Lima, Peru, pp. 115–124.

Tingey, W.M. 1991. Potato granular trichomes: defensive activity against inset attack. *In:* Naturally Occurring Pest Bioregulators. ACS Symp. Series 449. ACS Books, Wash., DC, pp. 126–135.

Trivedi, T.P. and D. Rajagopal. 1992. Distribution, biology, ecology, and management of potato tuber moth, *Phthorimaea operculella* (Zeller) (Lepidoptera: Gelechiidae): a review. Trop Pest Manage 38(3) 279–285.

US EPA, Office of Pesticide Programs. Revised Bt Crops Assessment. 2001. www.epa.gov/pesticides/biopesticides/otherdocs/bt—/reassess/5-Insect%20Resistance%20Management.pdf

Vaeck, M., A. Reynaerts, and H. Hofte. 1987. Transgenic plants protected from insect attack. Nature 328: 33–37.

Visser, R.G.F., E. Jacobsen, A. Hesseling-Meinders, M.J. Schans, B. Witholt, and W.J. Feenstra. 1989. Transformation of homozygous diploid potato with an *Agrobacterium tumefaciens* binary vector system by adventitious shoot regeneration on leaf and stem segments. Plant Mol Biol 12: 329–337.

Wenzler, H.C., G.A. Mignery, L.M. Fisher, and W.D. Park. 1989. Analysis of a chimeric class-I patatin-GUS gene in transgenic potato plants: High-level expression in tubers and sucrose-inducible expression in cultured leaf and stem explants. Plant Mol Biol 12: 41–50.

Westedt, A.L., D.S. Douches, W. Pett, and E.J. Grafius. 1998. Evaluation of natural and engineered resistance mechanisms in *Solanum tuberosum* for resistance to *Phthorimaea operculella* (Lepidoptera: Gelechiidae). J Econ Entomol 91(2): 552–556.

Whalon, M.E., D.L. Miller, R.M. Hollingworth, E.J. Grafius, and J.R. Miller. 1993. Laboratory selection of a resistant Colorado potato beetle (Coleoptera: Chrysomelidae) strain to the CRY IIIA coleopteran specific delta endotoxin of *Bacillus thuringiensis*. J Econ Entomol 86: 226–233.

Wierenga, J. and R.M. Hollingworth. 1992. Inhibition of insect acetylcholinesterase by the potato glycoalkaloid-chaconine. Nat Toxins 1: 96–99.

Wierenga, J.M. and R.M. Hollingworth. 1993. Inhibition of altered acetylcholinesterases from insecticide-resistant Colorado potato beetles (Coleoptera: Chrysomelidae). J Econ Entomol 86: 673–679.

Wiese, M., J. Guenther, A. Pavlista, J. Wyman and J. Sieczka. 1998. Use, target pests, and economic impact of pesticides applied to potatoes in the United States. USDA NAPIAP Report 2-CA-98.

Wolfenbarger L.L. and P.R. Phifer. 2000. The ecological risks and benefits of genetically engineered plants. Science 290: 2088–2093.

Wright, R.J. 1984. Evaluation of crop rotation for control of Colorado potato beetles (Coleoptera: Chrysomelidae) in commercial potato fields on Long Island. J Econ Entomol 77: 1254–1259.

Wünn, J., A. Klöti, P.K. Burkhardt, G.C. Ghosh, K. Launis, V. Iglesias, and I. Potrykus. 1996. Transgenic indica rice breeding line IR58 expressing a synthetic cryIA(b) gene from *Bacillus thuringensis* provides effective insect pest control. Biotech 14: 171–176.

Yadav, N.R. and M.B. Sticklen. 1995. Direct and efficient plant regeneration from leaf explants of *Solanum tuberosum* L. cv. Bintje. Plant Cell Rep 14: 645–647.

Yencho, C.G. and W.M. Tingey. 1994. Glandular trichomes of *Solanum berthaultii* alter host preference of the Colorado potato beetle, *Leptinotarsa decemlineata*. Entomol Exp Appl 70: 217–225.

Zhao, J., E. Grafius, and B. Bishop. 2000. Inheritance and synergism of resistance to imidacloprid in the Colorado potato beetle (Coleoptera: Chrysomelidae). J Econ Entomol 93: 1508–1514.

Zhou, J., H.L. Collins, J.D. Tang, J. Cao, E.D. Earle, R.T. Roush, S. Herrero, B. Escriche, J. Ferre, and A.M. Shelton. 2000. Development and characterization of Diamond back moth to transgenic broccoli expressing high levels of Cry1C. Appl Environ Microbiol 66(9): 3784–3789.

Zhu, K.Y. and J.M. Clark. 1995. Comparisons of kinetic properties of acetylcholinesterase purified from azinphosmethyl-susceptible and resistant strains of Colorado potato beetle. Pestic Biochem Physiol 51: 57–67.

12

Breeding for Resistance to *Meloidogyne* Species and Trichodorid-Vectored Virus

CHARLES R. BROWN AND HASSAN MOJTAHEDI
USDA/ARS, 24106 N. Bunn Rd, Prosser, WA 99350, USA, Tel: (509) 786-9525; Fax: (509) 786-9277; e-mail: cbrown@pars.ars.usda.gov

INTRODUCTION: NEMATODE-MEDIATED DISEASES OF POTATO

Two important diseases of potato (*Solanum tuberosum* L.) that impact the quality of the crop are induced by the action of nematodes, namely root-knot and corky ringspot disease. The first disease is caused by several root-knot nematode species, while the second is caused by tobacco rattle virus (TRV) vectored by stubby root nematode species. Both diseases damage the quality of the tuber and render it unmarketable. Both diseases are managed by controlling the nematodes. The damage threshold for both nematodes is very low. In the Pacific Northwest (PNW) of the USA where the Columbia root-knot nematode (CRN) *M. chitwoodi* (Golden et al. 1980), and the northern root-knot nematode (NRN) *M. hapla* (Chitwood 1949) threaten potato quality. Santo et al. (1981) estimated that one CRN and 50 NRN nematodes per 250 cm^3 soil at planting, may lead to crop failure. Mojtahedi et al. (2001) demonstrated that 3 viruliferous stubby root nematodes per 250 cm^3 soil suffice to cause visual viral symptoms on tubers. In the (PNW), a potato consignment with 10 or more percent tubers blemished due to corky ringspot (CRS) disease or any other injuries that may be considered by the processors as unmarketable. A tuber is considered a cull if six or more *M. chitwoodi* spots or one corky ringspot blemish are scored. Thus, a strict control of these nematodes before planting potato is essential for crop safety. Presently, soil fumigation with 1, 3 dichloropropene at 180-225 ha^{-1} is the only method labeled by the manufacturer to control CRN and stubby

Corresponding author: Charles R. Brown

root nematodes in the PNW (Santo et al. 1997; Ingham et al. 2000). This fumigant must be shanked 45 cm deep, costs the grower \$300-500 ha^{-1}, and may not even be environmentally sound. Contrary to PNW, control of CRS in the potato fields of northeastern Florida (Weingartner et al. 1983) is not achieved by fumigation. Alternative control measures such as green manure and organic amendments have reduced soil populations of root-knot nematodes, but the residual population was high enough to require additional augmentation with some sort of nematicidal chemical treatment to achieve an economic control (Mojtahedi et al. 1993). Stubby root nematodes are less sensitive to soil amendments, than root-knot nematodes (Mojtahedi unpubl. data). The root-knot and stubby root nematode species associated with potato have a wide host range and crop rotation has not provided a reliable method of control. Rotating potato with alfalfa, wheat or corn in PNW has not reduced populations of CRN and NRN nematode species enough to obviate the need to fumigate. Host races of CRN increase on alfalfa rendering it an unsuitable choice for rotation. Rotating potato with corn or wheat to reduce the impact of CRN also proved futile (Mojtahedi et al. 2002). Although alfalfa as a rotational crop may reduce CRS incidence on potato (Thomas et al. 1999), the presence of certain weeds may negate the benefit of such a practice (Boydston et al. 2002). Time of planting and early maturing varieties may provide relief for certain growers with root-knot nematode problems. In the PNW, early maturing varieties, such as Shepody and White Rose, may escape northern root-knot nematode damage especially if the initial soil population is low. In the Central Valley of California, by early planting, potato may escape damage by the southern root-knot nematode (SRN), *M. incognita* (Chitwood 1949). In tropical and subtropical regions, where SRN produces 12 generations per year, escaping the impact of nematodes is impossible. Trichodorid nematodes introduce TRV to potato tubers immediately after they are formed (van Hoof 1964), and therefore escape from CRS is unlikely. Based on information presented so far, it is clear that potato growers in hot production zones where SRN is a problem and those in temperate regions (such as the PNW and Europe) where CRN and *M. fallax* are problems, would benefit greatly by having varieties resistant to root-knot nematode and corky ringspot.

BIOLOGY AND AGRONOMIC FACTORS OF *MELOIDOGYNE* SPECIES

Root-knot nematodes (*Meloidogyne* spp,) are sedentary endpoparasitc organisms of the family *Heteroderidae.* The female is white, round to pyriform, with a protruding neck that is embedded in the host tissue. The male, if present, is vermiform and free living.

More than 80 species have been described (Karssen and Van Hoenselaar 1998). The genus is easy to recognize, but identification at the species level is difficult due to relatively small intra-specific morphological variations. Various morphological features [perineal patterns (Chitwood 1949), tail morphology of second stage juveniles (Nyczepir et al. 1982), numerical taxonomy (Hewlett and Tarjan 1983)], physiological characters [esterase and malate dehydrogenase patterns (Esbenshade and Triantaphyllou 1990)], and molecular techniques (Cenis 1993; Castagnone-Sereno et al. 1999) have been utilized to set root-knot nematode species apart. The occurrence of host races and pathotypes (Roberts 1995) in several species of root-knot nematode (Table 12.1) further complicates the task of breeders engaged in breeding for resistance (Brown et al. 1996). Unfortunately, none of the modern techniques have facilitated the identification of these host races and pathotypes (Powers and Harris 1993). But luckily, only half a dozen of these species have a world-wide distribution, and not more than seven species have been associated with potato. These species include *M. arenaria* [Chitwood 1949 (peanut root-knot nematode, PRN)]; *M. chitwoodi* [Golden et al. 1980 (Columbia root-knot nematode, CRN)]; *M. hapla* [Chitwood 1949 (northern root-knot nematode, NRN)]; *M. fallax* (Karssen and Van Hoenselaar 1998); *M. incognita* (Chitwood 1949).

Table 12.1 *Host races and pathotypes (Roberts 1995) of three root-knot nematode species associated with potato and the host differentials responsible for setting them apart (after Hartman and Sasser 1985, Mojtahedi et al. 1988, and Mojtahedi et al. 1994)*

Species & host races	*Differential hosts*								
	TOB	*W*	*P*	*A*	*C*	*CO*	*WM*	*SB 22*	*TO*
M. hapla (no host race)	+	–	+	+	+	–	–	+	+
M. chitwoodi, host race 1	–	+	–	–	+	–	–	–	+
M. chitwoodi, host race 2									
Pathotype 1	–	+	–	+	–	–	–	–	+
Pathotype 2	–	+	–	+	–	–	–	+	+
M. incognita									
Host race 1	–	+	+	+	+	–	+	nt	+
Host race 2	+	+	+	+	+	–	+	nt	+
Host race 3	–	+	+	+	+	+	+	nt	+
Host race 4	+	+	+	+	+	+	+	nt	+

Host differentials include tobacco, NC 95 (TOB); wheat, cv New Gaines (or cv Stevens) (W); pepper, cv California Wonder (P); alfalfa, cv Thor (A); carrot, cv Chantenay (C); Cotton, cv Delta Pine (CO); watermelon, cv Charleston Grey, *Solanum bulbocastanum* clone 22 (SB22); tomato, cv Rutgers (TO).

(–) indicates a nonhost status (reproductive factor (RF): $P_f \div P_i < 1$); (+) indicates suitable host status (RF > 1); nt indicates that a particular host was not tested for this nematode host race.

(southern root-knot nematode, SRN)]; *M. javanica* [Chitwood 1949 (Javanese root-knot nematode, JRN)]; and *M. thamesi* [Chitwood 1944 (Thames' root-knot nematode, TRN). The potato industry and potato researchers focus mainly on three nematode species: CRN (cool climate species), NRN (cool-warm climate species), and lastly SRN (warm climate species).

In the PNW states (Washington, Oregon, and Idaho), the two species CRN and NRN are widespread in potato-growing regions. Both species are well adapted to cool Northern production zones and also include Colorado (Pinkerton and McIntyre 1987), Utah (Griffin 1988), northern California and northern Nevada (Nyczepir et al. 1982), or in higher elevations of warmer production states of Texas (Szalanski et al. 2001) and New Mexico (Thomas et al. 2001).

In Europe, NRN is undoubtedly the most common and widely distributed root-knot nematode. CRN, however, was only reported from the Netherlands, Belgium, Germany, and Portugal. A closely related and newly discovered species, *M. fallax,* was reported from the Netherlands, Belgium, Germany (Karssen and Van Hoenselaar 1998).

SRN is adapted to warmer temperatures and may not survive the freezing cold winter of northern states of USA. In cooler climates like Europe it is frequently encountered in greenhouses. In tropical and subtropical countries SRN is a dominant species and causes serious damage to potato. Comments about SRN apply as well to *M. arenaria* and *M. javanica.*

PATHOGEN LIFE CYCLE AND BIOLOGY

The first stage juvenile (J1) of root-knot nematode species molts inside the egg and the second stage juvenile (J2) emerges as a free living vermiform larva from the egg. J2 is the infective stage and penetrates the roots in the zone basal to the growing meristematic tips. This induces giant cell formation for nourishing the nematode until maturity. Within 3-4 weeks, by molting three more times, J2 passes through the third and fourth ecdyses to become a pyriform female or a vermiform male. The female may/may not copulate before depositing 200-1000 eggs in a gelatinous matrix that protrudes out of the root surface. This matrix holds and protects the eggs from adverse effects of the soil environment. J2 may exit the matrix to search for new infection sites and to start a new life-cycle. In cooler growing regions of the US, this cycle may be repeated 2-5 times on the potato root system, depending on soil temperature and base-temperature requirement of a given nematode species (the lowest temperature at which life activity of the species begins). The base temperature is 6°, 10°, and 12-18°C for CRN (Inserra et al. 1983), NRN (Inserra et al. 1983), and SRN (Goodbell and Ferris 1989), respectively.

J2 may penetrate tubers only after the stomata on the skin develop into lenticels (Pinkerton et al. 1991). Tuber resistance to nematode infection might be independent from root resistance (Santo et al. 1994). Since CRN and NRN cause little disturbance of root growth, the important consideration is tuber infection which corresponds to completion of the first generation on the potato root system in the PNW (Pinkerton et al. 1991) (Fig. 12.1). Tuber infection in terms of degree-days (DD*) was determined for CRN and NRN to be 1,000-1,100 DD from time of planting to completion of the first generation, and 500-600 DD for the subsequent generations (Pinkerton et al. 1991). The difference in base temperatures usually makes NRN a minor factor in the PNW while for CRN it is a despoiler of tuber quality of considerable proportions. The damage threshold of CRN on potato at planting was estimated to be one nematode per 250 cm^3 soil and for NRN 50 cm^3 (Santo et al. 1981).

In contrast to cooler climates, in tropical and subtropical regions, SRN may go through 12 generations per growing season on potato and therefore escape from nematode damage is not possible. In the Mediterranean climate of California, SRN is of little consequence and is primarily controlled by planting potato when the average soil temperature is below 15°C in the hottest growing areas of the Central Valley.

Nematodes that penetrate tubers cause blemishes when the female deposits eggs, a process that may continue even in storage (Santo 1989). Thus, a harvested crop without apparent symptoms, if stored at 4 to 7°C might well reach an unmarketable state after 2-3 months.

DISEASE SYMPTOMS

No distinct above-ground symptoms are associated with root-knot nematode-infected potato plants. At high population density, the plants may show varying degrees of stunting, chlorosis, or even wilting. The root symptoms, however, are distinct. SRN causes large galls while NRN induces small galls plus excessive branching, giving the root system a wiry appearance. CRN does not cause galls but the infection sites swell. Infected tubers of all three types also exhibit characteristic symptoms. SRN causes large warty and CRN pimple-like galls on the tubers while NRN does not cause distinct galls but with severe infection a general swelling is discernible. All the species produce typical necrotic spots in the region between the tuber surface and the vascular ring. The necrotic spots are a tuber tissue reaction to an egg deposit (Fig. 12.1). Resistance to SRN is commonly judged by subjective scales of root and tuber galling, while

*DD (daily maximum temperature at 15-20 cm from the soil surface + minimum temperature) ÷ 2, the base temperature = accumulated degree days.

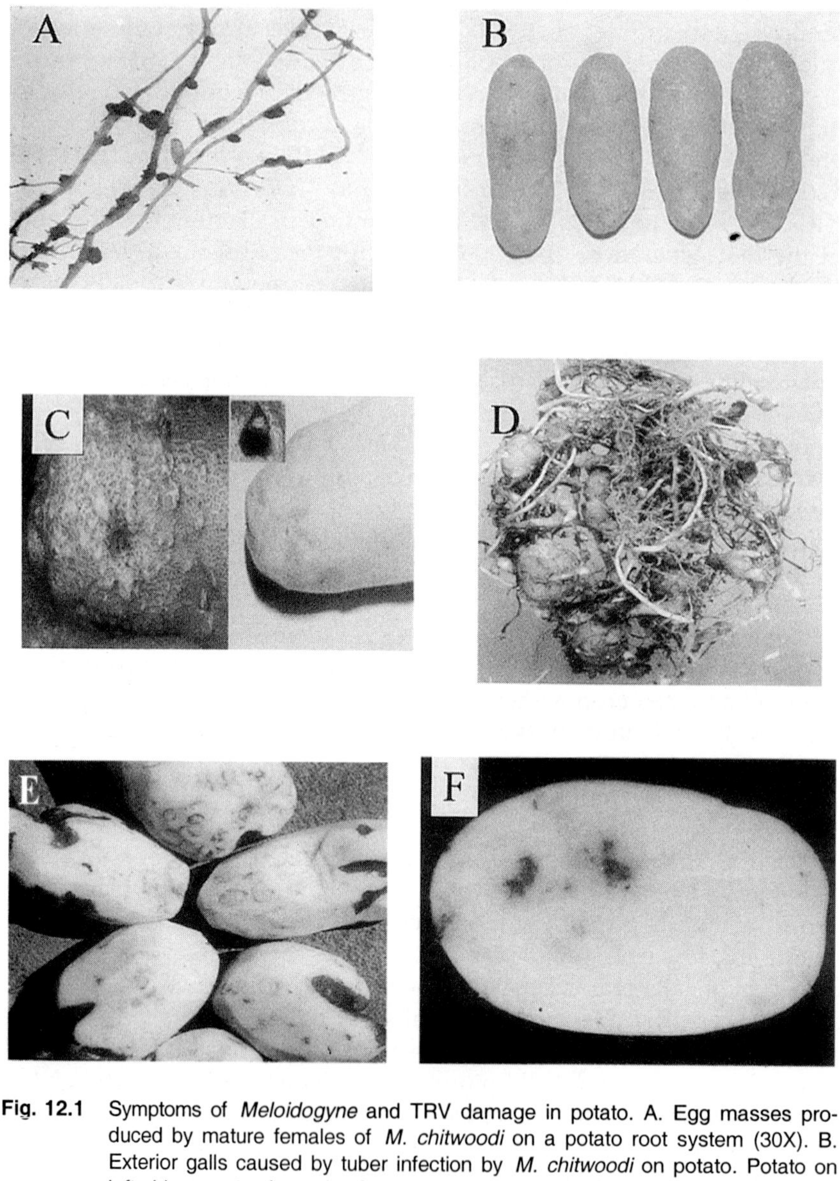

Fig. 12.1 Symptoms of *Meloidogyne* and TRV damage in potato. A. Egg masses produced by mature females of *M. chitwoodi* on a potato root system (30X). B. Exterior galls caused by tuber infection by *M. chitwoodi* on potato. Potato on left side symptomless; the three potatoes to the right show progressively more tuber galling. C. Close-up of tuber galling on left, females in tuber flesh after egg mass formation on right, and inset, female nematode with egg mass dissected from tuber flesh (60X). D. Root and tuber galling on potato caused by *M. incognita*. E. Necrotic arcs and spots caused by infection by TRV in potato tubers. F. Internal necrotic blotches caused by isolate of TRV present in the Columbia Basin of the Pacific Northwest USA.

NRN and CNR have been assessed by counting egg masses or extracted eggs. The standardized host test procedure for greenhouse experiments has been described by Hartman and Sasser (1985).

RESISTANCE TO *MELOIDOGYNE* SPECIES

With the advent of scientific breeding in the twentieth century and accessing of wild and cultivated germplasm, resistance to *Globodera* species, another important nematode pathogen, became necessary in traditional cool growing areas. With the introduction of potato culture to warmer growing areas, including the subtropics and tropics, consideration of the threat posed by root-knot nematodes naturally increased and warm-climate root-knot nematode species were originally the only breeding target. However, identification of a new species of root-knot, *Meloidogyne chitwoodi,* adapted to much cooler soil temperatures than other *Meloidogyne* species previously associated with potato, lent a new focus to root-knot resistance research. The importance of the *M. chitwoodi* became paramount with its ever-expanding geographic distribution, most notably by its discovery in the Netherlands and the description of a similar species, *M. fallax.*

Germplasm Surveys

Nirula et al. (1967) reported that *Solanum vernei* showed the highest level of resistance, while other species developed resistance through clonal selections: *S. acaule, S. acrosopicum, S. boliviense, S. cardiophyllum, S. chacoense, S. fendleri, S. ochrantum, S. raphanifolium, S. sparsipilum, S. spegazzinii,* and *S. stoloniferum.* This was a pot test and resistance was subjectively rated by degree of galling. Using a root system initiated adventitiously on leaf petioles, Nirula et al. (1967) found resistance to root galling in these wild species: *S. ajanhuiri, S. bulbocastanum, S. gandarillasii, S. lignicaule, S. spegazzinii,* and *S. vernei,* and in the cultivated *S. tuberosum* ssp. *andigena.* Brücher (1967) identified resistance to four species of *Meloidogyne* in pot experiments testing both wild and cultivated species, and found interestingly examples of resistance in the native cultivars (*Solanum tuberosum* ssp. *andigena*) of the Jujuy Province of Argentina (Table 12.2). González and Accatino (1974) failed to identify any resistance to galling by *M. incognita acrita* among the named varieties and advanced breeding clones derived from the Chilean National Breeding Program. Hoyman (1974) reported that several wild species, cultivars, and breeding clones showed reduced root galling when inoculated with *M. hapla.* It is likely, however, that his nematode cultures were mixed with *M. chitwoodi* and therefore evaluation for galling was inappropriate. Evaluation of resistance to *Meloidogyne* on adventitiously induced roots on nodal stem cuttings was employed at the International Potato Center

Table 12.2 Literature citations of cultivated and wild species resistant to Meloidogyne spp.

Species	*Evaluation Method*	*Nematode species*	*Reference*
S. acaule	EM	*Mh*	Janssen et al. 1996b
S. acaule	RG	*Mi*	Nirula et al. 1967
S. acrosopicum	RG	*Mi*	Nirula et al. 1967
S. ajanhuiri	RG	*Mi*	Nirula et al. 1969
S. arnezii	EM	*Mh*	Janssen et al. 1996b
S. bjugum	RG	*Mh, Mj, Mi, Mt*	Brücher 1967
S. boliviense	EM	*Mh*	Janssen et al. 1996b
S. boliviense	RG	*Mi*	Nirula et al. 1967
S. brachistotrichum	EM	*Mf*	Janssen et al. 1996b
S. brachistotrichum	EM	*Mh*	Janssen et al. 1996b
S. bulbocastanum	RF	*Mc*	Brown et al. 1989
S. bulbocastanum	EM	*Mc, Mf, Mh*	Janssen et al. 1996b
S. bulbocastanum	RG	*Mi*	Nirula et al. 1969
S. cardiophyllum	EM	*Mc, Mh*	Janssen et al. 1996b
S. cardiophyllum	RG	*Mi*	Nirula et al. 1967
S. chacoense	RG	*Mi*	Iwanaga et al. 1989
S. chacoense	EM	*Mc, Mh*	Janssen et al. 1996b
S. chacoense	RG	*Mi*	Nirula et al. 1967
S. chacoense	RG	*Mh, Mj, Mi, Mt*	Brücher 1967
S. commersonii	RG	*Mh, Mj, Mi, Mt*	Brücher 1967
S. fendleri	RF	*Mc*	Brown et al. 2002
S. fendleri	EM	*Mc*	Janssen et al. 1996b
S. fendleri	RG	*Mi*	Nirula et al. 1967
S. gandarillasii	RG	*Mi*	Nirula et al. 1969
S. gourlayi	EM	*Mc, Mh*	Janssen et al. 1996b
S. hougasii	RF	*Mc*	Brown et al. 1991, 1999
S. hougasii	EM	*Mf, Mh*	Janssen et al. 1996b
S. lignicaule	RG	*Mi*	Nirula et al. 1969
S. microdontum	EM	*Mh*	Janssen et al. 1996b
S. ochrantum	RG	*Mi*	Nirula et al. 1967
S. raphanifolium	RG	*Mi*	Nirula et al. 1967
S. ruiz-lealii	RG	*Mh, Mj, Mi, Mt*	Brücher 1967
S. schenkii	RG	*Mc*	Berthou et al. 1996a
S. seaforthium	RG	*Mi*	Shetty and Reddy, 1985
S. sparsipilum	RG	*Ma, Mi, Mj*	Berthou et al. 1996a
S. sparsipilum	RG	*Mf*	Berthou et al. 1999
S. sparsipilum	RG	*Mi, Ma, Mj*	Gomez-Cuervo 1982; Gomez et al. 1983
S. sparsipilum	RG	*Mi*	Gutierrez-Deza 1984
S. sparsipilum	RG	*Mi*	Iwanaga et al. 1989
S. sparsipilum	EM	*Mh*	Janssen et al. 1996b
S. sparsipilum	RG	*Mi*	Nirula et al. 1967
S. spegazzinii	EM	*Mh*	Janssen et al. 1996b
S. spegazzinii	RG	*Mi*	Nirula et al. 1967

(Contd.)

(*Contd.*)

S. spegazzinii	RG	*Mi*	Nirula et al. 1969
S. stoloniferum	EM	*Mf*	Janssen et al. 1996b
S. stoloniferum	RG	*Mi*	Nirula et al. 1967
S. sucrense	EM	*Mh*	Janssen et al. 1996b
S. tarijense	EM	*Mh*	Janssen et al. 1996b
S. tascalense	RG	*Mh, Mj, Mi, Mt*	Brücher 1967
S. torvum	RG	*Mi*	Shetty and Reddy 1985
S. tuberosum ssp. *andigena*	RG	*Mh, Mj*	Brücher 1967
S. tuberosum ssp. *andigena*	RG	*Mi*	Raj and Sharma 1987
S. tuberosum ssp. *tuberosum*	RG	*Mi*	Raj and Sharma 1987
S. vernei	RG	*Mi*	Nirula et al. 1969

RG = Assessment of root galling
EM = Counting of stained egg masses
RF = Extraction and counting of eggs
Ma = *Meloidogyne arenaria*
Mc = *M. chitwoodi*
Mf = *M. fallax*
Mh = *M. hapla*
Mi = *M. incognita*
Mj = *M. javanica*
Mt = *M. thamesii*

(CIP), Lima, Peru (Franco, 1974). Shetty and Reddy (1985) noted reduced invasion, galling, and reproduction in *S. torvum,* and *S. seaforthium.* Gomez-Cuervo (1982) and Gomez et al. (1983) reported resistance to *M. incognita, M. arenaria,* and *M. javanica* derived from diploid *S. sparsipilum,* introgressed into the tetraploid breeding pool. While tetraploid progeny showed resistance to *M. arenaria,* the same could be observed for other pests, *M. javanica* and *M. incognita.* Inheritance could be explained by oligogenic control for all three species of nematode, with the most useful phenotype being the multiple resistances not disrupted during recurrent selection. Gutierrez-Dega (1984) reported in a diallel cross, with reciprocals at the diploid level, that resistance to *M. incognita* from *S. sparsipilum* displayed a large proportion of variation as general combining, accompanied by a narrow-sense heritability of h^2 = 0.78. Significantly higher levels of resistance were found where *S. sparsipilum* was the female parent. A separate study reported that the introduction of *Meloidogyne* spp. resistance into the tetraploid breeding pool could be successfully carried out by "sexual polyploidiza-tion" (Iwanaga et al. 1989). This involves the identification of 2n gamete producing resistant parents crossed with a tetraploid as 4x-2x or 2x-4x crosses. It was also shown that 2n gametes transmitted resistance more effectively than n gametes, producing 25 versus 11% resistant progenies respectively. This study failed to find the maternal effect noted by Gomez-Cuervo (1982), Gomez et al. (1983), and Gutierrez-Dega (1984). In fact, *S. sparsipilum* as source of resistance emerged from studies at the (CIP) (Mendoza and Jatala 1985). Although resistance in breeding materials did not emerge directly from CIP, this source of resistance served as the basis, upon exchange of germplasm, for work at the Institut National de la Rechérche Agronomique (INRA) in France. Hypersensitive resistance to SRN extracted from *S. sparsipilum* by sexual polyploidization, possibly under simple (monogenic) genetic control, was incorporated into advanced tetraploid breeding clones (Berthou et al. 1996a, b). In addition, researchers at INRA discovered resistance to CRN and *M. fallax* in *S. sparsipilum* and *S. schenkii* (Berthou et al. 1999). Raj and Sharma (1987) noted complete resistance to root galling in several *S. tuberosum* ssp. *andigena* and partial resistance to galling in ssp. *tuberosum* breeding clones. Evaluation of tuber infection rarely showed resistance in normal breding clones. One instance was the clone A8292-5 that emerged from the breeding programs at USDA/ARS, Aberdeen, ID, and Prosser, WA, which was selected by Mark Martin and showed resistance to tuber infection while paradoxically serving as a good host in the root system (Santo et al., 1994).

Sasser et al. (1987) conducted extensive germplasm screening for resistance to tropical *Meloidogyne* ssp. This survey led to more detailed research that revealed *S. bulbocastanum* as a source of resistance to *M. chitwoodi* (Brown et al. 1989). This trait was introduced to the cultivated tetraploid breeding pool by protoplast fusion (Austin et al. 1993). Rivera-Smith et al. (1991) studied host suitability of various sources of resistance, including *S. sparsipilum* in an *in-vitro* excised root assay as well as in pot culture. They found a similarity between pot and *in vitro* results; however, most of the genetic materials seemed suitable hosts when tested against CRN, NRN, and SRN. Janssen et al. (1996b) surveyed a number of *Solanum* species for resistance to CRN, *M. fallax,* and NRN (see Table 12.2). Resistance to CRN and *M. fallax* was found to be similar to previous studies in *S. bulbocastanum* and *S. hougasii.* New discoveries included resistance in *S. brachistotrichum, S. cardiophyllum,* and *S. fendleri.* Resistance to NRN was found in a larger group of species (see Table 12.2). Resistance was assessed by counting egg masses on the root systems instead of extraction and counting of eggs. *M. fallax* appeared to match the pathogenicity pattern of CRN, race 1.

Genetics of Resistance

Resistance was also confirmed in *Solanum hougasii* (Brown et al. 1991) and *Solanum fendleri* (Brown et al. 2002). It is important to note that mapping studies have identified the location of resistance factors to CRN, $R_{MC\ 1\ (hou)}$ and $R_{Mc1(blb)}$, on chromosome 11 for both *S. hougasii* and *S. bulbocastanum* (Brown et al. 1996, 2003). Furthermore, in both species mapping populations showed that resistance to race 1 of CRN was easily identifiable as a single factor segregating in a 1:1 fashion while resistance to race 2 was less easily attributed to such simple inheritance. Resistance to race 1 derived from *S. bulbocastanum* was introduced to the cultivated breeding pool by protoplast fusion followed by backcrossing to recurrent and non recurrent parents drawn out of advanced breeding materials (Fig. 12.2). During the breeding process, transmission was associated with erratic meiosis and aberrant segregation of sterile pollen perpetuated by the hybrid cytoplasm derived from the fusion (Masuelli et al. 1995). However, expression of the trait was undiminished after four backcrosses. A comparison of the reproductive factor (Rf = Pf/Pi) and tuber damage, as assessed from counting the number of infestation sites on a peeled tuber revealed that resistance to reproduction on the root system translated well into reduction of tuber infestation sites ($r = 0.77$, $P < 0.01$) (Fig. 12.3). Two backcross populations were sent to the Netherlands and a Site Characterized Amplified Region (SCAR) marker was developed to facilitate Marker Assisted Selection (MAS). Confirmation of the chromosomal position on the upper arm of chromosome 11 was simultaneously obtained

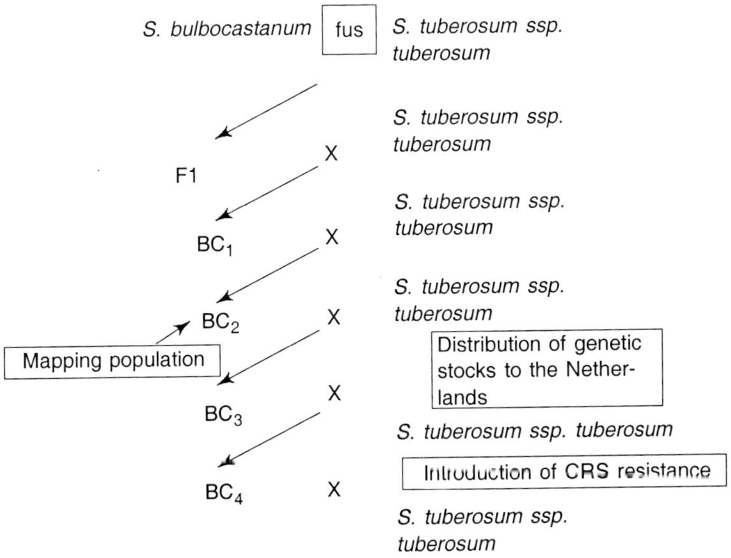

Fig. 12.2 Breeding program carried out at USDA/ARS, Prosser, Washington, USA. *Meloidogyne chitwoodi* resistance was introduced to tetraploid breeding by protoplast fusion with *Solanum bulbocastanum*. Corky ringspot resistance was subsequently inserted from advanced breeding clones. Backcrossing and selection occurred in a recurrent fashion.

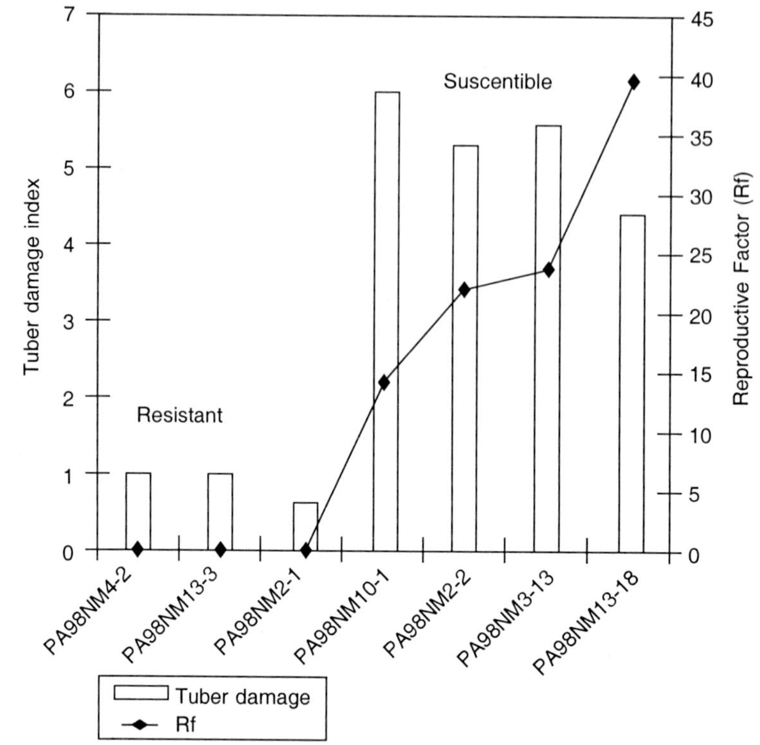

Fig. 12.3 Reproductive factor and tuber damage index of BC_4 progeny. Nonhost Rf is accompained by very low tuber damage in the field.

(Roupe van der Voort et al. 1999). Janssen et al. (1997b) succeeded in introgressing North American 2 EBN allotetraploids by crossing with 4 EBN *Solanum tuberosum* ssp. *tuberosum,* although the crosses were difficult and only a few seeds were obtained. In studies on the introgression of several species, evidence of monogenic inheritance was found in the sources of resistance *S. fendleri, S. hougasii,* and *S. stoloniferum* (Janssen et al. 1997a). Brown et al. (1996 and 1999) proposed the nomenclature for the monogenic resistances to race 1 of CRN from *S. bulbocastanum* and *S. hougasii* as $R_{Mc1\ (blb)}$ and $R_{Mc1\ (hou)}$ respectively, while Janssen et al. (1997a) proposed the gene symbol R_{Mc2} for resistance derived from *S. fendleri.*

Nature of Resistance

Growing wild species sources of resistance in infested fields showed that the accessions previously described as resistant in greenhouse experiments were also resistant to NRN and *M. fallax* in the field. This resulted in a decrease in field populations of the nematodes after a single growing season (Janssen et al. 1996a). CRN increased in virulence after several selections of *S. fendleri* during repeated passages of the inoculum, suggesting a breakdown of resistance (Janssen et al. 1998). Indeed Janssen and coworkers (1996b) called attention to the possible short-lived durability of resistance genes of CRN and *M. fallax,* emphasizing the need to identify distinct sources of resistance and pyramid them by breeding in future varieties.

A survey was done on the host suitability of commercial varieties in Holland when challenged by *M. fallax,* CRN, and NRN. Evaluating the egg mass number and reproductive factor based on number of juveniles present in the soil, interactions between nematodes and cultivars were recorded (Van der Beek et al. 1998) and levels of "resistance" found to be not useful for commercial production. Neither did any of the varieties serve as a genetic source of resistance. Canto-Saenz and Brodie (1986a, b) analyzed the resistance derived from *S. tuberosum* ssp. *andigena* and *S. sparsipilum* in potato clones incorporating these sources through breeding. Several races of *M. incognita* were used. None of the plants tested was consistently unsuitable as hosts for *M. incognita* at higher temperatures (>31°C). At lower temperatures *S. sparsipilum* as a source of resistance to *M. incognita* showed reduced galling and pest reproduction, resulting in higher yield even at a higher inoculum infestation level than in susceptible genotypes. Canto-Saenz (1982) and Canto-Saenz and Brodie (1987) described aspects of resistance in breeding clones deriving resistance from *S. sparsipilum* when challenged with SRN, PRN, and JRN. Resistance was characterized by a hypersensitive response, the root cells adjacent to the nematode becoming necrotic, accompanied by the death and disappearance of the nematode. Tuber penetration was reduced and no further development of the nematode in the tuber occurred. Host suitability increased dramatically at temperatures above 31°C. Mojtahedi et al. (1995) examined the nature of resistance in somatic hybrids between *Solanum tuberosum* ssp. *tuberosum* and *S. bulbocastanum* when challenged by race 1 of CRN. Resistance reduced the number of juveniles penetrating the root tip and lead to the virtually complete failure of ingressed nematodes to induce a feeding site (Fig. 12.4). Instead juveniles remained almost entirely in the vermiform stage and surrounding cells showed a buildup of brown pigment, suggestive of the necrotic response described by Canto-Saenz (1982). CRN does not incite galling of the root systems of potatoes or tomatoes. Thus a scoring system based on assessment of root damage due to galling is useless. Tubers, however, are penetrated by juveniles, which results in superficial galls and brown spots visible on the surface of peeled tubers (see Fig. 12.1). Resistance based on failure of reproduction on the root system appears to be effective in prevention of tuber damage. Resistance to SRN derived from *S. sparsipilum* introgressed into advanced breeding materials, has been described as a hypersensitive response by Berthou and coworkers (1996b).

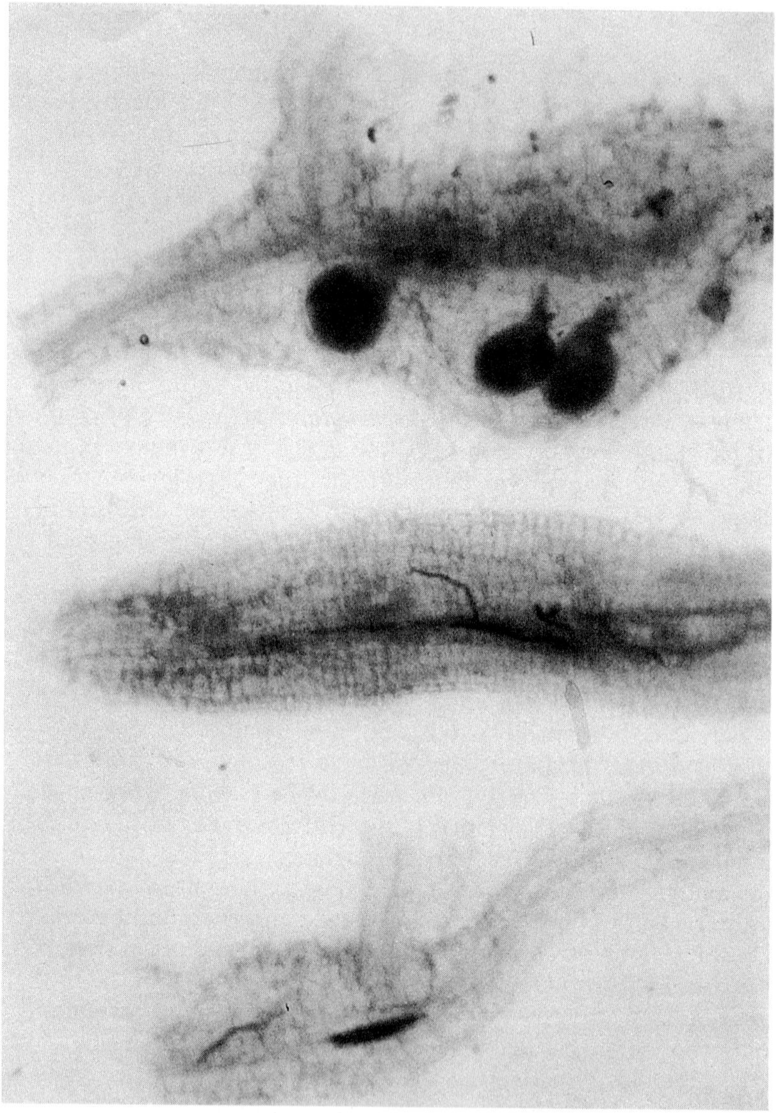

Fig. 12.4 Reaction of roots and penetrated *M. chitwoodi* juveniles in resistant *S. hougasii* (left), *S. bulbocastanum* (middle), and susceptible potato (right) 3 weeks after inoculation. Resistant potato roots do not permit establishment of feeding sites and juvenile development while susceptible roots show stage 3 juveniles.

CORKY RINGSPOT DISEASE (SPRAING)

Corky ringspot disease (CRS) or "spraing" of potato (*Solanum tuberosum* L.) is caused by the tobacco rattle virus (TRV) vectored by trichodorid nematode species (Harrison and Robinson, 1989). The virus is worldwide in distribution (Brunt et al. 1996) but the disease in potato tends to be associated with sandy soils where the vectors thrive. The disease is characterized by arcs, concentric rings, or diffuse extensive browning of tuber flesh that later dries into cork-like tissue. Sometimes necrotic rings develop on the tuber surface (see Fig. 12.2). The blemished tubers are categorized as culls and are unmarketable. In the US the disease has impacted the potato industry in northeastern Florida and to some extent in the PNW (USA). In Florida, approximately one-third of the northeastern potato farms are affected to varying degrees with TRV (Weingartner and Shumaker 1990). In the state of Washington alone, it was estimated that 3% of potato fields are at risk of severe losses due to CRS (Mojtahedi et al. 2000). Similar risks may threaten potato growers of Idaho and Oregon (Mojtahedi et al. 2000). Loss to growers is experienced in the form of culled potatoes, reduction in contract prices due to exceeding maximum incidence standards, and the cost of fumigating potato fields to avoid risk of CRS.

In many potato-growing countries of Europe, the disease is considered a serious threat to potato quality. In England, the disease is called spraing, and additional shoot symptoms have been ascribed to infected potato plants (Harrison and Robinson 1989). Either leaves and a few stems of the infec-ted potato are mottled, or the leaves puckered, small, and deformed. They may also develop bright yellow stripes shaped like chevrons, or arclike and ringlike yellow patches. Tubers from mottled stems are more likely to produce infected plants with infected daughter tubers rather than tubers with severe necrosis (Harrison and Robinson 1989). Those symptoms may help a commercial grower to rogue the infected plant in his field, but for a seed producer the presence of TRV is an unacceptable situation. Generally US seed certification programs have a zero tolerance for TRV. Reports of the disease indicate a worldwide distribution, but most of the research is focused on European and North American potato production areas.

Virus

TRV is the type member of genus *Tobravirus* consisting of two rod-shaped particles, short and long, measuring 50-115 and 190 nm respectively. Normal viral isolates (M-type) consist of two genomic RNAs (RNA-1 and RNA-2) and are readily transmitted by mechanical inoculation and by nematode vectors. Isolates which do not produce viral particles (NM-types) have only RNA-1 and are not nematode transmitted, but sap-transmitted with difficulty. Genomic RNA-1 is highly conserved among different isolates and encodes for viral replicase and proteins for symptom expression and cell-to-cell spread of the virus (Harrison and Robinson 1989). NM-type TRV can be detected in potato tissue by the polymerase-chain-reaction-based (PCR) technique but not by enzyme-linked immunosorbent-assay (ELISA) methods. On the other hand, genomic RNA-2 is variable among TRV isolates, encodes the coat protein plus 2 or 3 nonstructural proteins, and is involved in virus-vector specificity (Hernandez et al. 1997). In Europe, where numerous species of trichodorid nematodes are present in potato-growing countries, different serotypes of TRV have been recognized which are transmitted by distinct species of the vector, or by a single species that transmits different serotypes. If the titer of M-type virus is high enough in the infected tissue, it can be detected by ELISA. Through repeated mechanical inoculation in order to achieve omission of nematode transmission of normal M-type, TRV may be converted to NM-type that is no longer nematode transmissible (Hernandez et al. 1996). Also, differences in transmission efficiency of nematodes occur which are not related to origin of the vector or virus (Ploeg and Brown 1989). Thus method of inoculation, technique employed to detect the virus, and the environment in which the experiment is run play an important role in obtaining dependable results in identifying resistant breeding clones.

Vectors

Trichodorid nematodes (*Allotrichodorus, Monotrichodorus, Paratrichodorus spp.,* and *Trichodorus spp.*) are root ectoparasites (Decreamer 1995), and possess a long arcade stylet, with which they pierce plant cells. Trichodorid nematodes thrive best in sandy soil and are sensitive to low soil moisture (Mojtahedi and Santo 1999). They move downward in the soil and may escape the adverse impact of chemical control measures by migrating to depths at which fumigants have a reduced effect (Weingartner et al. 1983). Trichodorids are polyphagous and hence no crop rotations commonly adopted by potato growers are known to reduce their numbers without chemical control (Mojtahedi et al. 2002). Furthermore, the damage threshold for trichodorid nematodes as a vector is so low (3 per 250 cm^3) (Mojtahedi et al. 2001) that rotating potato with crops that are poor or nonhosts may not cleanse infested soil adequately to grow potato safely. Trichodorids complete their life-cycle in 15-30 days depending on soil temperature.

The role of *P. minor* (Colbran, 1956*) Siddiqi 1974* as a plant pathogen was first demonstrated in 1951 (Christie and Perry) and research interest rapidly increased after the discovery by Sol et al. (1961) that *P. pachydermus* (Seinhorst 1954) Siddiqi 1974* could transmit *Tobravirus* tobacco rattle virus (TRV), the causal agent of potato spraing disease (= corky ringspot

*part of the "attribution" that accompanies binomial genus species with no citation in bibliography.

= CRS). Subsequently, other researchers showed that pea early browning (PEBV) and pepper ringspot (PRSV), both *Tobraviruses,* were also transmitted by trichodorid nematode species (van Hoof 1962; Salomao 1973). It was demonstrated that viral particles attach to the cuticle lining of the posterior tract of the pharyngeal lumen of the vector (Taylor and Robertson 1970; Karanastasi et al. 2000) and are released into the plant cell upon probing by nematodes.

Today, eleven known species of trichodorids have been properly documented (Brown et al. 1989; Ploeg and Brown 1989) as viral vectors (Table 12.3). There is a substantial degree of specificity between trichodorid vector species and *Tobravirus* isolates. This specific relationship is more apparent with association between *Paratrichodorus* vector species than with those between *Trichodorus* species.

Table 12.3 *Virus vector trichodorid nematode species reported from various continents*

Vectors	*Continents*					*Virus*	*First reported by:*
	Europe	*N America*	*Africa*	*S & C America*	*Asia*		
Paratrichodorus allius	X	X	X	X	X	TRV	Jensen and Allen (1964)
P. anemones	X	X				TRV	Ploeg et al. (1993) Komuro et al. (1970)
P. minor	X	X	X	X	X	PRSV	Salomao (1973)
P. nanus	X	X	X			TRV	Cooper and Thomas (1971)
P. pachydermus	X	X			X	TRV	Sol et al. (1961)
P. teres	X	X	X			TRV	van Hoof (1962)
P. tuneisiensis	X		X			TRV	Roca and Rana (1981)
Trichodorus cylindricus	X					TRV PEBV	van Hoof (1968) van Hoof (1962)
T. primitivus	X	X		X	X	TRV	MacFarlane et al. (1999)
T. similes	X	X				TRV	Cremer and Schenk (1967)
T. viruliferous	X	X				PEBV TRV	Gibbs and Harrison (1964); van Hoof et al. (1966)

Except for *P. minor* and *T. primitivus* none of the other virus vector trichodorid species have been reported from Oceania.

According to Decraemer (1995), these species were not natives of the continents.

Two trichodorid species that have major impact on potato production in the USA are *P. allius* and *P. minor*. The first was originally described from onion fields in Oregon and is widespread in the potato production area of the PNW (Mojtahedi et al. 2000). This nematode is readily controlled by costly soil fumigation. The second species, a major concern to potato growers in northeastern Florida (Weingartner et al. 1975), is not readily controlled by soil fumigation (Weingartner et al. 1983). Mojtahedi and Santo (1999) speculated that the difference between the two vectors is that *P. minor* is more mobile than *P. allius* and after initial escape, can reenter the treated zone after the fumigant has dissipated. *P. minor* has been also reported from potato fields of New York (Brodie 1976) and Wisconsin (Walkinshaw et al. 1961). Hafez et al. (1992) recently reported that *P. minor* was present in sugar beet fields in Idaho.

Breeding for Resistance to Corky Ringspot Disease

Despite the complications of different species in the Trichodoridae present in distinct geographic areas, and obvious diversity in tobacco rattle virus, a high level of resistance has been noted in varieties and breeding clones. All of these have been determined by slicing the tubers and scoring symptoms after field or pot culture in the presence of viruliferous nematodes.

Germplasm Description

Harrison (1968) described resistance in British varieties and Bintje. After mechanical inoculation the virus could be detected in both susceptible and resistant varieties. Varieties with severe tuber symptoms rarely showed foliar symptoms (called stem mottle), while varieties with mild tuber expression showed a higher incidence of foliar expression. TRV appeared to be self-eliminating, and therefore not likely to be perpetuated in seed, but elimination proceeded more slowly in varieties with mild tuber symptom expression versus those with severe tuber symptoms. At the Potato Research Center (IHAR) in Mlochow, Poland, breeding clones are assessed by exposure to viruliferous nematodes both in the field and in pots (Borejko 2001). In addition, scions of breeding clones are grafted onto TRV-infected tobacco and the presence of virus tested in the potato tissues as well as in a top-graft of healthy tobacco. Assessment is done by direct observation of symptoms in the potato tissue and by ELISA detection of virus in potato foliage and tobacco top-graft (Muchalski, 2001). Clones and cultivars with high levels of resistance have been noted in Great Britain, Poland, and the US (Richardson 1970; Anonymous 1996; Dale 1989; Dale and Solomon 1988a, b; Shumaker et al. 1984; Weingartner and McSorley 1994; Brown et al. 2000). Thirteen of 93 varieties grown in Europe and North America were designated as having resistance to corky ringspot disease (Swiezynski et al. 1998). Eighteen of 59 Polish varieties had high resistance to corky ringspot (Anonymous

1996). The occurrence of resistance in cultivars (20.4% of varieties listed) reveals that it is common in modern varieties and there are many resistant parental materials from which to choose for breeding purposes.

Inheritance of resistance appears not to be simple, complicated by the fact that some genotypes react with mild symptoms, and certain resistant genotypes have been found to be susceptible to some viral isolates and not to others. Furthermore, it is not clear whether the economically desirable response (i.e., lack of symptoms) is a state of resistance or susceptibility (Dale and Solomon 1988b). Brown and coworkers (2000) showed that resistance does not correlate with suitability of the potato genotype as a host to the stubby root nematode vector, *Paratrichodorus allius,* in the Pacific Northwest of the US, thereby suggesting that resistance to TRV *per se* is the main component of resistance. Maas (1975) and Brown et al. (2000) found that some genotypes were resistant in one infested field but not another. Dale and Solomon (1988b) and Dale (1989) studied inheritance based on a diallel mating design. They found evidence of heritable resistance in some cases that could be cumulative due to a single major gene modified by genes of minor effect. Both general and specific combining ability were also significant. Umaerus (1979) found marked correlations of parent and progeny resistance levels resulting from their crosses. Mojtahedi and Santo (1999) found definite differences in virulence in the most important cultivars, e.g., "Russet Norkotah," and "Russet Burbank," of different nematode-virus isolates collected from geographically different fields in the Columbia Basin of the US. It is conceivable that lack of necrotic symptoms in the tuber is a systemic latent susceptibility of the virus (Robinson and Dale 1994). If this is true, it could be expected that resistant clones would be latent carriers from which resident trichodorid populations might possibly acquire virus, resulting in severe loss when susceptible cultivars are subsequently grown in this field.

Infection Detection Methodology

The development of a TRV detection technique by a reverse transcriptase polymerase chain reaction (rt-PCR) in tissues of potato tubers (Crosslin and Thomas 1995) improved detection for sensitivity. Using rt-PCR, evidence of symptomless infection in potatoes has emerged (Crosslin et al. 1999; Xenophontos et al. 1998). Occurrence of TRV in symptomless field-grown tubers of "resistant" potato genotypes has been noted, although it is rarer than detection in symptomless "susceptible" clones. Richardson (1970) indicated that seedlots of the susceptible Pentland Dell showed a 5% incidence after one year's exposure, which decreased to 1% after several years of multiplication in soil free of TRV, suggesting self-elimination. Dale et al. (2000) found that seedstock of the cultivar Wilja infected latently with TRV was inferior to healthy Wilja in plant vigor, dry matter, fry color, after-cooking darkening, tuber size, and percentage of culls.

Transgenic Resistance

Potato is easily transformed and a number of research endeavors have successfully enhanced resistance to several viruses. In the case of TRV, results have been encouraging but inconclusive. MacFarlane and Davies (1992) transformed *Nicotiana benthamiana* with the putative replicase cistron of *Tobravirus* pea early browning virus (PEBV), a close relative of TRV. Although most plants were resistant to mechanical inoculation of PEBV, they were not resistant to TRV and two variants of PEBV also broke resistance. Van Dun and Bol (1988) showed that expression of TRV coat protein in transformed plants of tobacco conferred resistance to one strain of TRV (TCM) but not to the other strain (PLB). The two strains share only 39% amino acid homology in the coat protein. Plants resistant to TRV-TCM were also resistant to a PEBV isolate that likewise shared greater amino acid homology in the coat protein cistron of TCM. Angenent et al. (1990) demonstrated a transformed tobacco separately expressing the coat proteins of TCM and PLB. Strains of TRV were resistant to the virions of the same coat protein type but susceptible to the other type. However, all transformants were susceptible to infection when viral RNA was used as inoculum. Ploeg et al. (1993) demonstrated that transformed tobacco plants expressing ceoat protein were resistant to mechanical transmission on the leaves but susceptible to nematode-vectored transmission. Further research into transgenic resistance to TRV appears warranted since results to date do not point the way to a commercially viable approach.

REFERENCES

Angenent, G.C., J.M.W. Van Den Ouweland, and J.F. Bol. 1990. Susceptibility to virus infection of transgenic tobacco plants expressing structural and nonstructural genes of tobacco rattle virus. Virology 175: 191–196.

Angenent, G.C., A. Mathis, J.F. Bol, D.J.F. Brown, and D.J. Robinson. 1993. Susceptibility of transgenic tobacco plants expressing tobacco rattle virus coat protein to nematode-transmitted and mechanically inoculated tobacco rattle virus. J Gen Virol 74: 2709–2715.

Anonymous. 1996. Characterystyka zrejonizowanych odmian ziemniaka [Characteristics of Registered Potato Cultivars]. Instytut Zemniaka, Wydanie iv. Bonin, Poland, (in Polish).

Austin, S., J.D. Pohlman, C.R. Brown, H. Mojtahedi, G.S. Santo, D. Douches, and J.P. Helgeson. 1993. Interspecific somatic hybridization between *Solanum tuberosum* L. and *S. bulbocastanum* DUN, as a means of transferring nematode resistance. Amer Potato J 70: 485-495.

Berthou, F., P. Rousselle, and D. Mugniéry, 1996a. Ex-*Solanum sparsipilum-chacoense* heat-stable resistance to*Meloidogyne arenaria* in*Solanum tuberosum*. III° Congrés International de Nématologie, 7-12/7/96, La Guadeloupe. Program and Abstract, pp. 101-102.

Berthou, F., P. Rousselle, and D. Mugniéry. 1996b. Thermo-instability of the resistance expressed in potato roots by hypersensitive reactions to *Meloidogyne arenaria* and selection for stability after short exposures to increased temperatures during invasion by juveniles. Eucarpia, Meeting on Tropical Plants, March 11-15, 1996, Montpellier, France. Communications and Posters, pp. 258–259.

Berthou, F.,D. Mugniéry, D. Ellisséche, and J.P. Dantec. 1999. Résistance à *M. chitwoodi* dans le genre *Solanum*. 14th Triennal Conf. European Assoc. Potato Research. May 2-7-1999, Sorrento, Italy. Abstracts of Conference Papers. Posters and Demonstrations, 325–326.

Borejko, T. 2001. The evaluation of resistance to tobacco rattle virus (TRV) in potato cultivars and breeders selections. *In:* E. Zimnoch-Guzowska, J. Syller, and M. Sieczka (eds.), Methods of Evaluation and Selection Applied in Potato Research and Breeding. Monografic I Rozprawy Naukowe 10a/2001. Plant Breeding and Acclimatization Institute (IHAR), Radzików, Poland, pp. 53–54.

Boydston, R.A., P.E. Thomas, H. Mojtahedi, G.S. Santo, and J.M. Crosslin. 2002. The role of weeds in corky ringspot disease persistence in crop rotations. WSSA Abstracts, 2002 Meeting of WSSA, vol. 42, p. 59.

Brodie, B.B. 1976. Vertical distribtuon of three nematode species in relation to certain soil properties. J Nematol 8: 234-247.

Brown, C.R., Mojitahedi, H., and G.S. Santo. 1989. Comparison of reproductive efficiency of Meloidogyne chitwoodi on Solanum bulbocastanum in soil and in virto tests. Plant Dis 73: 957-959.

Brown, C.R., H. Mojtahedi, and G.S. Santo. 1991. Resistance to Columbia root-knot nematode in *Solanum* ssp. and in hybrids of *S. hougasii* with tetraploid cultivated potato. Amer Potato J 68: 445–452.

Brown, C.R., H. Mojtahedi, and G.S. Santo. 1995. Introgression of resistance to Columbia and Northern root-knot nematodes from *Solanum bulbocastanum* into cultivated potato. Euphytica 83: 71–78.

Brown, C.R., H. Mojtahedi, and G.S. Santo. 1999. Genetic analysis of resistance to *Meloidogyne chitwoodi* introgressed from *Solanum hougasii* in cultivated potato. J Nematol 31: 264–271.

Brown, C.R., H. Mojtahedi, G.S. Santo, P. Hamm, J.J. Pavek, D. Corsini, S. Love, J.M. Crosslin and P.E. Thomas. 2000. Potato germplasm resistant to corky ringspot disease. Amer J Potato Res, 77: 23-27.

Brown, C.R., H. Mojtahedi, and G.S. Santo. 2003. Characteristics of resistance to Columbia root-knot nematode introgressed from several Mexican and American wild potato species. Acta Hortic 619: 117-125.

Brown, C.R., H. Mojtahedi, G.S. Santo, and S. Austin-Phillips. 1994. Enhancing resistance to root-knot nematodes derived from wild *Solanum* species in potato germplasm. *In:* G.W Zehnder, M.L. Powelson, R.K. Jansson, and K.V. Raman (eds.), Advances in Potato Pest Biology and Management. Amer Phytopath Soc. Minneapolis, MN, pp. 426–438.

Brown, C.R., C.P. Yang, H. Mojtahedi, G.S. Santo, and R. Masuelli. 1996. RFLP analysis of resistance to Columbia root-knot nematode derived from *Solanum bulbocastanum* in a BC_2 population. Theor Appl Genet 92: 572–576.

Brücher, H. 1967. Rootknot-eelworm resistance in some South American tuber-forming *Solanum* species. Amer Potato J 44: 370–375.

Brunt, A. A., K. Crabtree, M.J. Dallwitz, A.J. Gibbs, and L. Watson (eds.). 1996. Viruses of Plants, Description and Lists from the VIDE Database. CAB Intl, Wallingford, UK.

Canto-Saenz, M.A. 1982. Variability of root-knot nematodes and the nature of potato resistance to these nematodes. Ph.D. thesis, Cornell University, Ithaca, NY, USA.

Canto-Saenz, M. and B.B. Brodie. 1986a. Host efficiency of potato to *Meloidogyne incognita* and damage threshold densities on potato. Nematropica 16: 109–116.

Canto-Saenz, M. and B.B. Brodie. 1986b. Factors affecting host efficiency of potato to *Meloidogyne incognita*. Nematropica 16: 117–124.

Canto-Saenz, M. and B.B. Brodie. 1987. Comparison of compatible and incompatible response of potato to *Meloidogyne incognita*. J Nematol 19: 218–221.

Castagnone-Sereno, P., F. Leroy, M. Bongiovanni, C. Ziklstra, and P. Abad. 1999. Specific diagnosis of two root-knot nematodes, *Meloidogyne chitwoodi* and *M. fallax*, with satellite DNA probes. Phytopathology, 89: 380-384.

Cenis, J.L. 1993. Identification of four major*Meloidogyne* spp. by random amplified polymorphic DNA (RAPD-PCR). Phytopath 83: 76–80.

Chitwood, B.G. 1949. Root-knot nematodes. I.A revision of the genus *Meloidogyne* Goeldi, 1887. Proc Helminth Soc Washington 16: 90–104.

Christie, J.R. and U.G. Perry. 1951. A root disease of plants caused by a nematode of the genus *Trichodorus*. Science 4: 123-150.

Cooper, J.I. and P.R. Thomas. 1971. Chemical treatment of soil to precent transmission of tobacco rattle virus to potato by *Trichodorus* spp. Ann. Appl. Biol. 69: 23-34.

Cremer, M.G. and P.K. Schenk. 1967. Notched leaf in *Gladiolus* spp. caused by viruses of the tobacco rattle virus group. Netherlands J Plant Path 73: 33-48.

Crosslin, J.M. and P.E. Thomas. 1995. Detection of tobacco rattle virus in tubers exhibiting symptoms of corky ringspot by polymerase chain reaction. Amer Potato J 72: 606–609.

Crosslin, J.M., P.E. Thomas, and C.R. Brown. 1999. Distribution of tobacco rattle virus in tubers of resistant and susceptible potatoes and systemic movement of virus into daughter tubers. Amer J Potato Res 76: 191–197.

Dale, M.F.B. 1989. Advances in the breeding for tobacco rattle virus insensitivity in potatoes. Aspects Appl Biol 22: 87–92.

Dale, M.F.B. and R.M. Solomon. 1988a. A glasshouse test to assess the sensitivity of potato cultivars to tobacco rattle virus. Ann Appl Biol 113: 225–229.

Dale, M.F.B and R.M. Solomon. 1988b. Inheritance of insensitivity to tobacco rattle virus in potatoes. Ann App Biol 112: 225–229.

Dale, M.F.B., D.J. Robinson, D.W. Griffiths, D. Todd, and H. Bain. 2000. Effects of tuber-borne M-type strain of tobacco rattle virus on yield and quality attributes of potato tubers of the cultivar Wilja. Europ J Plant Path 106: 275–282.

Decraemer, W. 1995. The family *Trichodoridae*: Stubby root and virus vector nematodes. Kluwer Academic Publishers, Netherlands 360 pp.

Esbenshade, P.R. and A.C. Triantaphyllou. 1990. Isozyme phenotypes for the identification of *Meloidogyne* species. J Nematol 22: 10–15.

Franco, P. J. 1974. Ensayos preliminares de enraizamiento de hojas y foliolos de papa. Investigaciones Agropecuarias (Perú) 4(1-2): 56–61.

Gibbs, A.J. and B.D. Harrison. 1964. A form of Pea early-browning virus found in Britain. Ann. of App. Biol. 54: 1–11.

Gomez, P.L., R.L. Plaisted, and H.D. Thurston. 1983. Combining resistant *Meloidogyne incognita, M. javanica, M. arenaria,* and *Pseudomonas solanacearum* in potatoes. Amer Potato J 60: 353-360.

Gomez-Cuervo, P.L. 1982. Inheritance of the resistance to root-knot in potatoes and its combination with bacterial wilt. Ph.D. thesis, Cornell University, Ithaca, NY, USA.

González, H. and P. Accatino L. 1974. Estudio preliminar del comportamiento y reacción de diferentes variedades de papas frente al "Nematodo de la raíz" (*Meloidogyne incognita acrita* Chitwood, 1949). Agricultura Técnica (Chile) 34(3): 177–181.

Goodbell, P.B. and H. Ferris. 1989. Influence of environmental factors on the hatch and survival of *Meloidogyne incognita*. J Nematol 21: 328-334.

Griffin, G.D. 1988. The Columbia root-knot nematode, *Meloidogyne chitwoodi,* discovered in the state of Utah. Plant Disease, 72: 363.

Gutierrez-Deza, L.I.R. 1984. Herencia de la resistancia al nematodo del nudo de la raíz, *Meloidogyne incognita*, en papas diploides. M.S. thesis, Universidad Nacional Agraria, La Molnia, Perú.

Hafez, S.L., A.M. Golden, F. Rashid, and Z. Handoo. 1992. Plant-parasitic nematodes associated with crops in Idaho and eastern Oregon. Nematropica 22: 193-204.

Harrison, B.D. 1968. Reactions of some old and new British potato cultivars to tobacco rattle virus. Europ Potato J 11: 165–176.

Harrison, B.D. and D.J. Robinson. 1989. Tobraviruses. *In* L M.H.V. Van Regenmortel and H. Fraenkel-Conrat (eds.), The Plant Viruses, Volume 2: The Rod-shaped Plant Viruses. Plenum Press, NY. pp. 339-369.

Hartman K.M. and J.N. Sasser. 1985. Identification of *Meloidogyne* species on the basis of differential host test and perineal-pattern morphology. *In:* K.R. Barker et al. (eds.), An Advanced Treatise on *Meloidogyne,* vol. II: Methodology. North Carolina State Univ Graphics, Raleigh, NC, pp. 69–77.

Hernandez, C., J.E. Carette, J.D.F. Brown, and J.F. Bol. 1996. Serial passage of tobacco rattle virus under different selection conditions results in deletion of structural and nonstructural genes in RNA 2. J Virol 70: 4933–4940.

Hernandez, C., P.B. Visser, D.J.F. Brown, and J.F. Bol. 1997. Transmission of tobacco rattle virus isolate PpK20 by its nematode vector requires one of the two non-structural genes in the viral RNA 2. J Gen Virol 78: 465–467.

Hewlett, T.E. and A.C. Tarjan. 1983. Synopsis of the genus *Meloidogyne* Goeldi, 1887. Root-knot nematode, systematics, numerical taxonomy. Nematropica 13: 78-102.

Hoyman, Wm. G. 1974. Reaction of *Solanum tuberosum* and *Solanum* species to *Meloidogyne hapla.* Amer Potato J 51: 281–286.

Ingham, R.E., P.B. Hamm, R.E. Williams, and W.H. Swanson. 2000. Control of *Paratrichodorus allius* and corky ringspot disease of potato in the Columbia Basin of Oregon. J Nematol 32 (4S): 566–575.

Inserra, R.N., G.D. Griffin, and D.V. Sisson. 1983. Effect of temperature and root leachate on embryonic development and hatching of *Meloidogyne chitwoodi* and *M. hapla* J nematol 15: 123-127.

Iwanaga, M., P. Jatala, R. Ortiz, and E. Guevara. 1989. Use of FDR 2n pollen to transfer resistance to root-knot nematodes into cultivated 4x potatoes. J Amer Soc Hort Sci 114: 1008–1013.

Janssen, G.J. W.,R. Janssen, A. van Norel, B. Verkerk-Bakker, and C.J. Hoogendoorn. 1996a. Expression of resistance to the root-knot nematodes, *Meloidogyne hapla* and *M. fallax,* in wild *Solanum* spp. under field conditions. Europ J Plant Path 102: 859–865.

Janssen, G.J.W., A. van Norel, B. Verkerk-Bakker, and R. Janssen. 1996b. Resistance to *Meloidogyne chitwoodi, M. fallax,* and *M. hapla* in wild tuber-bearing *Solanum* spp. Euphytica 92: 287–294.

Janssen, G.J.W., A. van Norel, R. Janssen, and J. Hoogendoorn. 1997a. Dominant and additive resistance to the root-knot nemotodes *Meloidogyne chitwoodi,* and *M. fallax* in Central American *Solanum* species. Theor Appl Genet 94: 692–700.

Janssen, G.J. W., A van Norel, B. Verkerk-Bakker, R. Janssen, and J. Hoogendoorn. 1997b. Introgression of resistance to root-knot nematodes from wild Central American *Solanum* species into *S. tuberosum* ssp. *tuberosum.* Theor Appl Genet, 95: 490–496.

Janssen, G.J.W., O.E. Scholten, A. van Norel, and C.J. Hoogendoorn. 1998. Selection of virulence in *Meloidogyne chitwoodi* to resistance in the wild potato *Solanum fendleri.* Europ J Plant Path 104: 645–651.

Jensen, H.J. and T.C. Allen, Jr. 1964. Transmission of tobacco rattle virus by a stubby-root nematode, *Trichodorus allius.* Plant Dis Reporter 48: 333–334.

Karanastasi, E., N. Vassilakos, I.M. Roberts, S.A. MacFarlane, and D.J.F. Brown. 2000. Immunogold localization of tobacco rattle virus particles within *Paratrichodorus anemones.* J Nematol 32: 5–12.

Karssen, G. and T. Van Hoenselaar. 1998. Revision of the genus *Meloidogyne* Goeldi, 1892 (Nematoda: *Heteroderidae*) in Europe. Nematologica 44: 713–788.

Komuro, Y., M. Yoshino, and M. Ichinohe. 1970. Tobacco rattle virus isolated from aster showing ringspot syndrome and its transmission by *Trichodorus minor* Colbran (in Japanese) Annals of Phytopathological society, Japan 36: 17-36.

MacFarlane, S.A. and J.W. Davies. 1992. Plants transformed with a region of the 201-kilodalton replicase gene from pea early browning virus RNA-1 are resistant to virus infection. Proc Natl Acad Sci 89: 5829–5833.

MacFarlane, S.A., N. Vassilakos, and D.J.F. Brown. 1999. Similarities in the genome organization of tobacco rattle virus and pea early-browning virus isolates that are transmitted by the same vector nematode. J Gen. Virol 80: 273-276.

Mass, P.W.Th. 1975. Soil fumigation and crop rotation to control spraing disease in potatoes. Netherlands. J Plant Path 81: 138–143.

Masuelli, R., E. Tanimoto, C.R. Brown, and L. Comai. 1995. Erratic meiosis in a somatic hybrid between *S. bulbocastanum* and *S. tuberosum* detected by species-specific markers and cytological analysis. Theor Appl Genet 91: 401–408.

Mendoza, H.A. and P. Jatala. 1985. Breeding potatoes for resistance to the root-knot nematode *Meloidogyne* species. *In:* J. N. Sasser and C. C. Carter (eds.), An Advanced Treatise on *Meloidogyne,* vol. I: Biology and Control. North Carolina State Univ Graphics, Raleigh, NC, USA, pp. 217–224.

Mojtahedi, H. and G.S. Santo. 1999. Ecology of *Paratrichodorus allius* and its relationship to the corky ringspot disease of potato in the Pacific Northwest. Amer J Potato Res 76: 273–280.

Mojtahedi, H., G.S. Santo, and J.H. Wilson. 1988. Host test to differentiate *Meloidogyne chitwoodi* races 1 and 2 and *M. hapla.* J Nematol 20: 468–473.

Mojtahedi, H., C.R. Brown, and G.S. Santo. 1995. Characterization of resistance in a somatic hybrid of *Solanum bulbocastanum* and *S. tuberosum* to *Meloidogyne chitwoodi.* J Nematol 27: 86–93.

Mojtahedi, H., G.S. Santo, J.H. Wilson, and A.N. Hang. 1993. Managing *Meloidogyne chitwoodi* on potato with rapeseed as green manure. Plant Dis 77: 42–46.

Mojtahedi, H., G.S. Santo, C.R. Brown, H. Ferris, and V. Williamson. 1994. A new host race of *Meloidogyne chitwoodi* from California. Plant Dis 78: 1010.

Mojtahedi, H., G.S. Santo, J.M. Crosslin, C.R. Brown, and P.E. Thomas. 2000. Corky ringspot: review of the current situation. Proc 39th Washington State Potato Conf Trade Fair. Moses Lake, WA, pp. 9–13.

Mojtahedi, H., J.M. Crosslin, G.S. Santo, C.R. Brown, and P.E. Thomas. 2001. Pathogenicity of Washington and Oregon isolates of tobacco rattle virus on potato. Amer J Potato Res 77: 183–190.

Mojtahedi, H., G.S. Santo, Z. Handoo, J.M. Crosslin, C.R. Brown, and P.E. Thomas. 2000. Distribution of *Paratrichodorus allius* and tobacco rattle virus in Pacific Northwest potato fields. J Nematol 32: 447 (abstract).

Mojtahedi, H., J.M. Crosslin, P.E. Thomas, G.S. Santo, C.R. Brown, and J.H. Wilson. 2002. Impact of wheat and corn as rotational crops on corky ringspot disease of Russet Norkotah potato. *Amer J Potato Res* 79: 339-344.

Muchalski, T. 2001. Laboratory methods used for selection of the potato genotypes highly resistant to tobacco rattle virus (TRV). *In:* E. Zimnoch-Guzowska, J. Syller, and M. Sieczka (eds.), The Methods of Evaluation and Selection Applied in Potato Research and Breeding. Monografie I Rozprawy Naukowe 10a/2001. Plant Breeding and Acclimatization Institute (IHAR), Radzików, Poland, pp. 55–57.

Nirula, K.K., C.L. Khushu, and B.T. Raj. 1969. Resistance in tuber-bearing *Solanum* species to root-knot nematode, *Meloidogyne incognita.* Amer Potato J 46: 251–253.

Nirula, K.K., N.M. Nayar, K.K. Bassi, and G. Singh. 1967. Reaction of tuber-bearing *Solanum* species to root-knot nematode, *Meloidogyne incognita.* Amer Potato J 44: 66–69.

Nyczepir, A.P., J.H. O'Bannon, G.S. Santo, and A.M. Finely. 1982. Incidence and distinguishing characteristics of *Meloidogyne chitwoodi* and *Meloidogyne hapla* in potato from the northwestern United States. J Nematol 14: 347–353.

Pinkerton, J.N. and G.A McIntyre. 1987. The occurrence of *Meloidogyne chitwoodi* in potato fields in Colorado. Plant Dis 71: 192.

Pinkerton, J.N., G.S. Santo, and H. Mojtahedi. 1991. Population dynamics of *Meloidogyne chitwoodi* on Russet Burbank potatoes in relation to degree-day accumulation. J Nematol 23: 283–290.

Ploeg, A.T. and D.J.F. Brown. 1989. Factors affecting problems caused by tobacco rattle virus. Aspects Appl Biol 22: 67–72.

Ploeg, A.T., A. Mathis, J.F. Bol, and D.J.F. Brown. 1993. Susceptibility of transgenic tobacco plants expressing tobacco rattle virus coat protein to nematode-transmitted and mechanically inoculated tobacco rattle virus. J Gen Virol 74: 2709–2715.

Powers, T.O. and T.S. Harris. 1993. A polymerase chain reaction method for identification of five major *Meloidogyne* species. J Nematol 25: 1–6.

Raj, D. and R.K. Sharma. 1987. Reaction of some potato cultivars to root-knot nematode infection. Intl Nematol Network Newsl 4: 16–18.

Richardson, D.E. 1970. The assessment of varietal reactions to spraing caused by tobacco rattle virus. J Natl Inst Agricul Bot 12: 112–118.

Rivera-Smith, C.E., Ferris, H., and Voss, R.E. 1991. The application of an excised root assay for the determination of susceptibility or resistance to root-knot nematodes (*Meloidogyne* spp. Goeldi) in potatoes. Amer Potato J 68: 133–142.

Roberts, P.A. 1995. Conceptual and practical aspects of variability on root-knot nematodes related to host plant resistance. Annl Rev Phytopath 33: 199–221.

Robinson, D.J. and B.D. Harrison. 1989. Tobacco rattle virus. CAB Descriptions of Plant Viruses No. 346 (no. 12 revised).

Robinson, D.J. and M.F.B. Dale. 1994. Susceptibility, resistance and tolerance of potato cultivars to tobacco rattle virus infection and spraing disease. Aspects Appl Biol 39: 61–66.

Roca, F. and G.L. Rana. 1981. *Paratichodorous tunisiensis* (*Nematoda: Trichodoridae*), a new vector of tobacco rattle virus in Italy. Nematologia Mediterranea 9: 217–219.

Roupe van der Voort, J.N.A. M., G.J. W. Janssen, H. Overmars, P.M. Zandvoort, A. van Norel, O.E. Scholten, R. Janssen, and J. Bakker. 1999. Development of a PCR-based selection assay for root-knot nematode resistance (*Rmc1*) by a comparative analysis of the *Solanum bulbocastanum* and *S. tuberosum* genome. Euphytica 106: 187–195.

Salomao, T.A. 1973. Soil transmission of artichoke yellow band virus. 2nd Congress International Studi Carciofo, Bari, Italy, 21-24 Nov, pp. 831–854.

Santo, G.S. 1989. The role of crop rotation systems on non-cyst nematodes affecting potatoes in the temperate zones. *In:* J. Vos, C.D. Van Loon and G.J. Bollen (eds), Effects of Crop Rotation on Potato Production in the Temperate Zones. Kluwer Academic Publishers, London, pp. 121–130.

Santo, G.S., J.H. Wilson, and H. Mojtahedi, 1997. Management of the Columbia root-knot nematode and corky ringspot disease on potato. Proc 36th Annual Washington State Potato Conf Moses Lake, WA, USA, pp. 17–25.

Santo, G.S., J.H. O'Bannon, A.P. Nyczepir, and R.P. Ponti. 1981. Ecology and control of root-knot nematodes on potato. Proc 20th Annual Washington State Potato Conf Trade Fair. Moses Lake, WA, USA, pp. 135–139.

Santo, G.S., H. Mojtahedi, M.W. Martin, D.C. Hane, C.R. Brown, J.J. Pavek, and J.H. Wilson. 1994. Tuber resistance in a potato breeding clone that is a suitable host for *Meloidogyne chitwoodi*. J Nematol 26: 565 (abstract).

Sasser, J.N., C.C. Carter, and K.M. Hartman. 1984. Standardization of host suitability studies and reporting of resistance to root-knot nematodes. Crop Nem. Res. Control Proj., NCSU/USAID, Dept. Plant Path., NCSU, Box 7616, Raleigh, NC 27695, USA, 7 pp.

Sasser, J.N., K.M. Hartman, and C.C. Carter. 1987. Summary of preliminary crop germplasm evaluations for resistance to root-knot nematodes. Crop Nem. Res. Control Proj., NCSU/USAID, Dept. Plant Path., NCSU, Box 7616, Raleigh, NC 27695, USA, 88 pp.

Sasser, J.N., K.M. Hartman, and C.C. Carter. 1987. Summary of preliminary crop germplasm evaluations for resistance to root-knot nematodes. Crop Nema. Res. and Control Proj., NCSU/USAID, Dept. of Plant Path., NCSU, Box 7616, Raleigh, NC 27695, USA, 88 pp.

Shetty, K.D. and D.D.R. Reddy. 1985. Resistance in *Solanum* species to root-knot nematode, *Meloidogyne incognita*. Indian J Nematol 15: 230.

Shumaker, J.R., D.P. Weingartner, D.R. Hensel, and R.E. Webb. 1984. Promising russet-skin potato cultivars for North Florida. Soil Crop Sci Soc Florida 43: 166–169.

Sol, H.H., J.C. van Heuven, and J.W. Seinhorst. 1961. Transmission of rattle virus and Atropa belladonna virus by nematodes. Tijdschrift Plantenziekten, 66: 228-231.

Swiezynski, K.M., M.T. Sieczka, I. Stypa, and E. Zimnoch-Guzowska. 1998. Characteristics of major potato varieties from Europe and North America. Plant Breed Seed Sci, suppl 42: 1–44.

Szalanski, A.L, P.G. Mullin, T.S. Harris, and T.O. Powers. 2001. First Report of Columbia Root Knot Nematode (*Meloidogyne chitwoodi*) in potato in Texas. Plant Dis 85: 442.

Taylor, C.E. and W.M. Robertson. 1970. Location of tobacco rattle virus in the nematode vector *Trichodorus pachydermus* Seinhorst. J Gen Virol 6: 179–182.

Thomas, P.E., H. Mojtahedi, J.M. Crosslin, and G.S. Santo. 1999. Eradication of tobacco rattle virus from soils by growing weed-free alfalfa. Proc VII Intl Plant Virus Epidemiology Symp. Aguadulce (Almeria), Spain (abstract), pp. 164–165.

Thomas, S.H., S.A. Sanderson, and Z.A. Handoo. 2001. First report of Columbia root-knot nematode (*Meloidogyne chitwoodi*) in potato in New Mexico. Plant Dis 85: 924.

Umaerus, M. 1979. Hur viktg är rostringsresistens i potatisförädlingen? (How important is corky ringspot resistance in potato breeding?) Sveriges Utsädesförenings Tidskrift 89: 209–216 (in Swedish).

Van Dun, C.M.P. and J.F. Bol. 1988. Transgenic tobaco plants accumulating tobacco rattle virus coat protein resist infection with tobacco rattle virus and pea early browning virus. Virology 167: 649–652.

Van der Beek, J.G., P.F. G. Vereijken, L.M. Poleij, and C.H. Van Silfhout. 1998. Isolate-by-cultivar interaction in root-knot nematodes *Meloidogyne hapla, M. chitwoodi,* and *M. fallax* on potato. Can J Bot 76: 75–42.

Van Hoof, H.A. 1962. *Trichodorus pachydermus* and *T. teres*, vectors of early browning virus of peas. Tijdschrift Planteziekten 68: 391-396.

Van Hoof, H.A. 1964. Het Tijdstip van infectie en veraderingen in de concentratievan ratelvirus (Kringerheld) in de Aardappelknol. Meded. Landb. Mogesch. Opoekingster. Staat Gent 29: 944-955.

Van Hoof, H.A. 1968. Transmission of tobacco-rattle virus by *Trichodorus*-species. Nematologica 14: 20–24.

Van Hoof, H.A., D.Z. Maat, and J.W. Seinhorst. 1966. Viruses of the tobacco rattle virus grouping in northern Italy: their vectors and serological relationships. Netherlands J Plant Path 72: 253-258.

Walkinshaw, C.G., G.D. Griffin, and R.H. Larson. 1961. *Trichodorus christei* as a vector of potato corky ringspot (tobacco rattle virus). Phytopath 51: 806–808.

Weingartner, D.P. and J.R. Shumaker, G.C. Smart, Jr., and D.W. Dixon. 1975. A new nematode control program for potato grown in northeast Florida. Proc Florida State Hort Soc 89: 175-182.

Weingartner, D.P. and J.R. Shumaker. 1990. Effects of soil fumigant and aldicarb on corky ringspot disease and trichodorid nematodes on potato. J Nematol (suppl) 22: 665–671.

Weingartner, D.P. and R. McSorley. 1994. Management of nematodes and soil-borne pathogens in subtropical potato production. *In:* G.W. Zehnder, M.L Powelson, R.K. Jansson, and K.V. Raman (eds.), Advances in Potato Pest Biology and Management. Amer Phytopath Soc, Minneapolis, MN, USA, pp. 202–213.

Weingartner, D.P., J.R. Shumaker, and G.C. Smart Jr. 1983. Why soil fumigation fails to control potato corky ringspot disease in Florida. Plant Dis 67: 130–134.

Xenophontos, S., D.J. Robinson, M.F.B. Dale, and D.F.B. Brown. 1998. Evidence for persistent, symptomless infection of some potato cultivars with tobacco rattle virus. Potato Res 41: 255–265.

13

Resistance to Viruses

RAMONA THIEME[1] AND THOMAS THIEME[2]
[1]*Bundesanstalt für Züchtungsforschung an Kulturpflanzen*
Institut für Landwirtschaftliche Kulturen, Rudolf-Schick-Platz 3a
18190 Gross Lüsewitz, Germany, e-mail: r.thieme@bafz.de
Fax (+49) (0) 38209-45-222
[2]*Bio-Test Labor (GmbH) Sagerheide Birkenalle 19, 18184 Sagerheide, Germany*
e-mail: tt@btl. hro. uunet.de; Fax: (+49)(0) 38204-12-980

INTRODUCTION

Matthews (1992) defined a virus as a set of one or more nucleic acid molecules, normally encased in a protective coat of protein of lipoprotein, which is able to meditate its own replication only within suitable host cells and is dependent on the host's protein-synthesizing machinery.

Potato viruses are prevalent world-wide and regarded as important pathogens causing substantial economic losses. Although in developed countries these losses in yield have decreased because of the use of seed potatoes free of virus, increased viral resistance in potato and intensive applications of insecticides to control viral vectors, the virus problem is still unresolved. In spite of extensive investigations into the mechanisms of the virus life cycle, the biochemical and physiological processes of the interactions between viruses and plants are still poorly understood. Research in this area has been reviewed by Ross (1986), Kegler and Friedt (1993), Świeżyński (1994), Valkonen (1994), Valkonen et al. (1996), and Solomon-Blackburn (2001a, b). These authors summarized the information available and suggested strategies for breeding viral resistance into potato. This knowledge could result in the production of commercial potato clones and cultivars with multiple viral resistance.

Viruses are transmitted to successive potato generations through tubers. Pesticides have been used to control the spread of viruses but this strategy has not proven successful (Jones et al. 1982; De Bokx and Van der Want 1987; Valkonen et al. 1996). Planting of PVY- (potato Y *potyvirus*) or PLRV- (potato leafroll *luteovirus*) infected tubers of susceptible cultivars can cause yield reductions of as much as 80% (Banttari et al. 1993). According to

Corresponding author: Ramona Thieme.

Hull (1984), potato yield losses resulting from the most important viral strains in the UK were estimated at £ 30-50 million in a year of average infection (1982 prices). Yield losses attributable to PLRV worldwide have been estimated at 20 million tons per year (Kojima and Lapierre 1988). These losses were calculated as total cost of reduced yield, downgrading of seed crops, plus the costs of control measures. Currently there are three main categories of control measures for preventing virus-induced crop losses. The first aims to remove viral sources (for example, by producing virus-free seed potatoes), the second to reduce virus spread (for example, by killing viral vectors), and the third to grow more virus-resistant plant varieties (Corsini et al. 1994; Valkonen et al. 1994b). Use of insecticides to control viral vectors is not cheap. Furthermore, they are also undesirable given the increasing insecticidal resistance in aphids, occurrence of new genotypes through expansion of their distribution, growing concern about pesticide residues in food and their environmental impact. Hence resistant cultivars would seem to be the only economical and environmentally friendly way of controlling viral diseases of potato.

POTATO VIRUSES

Full exploitation of resistance in crop protection requires understanding the nature of plant-virus interaction. Various viruses infect potato cultivars and wild species of *Solanum*. The nomenclature of these pathogens was so confusing that in 1991 the International Committee on Taxonomy of Viruses (ICTV) started a universal virus database (ICTVdB). This include not only the characters of each virus but also a decimal numbering system (Brunt et al. 1996; Dallwitz 1980; Dallwitz et al. 1993). The ICTVdB terminology is followed herein.

The major viral pathogens found in *Solanum tuberosum* are listed in Table 13.1. Potato leafroll *luteovirus* (PLRV) was first reported in *Solanum tuberosum* from the Netherlands (Peters 1967). Strains of PLRV differ in severity of symptoms they induced in tested potato plants (Beemster and De Bokx 1987). For a fuller description of the virus see Peters (1967).

Potato Y *potyvirus* (PVY) was first reported in *S. tuberosum* from the UK (Smith 1931). Different strains are characterized according to the local and systemic infection each strain induces in diagnostic plants (see Jones 1990; Kerlan et al. 1999; Table 13.2). There are three main strains of PVY in potato: common or ordinary strain (PVY^{O}), tobacco veinal necrosis strain (PVY^{N}), and stipple streak strain (PVY^{C}). PVY^{C} has not spread worldwide and is less economically important than PVY^{O} and PVY^{N}. More recently, new variants have been reported: PVY^{NTN}, $PVY^{N}W$, and PVY^{Z}. PVY^{NTN} causes the potato tuber necrosis ringspot disease, now present in most potato-producing areas. $PVY^{N}W$ causes a typical vein necrosis in tobacco and was first reported infesting potato in cv "Wilga" in Poland in 1984 (Chrzanowska 1987) but is now rather widespread (Kerlan et al. 1999).

Table 13.1 *The major viral pathogens of potato worldwide (Beemster and De Bokx 1987; Burton 1989; Brunt et al. 1996)*

Virus (abbreviation)	*ICTV decimal code*	*Symptoms*	*Transmission by*
Potato leafroll *luteovirus* (PLRV)	39.0.1.0.012	persist	• potato colonizing aphids in a persistent manner • grafting
Potato Y *potyvirus* (PVY)	57.0.1.0.058	persist	• vectors in a nonpersistent manner • mechanical inoculation • grafting
Potato A *potyvirus* (PVA)	57.0.1.0.056	persist	• vectors in a nonpersistent manner • mechanical inoculation • grafting
Potato V *potyvirus* (PVV)	57.0.1.0.057	persist	• vectors in a nonpersistent manner • mechanical inoculation • grafting
Potato X *potexvirus* (PVX)	56.0.1.0.018	persist	• mechanical inoculation • contact between plants
Potato M *carlavirus* (PVM)	14.0.1.0.025	persist	• aphids in a nonpersistent manner (but not all isolates) • mechanical inoculation • grafting
Potato S *carlavirus* (PVS)	14.0.1.0.026	persist	• aphids in a nonpersistent manner (most isolates) • mechanical inoculation • grafting

Studying several PVY isolates, Glais et al. (2002) found evidence that $PVY^{N}W$ and PVY^{NTN} variants are single to multiple recombinants between PVY^{O} and PVY^{N}. Glais and co-workers performed a restriction fragment length polymorphism study (RFLP) of each gene of several isolates and sequenced the first 2,700 nucleotides of two $PVY^{N}W$ isolates, which showed either PVY^{O} or PVY^{N}-like sequences. Strangely, although $PVY^{N}W$ induces symptoms in *Nicotiana tabacum* typical of PVY^{N}, serological typing using monoclonal antibodies indicates PVY^{O}. Some extremely rare isolates of PVY^{O} which are capable of overcoming resistance in certain potato

Table 13.2 *Groups of PVY strains identified on the basis of hypersensitive response to infection in potato cultivars (Jones 1990; Valkonen et al. 1996; Kerlan et al. 1999; Schubert et al. unpubl. data)*

	Desirée Ny_{tbr}: *nc: nz* $Ny\ (+o/-c,\ n,\ z)_{tbr}$*	*Eersteling, King Edward* Ny_{tbr}: *Nc: nz* $Ny\ (+c/-n,\ o,\ z)_{tbr}$*	*Maris Bard* Ny_{tbr}: *Nc: Nz* $Ny\ (+c,\ o,\ z/-n)_{tbr}$*	*Nicotiana tabacum*
PVY^O	HR nll/sys n	s –/sys mo	HR nll/sys n	mosaic, vein lightening
PVY^C	s –/sys mo	HR nll/sys n	HR nll/sys n	mosaic, vein lightening
PVY^N	s –/mild mo	s –/mild mo	s –/mild mo	mosaic, vein lightening, vein necrosis
PVY^{NTN}		like PVY^N		mosaic, vein lightening, vein necrosis
PVY^NW	s chlsp/mild mo	s ?/sev mo	s ?/?	mosaic, vein lightening, light vein necrosis
PVY^Z	s –/sys mo	s –/sys mo	HR nll/sys mo	?
PVY^{ZE}	s –/sys mo	s –/sys mo	s –/sys mo	?

s—no symptoms on inoculated leaves; HR—hypersensitive response; inoculated leaves: nll—necrotic local lesions, sys—systemic, n—necrosis, mo—mottle, sev—severe, chlsp—chlorotic spots, ?—not tested; * nomenclature follows that of Valkonen et al. (1996).

cultivars were found and named PVY^Z (Jones 1990). Isolates of PVY^Z able to overcome the hypersensitive response in potato cultivars were designated PVY^{ZE} (Kerlan et al. 1999). Although the reliability of bioassays remains questionable (McDonald and Singh 1996), this method in which isolates of PVY are inoculated into sensitive potato cultivars, is still followed for distinguishing the main strains of PVY.

Currently there are not sufficient methods for identifying PVY^{NTN} and PVY^Z. In a recent paper, Boonham et al. (2002) reported an RT-PCR assay for the reliable detection of PVY^{NTN}, distinguishing all the main strains of PVY.

Potato A *potyvirus* (PVA) was first reported in *S. tuberosum* from Eire (Murphy and McKay 1932). For a fuller description of the virus see Brunt et al. (1996). Today PVA occurs worldwide with mild, moderate or severe strains, which differ in severity of symptoms produced in potato cultivars (Smith 1972; Beemster and De Bokx 1987). Valkonen et al. (1995a) divided PVA into strain groups that do (A^1) or do not (A^2) induce a hypersensitive response in the potato cv King Edward.

Potato X *potexvirus* (PVX) was first reported in *S. tuberosum* from the UK (Smith 1931). Strains were separated into four groups based on their ability to elicit dominant genes for the hypersensitive response (Cockerham 1955; Valkonen et al. 1996). PVX strains belonging to group $1(X^1)$ are not pathogenic to potato genotypes carrying the gene Nx_{tbr} or Nb_{tbr}. Strains of group 2 (X^2), commonly called B, are able to overcome the resistance provided by Nb_{tbr}. Strains of group 3 (X^3), the "common strain", are also able to overcome the resistance provided by Nb_{tbr}, and strains of group $4(X^4)$ provide resistance to both these genes.

An additional group is represented by the South American strain PVX_{HB}, which can overcome the resistance protecting the plant against other strains (Moreira et al. 1980).

Potato V *potyvirus* (PVV) was first reported in *S. tuberosum* from the Netherlands by Rozendaal et al. (1971). For a fuller description of the virus see Brunt et al. (1996). Today PVV occurs in Europe and South America and causes few symptoms. Strains have not been separated into groups. Phylogenetic analysis, performed with coat protein sequences (Oruetxebarria and Valkonen 2001), did not reveal geographic grouping. This was confirmed by a bioassay, which showed that the European and South American isolates belong to two strain groups. Only the European isolates induce a hypersensitive response in the potato cv Pentland Dell (Spetz et al. 2002).

Potato M *carlavirus* (PVM) was first reported in *S. tuberosum* from the USA by Schultz and Folsom (1923). For a fuller description of the virus see Brunt et al. (1996). Today PVM probably occurs worldwide and usually causes few symptoms. Strains have not been separated into groups, but a number of isolates were tested for their ability to induce symptoms in different potato cultivars (Chrzanowska and Kowalska 1978).

Potato S *carlavirus* (PVS) was first reported in *S. tuberosum* from the Netherlands by De Bruyn Ouboter (1952). For a fuller description of the virus see Brunt et al. (1996). Today PVS probably occurs worldwide and causes few or no symptoms, but decreases potato yield by up to 20%. Strains have not been grouped although isolates differ in their transmissibility by the vector, *Myzus persicae* (Bode and Weidemann 1970), and some plants show a strain-specific hypersensitive response (Santillan 1979; Bagnall 1972).

Another 25 viruses infect potatoes worldwide, mainly in the Andean region of South America, which is the center of origin of potato (Jones 1981; Jones et al. 1982; Valkonen 1994).

RESISTANCE TO VIRUS

Resistance is the ability of a plant to deter or control a disease. This ability is inherited. The terminology used to define various types and levels of resistance to virus is not always clear and comparable. There are numerous proposals for characterizing and cataloguing the responses of plants to viral infection (Cooper and Jones 1983; Kegler and Friedt 1993; Huth 2000) and the types of host resistance in potato (Ross 1986; Matthews 1992; Valkonen 1994; Valkonen et al. 1996; Solomon-Blackburn and Barker 2001a, b).

The responses of potato plants to attack by virus, mostly for host gene-mediated and transgenic resistance, are described below.

Plants with host gene-mediated extreme resistance (ER) either show no symptoms or limited necrosis. ER prevents virus multiplication at an early stage of infection and is characterized by an extremely low viral titre in the plant. Inheritance of *R* genes is monogenic dominant (Cockerham 1970), coding for extreme resistance to viruses, which is expressed as a highly reduced viral titre in inoculated plants/protoplasts, and operates against a broad spectrum of virus strains. These genes are not normally associated with cell death (Hämäläinen et al. 1997; Gilbert et al. 1998). *Ry* is a dominant gene that confers ER to several strains of PVY in potato (Cockerham 1943; Fernandez-Northcote 1983; Ross 1952; Mestre et al. 2000). *Ry*-mediated resistance is similar in several ways to the resistance conferred by the potato *Rx* gene to PVX; resistant plants do not develop visible symptoms when challenged with the virus, since resistance cannot be detected either by ELISA (Adams et al. 1986; Jones 1990) or RNA hybridization (Goulden et al. 1993). However, resistance is active at the protoplast level (Adams et al. 1986; Barker and Harrison 1984). These features differentiate ER from other types of viral resistance. According to Jones (1990), *R* genes are the obvious choice for incorporation into new cultivars because of their comprehensive action.

Hypersensitive resistance (HR) is coded by *N* genes, which are also monogenic in inheritance. This strain-specific hypersensitive response to viruses (Cadman 1942; Cockerham 1970; Jones 1990) is a rapid defense response that results in the death of a few cells (necrosis) at the site of inoculation, which prevents the infection from spreading further (Dixon and Harrison 1994).

Typically in HR the virus accumulates at the site of infection, but resistance is not active at the protoplast level (Nishiguchi et al. 1978). HR is a generalized resistance response, which follows from a highly specific recognition of an elicitor produced by a given virus (Hammond-Kosack and Jones 1997). Little is known about these gene products, but there are many dominant genes in plants that trigger hypersensitivity, while many virulent strains of viruses can overcome HR. This response can sometimes fail as it is affected by environmental conditions such as increased temperature, reduction in light intensity, and the physiology of the host plant (Ponz and Bruening 1986; Matthews 1991; Valkonen et al. 1998).

Ross (1958) reported that the hypersensitive response can occasionally occur in potato that shows ER, although the *R* gene usually dominates over the N gene. This may indicate a connection between ER and HR (Delhey 1975; Solomon-Blackburn and Barker 2001b). It is also suggested that ER may have a stronger expression for HR (Benson and Hooker 1960; Ross 1986; Valkonen 1994a).

Resistance to infection means a plant does not easily become infected because of reduced likelihood of its attack by vectors. This resistance character of plants is being utilized in different crops (Beekman 1987). For example an interesting resistance character is the presence of glandular hairs that trap insects (Tingey 1991). For viruses that have to be inoculated into the phloem (e.g. PLRV), such a resistance trait does not allow multiplication (Solomon-Blackburn and Barker 1993) or transport of PLRV (Wilson and Jones 1992).

Resistance to virus accumulation occurs when a virus is able to infect a plant but achieves a low multiplication in its host. This kind of resistance is known in *Solanum* species, and was reviewed by Solomon-Blackburn and Barker (2001a).

Resistance to viral movement occurs if the virus is localized in a particular plant part or organ. Viral movement between tubers of infected plants and daughter tuber plants is restricted to the respective part in potato genotypes resistant to PLRV (Barker et al. 1984; Barker 1987) or resistant to tobacco rattle *tobravirus* (TRV) (Muchalski 1997). Resistance to phloem transport (i.e., delayed PLRV systemic movement) is found in cv Bismark (Wilson and Jones 1992). Long distance movement of cucumber mosaic *cucumovirus* (CMV) is blocked in certain potato genotypes, although CMV generally replicates in cells and spreads from cell to cell (Celebi et al. 1998). Possibly the resistance in potato to CMV infection is due to restriction of localization of the virus at the point of entry. Analyzing PLRV movement in the phloem of PLRV-resistant potato clones, Derrick and Barker (1997) recorded little PLRV accumulation. PLRV was largely absent from external phloem bundles in the stem tissue of resistant clones and thus less pronounced in older tissues.

Some studies have shown that resistance to virus can also hamper short distance movement. For example, in *S. brevidens* PVX and PVY accumulate in the leaves, but few leaf cells are infected. Resistance of *S. brevidens* is associated with slow cell-to-cell spread (Valkonen et al. 1991; Valkonen 1992a, b).

HR could also operate via viral movement but it is arrested or impeded by cell death following the initial infection. Localization of a viral infection is often associated with the formation of callose near the necrotic zones caused by viruses, which might limit cell-to-cell movement (Allison and Shalla 1974; Wolf et al. 1991). Transition between the types of resistance is not always clear but fluid. Hinrichs et al. (1998) reported that necrotic formation three days after inoculation stopped the spread of a virus in resistant transgenic potato carrying Ry_{sto} from *S. stoloniferum* and seven days after inoculation accumulation of the virus ceased completely.

A special "response" in plants to viruses is immunity, a complete lack of infection, also referred to as an absolute state of exemption from infection by a virus (Cooper and Jones 1983). There are two kinds of immunity (Valkonen 1992a): either the plants do not permit viral particles or viral nucleic acid to undergo replication, or the plant inhibits the processing of a viral polyprotein, thereby preventing virus multiplication (Ponz et al. 1988). In both cases, plants do not develop symptoms or show an increase in viral titre (Ponz and Bruening 1986) and are immune and nonhosts to the virus. However, Valkonen et al. (1996) pointed out that immunity is unknown in potato. He supposed that reports describing immunity to viruses in *Solanum* spp. are cases of extreme resistance.

STRATEGIES FOR OBTAINING VIRUS-RESISTANT POTATO

Use of Natural Genes for Resistance to Virus

Potato breeding for virus resistance has been reviewed by Davidson (1980), Ross (1986), Świeżyński (1994), and Solomon-Blackburn and Barker (2001a). Sources of resistance in *S. tuberosum* include wild potato species, improved breeder clones, and existing cultivars. There are a number of major genes that confer resistance to virus (Table 13.3). They are used by breeders to incorporate virus resistance into new potato cultivars.

Table 13.3 Solanum *species with genes identified for resistance to potyviruses and PVX (modified from Solomon-Blackburn and Barker 2001a)*

Species	*Virus*	*Type of resistance*	*Gene*	*Reference*
S. tuberosum	PVY, PVA, PVA str A[1], PVA str C, PVV, PVX(1,3), PVX(2)	HR	*Ny*, Na_{tbr}, *Na*, Nc_{tbr}, *Nv*, Nx_{tbr}, Nb_{tbr}	Cadman (1942), Cockerham (1943, 1970); Davidson (1980); Fribourg and Nakashima (1984); Hutton (1951); Jones (1990); Valkonen et al. (1995b)
S. t. ssp. *tuberosum*	PVX(1, 2, 3, 4)	ER	*Rx*	Cockerham (1970); Fernandez-Northcote (1990); Mills (1965); Ross (1986)

(*Contd.*)

(Contd.)

S. t. ssp. *andigena*	PVA, PVY°,	HR	Ra_{adg}, Ny_{adg},	Hämäläinen et al. (1998);
	PVX(1, 2, 3, 4)		Rx_{acl}	Świeżyński (1994); Valkonen et al. (1994a)
	PVY[1)2)], PVA,	ER	Ry_{adg}, Ra_{adg},	Cockerham (1955; 1970);
	PVX(1, 2, 3, 4)[3)]		Rx_{adg}	Hämäläinen et al. (1998); Munoz et al. (1975); Munoz et al. (1994a)
S. sucrense	PVX(HB)[4)], PVX(2)[4)]	ER	Rx_{HB}, Rx_c	Brown et al. (1984)
S. stoloniferum	PVA, PVY, PVY[1)2)], PVV	ER, ER(Y), ER(V)/HR(V)	Ra, Ry_{sto}, $Ry_{sto}{}^{na}$, $Ry_{sto}{}^{rna}$, Na_{sto}	Barker (1996, 1997); Cockerham (1970); Jones (1990); Ross (1961)
	PVA, PVY	HR(A)	$Ry_{sto}{}^{na}$, $Ry_{sto}{}^{rna}$, $Ry_{sto}{}^{n1}$, $Ry_{sto}{}^{n2}$	Barker (1997); Cockerham (1970); Jones (1990); Ross (1961)
S. sparsipilum	PVX(1, 3)	HR	$Nx_{tbr}{}^{spl}$	Cockerham (1970)
S. phureja	PVX	HR	Nx_{phu}	Tommiska et al. (1998); Valkonen et al. (1995a)
S. microdontum	PVA	HR	Ny_{chc}	Cockerham (1970)
S. hougasii	PVY, PVA	ER	Ry_{hou}	Cockerham (1970)
S. demissum	PVY, PVA	HR	Ny_{dms}, Ry_{dms}	Cockerham (1958, 1970)
S. chacoense	PVY, PVX (1, 2, 3, 4)	HR	Ny_{chc}, Nx_{chc}	Cockerham (1970)
S. acaule	PVX (1, 2, 3, 4)	HR	Rx_{acl}	Cockerham (1958, 1970)
	PVX (1, 2, 3, 4)[3)]	ER	Rx_{acl}, X^i	Cockerham (1958, 1970); Ross (1954)

1) Cultivars with these genes showed ER or at least good resistance to PVY^{NTN} (Barker 1996; Le Romancer and Nedellec 1997).

2) Including PVY^N (Ry_{sto}: Barker 1996, Ry_{adg}: Hämäläinen et al. 1998); cv. Corine with $Ry_{sto}{}^{na}$ showed ER to PVY^N (Jones 1990). Ry_{adg} did not confer resistance to PVA (Ross 1986; Hämäläinen et al. 1998).

3) Including PVX group 4 according to Cockerham (1955, 1970) but not Valkonen (1994a) (no data or reference given).

4) Strains 1, 3, and 4 were not mentioned in the reference.
Numbers 1-4 and HB in parentheses after PVX are strains.

The modern European potato varieties, *S. tuberosum* ssp. *tuberosum*, were derived from introductions of cultivated *S. tuberosum* ssp. *andigena* from South America. Resistance to virus (PLRV) appears to have been derived from introgressions of South American *S. demissum*, a by-product of breeding for late blight resistance in the UK (Davidson 1980). In some modern German cultivars, resistance to PLRV can be traced back to *S. demissum* (Ross 1986) and PVY resistance to *S. stoloniferum* (Song et al. 2002).

Identification and transfer of resistance to cultivars is difficult because of single (monogenic) or several gene (polygenic) viral resistance (Cockerham 1970; Ross 1986; Jones 1990; Mendoza et al. 1990), and also strain-specific viral resistance (Ross 1978; Fernandez-Northcote and Brown 1981; Świeżyński 1983, 1994; Dziewońska 1986; Corsini et al. 1994; Jones 1990; Mendoza et al. 1996; Valkonen et al. 1996; Solomon-Blackburn 1998).

Plants with resistance genes are available in various collections around the world and may be used for screening resistance against viruses, other pathogens, and pests. This has resulted in the discovery of new resistance genes and new resistance specificities (Barker 1996; Hämäläinen et al. 1998; Kiru et al. 2002).

Recent taxonomic, phylogenetic, and molecular studies have identified 206 tuber-bearing wild, and seven primitive, cultivated *Solanum* spp. (Spooner and Hijmans 2001) as well as numerous nontuber-bearing species of the series Etuberosa. European potato breeders have used only some of these wild species for incorporating genes into cultivars. That means 90% of the wild species in collections and genebanks could be a source of new resistance.

Most of the accessions of a single species stored in genebanks can differ in genome. The resistance reported for a given accession may therefore be absent from other accessions. In screening for resistance against different pathogens, Ruiz-de-Galarreta et al. (1998) studied 98 accessions of 90 wild *Solanum* species. Interestingly, in six of the eight cases in which two accessions per species were analyzed, the response to a virus of the accessions differed: *S. demissum* and *S. pinnatisectum* differed in their response to PVY^N, and accessions of *S. gourlayi, S. stoloniferum,* and *S. verrucosum* differed in response to PLRV. These authors further reported that they found resistance to pathogens in accessions previously reported as susceptible. These finding thus demonstrate that within a gene pool of *Solanum* species still unexploited resistance is available, which could be used in potato breeding.

Use of Germplasm in Potato Breeding

Potatoes are notoriously difficult to breed due to their tetraploid nuclear state and heterozygous nature. It has been a challenge to accumulate many nonlinked genes in one plant because ratios do not follow the Mendelian model and it is difficult to predict the traits that will be inherited. Therefore progress in breeding for resistance has been slow and arduous, and so far has yielded only modest results. Introgression usually involves hybridization of the donor species with *S. tuberosum*, followed by repeated backcrossing and selection for the desired genes. However, not all potato species hybridize freely with *S. tuberosum*. The commonly cultivated *S. tuberosum* is a tetraploid ($2n = 4x = 48$). Wild species exist in a polyploid series from diploid to hexaploid, with 12 chromosomes as the basic number. Crossability of cultivated potato with wild species is hampered by pre- and postfertilization barriers. Some are genetically determined and some are environmentally related. Besides stylar barriers, ploidy barriers may exist, preventing hybridization in interploidy crosses. In addition, endosperm balance numbers (EBN) play a significant role in interspecific hybridizations (Bradshaw and Mackay 1994; Hermsen 1994). According to Johnston et al. (1980), for successful seed production the EBNs of the respective parents should be in a 2:1 maternal to paternal ratio in the endosperm. To overcome these problems two approaches have been adopted. One is to make bridging crosses and manipulate ploidy using intermediate species (Johnston and Hannemann 1982; Chavez et al. 1988). This approach is demanding of resources, time consuming, and limited in genetic efficiency (Hermsen 1987). A second approach could be somatic hybridisation via protoplast fusion (Barsby et al. 1984).

BIOTECHNOLOGICAL APPROACH

Somatic Hybridization

Shepard (1981) postulated the usefulness of protoplast fusion as an approach for inserting disease resistance into plants. Since the first successful production of interspecific, intergeneric somatic hybrids between *S. tuberosum* and *Lycopersicon esculentum* (Melchers et al. 1978), several cultivated, primitive or wild species have been hybridized with *S. tuberosum* by symmetric or asymmetric hybridization (Rokka 1998).

Fusion of protoplasts can be induced chemically by polyethylene glycol (PEG) or by electrofusion. In potato, both methods are used, but most laboratories prefer electrofusion because high numbers of protoplasts can be handled at the same time and it is less harmful to the cells (Bauer 1990).

During the 1980s and 1990s the development of more efficient methods for isolation, culture, fusion of protoplasts, and plant regeneration (Fig. 13.1; Shepard and Totten 1977; Haberlach et al. 1985; Austin et al. 1985a; Fish et al. 1988; Perl et al. 1988; Schilde-Rentschler et al. 1988; Deimling et al. 1988; Masson et al. 1988; Hunt and Helgeson 1989; Chaput et al. 1990; Möllers and Wenzel 1992; Menke et al. 1996; Thieme et al. 1996, 1997) as well as hybrid identification (Figs. 13.2 A and 13.2 B;

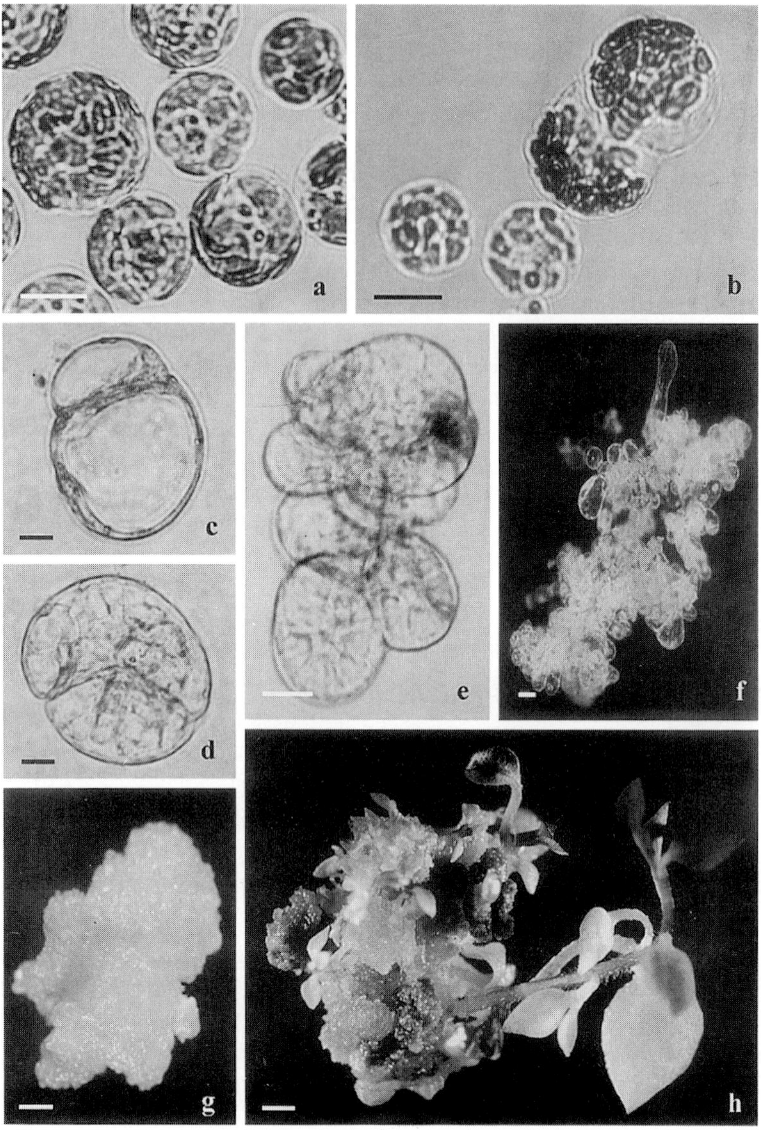

Fig. 13.1 Production of somatic hybrids—stages in plant regeneration after protoplast isolation and fusion between a wild species and *Solanum tuberosum* (photo by Dinu and Thieme 2001). a. Suspension of mixed parental mesophyll protoplasts; b. Electrofusion of two protoplasts; c.-d. First and second cell division; e. Cell colony; f.-g. Micro-and macrocalus; h. Plant regeneration. Magnification: a.-e.: 50x Bars represent 50 μm (a., b.), 30 μm (c., d.), 15 μm (e.), 20 μm (f.), 1 mm (g., h.)

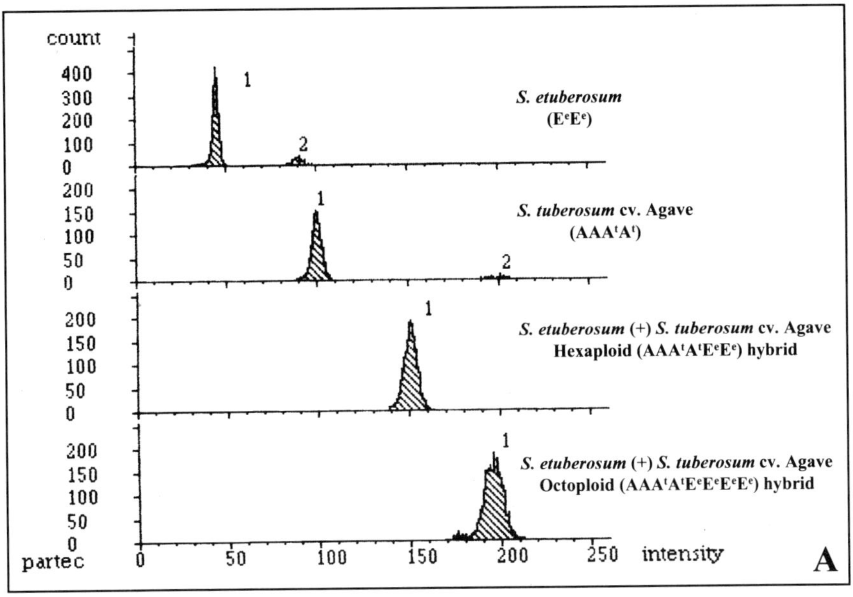

Fig. 13.2A Histograms of flow cytometric measurements of the diploid wild species *S. etuberosum*, the tetraploid cv. Agave and their somatic hybrids with different ploidy levels (Dinu and Thieme 2001).

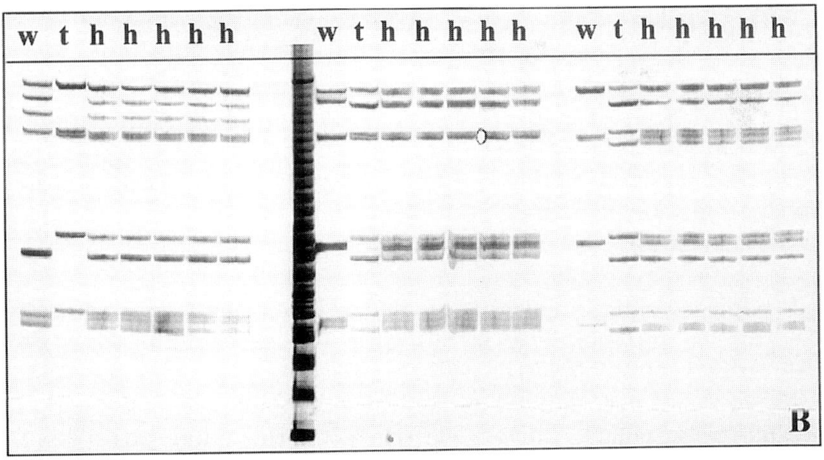

Fig. 13.2B Banding patterns on polyacrylamide gels for parental species (w-wild species, t-*tuberosum* partner) and hybrids (h) using three different microsatellite primer pairs.

Masson et al. 1989; Waara et al. 1989; Schweizer et al. 1990; Rasmussen and Rasmussen 1995; Stadler et al. 1995; Thieme and Hackauf 1998) resulted in improvement of somatic hybridization techniques, which made it possible to produce intra- and interspecific somatic hybrids of potato on an acceptable scale. The fusion method as well as culture conditions for certain genotypes of potato have to be optimized. In addition, there are clear differences in the suitability of clones as fusion partners, which could limit the yield of hybrids. Specific and general combining ability effects are known (Frei et al. 1998) and could be a bottleneck to widening the fusion procedure to include many other genotypes.

One attractive application of somatic hybridization is to combine asexually polygenic traits of different haploids and produce heterozygous interdihaploids (Wenzel et al. 1979). Therefore, protoplast fusion was applied in potato breeding at the diploid level as an alternative method to traditional potato breeding at the tetraploid level (Schilde-Rentschler et al. 1988; Möllers and Wenzel 1992; Möllers et al. 1994; Schwarzfischer 1994; Gavrilenko et al. 1999). The dominant characters of the dihaploid parental lines are expressed in the resultant heterozygous tetraploid somatic hybrids (Wenzel et al. 1979; Wenzel 1994). So protoplast fusion between dihaploid clones, containing a dominant allele for extreme resistance to PVX and/or PVY, can be used to combine resistances to different viruses. Most of these hybrids express resistance to both viruses, but a few deviant clones were also found, which showed phenotypic disappearance of the resistant alleles (Thach et al. 1993). Schwarzfischer et al. (1998) reported that in fusion combinations of two PVY resistant dihaploid clones mainly resistant hybrids were obtained after field trials, but in combinations between a resistant and a susceptible clone, high variation in the hybrids was the rule. The fusion hybrids were more resistant to PLRV than the fusion partners (Schwarzfischer et al. 1998).

Detailed analysis revealed genetic and phenotypic variation among intraspecific somatic hybrids derived from the same fusion combination (Gavrilenko et al. 1999). These data indicate that fusion hybrids originating from the same parental lines show variability in quantitatively inherited traits. Novel hybrid cytoplasm and nuclear-cytoplasmic interactions also play an important role in this variability (Lössl et al. 1994, 1996, 1999; Schilde-Rentschler et al. 1995).

As there has been limited success in incorporating wild germplasm into European potato cultivars via sexual hybridization, protoplast fusion is an alternative method to introgress this genetic material in the *S. tuberosum* genepool, thus bypassing the sexual incompatibility and gene segregation in breeding programs (Millam et al. 1995). Symmetric fusion between cells of wild and cultivated species has successfully resulted in the integration of virus and aphid resistance into potato (Table 13.4). Of

Fig. 13.3A Flowers of parental genotypes. Morphological variation in interspecific somatic hybrids. a. *S. etuberosum* (2n = 2x = 24), b. *S. tuberosum*, breeding clone (2n = 2x = 24), c. tetraploid (2n = 4x = 48), and d. hexaploid (2n = 6x = 72) somatic hybrid.

Fig. 13.3B Development of field-grown tubers. a. tetraploid (4x; EEAA), b. hexaploid (6x; EEEEAA) and c. hexaploid (6x; EEAAAA) somatic hybrid (E—genome of *S. etuberosum*, A—genome of *S. tuberosum*).

special interest for this was the diploid nontuber-bearing wild species, *S. brevidens* (Austin et al. 1985b; Fish et al. 1987; Gibson et al. 1988; Pehu et al. 1990; Valkonen et al. 1994b).

Table 13.4 *Examples of the Solanum species used in interspecific and intraspecific somatic hybridization (and development of their sexual progenies) for transfer of virus and aphid resistance into potato* (S. tuberosum *L.)*

Solanum species	*Resistance*	*References*
S. berthaultii	Aphid	Serraf et al. (1991)
S. brevidens	PLRV, PVY, PVX	Austin et al. (1985b; Gibson et al. 1988); Pehu et al. (1990; Valkonen et al. 1994b);
S. cardiophyllum	PVY	Dinu and Thieme (2001)
S. chacoense	PVY, PVX, PVA	Butenko and Kuchko (1994)
S. commersonii	PVX	Parrella and Cardi (1999)
S. etuberosum	PLRV, PVY, Aphid	Novy and Helgeson (1994a, b); Thieme et al. (1999); Novy et al. (2002); Gavrilenko et al. (2003)
S. tarnii	PVY	Thieme et al. (2002)
S. verrucosum	PLRV	Carrasco et al. (2000)
S. tuberosum	PLRV, PVY, PVX, PVM, PVA	Thach et al. (1993); Ducreux et al. (1993); Schwarzfischer et al. (1998)

Depending on their genome composition somatic hybrids show variation in morphological traits (Figs. 13.3a and 13.3b) and in response to PVY transmitted artificially and by vectors (Thieme et al. 2000). Somatic hybrids have been successfully backcrossed to potato cultivars (Williams et al. 1990; Helgeson et al. 1993; Thieme et al. 2002). Somatic hybrids (Fig 13.4) of *S. etuberosum* (+) *S. tuberosum* and some of their BC_1 clones showed increased resistance to PVY after mechanical virus inoculation (Novy and Helgeson 1994b) and extreme PVY resistance after grafting under greenhouse conditions (Gavrilenko et al. 2003) (Fig. 13.5). In the BC_2 progeny of somatic hybrids [potato × *S. etuberosum* (+) *S. tuberosum* × *S. berthaultii* hybrids] there was increased resistance to PVY, PLRV, and aphids in greenhouse and field trials (Novy et al. 2002). The breeding value of about one hundred and fifty somatic hybrids between Hungarian potato varieties (+) *S. brevidens* and BC progenies was investigated. All hybrids and most of the BC progenies expressed a high level of resistance to PLRV and PVY as shown in the parents (Polgar et al. 2000).

An effective identification of alien genetic material in interspecific somatic hybridization is essential for monitoring the introgression of important characters of wild species. Alien chromosomes in somatic hybrids and their progenies were identified using GISH in combinations: *S. etuberosum* (+) potato (Dong et al. 1999; Gavrilenko et al. 2003), *S. brevidens* (+) potato (Gavrilenko et al. 2002), and GISH and FISH with chromo-

Fig. 13.4 Production of backcross (BC) populations from somatic hybrids.
A. Fertile somatic hybrid successfully pollinated with *Solanum tuberosum*.
B. BC_1 plant with berries for production of BC_2 progeny.

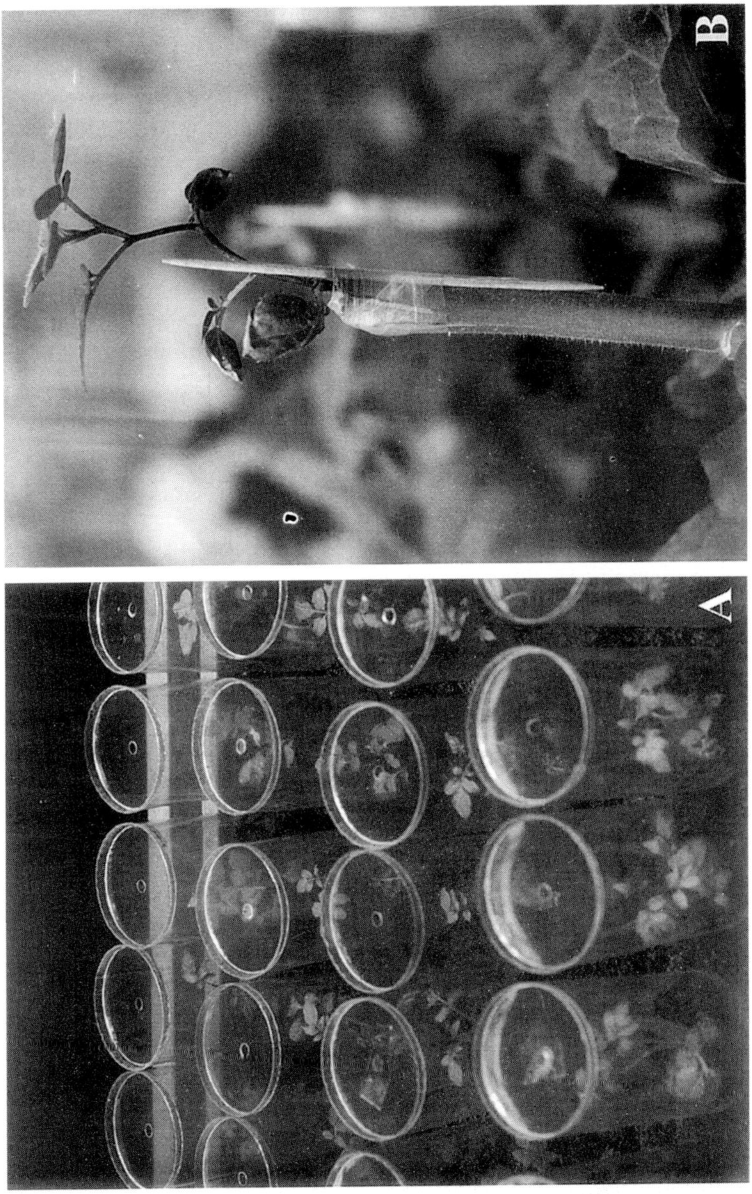

Fig. 13.5 Evaluation of somatic hybrids and progeny for resistance to PVY.
A. *In vitro* plants of hybrids and BC progeny transferred into greenhouse for mechanical PVY inoculation.
B. *In vitro* plantlet of a somatic hybrid grafted on greenhouse grown PVY infected tobacco plant.

some-specific cytogenetic DNA markers (CSCDMs) in the backcross progenies of *S. brevidens* (+) potato hybrids (Dong et al. 2000, 2001). The development of a potato bacterial artificial chromosome (BAC) library was reported by Song et al. (2000). RFLP and RAPD analyses of chromosomal segregation in backcross generations from somatic hybrids between *S. brevidens* and potato revealed intergenomic recombination (McGrath et al. 1994, 1996). Application of the new molecular and cytogenetic techniques will make it possible to localize, isolate, and clone the novel virus-resistant gene(s) from exotic wild *Solanum* species. Because of the great importance of virus and aphid infestations of potato crops and the adaptability of the pathogens, the introduction of "exotic" germplasm from wild species with novel types of resistance to virus is a promising method for increasing genetic diversity in plant breeding. Interspecific somatic hybrids and their progenies should result in the production of new clones breeding for high viral resistance (Polgar et al. 2000; Novy et al. 2002). Furthermore, use of such materials could facilitate the study of the mechanisms and genetic basis of viral resistance.

Transgenic Resistance

There are many reports of pathogen-derived resistance to potato viruses. The results of genetic engineering of plants for viral resistance are reviewed by Harrison (1992), Fitchen and Beachy (1993), Wilson (1993), Baulcombe (1994), Pierpoint (1996), Fuchs et al. (1997), Barker and Waterhouse (1999), and Solomon-Blackburn and Barker (2001a).

Use of coat protein-mediated resistance (CPMR) against PVX was one of the first examples of pathogen-derived resistance (Hemenway et al. 1988). In most cases when CP sequences are used, the transgenic plants are resistant to viral accumulation. Usually the precise reduction in viral concentration in infected transgenic plants varies depending on the cultivar used for the transformation. Besides resistance to viral accumulation, there is an example of a PLRV CP sequence conferring immunity to PLRV—the antisense CP gene (Palucha et al. 1998). The underlying mechanism for CPMR is not fully understood. Occasionally resistance is found although the CP gene is not detected (Van der Wilk et al. 1991). Possibly this is due to methodological difficulties in detecting very low amounts of CP or because the resistance affects viral multiplication at the RNA level.

In 1990, Monsanto scientists reported the development of transgenic Russet Burbank potatoes resistant to mixed viral infections. This was the first attempt to obtain coated protein-mediated protection against PVY and PVX (Villalobos and Rivera 1997).

There is an alternative strategy involving protein-mediated movement resistance. Potato was transformed with a PLRV sequence, which encodes for a phloem-specific movement protein (Tacke et al. 1996). When tested for protection against infection by PLRV, the transgenic lines showed a significant reduction in virus antigen; therefore, plants that accumulated these proteins were also resistant to infection by other potato viruses, PVY and PVX. Interestingly, the protection is not strongly strain specific.

A further strategy to establish resistance was reported by Golemboski et al. (1990) which involves the transfer of a region of the tobacco mosaic *tobamovirus* polymerase gene, which generates a replicase-mediated resistance, into potato. Subsequently, forms of the polymerase of other RNA viruses were transferred and similar replicase-mediated resistance to different viruses reported; e.g., PVX (Braun and Hemenway 1992; Longstaff et al. 1993), PVY (Audy et al. 1994). This resistance is characterized by strong inhibition of virus replication. Though active in protoplasts and against viral RNA, it seems to be strain specific. The mechanism of replicase-mediated resistance is still unclear. The replicase gene product does not seem to be involved in protection, as is observed with CPMR and potyviruses, and the highly resistant plants do not accumulate the replicase protein. This suggests RNA-mediated protection. Monsanto transferred a region of the PLRV polymerase gene into potato (Monsanto 1994). This transgenic line is able to express the whole polymerase and is strongly resistant to PLRV (Monsanto 1994).

Plant transformed with the untranslatable form of the tobacco etch *potyvirus* (TEV) CP transgene (Dougherty et al. 1994) are either highly resistant to the virus, or susceptible, with some recovering some weeks after the initial inoculation. This recovery in phenotype has also been reported for other potyviruses (Ling et al. 1991; Fang and Grumet 1993). Lindbo et al. (1993) reported that no virus could be detected in the plant tissues that recovered and that further inoculation failed. Dougherty et al. (1994) suggested a specific posttranscriptional degradation of the transgenic RNA occurs because the RNA level is reduced, although the transgene was transcribed. This posttranscriptional gene silencing (PTGS) can be induced not only by transgenes, but also by viruses (Baulcombe 1996), and seems to be a common defense strategy of plants and may explain many cases of pathogen-derived protection. PTGS is characterized by a highly specific degradation of introduced alien mRNA and target RNA. For PTGS to operate, the sequences of these two RNAs must be the same or highly homologous (English et al. 1996).

Studying PVY resistance in potato, Smith et al. (1994) found that a certain threshold level of foreign transcripts is necessary to activate PTGS. Interestingly, silencing correlated with an increase in RNA degradation intermediates (Metzlaff et al. 1997; Holtorf et al. 1999). Therefore, the subsequent decrease in mRNA levels is not due to a specific decrease in transcription (De Carvalho Niebel et al. 1995) but to degradation.

After it was shown that plant viruses could be targets of PTGS, viruses encoding proteins that inhibit PTGS were discovered (Voinnet et al. 2000). It seems the possible mechanism by which the viruses overcome resistance (Van Dyke 1994). Considering the step of PTGS they inhibit (Li and Ding 2001), these proteins may be grouped into (a) those that inhibit PTGS in all tissues of a plant, (b) those that block the spread of PTGS to newly developing tissues (Brigneti et al. 1998), and (c) those that prevent the spread of silencing, possibly by blocking the synthesis of the diffusible silencing signal (Voinnet et al. 2000).

Ecological Impact of Transgenic Potatoes

Gene-manipulated crops are a matter of great public concern (Malnoë et al. 1997; Stark 1997). The potential risk concerning resistance to viruses is discussed by Barker and Waterhouse (1999) under four headings: "(I) interaction of the transgene protein with an invading virus causing a new disease; (II) selection by the transgene-mediated resistance for the creation of a new resistance-breaking virus strain; (III) escape of the transgene into weed species, conferring on it greater fitness; and (IV) recombination between a transgene mRNA and the genome of an invading virus to create a novel virus and a new disease."

We shall not discuss these problems but it is likely that the recombination between the RNA of infecting viruses and the virus-derived transgene RNA poses the most important risk. Several studies report that recombinations between natural occurring isolates are common, but the likelihood of such an event is difficult to determine (Aaziz and Tepfer 1999), although Malnoë et al. (1997) reported it is not uncommon: The question is: are these recombinants able to compete with natural viruses?

In addition to these risks there are some practical disadvantages. It is possible to create combining resistance to different potato viruses by genetic manipulation, but the achieved level of resistance is incomplete or only acts against some of the virus strains occurring in the field (Schubert et al. 2002); this means such resistance is unlikely to persist for very long. It is also likely that a formerly resistant plant may become susceptible to other viruses after a single virus isolate overcomes its resistance.

Durability of Viral Resistance

In nature, there is an interaction between pathogens and plants which results in coevolution (Ehrlich and Raven 1964; Parlevliet 1986; Thompson 1994). Therefore, all resistance is temporary. The duration of resistance can be short or long, depending on genotype and environment. Although viruses may have limited (Andean potato latent virus) or very wide host ranges (tobacco mosaic virus, Singh et al. 1995), specificity does not appear to be related to the durability of viral resistance.

Obviously there are differences in the ability of viruses to overcome introduced resistance. Several forms of resistance are quite durable. Race specificity is not a good indicator of the durability of resistance.

The farming system can affect durability (Parlevliet 1993); sanitary and other activities reduce the possibility of a pathogen evolving new races. On the other hand, increase in area covered by a crop with a major resistance gene against a part of the pathogen population (the "boom"), increases the probability of a pathogen overcoming this resistance gene (the "bust"). The breakdown of genetic resistance is not caused by mutation of the resistance gene in the host, but by selection for pathogens better adapted to the resistant crop.

According to the model of the gene-for-gene (GFG) interaction between plants and pathogens, avirulence genes of pathogens produce elicitor molecules identified by specific receptors produced by resistance genes in plants. In the case of HR, both the infected plant cell as well as the pathogen die when a plant cell receptor recognizes a pathogen elicitor. Mutation from avirulence to virulence in the virus changes the elicitor and prevents identification by the host receptor. The evolution of a new virus population that breaks the resistance is caused by an increase in the virus strains that carry the virulent mutation. Gene-for-gene resistance is also called major-gene resistance, and acts only against those viruses that produce the elicitor. Resistance caused by other types of genetically encoded products (minor-gene resistance) does not follow the GFG pattern of the boom and bust cycle (McDonald and Linde 2002).

Different types of genetic variants causing the emergence of resistance-breaking virus variants occur in virus populations. Reviews of these genetic variants have been published by Gibbs et al. (1995), Harrison and Robinson (1999), Garcia-Arenal et al. (2001), and Harrison (2002).

A large proportion of viral genomic variation is caused by mutation. Because of their short replication cycles, high yield, and high mutation rates, RNA viruses develop many mutants, although most of them do not survive. A second type of genomic variation is recombination in several RNA plant viruses (MacFarlane 1997). Recombination is not only recorded in viral RNA, but also in host-RNA sequences, and occasionally in transcripts of virus-derived transgene sequences (Greene and Allison 1994). A third type of genomic variation is caused by reassortment of genomic segments. This pseudo-recombination is found in viruses with a segmented genome but many of these isolates are not positively selected.

Lastly, plant viruses are associated with extra nucleic acid components which may affect symptoms, levels of virus replication and/or vector transmissibility, but apparently not virus host range (Tamada et al. 1989; Liu et al. 1998; Mansoor et al. 1999; Garcia-Arenal et al. 2001; Harrison 2002).

Comparative studies of resistance-breaking and avirulent forms of a virus are used to identify the viral features involved in resistance. Analyses have shown that in some cases a specific point mutation is responsible for overcoming resistance, whereas in other cases at least two point mutations are needed. The potato genes *Nb* and *Nx*, inducing HR to PVX, are present in several potato cultivars. This resistance can be broken by a single point mutation. Whereas a specific mutation in the 25K movement protein gene is responsible for overcoming the resistance caused by *Nb* (Malcuit et al. 1999), a point mutation in the coat protein (CP) gene occurs in isolates breaking the resistance caused by *Nx* (Santa Cruz and Baulcombe 1993). The potato gene *Rx1* induces ER to PVX. Bendahmane et al. (1995) reported that *Rx1*-mediated resistance is broken by PVX isolates with two specific mutations in the coat protein gene. Whereas this mutant has not been positively selected for Europe, an *Rx1*-breaking form of PVX is found in South America, which has a different amino acid sequence in its coat protein.

The potato gene *Ry* confers ER to all strains of PVY. Until now no form of PVY able to break *Ry*-mediated resistance has been detected. Initially it was thought that genes causing ER encode inhibitors of viral accumulation. Recently, the virus-coded nuclear inclusion, (NIa) protease, was identified as the probable avirulence factor by Mestre et al. (2000). Harrison (2002) argued that because the polyprotein of the NIa protease cleaves at different sites when the viral RNA is translated during replication, malfunction at any of these cleavages would prevent the production of functional products and therefore it is unlikely to overcome changes in NIa protease, since changes at all cleavage sites in the viral polyprotein would be required at the same time.

VIRUS VECTORS

It is difficult to control a viral disease that involves transmission by aphids without some knowledge of how it is propagated and spread. Basic to understanding transmission is knowledge of the hosts, viruses, and vectors. As aphids feed on potato they can transmit and spread a virus in these crops. This has been the subject of extensive research. Earlier studies on the transmission of cucumber mosaic viruses by aphids were done by Doolittle (1916) and on spinach blight by McClintock and Smith (1918). From that time onward virus transmission by aphids has been under continuous study. One of the first papers on the spread of potato virus was published by Doncaster and Gregory (1948). Although the epidemiology of this and other viruses has been well studied, the spread of viruses is still a matter of concern. Viruses are transmitted by various species of aphid differing in efficiency. Efficiency is defined as the probability of an aphid transmitting a noncirculative virus or circulative virus after acquisition. This efficiency plays an important role in the rate of virus spread. Consequently, it mainly determines the prevalence of various viruses, strains, and isolates in potato-growing areas. Analyses of the virus transmission efficiency of a given aphid are usually done in laboratories and the results often differ. These differences can be related to the test plant species, virus strain or isolate, or species, host race, or genotype of aphid used. In addition, the methods used, in particular the conditions under which the vectors, virus source, and test plants are grown, and the age of the aphid colonies and host plants also affect the transmission rate.

Aphids Transmitting PLRV

For laboratory studies *Myzus persicae* is the favored vector of PLRV. Williams and Ross (1957) noticed that this species varied in efficiency to transmit PLRV. Aphids that were efficient vectors produced offspring which performed efficiently, whereas the offspring of aphids that performed poorly in transmitting a virus were also poor vectors. Hinz (1965, 1966) found differences in the transmission efficiency of six different taxa of this species. Because he had access to aphid cultures raised by a taxonomist, he was one of the first to use bionomically well-defined subspecies and races as vectors. The frequency with which efficient and inefficient vectors occur in the field is not known nor how well they perform.

PLRV transmission efficiency depends on the plant species or cultivar. In laboratory studies comparing *Physalis floridana* and *S. tuberosum* as source plants for PLRV, Kirkpatrick and Ross (1952) found *P. floridana* to be a better virus source than potato. Before registering potato varieties they are usually planted in the field alongside plants infested with an abundance of aphids to assess their resistance to PLRV and PVY. Resistance is assessed in terms of the preference of the aphids for the plants, infectivity of an aphid, and development of resistance in the mature plant.

Besides *M. persicae,* there are other potato-colonizing aphids that are less effective in transmitting viruses. Several authors found that *Aulacorthum solani* is poor at transmitting PLRV as could also be concluded from data compiled by Hinz (1970). Some authors failed to show transmission by this species (Heinze 1960). *Aphis nasturtii,* which is not always present but has sometimes been very abundant in potato crops of Central and Eastern Europe, is known to be only slightly effective at spreading PLRV (Gabriel et al. 1987; Kostiw 1980). Several authors found that *Macrosiphum euphorbiae* is a poor vector of PLRV (Hinz 1970). In their extensive studies on the transmission of PLRV, Robert and Maury (1970) used five strains of *M. persicae, A. solani,* and *M. euphorbiae.* They found that all strains of *M. persicae* and one of *A. solani* transmitted the virus efficiently, whereas four strains of *A. solani* and one of *M. euphorbiae* were moderately efficient while four strains of *M. euphorbiae* transmitted this virus only at a very low rate if at all. It was also confirmed by Sylvester (1980) that there are large differences in the transmission efficiency of strains of vector species.

Although it is the adults of the apterous morph (form) of aphids that tend to be used in virus transmission studies, some authors have reported that larval instars are more efficient at transmitting virus (Robert et al. 1969). Miyamoto and Miyamoto (1971) showed that larvae are as good as adults of *M. persicae* in transmitting PLRV. Generally, larvae are better than adults. This may be because larvae are born on a host plant whereas adults must first choose a plant. However, the opposite was reported by Robert (1971). The last larval stage of *A. solani* is very inefficient compared to other stages; age-dependent efficiency was likewise confirmed by studies on *M. euphorbiae* in which the last larval stage of the winged aphid was least efficient. Field observations also indicate *M. euphorbiae* is a poor vector of PLRV (Murphy and McKay 1929). In years when *M. euphorbiae* arrived early and became abundant the spread of PLRV was minimal (Woodford et al. 1983). However, some clones did transmit PLRV efficiently to *Nicotiana clevelandii* and potato test plants (Woodford 1992).

Aphids Transmitting PVY

The transmission efficiency of potato viruses in the nonpersistent mode has been studied by several authors, using both potato colonizing and noncolonizing aphids. Large differences in the transmission efficiency of PVY were found for each species tested. This is attributable to the use of different strains of virus and aphids and possibly also the experimental conditions. Laboratory experiments have shown that besides *M. persicae* and other potato-colonising species, a large number of aphid species can transmit strains of PVY (Van Hoof 1980; Karl 1971; De Bokx and Piron 1984; Sigvald 1984).

Although the above experiments produced some information on the transmission efficiency of various vectors, but it may be noted that the aphids used were prevented from making long distance flights and had perforce to probe potato plants in cages. After some probing *M. persicae* accepted these plants as hosts whereas noncolonizing species, such as *Rhopalosiphum padi* and *Acyrthosiphon pisum* moved from one plant to another searching for their host plant and therefore were more likely to transmit virus than *M. persicae*. These experiments were not made under natural conditions and could have resulted in a transmission rate too high for those aphid species that do not colonize potato. The same result was obtained when noncolonizing aphids were kept for a long period on test plants. To eliminate these factors, several authors analyzed the occurrence and abundance of vector species in the field. In the UK, flying aphids were trapped on a vertical net placed downwind of a plot of potato infected with PVY (Harrington et al. 1986). A total of 20 of the 119 species or species groups trapped were identified as vectors. The species that accounted for the majority of transmissions was *Brachycaudus helichrysi*, not mentioned as a vector by either Van Hoof (1980) or Sigvald (1984), but considered one by both Van Harten (1983) and Nemecek (1993). Extensive field studies in Sweden monitoring winged aphids over 15 years using yellow-water traps showed that about 25-30 of the species trapped were vectors, of which *R. padi*, *Aphis fabae*, *B. helichrysi*, and *A. pisum* are the most important (Sigvald 1995). Similar results were reported from Northern Germany, where 79% of the aphids transmitting PVY to test plants belonged to species that do not colonise potato (Thieme et al. 1998). In order to correlate the spread of PVY transmissions in the field with the time of flight and abundance of an aphid species, Van Harten (1983) and Sigvald (1984) calculated a relative efficiency factor. This factor is an estimate of the contribution of an aphid species to the spread of a given virus in the field. *M. persicae* was given the highest relative efficiency factor, viz. 1.0, and other species lower values. The relative efficiencies calculated for the different aphid species differs between these two authors. To obtain a more realistic transmission efficiency, the number of probes made by colonizing as well as by noncolonizing species on potato and the probability of the winged form of an aphid species transmitting virus per probe was considered. Transmission efficiency and number of aphids probing in a potato field determine infection pressure. This is supported by electrical penetration graphs (EPGs), which allow investigation of the stylet activities during probing and feeding by aphids and the importance of brief penetrations for the transmission of nonpersistent PVY. Puncture of a cell membrane by stylets, recorded as a potential drop (pd), was correlated with the acquisition and inoculation of PVY by *M. persicae* (Powell 1991). Electrically recorded stylet penetration activities clearly demonstrated that the vector efficiency of *B. helichrysi* is associated with the higher frequency of cell membrane punctures, whereas the nonvector, *Drepanosiphum platanoidis*, punctured membranes less frequently (Powell et al. 1992).

BREEDING FOR RESISTANCE TO APHIDS

Potato cultivars differ in resistance to aphids and in degree of aphid colonization. Aphids are important vectors of several economically important viruses. According to Sigvald (1989, 1990), resistance to aphids does not necessarily imply cessation of viral transmission by these insects. Some increased transmission rates of PVY were obtained for aphid-resistant plants (Oertel et al. 1980). This may be due to changes in the feeding behavior of the aphids: when they leave plants identified by probing as unsuitable, transmission of viruses carried by them from plant to plant is inevitable.

Potato breeding research has focused more and more over the years on aphid resistance of potatoes. Aphid resistance is based on a combination of several biochemical and morphological factors, in particular glycoalkaloid content and pubescence. Tingey et al. (1978) and Tingey and Sinden (1982), established the considerable importance of glycoalkaloids in the insect resistance of potatoes. Several studies have reported that foliar glycoalkaloids are harmful to leaf-feeding beetles and potato leafhoppers (Levinson 1976; Raman et al. 1979; Sanford et al. 1996). However, until now no correlation has been found between resistance to aphids and high levels of glycoalkaloids (Flanders et al. 1992).

Gibson (1971) studied the resistance of three *Solanum* species to different aphid species. Results indicated that resistance to aphids was basically nutritional. He also observed that resistance of each *Solanum* species was dependent on aphid species, the plant part colonized, and the physiological status of the leaves. Gibson therefore expected the mechanism of resistance to aphids to be nonspecific, viz. pubescence, to be more effective than a nutritional mechanism.

Some wild species of *Solanum* have two types of glandular hairs on their leaves: type A and type B. Type A has a short and four-lobed head filled with a water-soluble liquid. If the heads are ruptured the contents are released and form a viscous substance. Type B is a multicellular glandular trichome, with a longer stalk, exuding a sticky droplet at its end. This substance supposedly entangles aphids, which while trying to escape damage type A hairs and become physically trapped. The mobility of aphids is also reduced because parts of the insect body become covered with the sticky substance (Gibson and Pickett 1983).

Wild species of *Solanum* with pubescence are *S. berthaultii*, *S. polyadenium*, *S. wittmackii*, and *S. tarijense*. Sticky glandular hairs on the leaves of *S. berthaultii* reduce reproduction in *M. persicae* and disturb the takeoff of alate individuals. Furthermore, the pubescence influences aphid feeding behavior and decreases the number of probes and acquisition of PVY by *M. persicae* (Gunenc and Gibson 1980). Using two virus-susceptible accessions of *S. berthaultii*, Rizvi and Raman (1983) studied the effect of the two types of pubescence on the occurrence of aphids and viral transmission. One accession had hairs of type A, the other of both types A and B. The average number of aphids in the former was higher than on the latter. In the field, the percentage of plants with A and B type hairs with PLRV-infection was below that for plants with only type A hairs (22% against 84%), but all plants of both accessions were infected by PVY. These results show that a mixed pubescence of two types of hairs is more effective in decreasing the number of aphids and PLRV spread than type A pubescence alone.

The inheritance of type B hairs was studied by Gibson (1979). Within the species *S. berthaultii,* this type of trichome is a monogenic dominant. In crosses between *S. phureja* and *S. berthaultii,* however, at least two genes are needed to explain the observed segregation. Analysis over two generations of crosses between *S. berthaultii* and a potato cultivar were made by Mehlenbacher and Plaisted (1983). The heritability ratio of both A and B types of hairs and the size of the droplet on the type B hair, were calculated. They concluded it should be possible to enhance insect resistance of plants by recombination and selection of these characters.

Several other *Solanum* species were screened for resistance to aphids. Interesting results were reported for wild potatoes of the series Etuberosa. The resistance of *S. brevidens, S. etuberosum,* and *S. fernandezianum* to *M. persicae* was studied by Valkonen (1992a). Species of the Etuberosa group vary in their resistance to *M. persicae* depending on age of the plants. Whereas the young and mature leaves of *S. brevidens* were never colonized, those of *S. etuberosum* were colonized as they reached maturity. The senescent leaves of both *S. brevidens* and *S. etuberosum* also became infested. In contrast *S. fernandezianum* showed no resistance to *M. persicae.* The high mortality during development of *Aphis frangulae, A. solani,* and *Myzus nicotianae* reared on *S. etuberosum* suggested that its ER to virus is due in part to the unsuitability of the host for aphid vectors (Thieme and Thieme 1998). Besides high mortality of *M. persicae,* when reared on *S. etuberosum,* an abnormal feeding behavior of surviving aphids at the phloem and prephloem level was observed using the technique of recording EPGs (Thieme et al. 2000, 2002, 2004). Further studies of aphid feeding behavior on interspecific somatic hybrids and backcross progeny, should show if aphid resistance characteristics from the wild species were transferred to these plants. The resistance of *S. etuberosum* and its backcross progeny to *M. persicae* was also described by Novy et al. (2002). These authors found that resistance to the aphid involved a reduction in adult size and fecundity. In addition, they identified one BC_2 individual carrying a genetic factor detrimental to nymph survival. Although resistance to all aphids was not found, even a moderate level of resistance to aphids transferred into a cultivar might strongly reduce spread of viruses by aphids (Barker and Harrison 1986; Valkonen 1992a).

EFFECTIVENESS OF INSECTICIDE APPLICATIONS

In addition to the problems of virus-vector relationships, the effectiveness of insecticide applications depends on many factors, such as the pattern of migration, recovery (reinfestation in treated fields) and the number of aphids involved, size of aphid populations in adjacent fields, number of source plants inside and outside the treated potato fields, ratio of the surface area of foliage of sprayed and unsprayed fields, and mode of action and dosages of the active ingredient. These factors tend to vary depending on region and season. More detailed knowledge of these aspects might result in better timing of insecticide applications.

The effect of insecticides on the control of PLRV varies from one report to another. This may depend on the number of aphids in the surroundings of the experimental plots. If the experimental plots are small and surrounded by large untreated crops, the infection pressure will be equal to that in the surrounding fields by immigrating aphids. Therefore, the effect of an insecticide can become underestimated in experiments using small plots. PLRV can increase when infection pressure outside the experimental fields is high and the aphid population remains low in those fields (Bacon et al. 1976). Heimbach et al. (2002) observed that the treated plots used in their experiments were recolonised rapidly at the borders with untreated plots. Therefore the EPPO (European and Mediterranean Plant Protection Organization) guidelines for regi-stration of insecticides to control virus vectors demand a large size plot depending on the target virus.

Application of insecticides seems to have a much greater effect on the control of PLRV than of PVY. However, the spread of PVY may be reduced by controlling PLRV. The lack of large aphid populations can possibly affect the spread of PVY in summer since potato-colonizing aphids appear to be responsible for its spread. But on the other hand, there are indications that some insecticides stimulate the activity of treated aphids and therefore contribute to an increase in PVY infection in potatoes (Thieme and Heimbach 1998).

Since the discovery of Bradley (1963) that the transmission of PVY was impeded by coating plants with mineral oils, several authors have studied the effect of this treatment (Bradley et al. 1966; Shands 1977). Reduction in PVY transmission varied from at least 50% to over 90% expressed as infected tubers or plants. The underlying mechanism was not fully discovered. Although the prepenetration activities of aphids were longer on oil treated leaves, the inhibitory effect of the oil could not be attributed to differences in the duration of stylet penetration. Powell (1992) in his study on the effect of mineral oil on the feeding behavior of aphids found that both acquisition and inoculation of the PVY were reduced by the presence of oil on the plant surface, but these reductions could not be correlated to electrically recorded differences in plant penetration behavior; in particular stylet punctures of plant cell membranes were not reduced by mineral oil. Nonbehavioral reasons are suggested to explain the mode of action of the oil.

The use of oil has not become common practice in the production of seed potatoes. This may be because of the cost of oil, the number of sprays necessary, and the reduced infections and spread of PVY in recent years. Other disadvantages of using oil are that it becomes more difficult to control *Phytophthora infestans* and rogue virus-diseased plants.

In conclusion, a number of viruses attack potato and several methods have been proposed to protect potato against them. The most effective and durable way of protecting this crop, however, will be combining resistance genes (both natural and transgenic) for both virus and vectors.

REFERENCES

Aaziz, R. and M Tepfer. 1999. Recombination in RNA viruses and virus-resistant transgenetic plants. J Gen Virol 80: 1339–1346.

Adams, S.E., R.A.C. Jones, and R.H.A. Coutts. 1986. Expression of potato virus X resistance gene *Rx* in potato leaf protoplasts. J Gen Virol 67: 1341–1346.

Allison, A.V. and T.A Shalla. 1974. The ultrastructure of local lesions induced by potato virus X: a sequence of cytological events in the course of infection. Phytopath 64: 784–793.

Audy, P., P. Palukaitis, S.A. Slack, and M. Zaitlin. 1994. Replicase-mediated resistance to potato virus Y in transgenic tobacco plants. Mol Plant Micr Interact 7: 15–22.

Austin, S., M.A. Baer, M. Ehlenfeldt, P.J. Kazmierczak, and J.P. Helgeson. 1985a. Intra-specific fusions in *Solanum tuberosum*. Theor Appl Genet 71: 172–175.

Austin, S., M.A. Baer, and J.P. Helgeson. 1985b. Transfer of resistance to potato leafroll virus from *Solanum brevidens* into *Solanum tuberosum* by somatic fusion. Plant Sci 39: 75–82.

Bacon, O.G., V.E. Burton, D.L. McLean, R.H. James, W.D. Riley, K.G. Baghott, and M.G. Kinsey. 1976. Control of the green peach aphid and its effect on the incidence of potato leafroll virus. J Econ Entomol 69: 410–414.

Bagnall, R.H. 1972. Resistance to viruses M, S, X and the spindle tuber virus in tuber-bearing *Solanum* species. Amer Potato J 49: 342–348.

Banttari, E.E., P.J. Ellis, and S.M.P. Khurana. 1993. Management of diseases caused by viruses `and virus like pathogens. *In:* R.C. Rowe (ed.), Potato Health Management. Amer Phytopath Soc Press, St. Paul, MN, USA, pp. 127–133.

Barker, H. 1987. Multiple components of the resistance of potatoes to potato leafroll virus. Ann Appl Biol 111: 641–648.

Barker, H. 1996. Inheritance of resistance to potato virus Y and A in progeny obtained from potato cultivars containing gene Ry: evidence for a new gene for extreme resistance to PVA. Theor Appl Genet 93: 710–716.

Barker, H. 1997. Extreme resistance to potato virus V in clones of *Solanum tuberosum* that are also resistant to potato virus Y and A: evidence for a locus conferring broad-spectrum resistance. Theor Appl Genet 95: 1258–1262.

Barker, H. and B.D. Harrison. 1984. Expression of genes for resistance to potato virus Y in potato plants and protoplasts. Ann Appl Biol 105: 539–545.

Barker, H. and B.D. Harrison. 1986. Restricted distribution of potato leafroll virus antigen in resistant potato genotypes and its effect on transmission of the virus by aphids. Ann Appl Biol 109: 595–604.

Barker, H. and P.M. Waterhouse. 1999. The development of resistance to luteovirus mediated by host genes and pathogen—derived transgenes. *In:* H.G. Smith and H. Barker (eds.), The Luteoviridae. CAB Intl, Wallingford, UK. pp. 169–210.

Barker, H., I.M. Roberts, B.D. Harrison, and P.R. Massalski. 1984. Studies on potato viruses. Annual Report, Scottish Crop Research Institute, pp. 193–196.

Barsby, T.L., J.F. Shepard, R.J. Kemble, and R. Wong.1984. Somatic hybridization in the genus *Solanum, S. tuberosum* and *S. brevidens*. Plant Cell Rep 3: 165–167.

Bauer, R. 1990. Somatic plant hybridization—a review. Arch. Züchtungsforsch 20: 66–79.

Baulcombe, D.C. 1994. Novel strategies for engineering virus resistance in plants. Curr Opin Biotechnol 5: 117–124.

Baulcombe, D.C. 1996. RNA as a target and an initiator of post-transcriptional gene silencing in transgenic plants. Plant Molec Biol 32: 79–88.

Beekman, A.G.B. 1987. Breeding for resistance. *In:* J.A. de Bokx and J.P.H. van der Want (eds.), Viruses of Potatoes and Seed-potato Production. PUDOC, Wageningen, The Netherlands, pp. 162–170.

Beemster, A.B.R. and J.A. De Bokx. 1987. Survey of properties and symptoms. *In:* J.A. de Bokx and J.P.H van der Want (eds.), Viruses of Potatoes and Seed-potato Production. PUDOC, Wageningen, The Netherlands, pp. 84–113.

Bendahmane, A., B.A. Kohm, C. Dedi, and D.C. Baulcombe. 1995. The coat protein of potato virus X is a strain-specific elicitor of *Rx1*-mediated virus resistance in potato. Plant J 8: 933–941.

Benson, A.P. and W.J. Hooker.1960. Isolation of virus X from 'immune' varieties of potato, *Solanum tuberosum*. Phytopath 50: 231–234.

Bode, O. and H.L. Weidemann. 1970. Weitere Untersuchungenüber M-und S-Virus und itre übertragung durch Vektoren. *In:* P.M. Madec (ed.), Proc 4th Triennial Conf EAPR, Brest, France, pp. 224–226.

Boonham N., K. Walsh, S. Preston, J. North, P. Smith, and I. Barker. 2002. The detection of tuber necrotic isolates of potato virus Y, and the accurate discrimination of PVY^{O}, PVY^{N} and PVY^{C} strains using RT-PCR. J Virol Meth 102: 103–112.

Bradley, R.H.E. 1963. Some ways in which a paraffin oil impedes aphid transmission of potato virus Y. Can J Microbiol 9: 369–380.

Bradley, R.H.E., C.E. Moore, and D.D., Pond. 1966. Spread of potato virus Y curtailed by oil. Nature (London) 209: 1370–1371.

Bradshaw, J.E. and G.R. Mackay. 1994. Breeding strategies for clonally propagated potatoes. *In:* J.E.Bradshaw and G.R. Mackay (eds.), Potato Genetics. CAB Intl, Wallingford. UK, pp. 467–497.

Braun, C.J. and C.L. Hemenway. 1992. Expression of amino-terminal portions or full-length replicase genes in transgenic plants confers resistance to potato virus X infection. Plant Cell 4: 735–744.

Brigneti, G., O. Voinnet, W.X. Li, L.H. Ji, S.W. Ding, and D.C. Baulcombe. 1998. Viral pathogenicity determinants are suppressors of transgene silencing in *Nicotiana benthamiana*. EMBO J 17: 6739–6746.

Brown, C.R., E.N. Fernandez-Northcote, U. Jayasinghe, and L. Salazar. 1984. Breeding virus-resistant potato cultivars for developing countries. CIP Circular 1: 1–4.

Brunt, A.A., K. Crabtree, M.J. Dallwitz, A.J. Gibbs, L. Watson, and E.J. Zurcher (eds.). 1996. Plant Viruses Online: Descriptions and Lists from the VIDE Database. Version: 20th August 1996. URL *http://biology.anu.edu.au/Groups/MES/vide/*

Burton, W.G. 1989. The Potato (3rd ed.). Longman Tech. & Sci., Singapore.

Butenko, R.G. and A.A. Kuchko. 1994. Somatic hybridization in *Solanum tuberosum* × *S. chacoense*. *In:* Y.P.S. Bajaj (ed.), Biotechnology in Agriculture and Forestry, vol 27: Somatic Hybridization in Crop Improvement I. Springer-Verlag. Berlin-Heidelberg, Germany, pp. 183–195.

Cadman, C.H. 1942. Autotetraploid inheritance in the potato: some new evidence. J Genet 44: 33-52.

Carrasco, A., J.I. Ruiz de Galarreta, A. Rico, and E. Ritter. 2000. Transfer of PLRV resistance from *Solanum verrucosum* Schlechdt to potato (*S. tuberosum* L.) by protoplast electrofusion. Potato Res 43: 31–42.

Celebi, F., P. Russo, K. Watanabe, J.P.T. Valkonen, and S.A.Slack. 1998. Resistance of potato to cucumber mosaic virus appears related to localization in inoculated leaves. Amer J Potato Res 75: 195–199.

Chaput, M.H., D. Sihachakr, G. Ducreux, D. Marie, and N. Barghi. 1990. Somatic hybrid plants produced by electrofusion between dihaploid potatoes: BF15 (H1), Aminca (H6) and Cardinal (H3). Plant Cell Rep 9: 411–414.

Chavez, R., C.R. Brown, and M. Iwanaga. 1988. Transfer of resistance to PLRV titre buildup from *Solanum etuberosum* to a tuber-bearing *Solanum* gene pool. Theor Appl Genet 76: 129–135.

Chrzanowska, M. 1987. Nowe izolaty wirusa Y zagrazajace ziemniakom w Polsce. I Nasiennictwo 5-6: 8–11.

Chrzanowska, M. and A. Kowalska 1978. Reaction of potato cultivars to isolates of virus M. Biuletyn Instytutu Ziemniaka 22: 37–61.

Cockerham, G. 1943. Potato breeding for virus resistance. Ann Appl Biol 30: 105–108.

Cockerham, G. 1955. Strains of potato virus X. *In:* E. Streutgers, A.B. R. Beemster, and J.P.H. van der Want (eds.). Proc. 2nd Conf. Potato Virus Diseases, Lisse-Wageningen. H. Veenman & Zonen, Wageningen, The Netherlands, pp. 89–92.

Cockerham, G. 1958. Experimental breeding in relation to virus resistance. *In:* F. Quark, J. Dijkstra, A.B. R. Beemster, and J.P. H. van der Want (eds.), Proc. 3rd Conf. Potato Virus Diseases, Lisse-Wageningen. H. Veenman & Zonen, Wageningen, The Netherlands, pp. 199–203.

Cockerham, G. 1970. Genetical studies on resistance to potato virus X and Y. Heredity 25: 309–348.

Cooper, J.I. and A.T. Jones. 1983. Responses of plants to viruses, proposals for the use of terms. Phytopath 73: 127–128.

Corsini, D.L., J.J. Pavek, M.W. Martin, and C.R. Brown. 1994. Potato germplasm with combined resistance to leafroll virus and viruses X and Y. Amer Potato J 71: 377–385.

Dallwitz, M.J. 1980. A general system for coding taxonomic descriptions. Taxon 29: 41–46.

Dallwitz, M.J., T.A. Paine, and E.J. Zurcher. 1993. User's Guide to the DELTA System: a general system for processing taxonomic descriptions. (4th ed.). CSIRO Division of Entomology, Canberra, 136 pp.

Davidson, T.M.W. 1980. Breeding for resistance to virus disease of the potato (*Solanum tuberosum*) at the Scottish Plant Breeding Station. *In:* Scottish Plant Breeding Station 59th Annual Report, Dundee, UK, pp. 100–108.

De Bokx, J.A. and P.G.M. Piron.1984. Aphid trapping in potato fields in the Netherlands in relation to transmission of PVY^{N}. Meded. Facult. Landbouw. Rijksuniv Genet 49: 443–452.

De Bokx, J.A. and J.P.H. Van der Want. 1987. Viruses of Potatoes and Seed-potato Production. PUDOC, Wageningen, The Netherlands.

De Bruyn Ouboter, M.P. 1952. *In:* Proc. 1st Conf. Potato Virus Diseases, Lisse-Wageningen, The Netherlands.

De Carvalho Niebel, F., P. Frendo, M. Van Montagu, and M. Cornelissen. 1995. Post-transcriptional cosuppression of β-1,3-glucanase genes does not affect accumulation of transgene nuclear mRNA. Plant Cell 7: 347–358.

Deimling, S.C., J. Zitzlsperger, and G. Wenzel. 1988. Somatic fusion for breeding of tetraploid potatoes. Plant Breed. 101: 181–189.

Delhey, R. 1975. Zur Natur der extremen Virusresistenz bei der Kartoffel II. Das Y-Virus. Phytopath Z 82: 163–168.

Derrick, P.M. and H. Barker. 1997. Short and long distance spread of potato leafroll luteovirus, effects of host genes and transgenes conferring resistance to virus accumulation in potato. J Gen Virol 78: 243–251.

Dinu, I. and R. Thieme. 2001. Utilization of genetic resources in *Solanum* for potato breeding through biotechnological methods. Schriften zu Genetischen Ressourcen 16: 120–127.

Dixon, R.A. and M.J. Harrison. 1994. Early events in the activation of plant defense responses. Ann Rev Phytopath 32: 479–501.

Doncaster, J.P. and P.H. Gregory. 1948. The spread of virus diseases in the potato crop. Agric Res Counc Rep 7: 1–189.

Dong, F., R.G. Novy, J.P. Helgeson, and J. Jiang. 1999. Cytological characterization of potato—*Solanum etuberosum* somatic hybrids and their backcross progenies by genomic in situ hybridization. Genome 42: 987–992.

Dong, F., J.M. McGrath, J.P. Helgeson, and J. Jiang. 2001. The genetic identity of alien chromosomes in potato breeding lines revealed by sequential GISH and FISH analyses using chromosome-specific cytogenetic DNA markers. Genome 44: 729–734.

Dong, F., J. Song, S.K. Naess, J.P. Helgeson, C. Gebhardt, and J. Jiang. 2000. Development and application of a set of chromosome-specific cytogenetic DNA markers in potato. Theor Appl Genet 101: 1001–1007.

Doolittle, S.P. 1916. New infectious mosaic disease of cucumber. Phytopath 6: 145–147.

Dougherty, W.G., J.A. Lindbo, H.A. Smith, T.D. Parks, S. Swaney, and W.M. Proebsting. 1994. RNA-mediated virus resistance in transgenic plants: Exploitation of a cellular pathway possibly involved in RNA degradation. Molec Plant Micr Interact 7: 544–552.

Ducreux, G., I. Serraf, M.H. Chaput, R. Jadari, S. Tizroutine, L. Rossignol, and D. Sihachakr. 1993. Transfer of genes by protoplast electrofusion for improvement of the potato crops. *In:* B. Jouan (ed.), Abstracts 12th Triennal Conference EAPR, Paris, France, pp. 101–102.

Dziewońska , M.A. 1986. Development of parental lines for breeding of potatoes resistant to virus and associated research. *In:* A.G.B. Beckman, K.M. Louwes, L.M.W. Dellaert, and A.E.F. Neele (eds.), Potato Research for Tomorrow. PUDOC, Wageningen, Netherlands, pp. 96-100.

Ehrlich, P.R. and P.H. Raven. 1964. Butterflies and plants: A study in coevolution. Evolution 18: 586–608.

English, J.J., E. Mueller, and D.C. Baulcombe. 1996. Suppression of virus accumulation in transgenic plants exhibiting silencing nuclear genes. Plant Cell 8: 179–188.

Fang, G. and R. Grumet. 1993. Genetic engineering of potyvirus resistance using constructs derived from the zucchini yellow mosaic virus coat protein gene. Molec Plant Micr Interact 6: 358–367.

Fernandez-Northcote, E.N. 1983. Reaction of potato clones 'immune' to potato virus Y, to potato viruses A, X, and to a wide range of potato virus Y strains from the Andean region. Phytopath 73: 788.

Fernandez-Northcote, E.N. 1990. Variability of PVX and PVY and its relationship to genetic resistance. *In:* Control of Virus and Virus-Like Diseases of Potato and Sweet Potato. Rept 3rd Planning Conf. CIP, Lima, Peru, pp. 131–139.

Fernandez-Northcote, E.N. and C.R. Brown. 1981. Resistance in diploid *Solanum phureja, S. stenotomum*, and *S. berthaultii* intercrosses to potato virus Y. Phytopath 71: 873.

Fish, N., A. Karp, and M.G.K. Jones. 1987. Improved isolation of dihaploid *Solanum tuberosum* protoplasts and the production of somatic hybrids between dihaploid *S. tuberosum* and *S. brevidens*. *In Vitro* 23: 575–580.

Fish, N., A. Karp, and M.G.K. Jones. 1988. Production of somatic hybrids by electrofusion in *Solanum*. Theor Appl Genet 76: 260–266.

Fitchen, J.H. and R.N. Beachy. 1993. Genetically engineering protection against viruses in transgenic plants. Annu Rev Microbiol 47: 739–763.

Flanders, K.L., J.G. Hawkes, E.B. Radcliffe, and F.I. Lauer. 1992. Insect resistance in potatoes, sources, evolutionary relationships, morphological and chemical defenses, and ecogeographical associations. Euphytica 61: 83–111.

Frei, U., M. Stattmann, A. Lössl, and G. Wenzel. 1998. Aspects of fusion combining ability of dihaploid *S. tuberosum* L.: influence of the cytoplasm. Potato Res 41: 155–162.

Fribourg, C.E. and J. Nakashima. 1984. Characterisation of a new potyvirus from potato. Phytopath 74: 1363–1369.

Fuchs, M., S. Ferreira, and D. Gonsalves. 1997. Management of virus diseases by classical and engineered protection. Molec Plant Pathol. On-Line, 1997, 0116fuchs at *http://www.bspp.or-g.uk/mppol*.

Gabriel, W., M. Kostiw, L. Styszko, and M. Wislocka. 1987. Relationship between air-temperature, aphid incidence, cultivar resistance and potato infection with PVY and PLRV in Poland. Potato Res 30: 165.

Garcia-Arenal, F., A. Fraile, and J.M. Malpica. 2001. Variability and genetic structure of plant virus populations. Annu Rev Phytopath 39: 157–186.

Gavrilenko, T., R. Thieme, and H. Tiemann. 1999. Assessment of genetic and phenotypic variation among intraspecific somatic hybrids of potato, *Solanum tuberosum* L. Plant Breed 118: 205–213.

Gavrilenko, T., J. Larkka, E. Pehu, and V.-M Rokka. 2002. Identification of mitotic chromosomes to tuberous and non-tuberous of *Solanum* species (*Solanum tuberosum* and *Solanum brevidens*) by GISH (genomic *in situ* hybridization) in their interspecific hybrids. Genome 45: 442–449.

Gavrilenko, T., R. Thieme, U. Heimbach, and T. Thieme. 2003. Fertile somatic hybrids of *Solanum etuberosum* (+) dihaploid *Solanum tuberosum* and their backcrossing progenies: relationships of genome dosage with tuber development and resistance to potato virus Y. Euphytica 131: 323–332.

Gibbs, A.J., C.H. Calisher, and F. Garcia-Arenal (eds.). 1995. Molecular Basis of Virus Evolution. Cambridge Univ. Press, Cambridge, UK.

Gibson, R.W. 1971. The resistance of three *Solanum* species to *Myzus persicae, Macrosiphum euphorbiae* and *Aulacorthum solani* (Aphididae: Homoptera). Ann Appl Biol 68: 245–251.

Gibson, R.W. 1979. The geographical distribution, inheritance and pest-resisting properties of sticky-tipped foliar hairs on potato species. Potato Res 22: 223–236.

Gibson, R.W. and J.A. Pickett. 1983. Wild potato repels aphids by release of aphid alarm pheromone. Nature (London) 302: 608–609.

Gibson, R.W., M.G.K. Jones, and N. Fish. 1988. Resistance to potato leafroll virus and potato virus Y in somatic hybrids between dihaploid *Solanum tuberosum* and *S. brevidens*. Theor Appl Genet 76: 113–117.

Gilbert, J., C. Spillane, T.A. Kavanagh, and D.C. Baulcombe. 1998. Elicitation of *Rx*-mediated resistance to PVX in potato does not require new RNA synthesis and may involve a latent hypersensitive response. Molec Plant Micr Interact 8: 833–835.

Glais, L., M. Tribodet, and C. Kerlan. 2002. Genomic variability in Potato potyvirus Y (PVY): evidence that $PVY^{N}W$ and PVY^{NTN} variants are single to multiple recombinants between PVY^{O} and PVY^{N} isolates. Arch Virol 2: 363–378.

Golemboski, D.B., G.P. Lomonossoff, and M. Zaitlin. 1990. Plants transformed with tobacco mosaic virus nonstructural gene sequence are resistant to the virus. *In:* Proc. Natl. Acad. Sci., USA, pp. 6311–6315.

Goulden, M.G., B.A. Kohm, C.S. Santa, T.A. Kavanagh, and D.C. Baulcombe. 1993. A feature of the coat protein of potato virus X affects both induced virus resistance in potato and viral fitness. Virology (New York) 197: 293–302.

Greene, A.E. and R.F. Allison. 1994. Recombination between viral RNA and transgenic plant transcripts. Science 263: 1423–1425.

Gunenc, Y. and R.W. Gibson. 1980. Effects of glandular foliar hairs on the spread of potato virus Y. Potato Res 23: 345–351.

Haberlach, G.T., B.A. Cohen, N.A. Reichert, M.A. Baer, L.E. Towill, and J.P. Helgeson. 1985. Isolation, culture and regeneration of protoplasts from potato and several related *Solanum* species. Plant Sci 39: 67–74.

Hämäläinen, J.H., V.A. Sorri, K.N. Watanabe, C. Gebhardt, and J.P.T. Valkonen. 1998. Molecular examination of a chromosome region that controls resistance to potato Y and A potyvirus in potato. Theor Appl Genet 96: 1036–1043.

Hämäläinen, J.H., K.N. Watanabe, J.P.T. Valkonen, A. Arihara, R.L. Plaisted, E. Pehu, et al. 1997. Mapping and marker-assisted selection for a gene for extreme resistance to potato virus Y. Theor Appl Genet 94: 192–197.

Hammond-Kosack, K.E. and J.D.G. Jones. 1997. Plant disease resistance genes. Annu Rev Plant Physiol Plant Molec Biol 48: 575–607.

Harrington, R., N. Katis, and R.W. Gibson. 1986. Field assessment of the relative importance of different aphid species in the transmission of potato virus Y. Potato Res 29: 67–76.

Harrison, B.D. 1992. Genetic engineering of virus resistance, a successful genetical alchemy. Proc Roy Soc 99: 61–77.

Harrison, B.D. 2002. Virus variation in relation to resistance-breaking in plants. Euphytica 124: 181–192.

Harrison, B.D. and D.J. Robinson. 1999. Natural genomic and antigenic variation in whitefly-transmitted geminiviruses (begomoviruses). Annu Rev Phytopath 37: 369–398.

Heimbach, U., C. Eggers, and T. Thieme. 2002. Aphids in oil seed rape in autumn, possibilities to reduce virus transmission. IOBC/WPRS Bull 25: 123–131.

Heinze, K. 1960. Versuche zur Übertragung nichtpersistenter und persistenter Viren durch Blattläuse. Nachrichtenbl. Deut. Pflanzenschutzd 12: 119–121.

Helgeson, J.P., G.T. Haberlach, M.K. Ehlenfeldt, G. Hunt, J. Pohlman, and S. Austin. 1993. Sexual progeny of somatic hybrids between potato and *Solanum brevidens*: potential for use in breeding programs. Amer Potato J 70: 437–452.

Hemenway, C., R.F. Fang, W.K. Kaniewski, N.H. Chua, and N.E. Tumer. 1988. Analysis of the mechanism of protection in transgenic plants expressing the potato virus X coat protein or its antisense RNA. EMBO J 7: 1273–1280.

Hermsen, T.J.G. 1987. Efficient utilization of wild and primitive species in potato breeding. *In:* G.J. Jellis and D.E. Richardson (eds.), The Production of New Potato Varieties. Technological Advances. Cambridge Univ. Press, Cambridge, UK, pp. 172–185.

Hermsen, T.J.G. 1994. Introgression of genes from wild species, including molecular and cellular approaches. *In:* J.E. Bradshaw and G.R. Mackay (eds.), Potato Genetics. CAB Intl., Wallingford, UK, pp. 515–538.

Hinrichs, J., S. Berger, and J.G. Shaw. 1998. A hypersensitive response-like mechanism is involved in resistance of potato plants bearing the Ry_{sto} gene to the potyviruses potato virus Y and tobacco etch virus. J Gen Virol 79: 167–176.

Hinz, B. 1965. Beiträge zur Analyse der Vektoreignung einiger wirtschaftlich wichtiger Blattlausarten und –rassen. PhD thesis, Univ. Rostock, Germany.

Hinz, B. 1966. Beiträge zur Analyse der Vektoreignung einiger wirtschaftlich wichtiger Blattlausarten und –rassen. I. Versuche zur Ermittlung der Vektoreigenschaften für das Blattrollvirus der Kartoffel bei Rassen von *Myzus persicae* (Sulz.). Phytopath Z 56: 54–77.

Hinz, B. 1970. Zum Einfluß der Vektor- und Virus-Wirtspflanze auf den Infektionserfolg bei Virusübertragungen durch Blattläuse. Wiss. Z. Univ. Rostock 19, Math.-naturw. Reihe 8: 601–609.

Holtorf, H., H. Schob, C. Kunz, R. Waldvogel, and F. Meins, Jr. 1999. Stochastic and nonstochastic post-transcriptional silencing of chitinase and β-1,3-glucanase genes involves increased RNA turnover—possible role for ribosome-independent RNA degradation. Plant Cell 11: 471–484.

Hull, R. 1984. Rapid diagnosis of plant virus infections by spot hybridisation. Trends Biotech 2:2.

Hunt, G.J. and J.P. Helgeson. 1989. A medium and simplified procedure for growing single cells from the *Solanum* species. Plant Sci 60: 251–258.

Huth, W. 2000. Von der Notwendigkeit, Methoden zur Bestimmung der Resistenz gegenüber Pathogenen zu überdenken und zu standardisieren. Nachrichtenbl. Deut. Pflanzenschutzd 52: 304–308.

Hutton, E.M. 1951. Possible genotypes conditioning virus resistance in the potato and tomato. J Aust Inst Agric Sci 17: 132–138.
Johnston, S.A. and R.E. Hanneman. 1982. Manipulations of endosperm balance number overcome crossing barriers between diploid *Solanum* species. Science 217: 446–448.
Johnston, S.A., T.P.M. den Nijs, S.J. Peloquin, and R.E. Hanneman, Jr. 1980. The significance of genic balance to endosperm development in interspecific crosses. Theor Appl Genet 57: 5–9.
Jones, R.A.C. 1981. The ecology of viruses infecting wild and cultivated potatoes in the Andean region of South America. *In:* J.M. Thresh (ed.), Pests, Pathogens and Vegetation. Pitman Books, London, UK, pp. 89–107.
Jones, R.A.C. 1990. Strain group specific and virus specific hypersensitive reactions to infection with potyvirus in potato cultivars. Ann Appl Biol 117: 93–105.
Jones, R.A.C., C.E. Fribourg, and S.A. Slack. 1982. Potato virus and virus-like diseases. *In:* O.W. Barnett, and S.A. Tollin (eds.), Plant Virus Slide Series. Clemson Univ., Clemson, SC, USA.
Karl, E. 1971. Neue Vektoren für einige nichtpersistente Viren. Arch. Pflanzenschutz 7: 337–342.
Kegler, H. and W. Friedt (eds.). 1993. Resistenz von Kulturpflanzen gegen pflanzenpathogene Viren. G. Fischer Verlag, Jena, Germany.
Kerlan, C., M. Tribodet, L. Glais, and M. Guillet. 1999. Variability of potato virus Y in potato crops in France. J Phytopath 147: 643–651.
Kirkpatrick, H.C. and A.F. Ross. 1952. Aphid transmission of potato leafroll virus to solanaceous species. Phytopath 42: 540–546.
Kiru, S.D., S.V. Palekha, S.A. Makovskaya, and L.P. Evstratova. 2002. New genetic sources for breeding among several forms of South American cultivated *Solanum* species. *In:* G. Wenzel and I. Wulfert (eds.), Vorträge für Pflanzenzüchtung. Abstracts of Papers and Posters, EAPR-Conference, Potatoes Today and Tomorrow, Hamburg, Germany, WPR Communication Gmbh & Co. KG, Königswinter, p. 27.
Kojima, R. and H. Lapierre. 1988. Potato leafroll virus. *In:* I.M. Smith, V. Dunez, D.H. Philips, R.A. Leliot, and S.A. Archer (eds.), European Handbook of Plant Disease. Blackwell Sci. Publi., Oxford, UK, pp. 23–24.
Kostiw, M. 1980. Transmission of potato viruses by some aphid species. Tagungsbericht, Akad. Landwirtschaftswissenschaften DDR, Berlin 184: 339–414.
Le Romancer, M. and M. Nedellec. 1997. Effect of plant genotype, virus isolate and temperature on the expression of the potato tuber necrotic ringspot disease (PTNRD). Plant Path 46: 104–111.
Levinson, H.Z. 1976. The defensive role of alkaloids in insects and plants. Experientia 32: 408–411.
Li, W.X. and S.W. Ding. 2001. Viral suppressors of RNA silencing. Curr Opin Biotech 12: 150–154.
Lindbo, J.A., L. Silva-Rosales, W.M. Proebsting, and W.G. Dougherty. 1993. Induction of a highly specific antiviral state in transgenic plants: Implications for regulation of gene expression and virus resistance. Plant Cell 5: 1749–1759.
Ling, K., S. Namba, C. Gonsalves, J.L. Slightom, and D. Gonsalves. 1991. Protection against detrimental effects of potyvirus infection in transgenic tobacco plants expressing the papaya ringspot virus coat protein gene. BioTech 9: 752–758.
Liu, Y., D.J. Robinson, and B.D. Harrison. 1998. Defective forms of cotton leaf curl virus DNA-A that have different combinations of sequence deletion, duplication, inversion and rearrangement. J Gen Virol 79: 1501–1508.
Longstaff, M., G. Brigneti, F. Boccard, S. Chapman, and D. Baulcombe. 1993. Extreme resistance to potato virus X infection in plants expressing a modified component of the putative viral replicase. EMBO J 12: 379–386.

Lössl, A., U. Frei, and G. Wenzel. 1994. Interaction between cytoplasmic composition and yield parameters in somatic hybrids of *S. tuberosum* L. Theor Appl Genet 89: 873–878.

Lössl, A., U. Frei, and G. Wenzel. 1996. Genetic structures of the best fitting cytoplasms for somatic hybrids of potato. *In:* C.D. Van Loon (ed.). Abstracts Papers, Posters and Demonstrations 13th Triennal Conference EAPR, Veldhoven, Wageningen, The Netherlands, pp. 395–396.

Lössl, A., N. Adler, R. Horn, U. Frei, and G. Wenzel. 1999. Chondriome-type characterization of potato: mt $\alpha, \beta, \gamma, \delta, \varepsilon$ and novel plastid-mitochondrial configurations in somatic hybrids. Theor Appl Genet 98: 1–10.

MacFarlane, S.A. 1997. Natural recombination among plant virus genomes: evidence from tobraviruses. Semin Virol 8: 25–31.

Malcuit, I., M.R. Marano, T.A. Kavanagh, W. De Jong, A. Forsyth, and D.C. Baulcombe. 1999. The 25-kDa movement protein of PVX elicits *Nb*-mediated hypersensitive cell death in potato. Molec Plant Micr Interact 12: 536–543.

Malnoë, P., G. Jakab, E. Droz, and F. Vaistij. 1997. Risk assessment, the evolution of a viral population in transgenic plants. *In:* J. Wirz (ed.). Dialogue on Risk Assessment of Transgenic Plants, Scientific, Technological and Societal Perspectives. Proc. 3rd Ifgene Workshop, Dornach, Switzerland, Ifgene International Forum Genetic Engineering, Criccieth, UK.

Mansoor, S., S.H. Khan, A. Bashir, M. Saeed, Y. Zafar, K.A. Malik, R. Briddon, J. Stanley, and P.G. Markham. 1999. Identification of a novel circular single-stranded DNA associated with cotton leaf curl disease in Pakistan. Virology 259: 190–199.

Masson J., M. Lecerf, P. Rousselle, P. Perennec, and G. Pelletier. 1988. Plant regeneration from protoplasts of diploid potato derived from crosses of *Solanum tuberosum* with wild *Solanum* species. Plant Sci 53: 167–176.

Masson J., D. Lancelin, C. Bellini, M. Lecerf, P. Guerche, and G. Pelletier. 1989. Selection of somatic hybrids between diploid clones of potato (*Solanum tuberosum* L.) transformed by direct gene transfer. Theor Appl Genet 78: 153–159.

Matthews, R.E.F. 1991. Plant Virology. Acad. Press, New York, NY, USA.

Matthews, R.E.F. 1992. Fundamentals of Plant Virology. Acad. Press, San Diego, CA, USA.

McClintock, J.A. and L.B. Smith. 1918. True nature of spinach-blight and relation of insects to its transmission. J. Agric Res 14: 1–59.

McDonald, B. and C. Linde. 2002. The population genetics of plant pathogens and breeding strategies for durable resistance. Euphytica 124: 163–180.

McDonald, J.G. and R.P. Singh. 1996. Host range, symptomology and serology of isolates of potato virus Y (PVY) that shared properties with both the PVY^N and PVY^O strain groups. Amer Potato J 73: 309–315.

McGrath, J.M., S.M. Wielgus, and J.P. Helgeson. 1996. Segregation and recombination of *Solanum brevidens* synteny groups in progeny of somatic hybrids with *S. tuberosum*: intragenomic equals or exceeds intergenomic recombination. Genetics 142: 1335–1348.

McGrath, J.M., S.M. Wielgus, T.F. Uchytil, H. Kim-Lee, G.T. Haberlach, and J.P. Helgeson. 1994. Recombination of *Solanum brevidens* chromosomes in the second backcross generation from a somatic hybrid from *S. tuberosum*. Theor Appl Genet 88: 917–924.

Mehlenbacher, S.A. and R.L. Plaisted. 1983. Heritability of grandular trichome characteristics in a *Solanum tuberosum x Solanum berthaultii* hybrid population. *In:* Research for the potato in the year 2000. Proc. Intl. Cong. Celebration of 10th Anniv. CIP, Lima, Peru, pp. 128–130.

Melchers, G., M.D. Sacristan, and A.A. Holder. 1978. Somatic hybrid plants of potato and tomato regenerated from fused protoplasts. Carlsberg Res Comm 43: 203–218.

Mendoza, H.A., E.J. Mihovilovich, and F. Saguma. 1996. Identification of triplex (YYYy) potato virus Y (PVY) immune progenitors derived from *Solanum tuberosum* ssp. *andigena*. Amer Potato J 73: 13–19.

Mendoza, H.A.,E. Fernandez-Northcote, U. Jayasinghe, L.F. Salazar, R. Galvez, and C. Chuquillanqui. 1990. Breeding for resistance to potato viruses Y, X and leafroll: research strategy, selection procedures, and experimental results. *In:* Control of Virus and Virus-like Diseases of Potato and Sweet Potato. CIP, Lima, Peru, pp. 155–171.

Menke, U., L. Schilde-Rentschler, B. Ruoss, C. Zanke, V. Hemleben, and H. Ninnemann. 1996. Somatic hybrids between the cultivated potato *Solanum tuberosum* L. and the 1 EBN wild species *Solanum pinnatisectum* Dun.: morphological and molecular characterization. Theor Appl Genet 92: 617–626.

Mestre, P., G. Brigneti, and D.C. Baulcombe. 2000. An *Ry*-mediated resistance response in potato requires the intact active site of the Na proteinase from potato virus Y. Plant J 23: 653–661.

Metzlaff, M., M. O'Dell, P.D. Cluster, and R.B. Flavell. 1997. RNA-mediated RNA degradation and chalcone synthase A silencing in *Petunia.* Cell 88: 845–854.

Millam, S., L.A. Payne, and G.R. Mackay. 1995. The integration of protoplast fusion-derived material into a potato breeding programme—a review of progress and problems. Euphytica 85: 451–455.

Mills, W.R. 1965. Inheritance of immunity to potato virus X. Amer Potato J 42: 294–295.

Miyamoto, S. and Y. Miyamoto. 1971. Notes on aphid-transmission of potato leafroll virus 2. Transference of the virus to nymphs from viruliferous adults of *Myzus persicae* Sulz. Sci Rept, Fac Agric, Kobe Univ., Kobe, Japan, 9: 59–70.

Möllers, C. and G. Wenzel. 1992. Somatic hybridization of dihaploid potato protoplasts as a tool for potato breeding. Acta Bot 105: 133–139.

Möllers, C., U. Frei, and G. Wenzel. 1994. Field evaluation of tetraploid somatic potato hybrids. Theor Appl Genet 88: 147–152.

Monsanto 1994. Plants resistant to infection by PLRV. International patent application WO 94/18336.

Moreira, A., R.A.C. Jones, and C.E. Fribourg. 1980. Properties of a resistance-breaking strain of potato virus X. Ann Appl Biol 95: 93–103.

Muchalski, T. 1997. Evaluation of tobacco rattle tobravirus (TRV) infection of selected potato genotypes in a greenhouse test. Biuletyn Instytutu Ziemniaka 48: 107–115.

Munoz, F.J., R.L. Plaisted, and H.D. Thurston. 1975. Resistance to potato virus Y in *Solanum tuberosum* ssp. *andigena.* Amer Potato J 52: 107–115.

Murphy, P.A. and R. Mckay. 1929. The insect vectors of the leafroll disease of the potato. Sci Proc Roy Dubl Soc 19: 341–353.

Murphy, P.A. and R. Mckay. 1932. A comparison of some European and American virus diseases in potato. Sci Proc Roy Dubl Soc 20: 347–358.

Nemecek, T. 1993. The role of aphid behaviour in the epidemiology of potato virus Y: a simulation study. PhD thesis, Univ. Zürich, Zürich, Switzerland.

Nishiguchi, M., F. Motoyoshi, and N. Oshima. 1978. Behaviour of a temperature sensitive strain of tobacco mosaic virus in tomato leaves and protoplasts. J Gen Virol 39: 53–61.

Novy, R.G. and J.P. Helgeson. 1994a. Somatic hybrids between *Solanum etuberosum* and diploid, tuber bearing *Solanum* clones. Theor Appl Genet 89: 775–782.

Novy, R.G. and J.P. Helgeson. 1994b. Resistance to potato virus Y in somatic hybrids between *Solanum etuberosum* and *S. tuberosum* × *S. berthaultii* hybrid. Theor Appl Genet 86: 783–786.

Novy, R.G., A. Nasruddin, D.W. Ragsdale, and E.B. Radcliffe. 2002. Genetic resistances to potato leafroll virus, potato virus Y, and Green Peach Aphid in progeny of *Solanum etuberosum.* Amer J Potato Res 79: 9–18.

Oertel, H., U. Hamann, and K. Neitzel. 1980. Stand der Virusresistenzzüchtung bei Kartoffeln. Fortschrittsb. Land. u. Nahrungsg. 18: 505.

Oruetxebarria, I. and J.P.T. Valkonen 2001. Analysis of the P1 gene sequences and the 3'-terminal sequences of the single-stranded RNA genome of Potato virus V. Virus Genes 22: 335–343.

Palucha, A., W. Chrzanowska, and D. Hulanicka. 1998. An antisense coat protein gene confers immunity to potato leafroll virus in a genetically engineered potato. Eur J Plant Path 104: 287–293.

Parlevliet, J.E. 1986. Coevolution of host resistance and pathogen virulence: possible implications for taxonomy. *In*: A.R. Stone and D.L. Hawksworth (eds.). Coevolution and Systematics, Clarendon Press, Oxford, UK, pp. 19–34.

Parlevliet, J.E. 1993. What is durable resistance? A general outline. *In*: T. Jacobs and J.E. Parlevliet (eds.). Durability of Disease Resistance. Kluwer Acad. Publ., Dordrecht, The Netherlands, pp. 23–39.

Parrella, G. and T. Cardi. 1999. Transfer of a new PVX resistance gene from *Solanum commersonii* to *S. tuberosum* through somatic hybridization. J Genet Breed 53: 359–362.

Pehu, E., R.W. Gibson, M.G.K. Jones, and A. Karp. 1990. Studies on the genetic basis of resistance to potato leaf roll virus, potato virus Y and potato virus X in *Solanum brevidens* using somatic hybrids of *Solanum brevidens* and *Solanum tuberosum*. Plant Sci 69: 95–101.

Perl, A., D. Aviv, and E. Galun. 1988. Ethylene and in vitro culture of potato: suppression of ethylene generation vastly improves protoplast yield, plating efficiency and transient expression of an alien gene. Plant Cell Rept 7: 403–406.

Peters, D. 1967. Purification of potato leafroll virus from its vector *Myzus persicae*. Virology 31: 46.

Pierpoint, W.S. 1996. Modifying resistance to plant viruses. *In*: W.S. Pierpoint and P.R. Shewry (eds.). Genetic Engineering of Crop Plants for Resistance to Pests and Diseases. British Crop Protection Council, Farnham, UK, pp. 16–37.

Polgar, Z., J.P. Helgeson, S.M. Wielgus, I. Wolf, C. Pinter, D. Dudits, and S. Horvath. 2000. Breeding value of potato + *Solanum brevidens* somatic hybrids. *In*: B. Czech (ed.), Abstracts of Joint Conf. EAPR, Section Breeding and Varietal Assessment and EUCARPIA, Potato Section. Breeding Research for Resistance to Pathogens and for Quality Traits. VIHAR, Mlochow Research Center, Warsaw, Poland, pp. 42–43.

Ponz, F. and G. Bruening. 1986. Mechanisms of resistance to plant viruses. Annu Rev Phytopath 24: 355–381.

Ponz, F., C.B. Glascock, and G. Bruening. 1988. An inhibitor of polyprotein processing with the characteristics of a natural virus resistance factor. Molec Plant Micr Interact 1: 25–31.

Powell, G. 1991. Cell-membrane punctures during epidermal penetrations by aphids—consequences for the transmission of 2 potyviruses. Ann Appl Biol 119: 313–321.

Powell, G. 1992. The effect of mineral-oil on stylet activities and potato virus Y transmission by aphids. Entomol Exp Appl 63: 237–242.

Powell, G., R. Harrington, and N.J. Stiller. 1992. Stylet activities and potato virus Y vector efficiencies by the aphids *Brachycaudus helichrysi* and *Drepanosiphum platanoidis*. Entomol Exp Appl 62: 293–300.

Raman, K.V., W.M. Tingey, and P. Gregory. 1979. Potato glycoalkaloids: effect on survival and feeding behavior of the potato leafhopper. J Econ Entomol 72: 337–341.

Rasmussen, J.O. and O.S. Rasmussen. 1995. Characterization of somatic hybrids of potato by use of RAPD markers and isoenzyme analysis. Physiol Plantarum 93: 357–364.

Rizvi, S.A.H. and K.V. Raman. 1983. Effect of glandular trichomes on the spread of potato virus Y (PVY) and potato leafroll virus (PLRV) in the field. *In*: W.J. Hooker (ed.). Proc. Intl. Cong. Res. Potato in the Year 2000. CIP, Lima, Peru, pp. 162–163.

Robert, Y. 1971. Épidémiologie de l' enroulement de la pomme de terre: capacité vectrice de stades et de formes des puccrons *Aulacorthum solani* Ktb., *Macrosiphum euphorbiae* Thom. et *Myzus persicae* Sulz. potato Res 14: 130–139.

Robert, Y. and Y. Maury. 1970. Capacités vectrices comparées de plusieurs souches de *Myzus persicae* Sulz., *Aulacorthum solani* Kltb. et *Macrosiphum euphorbiae* Thom. dans l' étude de la transmission de l' enroulement de la pomme de terre. Potato Res 13: 199–209.

Robert, Y., Y. Maury, and J. Quéméner. 1969. Transmission du virus de l' enroulement de la pomme terre par différentes formes et stades d'une souche de *Myzus persicae* (Sulz.): Résultats comparés sur *Physalis floridana* Rydberg et *Solanum tuberosum* L. var. Claudia. Ann Phytopath 1: 167–179.

Rokka, V.-M. 1998. Androgenic haploidization and interspecific and intraspecific somatic hybridization in potato germplasm development. PhD thesis, Univ. Helsinki, Helsinki, Finland.

Ross, H. 1952. Studies on mosaic resistance in the potato. *In*: Proc. Conf. Potato Virus Diseases 1951, Wageningen-Lisse, Veenman & Zonen, Wageningen, The Netherlands. pp. 40–47.

Ross, H. 1954. Über die extreme Resistenz von *Solanum acaule* gegen das X-virus. Mitt. Biol. Bundesanst. Land-u. Forstwirtsch 80: 144–145.

Ross, H. 1958. Virusresistenzzüchtung an der Kartoffel. Eur Potato J 1: 1–9.

Ross, H. 1961. Über die Vererbung von Eigenschaften für Resistenz gegen das Y-und A- Virus in *Solanum stoloniferum* und die mögliche Bedeutung für eine allgemeine Genetik der Virusresistenz in *Solanum* sect. *Tuberarium*. *In*: Proc. 4th Conf. Potato Virus Diseases, Braunschweig, Veenman & Zonen, Wageningen, The Netherlands, pp. 40–49.

Ross, H. 1978. Methods for breeding virus resistant potatoes. *In*: Report Planning Conf. Developments in Control of Potato Virus Diseases. CIP, Lima, Peru, pp. 93–114.

Ross, H. 1986. Potato breeding—problems and perspectives *In*: J. Brandes, R. Bartels, J. Völk, and C. Wetter (eds.). Advances in Plant Breeding. J Plant Breed 13 (suppl.). Paul Parey, Berlin, Germany, pp. 64–75

Rozendaal, A., J. van Binsbergen, B. Anema, D.H.M. van Slogteren, and M.H. Bunt. 1971. Serology of deviating potato virus Y^{C} strain in the potato variety Gladblaadje. Potato Res 14: 241.

Ruiz-de-Galarreta, J.I., A. Carrasco, A. Salazar, I. Barrena, E. Iturritxa, R. Marquineż, F.J. Legorburu, and E. Ritter, 1998. Wild *Solanum* species as resistance sources against different pathogens of potato. Potato Res 41: 57–68.

Sanford, L.L., J.M. Domek, W.W. Cantelo, R.S. Kobayashi, and S.L. Sinden. 1996. Mortality of potato leafhopper adults on synthetic diets containing seven glycoalkaloids synthesized in the foliage of various *Solanum* species. Amer Potato J 73: 79–88.

Santa Cruz, S. and D.C. Baulcombe. 1993. Molecular analysis of potato virus X isolates in relation to the potato hypersensitivity gene *Nx*. Molec Plant Micr Interact 6: 707–714.

Santillan, F.W. 1979. Estudio comparativo de once aislamientos de virus S de la region Andina. MSc thesis, Univ. Nacional Agraria, Lima, Peru.

Schilde-Rentschler, L., G. Boos, and H. Ninnemann. 1988. Somatic hybridization of diploid potato lines, a tool in potato breeding. *In*: K.J. Puite, J.J.M. Dons, H.J. Huizing, A.J. Kool, M. Koornneef, and F.A. Krens (eds.). Progress in Plant Protoplast Research. Kluwer Acad. Publ., Dordrecht, The Netherlands, pp. 195–196.

Schilde-Rentschler, L., K. Kugler-Busch, A. Schweis, and H. Ninnemann. 1995. Somatic hybrids for the study of nuclear-cytoplasmic interaction in *Solanum tuberosum*. *In*: U. Kück and G. Wricke (eds.). Advances in Plant Breeding, vol. 18: Genetic Mechanisms for Hybrid Breeding. Blackwell Wissenschafs-Verlag, Berlin, Germany, pp. 37–48.

Schubert, J., D. Mattern, and J. Matoušek. 2002. Zur Stabilität transgener Virusresistenz am Beispiel der Kartoffel/PVY. Vortr. Pflanzenzüchtg 54: 349–352.

Schultz, E.S. and D. Folsom. 1923. Transmission, variation, and control of certain degeneration diseases of Irish potatoes. J Agric Res 25: 43.

Schwarzfischer, A. 1994. Use of somatic hybrids in an applied breeding programme. Potato Res 37: 346.

Schwarzfischer, A., J. Schwarzfischer, and L. Hepting. 1998. Resistance and quality breeding with somatic hybridisation. Beitr. Züchtungsfg 4(2): 165–172.

Schweizer, G., T. Stelzer, and V. Hemleben. 1990. RFLP-Analyse mit spezifischen Genomkomponenten der Kartoffel zur Identifizierung von symmetrischen und asymmetrischen Hybriden. Vortr. Pflanzenzüchtg 18: 178–192.

Serraf, I., D. Sihachakr, G. Ducreux, S.C. Brown, M. Allot, N. Barghi, and L. Rossignol. 1991. Interspecific somatic hybridization in potato by protoplast electrofusion. Plant Sci 76: 115–126.

Shands, W.A. 1977. Control of aphid-borne potato virus Y in potatoes with oil emulsion. Amer Potato J 54: 179–187.

Shepard, J.F. 1981. Protoplasts as sources of disease resistance in plants. Ann Rev Phytopath 19: 145–166.

Shepard, J.F. and R.E. Totten. 1977. Mesophyll cell protoplasts of potato. Isolation, proliferation and plant regeneration. Plant Physiol. 60: 313–316.

Sigvald, R. 1984. The relative efficiency of some aphid species as vectors of potato virus Y^O (PVY^O). Potato Res 27: 285–290.

Sigvald, R. 1989. Relationship between aphid occurrence and spread of potato virus Y^O (PVY^O) in field experiments in southern Sweden. J Appl Entomol 108: 35–43.

Sigvald, R. 1990. Aphids on potato foliage in Sweden and their importance as vectors of potato virus Y^O. Acta Agric Scand 40: 53–58.

Sigvald, R. 1995. Forecasting potato virus Y. Vaxtskyddsnotiser 59: 116–120.

Singh, R.P., U.S. Singh, and K. Kohmoto (eds.). 1995. Pathogenesis and Host specificity in Plant Diseases, vol III: viruses and Viroids. Elsvier Science, Oxford, UK.

Smith, H.A., S.L. Swaney, T.D. Parks, E.A. Wernsman, and W.G. Dougherty. 1994. Transgenic plant virus resistance mediated by untranslatable sense RNAs: expression, regulation, and fate of nonessential RNAs. Plant Cell 6: 1441–1453.

Smith, K.M. 1931. On the composite nature of certain potato virus diseases of the mosaic group as revealed by the use of plant indicators. Proc. Roy. Soc. London Ser. B 109: 251–267.

Smith, K.M. 1972. A Textbook of Plant Virus Diseases. Longman, London, UK (3rd ed.).

Solomon-Blackburn, R.M. 1998. Progress in breeding potatoes for resistance to virus diseases. Aspects Appl Biol 52: 299–304.

Solomon-Blackburn, R.M. and H. Barker. 1993. Resistance to potato leafroll luteovirus can be greatly improved by combining two independent types of heritable resistance. Ann Appl Biol 122: 329–336.

Solomon-Blackburn, R.M. and H. Barker. 2001a. A review of host major-gene resistance to potato viruses, X, Y, A and V in potato: genes, genetics and mapped locations. Heredity 86: 8–16.

Solomon-Blackburn R.M. and H. Barker. 2001b. Breeding virus resistant potatoes (*Solanum tuberosum*): a review of traditional and molecular approaches. Heredity 86: 17–35.

Song, J., F. Dong, and J. Jiang. 2000. Construction of a bacterial artificial chromosome library for potato molecular cytogenetics research. Genome 43: 199–204.

Song, Y.S., L. Hepting, L. Hartl, G. Schweizer, G. Wenzel, and A. Schwarzfischer. 2002. Development of genetic markers to potato virus Y-immunity from *S. stoloniferum* (Ry_{sto}) using primary dihaploid Assia population from anther culture. Vortr. Pflanzenzüchtg 54: 267–270.

Spetz, C.J., A.M. Taboada, L.F. Salazar, and J.P.T. Valkonen. 2002. Evidence for two strain groups of potato virus V. *In*: G. Wenzel and I. Wulfert (eds.). Vorträge für Pflanzenzüchtung, Abstracts of Papers and Posters, EAPR-Conference, Potatoes Today and Tomorrow. Hamburg, Germany, p. 295.

Spooner, D.M. and R.J. Hijmans. 2001. Potato systematics and germplasm collecting, 1989-2000. Amer J Potato Res 78: 237–268.

Stadler, M., T. Stelzer, N. Borisjuk, C. Zanke, L. Schilde-Rentschler, and V. Hemleben. 1995. Distribution of novel and known repeated elements of *Solanum* and application for the identification of somatic hybrids among *Solanum* species. Theor Appl Genet 91: 1271–1278.

Stark, D.M. 1997. Risk assessment and criteria for commercial launch of transgenic plants. *In*: J. Wirz (ed.), Dialogue on Risk Assessment of Transgenic Plants: Scientific, Technological and Societal Perspectives. Proc. 3rd Ifgene Workshop, Dornach, Switzerland, Ifgene Intl. Forum Genetic Engineering Criccieth, UK, pp. 28–35.

Świeżyński, K.M. 1983. Parental line breeding in potatoes. Genetica (Yugoslavia) 15: 243–256.

Świeżyński, K.M. 1994. Inheritance of resistance to viruses. *In*: J.E. Bradshaw and G.R. Mackay (eds.). Potato Genetics. CAB Intl., Wallingford, UK, pp. 339–364.

Sylvester, E.S. 1980. Circulative and propagative virus transmission by aphids. Annu Rev Entomol 25: 257–286.

Tacke, E., F. Salamini, and W. Rohde. 1996. Genetic engineering of potato for broad-spectrum protection against virus infection. Nature Biotech 14: 1597–1601.

Tamada, T., Y. Shirako, H. Abe, M. Saito, T. Kiguchi, and T. Harada. 1989. Production and pathogenicity of isolates of beet necrotic yellow vein virus with different numbers of RNA components. J Gen Virol 70: 3399–3409.

Thach, N.Q., U. Frei, and G. Wenzel. 1993. Somatic fusion for combining virus resistances in *Solanum tuberosum* L. Theor Appl Genet 83: 81–88.

Thieme, R. and B. Hackauf. 1998. Nutzung von Mikrosatellitenloci zur Identifizierung somatischer Kartoffelhybriden. Vortr. Pflanzenzüchtg 42: 149–151.

Thieme, R., K. Sonntag, T. Gavrilenko, and T. Tiemann. 1996. Intraspezifische somatische Hybridiseirung durch Protoplastenfusion und Charakterisierung der Hybriden bei der Kartoffel. Vortr. Pflanzenzüchtg 32: 154–157.

Thieme, R., U. Darsow, T. Gavrilenko, D. Dorokhov, and T. Tiemann. 1997. Production of somatic hybrids between *S. tuberosum* L. and late blight resistant Mexican wild potato species. Euphytica 97: 189–200.

Thieme, R., T. Gavrilenko, T. Thieme, and U. Heimbach. 1999. Production of potato genotypes with resistance to Potato Virus Y (PVY) by biotechnological methods. *In*: A. Alman, M. Ziv, and S. Izhar (eds.). Proc. 9th Intl Cong. IAPTC and Biotech on "Plant Biotechnology and *in vitro* Biology in the 21st Century". Jerusalem, Israel. Kluwer Acad. Publ., The Netherlands, pp. 557–560.

Thieme, R., T. Thieme, U. Heimbach, and T. Gavrilenko. 2000. Incorporation and testing of new genetic sources for virus resistance in potato. Potato, Global Res Devel 1: 271–278.

Thieme, R., T. Gavrilenko, U. Heimbach, and T. Thieme. 2002. Development and utilisation of genetic diversity in *Solanum* for breeding potatoes. *In*: G. Wenzel and I. Wulfert (eds.), Vorträge für Pflanzenzüchtung. Abstracts of Papers and Posters, EAPR-Conference, Potatoes Today and Tomorrow, Hamburg, Germany, WPR Communication Gmbh & Co. KG, Königswinter, p.28.

Thieme, R., U. Heimbach, and T. Thieme. 2004. Breeding for aphid and virus resistance in potatoes. *In:* J.C. Simon, C.-A. Dedryver, C. Rispe, M. Hulle (eds.), Aphids in a new millenium. JNRA Editions, Paris, pp. 507–511.

Thieme, T. and U. Heimbach. 1998. Wirksamkeit von Pflanzenschutzmitteln gegen virusübertragende Blattläuse in Kartoffelbau. Mitt. Biol. Bundesanst. Land-Forstwirtsch 357: 100–101.

Thieme, T. and R. Thieme. 1998. Evaluation of resistance to potato virus Y (PVY) in wild species and potato breeding clones of the genus *Solanum*. Aspects Appl Biol 52: 355–359.

Thieme, T., U. Heimbach, R. Thieme, and H.-L. Weidemann. 1998. Introduction of a method for preventing transmission of potato virus Y (PVY) in Northern Germany. Aspects Appl Biol 52: 25–29.

Thompson, J.N. 1994. The Coevolutionary Process. Univ. Chicago Press, Chicago, IL, USA.

Tingey, W.M. 1991. Potato glandular trichomes—defensive activity against insect attack. *In*: P.A. Hedin (ed.). Naturally Occurring Pest Bioregulators. Acs Symp. Series 449, pp. 126–135.

Tingey, W.M. and S.L. Sinden. 1982. Glandular pubescence, glycoalkaloid composition, and resistance to the green peach aphid, potato leafhopper, and potato flea-beetle in *Solanum berthaultii*. Amer Potato J 59: 95–106.

Tingey, W.M., J.D. MacKenzie, and P. Gregory. 1978. Total foliar glycoalkaloids and resistance of wild potato species to *Empoasca fabae* (Harris). Amer Potato J 55: 577–585.

Tommiska, T.J., J.H. Hamalainen, K.N. Watanabe, and J.P.T. Valkonen. 1998. Mapping of the gene Nx_{phu} that controls hypersensitive resistance to potato virus X in *Solanum phureja* IvP35. Theor Appl Genet 96: 840–843.

Valkonen, J.P.T. 1992a. Characterization of virus resistance in *Solanum brevidens* Phil. PhD thesis, Univ. Helsinki, Finland.

Valkonen, J.P.T. 1992b. Resistance mechanisms against different virus diseases. Nordisk Jordbruksforskning 74: 113–114.

Valkonen, J.P.T. 1994. Natural genes and mechanisms for resistance to viruses in cultivated and wild potato species (*Solanum* spp.). Plant Breed 112: 1–16.

Valkonen, J.P.T., M.G.K. Jones, and R.W. Gibson. 1991. Resistance in *Solanum brevidens* to both potato virus Y and potato virus X may be associated with slow cell-to-cell spread. J Gen Virol 72: 231–236.

Valkonen, J.P.T., V.-M. Rokka, and K.N. Watanabe. 1998. Examination of the leaf-drop symptom of virus-infected potato using anther culture-derived haploids. Phytopath 88: 1073–1077.

Valkonen, J.P.T., S.A. Slack, R.L. Plaisted, and K.N. Watanabe. 1994a. Extreme resistance is epistatic to hypersensitive resistance to potato virus Y^O in a *Solanum tuberosum* subsp. *andigena*—derived potato genotype. Plant Dis 78: 1177–1180.

Valkonen, J.P.T., Y.-S. Xu, V.-M. Rokka, S. Pulli, and E. Pehu. 1994b. Transfer of resistance to potato leafroll virus, potato virus Y and potato virus X from *Solanum brevidens* to *S. tuberosum* through symmetric and designed asymmetric somatic hybridization. Ann Appl Biol 124: 351–362.

Valkonen, J.P.T., R.A.C. Jones, S.A. Slack, and K.N. Watanabe. 1996. Resistance specificities to virus in potato: Standardization of nomenclature. Plant Breed 115: 433–438.

Valkonen, J.P.T., M. Orillo, S.A. Slack, R.L. Plaisted, and K.N. Watanabe. 1995a. Resistance to virus in F_1 hybrids produced by direct crossing between diploid *Solanum* series *Tuberosa* and diploid *S. brevidens* (series *Etuberosa*) using *S. phureja* for rescue pollination. Plant Breed 114: 421–426.

Valkonen, J.P.T., Ü. Puurand, S.A. Slack, K. Mäkinen, and M. Saarma. 1995b. Three strain groups of potato A potyvirus based on hypersensitive responses in potato, serological properties, and coat proteins sequences. Plant Dis 79: 748–753.

Van der Wilk, F., D.P.-L. Willink, M.J. Huisman, H. Huttinga, and R. Goldbach. 1991. Expression of the potato leafroll luteovirus coat protein gene in transgenic potato plants inhibits viral infection. Plant Molec Biol 17: 431–439.

Van Dyke, T.A. 1994. Analysis of viral-host protein interactions and tumorigenesis in transgenic mice. Semin Cancer Biol 5: 47–60.

Van Harten, A. 1983. The relation between aphid flights and the spread of potato virus Y^N (PVY^N) in the Netherlands. Potato Res 26: 1–15.

Van Hoof, H.A. 1980. Aphid vectors of potato virus Y^N. Neth J Plant Path 86: 159–162.

Villalobos, V.M. and B.R. Rivera. 1997. Field tests on transgenic potatoes in Mexico. *In*: 3rd JIRCAS Intl. Symp. "4th Intl. Symp. on Biosafety Results of Field Tests of Genetically Modified Plants and Microorganisms" 5, pp. 157–162.

Voinnet, O., C. Lederer, and D.C. Baulcombe. 2000. A viral movement protein prevents spread of the gene silencing signal in *Nicotiana benthamiana*. Cell 103: 157–167.

Waara S., H. Tegelström, A. Wallin, and T. Eriksson. 1989. Somatic hybridization between anther-derived dihaploid clones of potato (*Solanum tuberosum* L.) and the identification of hybrid plants by isoenzyme analysis. Theor Appl Genet 77: 49–56.

Wenzel, G. 1994. Tissue Culture. *In*: J.E. Bradshaw and G.R. Mackay (eds.), Potato genetics. CAB Intl., Wallingford, UK, pp. 173–195.

Wenzel, G., O. Schieder, T. Przewozny, S.K. Sopory, and G. Melchers. 1979. Comparison of single cell culture-derived *Solanum tuberosum* L. plants and model for their application in breeding programs. Theor Appl Genet 55: 49–55.

Williams, C.E., G.J. Hunt, and J.P. Helgeson. 1990. Fertile somatic hybrids of *Solanum* species: RFLP analysis of a hyrbid and its sexual progeny from cross with potato. Theor Appl Genet 80: 545–551.

Williams, W.L. and A.F. Ross 1957. Aphid transmission of potato leafroll virus affected by the feeding of non-viruliferous aphids on the test plants and by vector variability. Phytopath 47: 538.

Wilson, C.R. and R.A.C. Jones 1992. Resistance to phloem transport of potato leafroll virus in potato plants. J Gen Virol 73: 3219–3224.

Wilson, T.M.A. 1993. Strategies to protect crop plants against viruses: Pathogen-derived resistance blossom. Proc Natl Acad Sci USA 90: 3134–3141.

Wolf, S., C.M. Deom, R. Beachy, and W.J. Lucas. 1991. Plasmodesmatal function is probed using transgenic tobacco plants that express a virus movement protein. Plant Cell 3: 593–604.

Woodford, J.A.T. 1992. Virus transmission by aphids in potato crops. Neth J Plant Path 98: 47–54.

Woodford, J.A.T., B.D. Harrison, C.S. Aveyard, and S.C. Gordon. 1983. Insecticidal control of aphids and the spread of potato leafroll virus in potato crops in eastern Scotland. Ann Appl Biol 103: 117–130.

14

Resistance to Bacterial Pathogens

EWA ZIMNOCH-GUZOWSKA[1], EWA LOJKOWSKA[2] AND MICHEL PÉROMBELON[3]

[1]*Plant Breeding and Acclimatization Institute, Mlochow Research Centre, Platanowa 19, 05-831 Mlochow, Poland,*

[2]*University of Gdansk & Medical University of Gdansk, Dept. Plant Protection and Biotechnology, Kladki 24, 80-822 Gdansk, Poland*

[3]*Scottish Crop Research Institute, Invergowrie, Dundee DD2 5DA, Scotland, UK*

IMPORTANCE OF BACTERIAL DISEASES IN POTATO PRODUCTION

Bacterial diseases constitute the second major constraint to potato production worldwide (late blight is first). The main bacterial diseases affecting potatoes are caused by:

1. *Erwinia carotovora* subsp. *atroseptica* (van Hall) Dye (reclassified as *Pectobacterium carotovorum* subsp. *atrosepticum*); *Erwinia carotovora* subsp. *carotovora* (Jones) Bergey, Harrison, Breed, Hamer and Huntoon (reclassified as *Pectobacterium carotovorum* subsp. *carotovorum*); *Erwinia chrysanthemi* Burkholder, McFadden and Dimock (reclassified as *Pectobacterium chrysanthemi*)—causal agents of tuber bacterial soft rot and blackleg.
2. *Ralstonia solanacearum* Smith (formerly called *Pseudomonas solanacearum*)—causal agent of bacterial wilt or brown rot.
3. *Clavibacter michiganensis* subsp. *sepedonicus* (Skaptason and Burkholder) (formerly called *Corynebacterium sepedonicum*)—causal agent of bacterial ring rot.
4. *Streptomyces scabies* (Thaxter) Lambert and Loria, *S. acidiscabies* Lambert and Loria, and *S. turgidiscabies* Takeuchi—causal agents of common scab.

Jansky (2000) estimated the total annual losses in potato production worldwide to be 65 million tons, equivalent to 22% total production, which Ross (1986) attributed to diseases and pests. Data pertaining exclusively to the impact of bacterial diseases on potato production are not available. The extent of losses varies with the bacterial disease and between regions. Pérombelon and Kelman (1980) estimated global annual

Corresponding author: Ewa Zimnoch-Guzowska Tel.: +48 22 7299089; 7299248 ext. 206, Fax: +48 22 7299247, *e-mail: E.Zimnoch-Guzowska@ihar.edu.pl* losses caused by *Erwinia* spp. to be 50 to 100 million US$. Average losses in Poland caused by tuber soft rot from 1976 to 1979 were estimated by Pietkiewicz (1980) to be 6.7% while Hide et al. (1996) reported that blackleg reduced yield by 28%, although average disease incidence in temperate regions is ca. 2% (Elphinstone 1987). Bacterial wilt is a major constraint to solanaceous vegetables in general but potato cultivation in particular, both in the lowland and highland tropics. Estimates of potato yield losses due to the disease range from 15 to 95% (Javier 1994). Losses due to bacterial wilt increase at higher temperatures: when the pathogen is soil-borne, losses are lower with an average disease incidence of 10% in cool climates and 25% in warmer. In contrast, when tuber-borne, disease level ranges from 30 to 100% (French et al. 1998).

Attempts at chemical control of potato bacterial diseases have been unsuccessful. The general strategy for disease control relies on an integrated management approach, based on resistant or tolerant cultivars, proper seed certification program, careful handling of the crop at planting and harvest time as well as in storage by reducing the risk of wounding and cross-contamination, and improving general on-farm hygiene, cropping methods, and diagnostics of bacterial diseases (De Boer 1990; Elphinstone 1994; Pérombelon 2002). However, the cheapest and most reliable method of plant protection is cultivation of disease-resistant cultivars if available. Production of cultivars with durable resistance is not readily achieved because of the variability and adaptability of the pathogens. Genetic variations for resistance to bacterial diseases in potato cultivars and related wild as well as cultivated species have been evaluated. In general, resistance in a commercial popular cultivar is insufficient to protect the crop. Lack of progress in the development of resistant cultivars stems from a paucity of sources of high resistance, effect of environmental factors on the expression of resistance in different climatic regions, and difficulties in the utilization of identified wild sources of resistance (Elphinstone 1994). Nevertheless, new cultivars expressing an improved level of agronomic characters and resistance traits have been bred using conventional breeding techniques. Recent progress in diagnostic techniques has allowed the identification and detection of latent infections, then the elimination in breeding programs of undesired tolerant material known to be a symptomless source of inoculum, which can not only infect neighbouring susceptible crops, but also spread the pathogen to clean fields.

Several reviews of breeding strategies against bacterial diseases and associated aspects have been published recently (Tarn et al. 1992; Wastie et al. 1993; Elphinstone 1994; Ghislain and Golmirzaie 1998; Jansky 2000; Koppel 2000; Watanabe and Watanabe 2000). The purpose of this chapter is to pool information on various aspects of potato resistance to bacterial diseases and examine how progress could be achieved in this field.

BIOLOGY OF PATHOGENS

The above-mentioned species of genera *Clavibacter, Erwinia, Ralstonia,* and *Streptomyces* are able to infect both tuber and stem tissues. The bacteria generally enter plants via wounds or natural openings, where they can cause disease immediately or survive latently to develop at a later date when environmental conditions are favorable. Although they are both soil-and tuber-borne, usually one of the two plays a crucial role, depending on the pathogen. There are different venues for infection during plant growth, harvesting, and tuber handling in storage, but most rely on prior wounding of the tissue. Disease control relies largely on disease avoidance measures based on a knowledge of the etiology and epidemiology of the disease (including pathogen diversity), diagnostics, seed certification, and production of resistant cultivars, which involve studies on the interactions between pathogen and host at biochemical and molecular levels as well as the identification of novel sources of resistance. Application of the different control measures is dependent on the relative economic importance of the different diseases occurring in a given area and whether the crop is for seed or ware production.

Erwinia spp.

Pathogens

Three soft rot erwinias, *Erwinia carotovora* subsp. *atroseptica, E. carotovora* subsp. *carotovora,* and *E. chrysanthemi,* are associated with potatoes, causing soft rotting of tubers and blackleg, which affect the growing plant (Pérombelon and Kelman 1980). These bacteria have recently been reclassified in the genus *Pectobacterium* (Hauben et al. 1998), but the new nomenclature has yet to be widely accepted by plant pathologists. The bacteria are nonspore-forming facultative anaerobes capable of producing large quantities of pectic enzymes, which enable them to macerate parenchymatous tissues (Basham and Bateman 1975; Pérombelon 1982; Collmer and Keen 1986; Kotoujansky 1987).

These bacterial pathogens are worldwide in distribution but have contrasting host ranges which reflect their temperature characteristics: *E. carotovora* subsp. *atroseptica* is restricted solely to potato and mainly in temperate regions; *E. chrysanthemi* has a wider range of cultivated and noncultivated host plant species, mainly in tropical and subtropical areas, but some cold-adapted strains can affect certain crops including potato in temperate countries. *E. carotovora* subsp. *carotovora* strains are widely distributed in all climates and are pathogenic to a wide range of plants.

Diversity

Strain diversity within the different species tends to reflect their geographic and host range distributions. *E. carotovora* subsp. *carotovora* and

E. carotovora subsp. *atroseptica* are serologically related and divided into some 50 sero-groups, of which only nine describe *E. carotovora* subsp. *atroseptica* and, of these, 60-95% of the strains belong to sero-group I (De Boer and McNaughton 1987; Pérombelon 2002). Considerable serological diversity has been noted in *E. chrysanthemi* (Samson et al. 1989). Similarly, two RFLP groups were only obtained after analysis of the polymorphism of the *E. carotovora* subsp. *atroseptica* genes encoding pectate lyase (Darrasse et al. 1994; Śeledż et al. 2000) and recombinase A (Waleron et al. 2002). Phage and molecular typing (RFLP, ERIC-PCR, and RAPD) showed the highest diversity within *E. carotovora* subsp. *carotovora* (Mäki-Valkama and Karjalainen 1994; Toth et al. 1999a, b; Fessehaie et al. 2002; Seo et al. 2002; Śeledż 2002). A high level of diversity was observed in *E. chrysanthemi* in RFLP analysis of the sequences of genes encoding isoenzymes from the pectate lyase family (Darrasse et al. 1994; Nassar et al. 1996) and in RFLP analysis of the sequences of different genes (Helias et al. 1998; Toth et al. 2001; Waleron et al. 2002).

Disease Symptoms

Environmental factors, especially humidity, determine the type of symptoms on stems: a soft rot tends to develop under wet conditions whereas wilting and desiccation predominate under dry conditions regardless of the form of *Erwinia* involved (Pérombelon and Kelman 1987). Under wet conditions, blackleg symptoms progress from the rotting seed (mother tuber) and range from a small, black, water-soaked lesion restricted to the base of stem to extensive rotting of the whole stem. Under dry conditions, diseased stems are stunted and wilting, with chlorotic leaves and infected tissue black to light brown. Tuber soft rot symptoms develop in storage and can range from a small lesion often restricted to lenticels to rotting of the whole tuber. Affected tuber tissue is soft, cream-colored, granular, and odorless. At the margin of decayed tissue, brown or black pigmentation often develops.

Etiology and Epidemiology

The soft rot erwinias have been described as opportunistic pathogens with little or no host specificity, mainly because the disease tends to develop only when host resistance is impaired, especially under wet, anaerobic conditions (Pérombelon 2002). Under these conditions, bacterial growth occurs and the strongly pectolytic erwinias tend to predominate, some more effectively at higher than at lower temperatures. Blackleg develops when large numbers of the pathogen invade the stems after multiplication in the rotting mother tubers. The extent of disease development is related to the level of seed tuber contamination by the pathogen, i.e., the greater the level of seed contamination, the greater the incidence of blackleg (Bain et al. 1990). Of the three bacteria, only *Erwinia carotovora* subsp. *atroseptica* and *E. chrysanthemi* appear to cause blackleg symptoms but all three can cause tuber soft rot (Pérombelon et al. 1987). Their pathogenesis is temperature dependent; *E. carotovora* subsp. *atroseptica* tends to cause blackleg at temperatures <25°C and *E. chrysanthemi* at higher temperatures (Pérombelon et al. 1987). However, what are known as cold strains (*E. chrysanthemi*) have recently been found to cause blackleg in temperate countries. Occasionally when conditions allow *E. carotovora* subsp. *carotovora* to predominate in rotting mother tubers, it can cause blackleg. This bacteria produces a set of pectin-degrading enzymes, such as pectin methylesterase, pectin acetyl esterase, pectate lyase, pectin lyase, and polygalacturonase (Collmer and Keen 1986; Barras et al. 1994). The cell wall degrading enzymes are secreted into the external medium via a common secretion system, which is encoded by the *out* operon, essential for pathogenicity. Plant factors may influence the expression of genes involved in pathogenicity (Boccara et al. 1988; Beaulieu et al. 1993; Lojkowska et al. 1993, 1995; Dorel et al. 1996). In addition to the production of large quantities of extracellular enzymes, in particular pectic enzymes, several other pathogenicity-associated characters of lesser importance are involved with the establishment of the bacteira in plant tissues, e.g. flagella (motility), fimbriae (adhesion), siderophores (iron uptake), etc. (Pérombelon 2002). There is also another set of pathogenicity characters, linked with host specificity (*hrp* and possibly *avr* genes), present mainly in "frank" plant pathogenic bacteira with a restricted natural host range, which could also be involved, but their role in pathogenesis has not yet been established (Bauer et al. 1994; Mukherjee et al. 1997; Pérombelon 2002).

Survival of erwinias in soil is restricted to a few months at most, hence blackleg is primarily a seed-borne disease. A high proportion of tubers of most commercial seed stocks are contaminated with erwinias at varying levels. Several sources of the pathogens have been identified, of which the most important are the rotting mother tubers from which the bacteria are transmitted to the lenticels of daughter tubers through soil water and rotting tubers during pre- and post-harvest, from which the bacteria get transferred to wounds inflicted on tubers during mechanical tuber handling (Pérombelon and Kelman 1987).

Detection and Identification Methods

A pectate-based selective and diagnostic growth medium is commonly used to isolate soft rot erwinias (Pérombelon and Kelman 1987). Identification at species and subspecies level requires additional biochemical, physiological, and immunological tests (De Boer and McNaughton 1987; Gorris et al. 1994; Hyman et al. 1995). Serological tests are no longer considered reliable for identification purposes because of poor specificity. Instead, PCR-based methods are more accurate for identification and quantification of *Erwinia carotovora* subsp. *atroseptica* and *E. chrysanthemi*

(Darrasse et al. 1994; De Boer and Ward 1995; Smith et al. 1995; Nassar et al. 1996; Fréchon et al. 1998; Pérombelon et al. 1998; Toth et al. 2001). Toth et al. (1999b) described a one-step, 16S rDNA PCR-based method, for the detection of all soft rot *Erwinia* species. Recently, Waleron et al. (2002) described a method based on restriction analysis of the amplified fragment of *recA* gene (*recA* PCR-RFLP) that can be used to rapidly identify all species and subspecies of pectinolytic *Erwinia*.

Disease Control

As tuber contamination by one or more erwinias is extensive and rotting occurs when host resistance is impaired, disease control is based mainly on measures which avoid risks of wetting of tubers and development of anaerobic conditions in storage and in transit, i.e., nonfluctuating low temperatures to avoid condensation and bacterial multiplication and adequate ventilation. Blackleg being a seed-borne disease, its control relies mainly on planting seed with contamination level below the threshold level for disease development (Pérombelon 2000).

Ralstonia solanacearum

Pathogen

Ralstonia solanacearum is a rod-shaped, Gram-negative, nonspore-forming, noncapsulated, nitrate reducing, and aerobic bacterium responsible for causing bacterial wilt or brown rot of potato. The bacterium can affect a wide range of plants, mostly solanaceous species, in tropical and subtropical regions. Nonvirulent spontaneous mutant colonies readily develop when grown on tetrazolium chloride medium (Kelman 1954). Whereas virulent colonies appear as fluidal and opaque white with a pink center, the mutant forms frequently develop forming smaller transparent red colonies with flagellated cells.

Diversity

R. solanacearum has been divided into five races defined by host affinity and five biovars on the basis of biochemical properties, but there is no close correlation between the two classifications. Only Races 1 and 3 cause severe symptoms on potatoes, the former in the hot tropics and the latter in cooler areas. Race 1 has a wide host range while Race 3, which contains the same strains as biovar II, is highly homogeneous and specific to potato (Timms-Wilson et al. 2001). Results of diversity studies on *R. solanacearum* strains were obtained worldwide using FAMEs analysis (Janse 1996). RFLP analysis with DNA probes, which encode information required for virulence and hypersensitive reaction PCR-RFLP (Cook et al. 1989, 1991), analysis of the *hrp* gene region, amplified fragment length polymorphism (AFLP) and sequencing of 16S rDNA (Poussier et al. 2000) showed high variability. Genetic diversity among isolates of

R. solanacearum Race 3 biovar II of European origin was studied by two genomic fingerprinting methods: macrorestriction analysis of genomic DNA resolved by pulsed-field gel electrophoresis (Smith et al. 1995) and repetitive element sequence based (REP-PCR; Smith et al. 1998).

Disease Symptoms

Bacterial wilt or brown rot is a soil- and seed-borne disease. Infection can occur when bacteria in the soil penetrate roots, especially following wounding caused by nematode activity or when the seed-borne pathogen moves systemically in the vascular system from the seed or mother tubers to stems and daughter tubers. Under favorable weather conditions, especially at high temperatures, the bacteria multiply in the stems causing wilting, hence the name bacterial wilt, followed by yellowing, stunting and desiccation. Discoloration of the vascular system is characteristic of the disease. It is common for secondary infection of soft rot erwinias to result in stem rot. But in cooler (temperate) regions, the more common symptoms affecting tubers are a brown staining of the vascular tissues which later ooze bacterial slime from the heel end and the eyes, often ending as a rot; hence the name brown rot. It is characteristic of the disease under cool conditions that tuber infection may remain symptomless for some time.

Etiology and Epidemiology

Survival of the bacterium in soil is poorly understood. It is affected by several interacting physical, chemical, and biological factors. The more important ones, which appear to adversely affect survival, are flooding, desiccation (usually associated with low rainfall and high soil temperature), and antagonism by certain soil bacteria which may be favored by growing cereals. Monitoring studies on bacterial survival in soil of fields planted with infected potato crops in Australia, Sweden, and the UK showed that the pathogen could not be detected even after two years. However, infection of a relatively common aquatic solanaceous weed, *Solanum dulcamara*, growing on riverbanks, has been detected in several European countries and can result in large numbers of the bacteria being liberated in water especially during the warmer summer months. As the weed is perennial, the bacterium can survive for several years. The origin of this infection is not clear but it could have arisen from bacteria present in water draining from fields with a diseased potato crop or in waste water from potato-processing factories using infected potatoes imported from a country where the disease is endemic or present in domestic waste water used for washing diseased tubers and subsequently discharged into a river (Elphinstone et al. 1998). Infection of progeny tubers can also occur as a result of the spread of the bacteria from rotting mother tubers. The spread of Race 3 of *R. solanacearum* worldwide has been associated with its dissemination in symptomless latently infected seed tubers.

Pathogenicity of *R. solanacearum* has been extensively studied. The main virulence factors involved in infection include the production of cell wall-degrading polygalacturonases, cellulase, and extracellular polysaccharides (Husain and Kelman 1958; Cook and Sequeira 1991; Hayward 1991; Huang and Allen 2000). Extracellualar polysaccharides of *R. solanacearum* may act as a protective coating that prevents bacteria from agglutination by divalent cations in intracellular space or fixation onto plant cell walls through binding between liposaccharides and lectins (Araud-Razou et al. 1998). Disease severity is determined by the expression of the set of *hrp* genes in potato tissue (Aldon et al. 2000; Genin and Boucher 2002) and by the restricted movement of the pathogen within the plant attributed to formation of a mechanical barrier (Elphinstone 1994; Priou et al. 2001). Recent completion of genome sequencing of the pathogen should greatly enhance our understanding of its infective process (Salanoubat et al. 2002).

Detection and Identification Methods

Traditional methods for detecting *R. solanacearum* are visual inspection of planting material and selective planting followed by biochemical characterization. Because of international spread of this bacterium through latent infection, more sensitive and rapid detection methods are required to control spread of the pathogen to new areas as well as to test the soil and water for irrigation prior to planting.

Sensitive and specific detection methods based on specific antisera (Wolf et al. 1998; Lyons et al. 2001; Caruso et al. 2002), nucleic acid probes (Wullings et al. 1998), PCR tests (Seal 1995; Boudazin et al. 1999; Pastrik and Maiss 2000; Poussier and Luisetti 2000) have been developed. Current methods in routine use include: growing the bacteria on a semi-selective/diagnostic medium, serological testing (immunofluorescence staining and ELISA tests on enriched or nonenriched material), molecular assay (PCR), and pathogenicity testing on a susceptible host (Elphinstone et al. 2000). All have certain advantages and disadvantages regarding not only specificity and sensitivity but also speed and user-friendliness for large-scale use. The sensitivity of any of the aforesaid methods is theoretically sufficient to detect latent symptomless infection in tubers because of number of the bacteria usually present. Initially, only two or three simple and rapid screening methods are routinely used, e.g. growth on selective medium and ELISA. A putative positive result can be confirmed by applying more specific but slower tests, such as FAP or PCR assay and pathogenicity testing. The methods need to be modified for the detection of the pathogen in complex substrates, such as soil and waste effluent during potato processing and sewage treatment. Recently, evaluation of several optimized procedures (IFAS, ELISA, and PCR) showed that IFAS is the most reliable (Elphinstone et al. 2000). Further, to eliminate the need for post-PCR gel electrophoresis and ethidium-bromide staining of target DNA, Weller et al. (2000) advocated use of quantitative, multiplex, real-time fluorogenic PCR (TaqMan) for detection of this bacterium in potato tubers. Recently, van der Wolf et al. (2002) applied flow cytometry (FCM) and Bentsink et al. (2002) the NASBA (Nucleic Acid Sequence Based Amplification) method to detect *R. solanacearum* and assess cell viability. Furthermore to the specificity and sensitivity of the detection methods used, detection also depends on the tuber sampling strategy.

Disease Control

In the absence of highly resistant cultivars, only disease evasion measures are feasible in regions where the pathogen is endemic. These include use of clean seed, non- or hardly contaminated fields, crop rotation which excludes a susceptible alternative host crop, use of nematicides to control nematodes that favour infection, care against cutting of seed tubers, and use of noncontaminated irrigation water. Where the pathogen is not endemic, it is treated as a quarantine material and like plant material subjected to strict regulations.

Current strategies to control brown rot in Europe fall into three categories: (i) preclusion of importation of potatoes from pathogen-infested areas, (ii) detection of infection in imported potatoes, and (iii) elimination if possible of the pathogen in seed stocks, surrounding environment.

Clavibacter michiganensis subsp. *sepedonicus*

Pathogen

Clavibacter michiganensis subsp. *sepedonicus* is a Gram-positive, nonmotile, obligate aerobic, pleomorphic rod-shaped bacterium causing bacterial ring rot (Skaptason and Burkholder 1942). Growth on agar medium is slow, taking several days to form small 1 mm diameter mucoid or nonmucoid colonies depending on the strain. Although when artificially inoculated it can infect several plant species, mostly *Solanum*, natural infection occurs only on potato.

Diversity

Genomic fingerprinting of different subspecies of *C. michiganensis* using restriction endonucleases and REP-PCR in conjunction with primers corresponding to conserved repetitive motifs (REP, BOX, ERIC), has generated different fingerprints for bacteria belonging to different subspecies of *C. michiganensis* (Louws et al. 1998; Brown et al. 2002). This technique enabled establishing the characteristic pattern for five tested subspecies of *C. michiganensis* (*sepedonicus, michiganensis, insidiosus, tessellarius*, and *nebraskensis*); however, it was not possible to identify differences within the two most common subspecies, *sepedonicus* and *michiganensis*.

Symptoms

When severely infected seed tubers are planted, they may fail to sprout or sprouted shoots die soon. Otherwise emerging shoots are stunted with yellowing and curling of the leaves. Affected stems eventually wilt and die, generally sooner under dry than in wet conditions. Milky bacterial ooze is seen when diseased stems are cut and squeezed. Tubers are infected *via* the stolons and, when sectioned, a yellowish ring along the vascular tissue is visible. Upon squeezing, a creamy mass of macerated tissue emerges from the vascular area.

Etiology and Epidemiology

Clavibacter michiganensis subsp. *sepedonicus* is primarily tuber borne and is spread during seed cutting and planting. Insect pests (Colorado potato beetle, potato flea beetle, and aphids) can also act as vectors under field and greenhouse conditions. Under cool environment with low relative humidity, the pathogen can survive on the surface of transport and storage equipment for up to five years. It can also survive in the field in diseased plant debris until the plant tissues have completely rotted. Dry soil and temperatures of 24-32°C are favorable for disease development.

The bacteria enter, multiply inside, and move through the xylem vessels of the potato plants. They interfere with translocation of water and nutrients. This results in wilting and death of the plants. After invasion of the xylem vessels, the bacteria produce mucoid exopolysaccharides that plug the vessels (Westra and Slack 1992). In addition, glycoproteins are formed that can affect permeability of cell membranes (Strobel 1970; Romanenko et al. 1997) and extracellular cellulases which degrade xylem cell walls (Baer and Gudmestad 1993; Metzler et al. 1997).

Detection and Identification Methods

Traditional methods of disease detection in the field are visual inspection followed by serological tests to identify *C. michiganensis* subsp. *sepedonicus*. ELISA and IFAS are commonly performed using mono- or polyclonal antibodies (De Boer and Wieczorek 1984; De Boer and Mc Naughton 1986; De Boer and Hall 2000b). Recently, molecular detection methods have been developed, such as genomic fingerprinting, DNA probe, and PCR (Schneider et al. 1993; Firrao and Lozzi 1994; Mills et al. 1997). The primers were derived from DNA fragments unique to *C. michiganensis* subsp. *sepedonicus* (Mills et al. 1997), from repetitive BOX sequence (Smith et al. 2001), and from the intergenic transcribed region (ITS) of the 16S-23S rDNA (Pastrik 2000). Detection of the bacterium in potato tubers by BIO-PCR and an automated real-time PCR detection system was recently described (Schaad et al. 1999).

Disease Control

C. michiganensis subsp. *sepedonicus* is a systemic pathogen and therefore requires quarantine measures in most countries to check for its presence in seed stocks. Control measures rely on seed certification based on visual field examination aided by laboratory-based methods for its detection as outlined above. In addition, agronomic measures, such as crop rotation and disinfection of knives when cutting seed tubers prior to planting and also of equipment, are essential to reduce risks of spread.

Streptomyces spp.

Pathogens

Three *Streptomyces* spp., namely *S. scabies, S. acidiscabies,* and *S. turgidiscabies* are the main causal agents of potato scab. *Streptomyces* are atypical bacteria that resemble fungi by producing mycelium with aerial filaments from which spores develop through fragmentation. The three species are distinguished by pigmentation, color and shape of their sporulating mycelium, and their ability to utilize specific sets of sugars (Lambert and Loria 1989). The host range of *S. scabies* includes beet, cabbage, carrot, eggplant, turnip, parsnip, and radish, but their economic importance relates mainly to potato. The pathogens are both soil- and seed-borne. Whereas *S. scabies* occurs worldwide, the other two species are restricted to northeastern USA and Japan. *S. acidiscabies* and *S. turgidiscabies* have a lower limiting pH (<5.0) for infection in soil than *S. scabies*.

Diversity

Phenotypic, genotypic, and pathogenic variations in strains of *Streptomyces* spp. examined by 16S rDNA analysis have allowed differentiation of several genotypes, but no correlation was found with their ability to cause common scab symptoms on potato (Doering-Saad et al. 1992; Bramwell et al. 1998; Conn et al. 1998). Diversity within species of bacteria causing common scab was also studied by RAPD and REP-PCR (Doering-Saad et al. 1992; Bramwell et al. 1998).

Disease Symptoms

Scab lesions on tubers are characterized by dark, irregular concentric cork layers developed as a reaction to infection around the central depression of the tuber and can be divided into three types: common raised scab, pitted scab, and russet scab. Their precise appearance, however, can be affected by the particular species or strain of the pathogen, cultivar, and or edaphic factors (Lambert and Loria 1989).

Etiology and Epidemiology

Growing potato tubers are infected through natural openings (lenticels) and mostly during their early development in wet soils. Their development may also be suppressed by soil microorganisms under still ill-defined conditions.

Streptomyces spp. causing common scab produce phytotoxins (thaxtomins) that can induce symptoms of disease, including cell hypertrophy and later cell death (Loria et al. 1995). The amount of thaxtoxin A produced by *S. scabies* strains is a good indicator of their pathogenicity (Kinkel et al. 1998).

Detection and Identification Methods

As previously mentioned, the three pathogenic species of *Streptomyces*—*S. scabies*, *S. acidiscabies*, and *S. turgidiscabies*—are distinguished from each other by color and shape of their sporulating filaments and ability to utilize a specific group of sugars. *Streptomyces* spp. can be isolated from infected tissue by dilution plating on water agar. However, for positive identification it is necessary to grow the bacteria up to 10 days, for bacterial filament and pigment production in order to allow differentiation of the pathogens (Loria et al. 2001). Pathogenicity tests may be necessary for confirmation. Detection and identification of pathogenic *Streptomyces* strains can be achieved by PCR-based tests using primers designed for genes coding for putative pathogenicity factors (Cullen et al. 2000).

Disease Control

Although disinfection of seed tubers can reduce infection, the main control measure is use of scab-free seed together with long crop rotation with cereal or grass crops. Acidification of soil by application of ammonium sulphate can help.

NATURE OF DISEASE RESISTANCE

Plant-Bacteria Interaction at Biochemical and Molecular Level

Plant-bacteria interaction can be divided into two classes, biotrophic or necrotrophic. Infection with a biotroph pathogen (*Streptomyces* spp.) does not result in plant cell mortality, whereas in the case of necrotroph pathogens, plant cell death occurs directly after invasion (*Erwinia*, *Ralstonia* and *Clavibacter*) and not during latent infection.

Resistance or susceptibility of the plant to infection depends on a wide range of different molecular interactions between molecules produced by plant pathogenic bacteria and by the host plant. One type of plant-microbe interaction, known as gene-for-gene, involves the avirulence genes of the pathogen genome and resistance genes of the host plant. The interaction is specific for the race of the pathogen and a plant cultivar. A second type of interaction concerns general host resistance which is determined by host genes that are equally effective against all races of the pathogen.

The products of avirulent genes (*avr*) determine host specificity as wells as compatible or incompatible interactions. Plants possessing resistance genes, which are pathogen-specific, interact specifically with pathogens with corresponding *avr* genes. This interaction involves a particular molecular signaling reaction between *avr* gene products and receptors (resistance gene products) on the surface of plant cells (Staskawicz et al. 1995, 2001; Aldon et al. 2000). *R. solanacearum* resistance gene products are similar to proteins involved in protein-protein binding, e.g. membrane-spanning regions and leucine-rich repeats (LRR; Gueneron et al. 2000).

Plant pathogenic bacteria (*Erwinia* sp., *Ralstonia solanacearum*) also have a set of genes (*hrp*) whose products are associated with induction of the hypersensitive response (HR) in plant tissue (Bauer et al. 1994; Mukherjee et al. 1997; Aldon et al. 2000). In this case, plant resistance is associated with an oxidative burst with the production of free radicals and cross-linking of proteins in plant cell walls (Aldon et al. 2000; Nissinen et al. 2001). At the same time, a flux of intracellular calcium occurs and is probably involved in intracellular signaling. The later steps of HR response are cell necrosis and phytoalexin accumulation within the infected cells and in neighboring cells. Phytoalexins are toxic to bacteria and there is no evidence that bacteria can degrade phytoalexins. The end result is programmed cell death—apoptosis (Dangl et al. 1996). It is now clear that plants and animals use similar mechanisms to invade their hosts. The type III secretion systems encoded by the *hrp* genes of plant pathogenic bacteria (e.g. *R. solanacearum, Erwinia amylovora*) and by the related *ysc* (Yop secretion) genes of animal pathogens in the genus *Yersinia,* have the ability to secrete virulence proteins and operate fundamentally in the same way (Alfano and Collmer 1996).

An important factor determining virulence of necrotroph bacteria is the production of a wide range of extracellular enzymes, many of which can degrade plant cell wall components. In contrast, biotrophs produce fewer extracellular enzymes and in smaller amounts. Necrotrophic *Erwinia* spp. are able to produce several isoenzymes of the more important extracellular enzymes, e.g. polygalacturonases, pectate lyases and cellulases, which allow rapid degradation of cell wall components, thereby preventing induction of any resistant reaction. Production and secretion of extracellular enzymes are controlled by a complex regulatory system, the most common signaling molecules being N-acyl derivatives of homoserine lactone (acyl-HSL), commonly produced by *E. carotovora, E. chrysanthemi,* and *R. solanacearum* (Jones et al. 1993). Modulation of the physiological processes controlled by acyl-HSL occurs in a cell density and growth phase-dependent manner, quorum-sensing (Pirhonen et al. 1993; Salmond et al. 1995).

The action of pectinases can induce release of oligogalacturonides which are able to elicit a resistant reaction through phytoalexin accumulation only under aerobic conditions (Vayda and Schaffer 1988). Specific receptors to these oligogalacturonides are present on the plasmalemma. On the other hand, plants also produce proteins able to inhibit cell-wall-degrading enzymes, e.g. a family of polygalacturonase-inhibiting proteins (PGIP) isolated from different plant species, which are able to inhibit different polygalacturonase isoenzymes (Darvill et al. 1994; De Lorenzo and Ferrari 2002). Bacterial proteases can degrade plant cell wall proteins associated with resistance, e.g. proline and hydroxyproline-rich proteins. Plant pathogenic bacteria can also produce phytotoxins, resulting in host cell death (Loria et al. 1997).

Plant tissue reaction to infection can be local (local acquired resistance, LAR) or systemic (systemic acquired resistance, SAR; Ross 1961; Rayls et al. 1996; Sticher et al. 1997). Systemic accumulation of salicylic and jasmonic acids in resistant tissue was observed following infection with the pathogen (Rasmussesn et al. 1991; Delanay et al. 1994). An application of salicylic acid can also induce a resistance reaction of the whole plant to infection caused by *E. carotovora* (Palva et al. 1994; Lopez et al. 2001).

Potato cells have the ability to produce antimicrobial peptides, essential effectors of innate immunity, a nonspecific defense mechanism occurring in plants and animals (Garcia-Olmedo et al. 1995; Broekaert et al. 1997). At least eight families of antimicrobial peptides, ranging in size from 2 to 9 kD, have been identified in plants (several in potato). These are thionins, defensins, so-called lipid transfer proteins, hevein-and knottin-like peptides, and snakins (Garcia-Olmedo et al. 1998). All of them have compact structures that are stabilized by 2-6 disulfide bridges. Antimicrobial peptides can affect the ability of the pathogen to survive and spread. *E. chrysanthemi sap* mutant (sensitive to antimicrobial peptides) was more sensitive than the wild type to alpha-thionin and snaking-1 and also less virulent than the wild-type strain in potato tubers; the lesion area was 37% of control and growth rate two orders of magnitude lower (Lopez-Solanilla et al. 1998). Pathogen sensitivity to host-produced peptides is mediated by two kinds of bacterial proteins: those affecting the overall permeability of the pathogen's extracellular matrix and those involved in peptide import (Titarenko et al. 1997). Mutation in the genes encoding the first group of proteins render the pathogen more susceptible to a range of general host defense mechanisms, whereas mutation in the second group of genes specifically affects the pathogen's ability to withstand antimicrobial peptides (Molina and Garcia-Olmedo 1997; Titarenko et al. 1997). The role of antibacterial peptides in the innate immunity system of plants is well established. These peptides are part of developmentally regulated, preexisting defense barriers and may be accumulated as a result of the expression of the corresponding genes upon infection (Garcia-Olmedo et al. 1998). Several genes encoding antimicrobial peptides were used in potato transformation toward enhanced resistance to bacterial diseases.

Genetic Determination of Potato Resistance to Bacterial Pathogens

Resistance to Tuber Soft Rot and Blackleg

Genetic variations for resistance to tuber soft rot and blackleg have been found in both wild and cultivated *Solanum* species (for references, see Table 14.1). Higher frequencies of resistant progeny to both diseases were found in crosses between tetraploid resistant than in susceptible parents (Dobias 1976; Hidalgo and Echandi 1982; Vlasov 1983; Pawlak et al. 1987). Similar results were obtained with diploid hybrids (Bushkova and Vlasov 1983; Lebecka et al. 1998). Resistance to tuber soft rot and blackleg is not closely related. Zadina and Dobias (1976), Munzert (1984) and Zimnoch-Guzowska et al. (1999a) found a correlation coefficient between the two characters of $r = 0.65$, $r = 0.57$, and $r = 0.725$ respectively. Such results support Lyon's statement (1989) that resistance to *Erwinia* spp. in potato tubers and stems is under partially independent genetic control. Similarly, Allefs (1995) found resistance of stored tubers did not correlate with resistance to blackleg. Ranking of genotypes depends on the screening methods applied and environmental conditions (Wastie and Mackay 1985).

Resistance to erwinias is inherited polygenically. Lellbach (1978), based on tuber soft rot resistance in progenies from diallel crosses among three cultivars, suggested that minor genes with additive effect determined the studied resistance. General Combining Ability (GCA) was the most important source of expressing resistance. Moderate broad-sense heritability estimates of resistance to tuber soft rot were found in diploid hybrid populations *S. phureja, S. stenotomum* after recurrent selection cycles (h^2 = 0.43-0.59) by Wolters and Collins (1991). After a three-year study at IHAR, (Poland) of 10 unselected diploid progenies, in which resistance originated from *S. chacoense, S. yungasense,* and *S. phureja,* significance of variance due to GCA of seed and pollen parents, as well as SCA effects, were shown in ANOVA of this factorial design; however, GCA effects dominated over SCA effects (Lebecka et al. 1998). When selected diploids from this program were used in 4x-2x crosses, similar results were obtained after three-year testing of 24 tetraploid progenies. Significant GCA and SCA effects in genetic determination of tuber resistance to soft rot were found. (Lebecka and Zimnoch-Guzowska 2001). The significant effects of SCA and GCA found in diploid and tetraploid progenies indicate involvement of both additive and nonadditive gene action in inheritance of tuber soft rot resistance. Maternal effect was not noticed in reciprocal crosses among two sets of diploids hybrids (Lebecka et al. 1998). Lack of cytoplasmic inheritance of resistance to tuber soft rot was also stated by McGrath et al. (2002) in reciprocal crosses of BC_4 generation of somatic hybrids *Salanum tuberosum + S. brevidens.*

The significant influence of year, such as interactions between a year and parents, indicated the relative importance of environment on

Table 14.1 *Solanum* species used in potato breeding programs as sources of resistance to bacterial diseases

Solanum species	*Soft rot*	*Blackleg*	*References*	*Bacterial wilt*	*References*	*Ring rot*	*References*	*Common scab*	*References*
acaule						+	Kurowski and Manzer (1992); Kriel et al. (1995)		
brevidens	+	+	Austin et al. (1988)						
chacoense	+	+	Zimnoch-Guzowska and Lojkowska (1993); Capo et al. (2002)	+	Schmiediche (1986)			+	Dionne and Lawrence (1961); Hawkes (1990)
commersonii	+		Carputo et al. (1997)					+	Hawkes (1990)
microdontum				+	Schmiediche (1986)				
phureja	+	+	Iwanaga (1985); De Maine et al. (2000)	+	Rowe and Sequeira (1970)			+	De Maine et al. (1993)
raphanifolium				+	Schmiediche (1986)				
sparsipilum				+	Schmiediche (1986)				
stenotomum	+	+	Iwanaga (1985)	+	Fock et al. (2000)				
tuberosum								+	Clark et al. 1938; Zadina (1958)
tuberosum subsp. *andigena*	+	+	Huaman et al. (1998)						
yungasense	+		Zimnoch-Guzowska and Lojkowska (1993)					+	Hawkes (1990)

expression of resistance to erwinias in potato (Lellbach 1978; Hidalgo and Echandi 1983; Allefs 1995; Zimnoch-Guzowska et al. 1999a; Lebecka and Zimnoch-Guzowska 2001). Potato resistance to each *Erwinia* strain pathogenic to potato was found to be highly correlated (Wolters and Collins 1994; Zimnoch-Guzowska et al. 1999b).

DNA markers are excellent tools for analysis of Quantitative Trait Loci (QTL) controlling genetic resistance to erwinias. QTLs of tuber and leaf resistance to *E. carotovora* subsp. *atroseptica* were studied by Zimnoch-Guzowska et al. (2000) based on a cross between diploid hybrids originated from *S. chacoense* and *S. yungasense* which differed in tuber and detached leaf resistance to this pathogen. A linkage map was constructed with application of AFLP and RFLP markers including three resistance-gene-like (RGL) loci. The QTL analysis confirmed the truly polygenic mode of inheritance of resistance to *E. carotovora* subsp. *atroseptica* as markers for resistance were found on all the twelve chromosomes of potato. QTLs for tuber resistance were found only on ten chromosomes, with larger QTL located on chromosome I, which explained 19% of the variation in resistance. This QTL was linked to RGL locus. It was the most prominent QTL for possible marker-assisted selection. Other reproducible QTLs for tuber resistance were found on chromosomes VI, XI and XII. Putative QTLs for leaf resistance found on 10 chromosomes were less reproducible than those for tuber resistance. Localization of QTLs for tuber and leaf resistance was largely independent. Several QTLs were mapped to similar positions as were other QTLs or major genes of resistance to pathogens in other solanaceous crops.

Resistance to Bacterial Wilt

Resistance to bacterial wilt was found to be strain-specific and temperature dependent (Thurston and Lozano 1968). The first report on inheritance of resistance to bacterial wilt assumed a simple model of dominant independent three gene actions found in *S. phureja* (Rowe and Sequeira 1970), extended in later studies by an additional function of modifying factors.

Resistance to bacterial wilt was found to be a quantitatively inherited trait by Watanabe et al. (1992). When 4x-2x crosses were used for transfer of resistance to bacterial wilt from diploid to tetraploid progeny via 2n FDR gametes, the female × male interaction was a significant element. Properly selected, both 4x and 2x parents were important for obtaining a high frequency of resistant progeny.

Inheritance of bacterial wilt resistance in tetraploid potato clones was investigated in the Philippines and Vietnam by Tung et al. (1993). The progenies segregated for the trait since resistance was derived from different species sources with different types of adaptations. The best parent was a hybrid AVRDC-1287.19 selected by CIP, having in its pedigree *Solanum chacoense* and *S. raphanifolium*. The results indicated partial dominance for resistance, with significant effects of GCA and SCA, and showed that both additive and nonadditive gene actions (epistasis) were important for resistance expression. The given results suggested that additive gene effects might be of significance in parents with a similar type of adaptations, whereas nonadditive gene effects are of importance only when parents differ in adaptability type. In the case of resistance in tomato to bacterial wilt, both GCA and SCA effects were found to be significant for inheritance of this trait; however, GCA was six times larger than SCA, indicating the predominance of additive gene effects in bacterial wilt resistance (Osiru et al. 2001).

The significance of SCA for inheritance of bacterial wilt resistance in 4x-2x seedling progenies of the CIP program was found by Chakrabarti et al. (1994) in seedling tests performed in a greenhouse. The authors state that resistance in seedling and clonal generations is probably governed by various gene systems.

Gang et al. (2000) used bulk segregant analysis to identify RAPD, SSR, and AFLP markers linked to resistance to bacterial wilt in diploid hybrid populations of the cross between susceptible and resistant parents (with *Solanum phureja* and *S. vernei* in pedigree). Three RAPD markers, OPG05940, OPR11800, and OPO13770, and one SSR marker, STM0032, were found to be linked with resistance to bacterial wilt. The STM0032 marker was mapped on the chromosome XII. Seven AFLP markers were considered associated with resistance.

Resistance to Bacterial Ring Rot

Genetic variation was found among potato relatives in respect of resistance to bacterial ring rot. However, among cultivars of *Solanum tuberosum,* extreme resistance (ER) was not found. Resistant cultivars were mainly identified on the basis of resistance to expression of symptoms rather than infection (De Boer and McCann 1990). However, symptomless (tolerant) cultivars are discouraged as they might be a source of inoculum. Limited studies were done on the genetic control of resistance to ring rot disease. For breeding purposes, use of an ER source is recommended because of the zero tolerance policy for this pathogen in the USA, Canada, and most European countries.

One of the identified sources of ER to ring rot was *Solanum acaule*. ER forms of *S. acaule* were tested by Kriel et al. (1995) in 16 F_1 progenies obtained by matings among ER and non-ER forms in various directions. Over 50% of the progeny expressed ER to inoculation. The expression of two dominant alleles at two loci, in addition to minor modifying factors, contributed to ER to ring rot found in this source. Kriel and coworkers suggested that transfer of ER into the cultivated potato would be relatively easy.

To use molecular tools for identification and selection of germplasm extremely resistant to ring rot, DNA polymorphism was assessed in relation to stem resistance to bacterial ring rot and supplemented by IFAS detection of *Clavibacter michiganensis* subsp. *sepedonicus* (Ishimaru et al. 1994). In a set of six cultivars and 22 accessions of wild species, two accessions—*Solanum acaule* #7-8 and *S. phureja* AD29-1—exhibited ER and were free from pathogen after inoculation. Polymorphism of DNA was detected by RFLP and RAPD markers among tested accessions and gene distances were defined among 23 accessions. Mapping populations to localize genes responsible for ER to ring rot would be helpful to find suitable parental forms.

Resistance to Common Scab

Information on the inheritance of potato to common scab is limited. Sources of high resistance were identified at the beginning of the twentieth century and widely used in cultivar breeding (see next section). Inheritance to common scab was shown to be controlled in a simple manner by a single gene and the level of resistance to scab was related to the condition of alleles of this gene. Quadruplexes were highly resistant, e.g., cultivar Jubel, a known ancestor of scab-resistant cultivars (Kranz and Eide 1941). A similar simple model was suggested by Lauer and Eide (1963), with a duplex level for respective levels of resistance to scab; however, the authors indicated that the type of lesion caused by the pathogen is more informative for proper selection than the surface area covered by the scab.

The maternal effect on the inheritance of resistance was demonstrated by Zadina (1958) in reciprocal crosses. A higher proportion of resistant progeny was found in the cross between resistant and susceptible forms when the seed parent was resistant.

O'Keefe (1965) field tested the heritability of common scab resistance to two locations for 198 clones and found it to be $h^2 = 0.69$ due to additive and dominant epistatic effects. Variance according to location was highly significant. In a three-year evaluation of 23 clones, the genotype × year interaction was significant.

Studies by Cipar and Lawrence (1972), and Howard (1978), indicated that more than one gene is involved in determination of resistance reaction to scab; however, the mode of resistance operation was still assumed to be simple. Alam (1972) proposed two loci model for inheritance to *Streptomyces* sp., while polygenic inheritance of this resistance was stated by Pfeffer and Ettmert (1985) and Haynes et al. (1997). Haynes et al. (1997) estimated broad-sense heritability for surface area covered by scab and for type of lesion on a mean basis of scab resistance in 4x *S. tuberosum* populations. In a two-year field testing at two locations, heritabilities were very high: for surface area covered $h^2 = 0.89$ and for type of lesion $h^2 = 0.93$. Both measurements showed significant effects of the interaction between genotype and environment. Such high heritabilities indicate a relatively simple model for genetic determination of resistance to this pathogen. Also, an effective transmission of scab resistance from diploid onto tetraploid level via 4x-2x interploid crosses argues for simple inheritance of this trait (Murphy et al. 1995).

Pietkiewicz (1980) tested 21 cultivars for scab resistance over three years at two to four locations and found significant genetic variation among them for resistance to common scab based on weakly modifying environmental effects.

Sources of Resistance within Solanaceae

At present, many potentially useful sources of resistance lie in germplasm collections of wild species and commercial cultivars across Europe and elsewhere in the world. However, sources which have been successfully utilized in breeding programs are rather limited (Table 14.1).

Blackleg and Tuber Soft Sot

Erwinia carotovora subsp. *atroseptica, E. carotovora* subsp. *carotovora,* and *E. chrysanthemi* are the causal agents of two of the most important bacterial diseases of potato worldwide: blackleg of stem in the growing season and tuber soft rot in storage (Pérombelon and Kelman 1980). Because there is no practical means of chemical control against these two diseases, breeding for resistance is an important issue (Weber 1990; Elphinstone 1994).

Cultivars Extreme resistance to *Erwinias* has not been found (Allefs 1995; Dorel et al. 1996). Resistance to blackleg and tuber soft rot does not show a high correlation and hence data on resistance to these two diseases have yet to be collected (Zimnoch-Guzowska et al. 1999a). Among the 137 economically important European and North American cultivars, only two cultivars bred in Belarus (Belorusky 3 and Loshitsky)* scored an 8 in a 1-9 scale of increasing resistance to blackleg (Swiezynski et al. 1998). In several German publications from the early 1950s, a set of cultivars expressing resistance to blackleg is mentioned: Carnea, Flava, Johanna, Prisca, Robusta, Sickingen, Anemone, Amsel, Agro, Bliss Triumph, Essex, Katahdin (after Schick and Hopfe 1962). Variability in cultivar resistance to blackleg was studied by Bushkova (1983), Lewosz (1984), Vlasov et al. (1987), Bains et al. (1999). Screening of cultivars for tuber resistance to soft rot was reported more often (Zadina and Dobias 1976; Bourne et al. 1981; Hidalgo and Echandi 1982; Krause et al. 1982; Ciampi-Panno and Andrade-Soto 1984; Tzeng et al. 1990; Allefs 1995; Reeves et al. 1999).

In the European database of cultivars (www.europotato.org), 656 taxa have been described for resistance to blackleg. It is noteworthy that nine cultivars were classified in the highest class (9) of blackleg resistance (Realta, Palma, Reina, Brodie, Karama, Picnic, Golden Millenium, Osprey,

---

*(acc. Swiezynski); Belorusky 3 and Loshitsky
(acc. world catal. potato varieties 1999)

Sebastian). The last four cultivars were released in the last 2-3 years. Among the 268 taxa assessed for resistance to soft rot, only two were classified in the highest class (cv. Record, released in 1937 and cv. Mensa, in 1967).

Wild and Primitive Cultivated Species Sources of high resistance to *Erwinia* spp. were discovered in wild and primitive cultivated *Solanum* species (Dobias 1978; Van Soest 1983; Huaman et al. 1989; Lojkowska and Kelman 1989; Rousselle-Bourgeois and Priou 1995). Among the 500 accessions of *Solanum* most of the lines resistant to stem rot were found in *S. berthaultii, S. bulbocastanum, S. chacoense, S. stoloniferum, S. tarijense* and *S. tuberosum* subsp. *andigena* by Lojkowska and Kelman (1989). Dobias (1978) tested 45 species for blackleg resistance and identified high resistance to this disease, surpassing that of cultivars Sickingen and Flavia in *S.boergeri, S. horovitzii, S. subtilius, S. graciae, S. laplaticum, S. chacoense, S. antipovitzii, S. schickii.* Rousselle-Bourgeois and Priou (1995) tested soft rot resistance in tubers of 11 *Solanum* species represented by 100 accessions. Confirmed resistances were identified by them in clones of *S. tuberosum* subsp. *andigena, S. chacoense, S. phureja, S. berthaultii, S. verneii,* and *S. sparsipilum.*

Watanabe and Watanabe (2000) listed 35 *Solanum* species, with various ploidy levels, in which resistance to erwinias was identified by various researchers. However, it is noteworthy that species utilized in breeding programs and already present in advanced breeding pools comprise *S. chacoense, S. phureja, S. brevidens, S. commersonii, S. tuberosum* subsp. *andigena, S. stenotomum,* and *S. yungasense* (see Table 14.1).

Bacterial Wilt

This disease is reported worldwide in regions of warm temperate and tropical climate. Host plants are represented by 35 (Kelman 1953) or 50 (Hayward 1994) families, with most genera in family Solanaceae; however, among the true host group 33 families were listed by Kelman and an additional 20 families (not considered true hosts) listed by Hayward. Potato is one of the major host plants for *Ralstonia solanacearum*. Thus the program of resistance breeding against bacterial wilt has been developed by CIP, Peru, and institutions focused on breeding potatoes in developing countries. Potato is affected in most cases by Race 3 biovar 2A of the pathogen, which is adapted to cool temperature. The pathogen is principally transmitted by tuber latent infection. Compared to other pathogens, the sources of resistance are very limited at both cultivar and wild species level.

Cultivars A reasonable level of resistance to *R. solanacearum* among tested potato cultivars is very rare. French et al. (1998) indicated two resistant cultivars from those tested by Nielsen and Haynes Jr. (1960), Jackson et al. (1979), and Bicamumpaka and Devaux (1984): cv. Prisca bred in

Germany, resistant to Race 1, and cv. Cruza 148 bred in Mexico, with confirmed resistance in East Africa, Brazil, and Peru. Cultivar Cruza 148 expressed besides resistance to wilt, some resistance to latent infection. Three cultivars originating from cultivated *S. phureja* were selected from a breeding program for bacterial wilt resistance of Sequeira and Rowe: Caxamarca, Molinera, and Amapola (after French et al. 1998).

From the Gross Lusewitz collection of cultivars tested for resistance to bacterial wilt, five were mentioned as having some resistance to this disease: Aquila, Epoka, Greta, Marilenem, and Unzen.

The European database of cultivars for bacterial wilt resistance shows 6 taxa only. Among cultivars and clones assessed in Brazil for resistance to *R. solanacearum* by growing them in a field naturally infested with Race 1 biovar I, the German cultivar Achat (very popular in Brazil) performed better than the other cultivars submitted for certification (Lopes et al. 1993).

Wild and Primitive Cultivated Species A few *Solanum* species have been selected as possible sources of resistance to bacterial wilt, including *S. phureja, S. chacoense, S. sparsipilum,* and *S. multidissectum* (Thurston and Lozano 1968; Rowe and Sequeira 1970; Schmiediche 1983; French et al. 1998). Hartman and Elphinstone (1994), however, adduced a longer list of *Solanum* species based on several publications: *S. acaule, S. acroscopicum, S. berthaultii, S. blanco-galdosii, S. boliviense, S. brachycarpum, S. chacoense, S. chomatophilum, S. commersonii, S. demissum, S. microdontum, S. phureja, S. pinnatisectum, S. polytrichon, S raphanifolium, S. sparsipilum, S. stenotomum, S. stoloniferum,* and *S. sucrense.* The sources of *S. phureja* were utilized by Rowe and Sequeira in their breeding program (1970).

The breeding program initiated in the 1960s at the CIP concerned with improving resistance to diseases, including bacterial wilt, explored several species of *Solanum*: *S. chacoense, S. sparcipilum, S. phureja, S. microdontun and S. raphanifolium* (Schmiediche 1986).

Ring Rot

This quarantine disease in potatoes caused by *Clavibacter michiganensis* subsp. *sepedonicus* is spread mainly via infected seed potatoes. Seed cutting is considered a major factor in ring rot spread in North America (De Boer and Slack 1984). Breeding for resistance to bacterial ring rot in the past involved selection for resistance based on low expression or lack of foliar and tuber symptoms. Such procedure favored tolerant forms, which were often the source of inoculum for cultivated potatoes. Thus in some countries, e.g. the USA, breeding strategy focused on selection of highly susceptible cultivars to reduce probability of introducing latently infected tubers in seed and ware production (Ishimaru et al. 1994). Genetic variability in resistance to this pathogen was found both among cultivars and wild species however, resistant forms are rare and ER to this disease has not been found in potato germplasm.

Cultivars In the 1940s the search for sources of ER to ring rot was expanded in the USA consequent to large-scale losses sustained by infected potato crops in the 1930s. Several cultivars were identified as expressing resistance to bacterial ring rot, viz. Frisco, Furore, Merrimack, President, Saranac, and Teton were selected in the 1940s and 50s based on symptomless behavior after infection with *C. michiganensis* subsp. *sepedonicus* (Bonde et al. 1947; Riedl et al. 1946; Lansade 1950; Akeley et al. 1955; De Boer and McCann 1990).

Expression of foliar and tuber bacterial ring rot symptoms in 108 cultivated potato genotypes was examined by Kawchuk et al. (1998) in a ten-year field experiment. Considerable variation in incidence of these symptoms was observed in all cultivars, ranging from no symptoms detectable to strong total necrosis of the leaves, or to breakdown of the vascular ring extending throughout the tuber. Neither cultivar Teton nor clone ND2042-2 expressed foliar symptoms of the disease, while the rest of the cultivars tested were rather susceptible in foliage. No significant year relationship was found between severity of foliar and tuber symptoms.

In studies by De Boer and McCann (1990), tubers vacuum inoculated with strain CS3 of *C. michiganensis* subsp. *sepedonicus* such as cultivars BelRus, Desirée, Kathadin, Rose Gold, Teton, and Urgenta, were classified in the group with very low frequency of tuber and foliage symptoms; however, symptomless samples tested by IFAS showed bacterial presence, sometimes at a high concentration.

Some cultivars (e.g. Rezerv and Lyubimets) were nonresistant to either strain of the pathogen, whereas other cultivars (e.g. Stolovyj 19 and Gatchinskij) were resistant to one strain and moderately resistant to the other strain of ring rot pathogen in studies conducted by Popkova et al. (2000).

In the European potato database, data for resistance level to bacterial ring rot are given for only 16 taxa. The highest resistance score was achieved by cv. Prof. Wohltmann bred in 1895 in Germany; highly resistant cvs. were the Irish Cobbler bred in 1876 in the USA, Uljanovskij bred in 1943 in Russia, Sotka likewise bred in Russia in 1977, and Lady Rosetta bred in 1988 in the Netherlands.

Wild and Primitive Cultivated Species Given the absence of ER to bacterial ring rot among cultivars, some investigations focused on wild *Solanum* species. A putative source of extremely high resistance to *C. michiganensis* subsp. *sepedonicus* was identified in several accessions of *S. acaule* by Kurowski and Manzer (1992), Ishimaru et al. (1994), and Kriel et al. (1995). The bacteria in symptomless plants previously artificially infected with ring rot agent were tested with IFAS serology in accessions of 15 species including *S. acaule, S. candolleanum, S. gourlayi, S. infudibuliforme,* and *S. vernei.* Some plants showing zero IFAS tests proved to be extremely resistant when repeated inoculations were carried out (Kurowski and Manzer 1992).

Common Scab

Breeding for scab resistance has a relatively long tradition since this trait is closely linked to tuber quality degradation caused by corky lesions. Strongly infected tubers are not marketable at all. The pathogen is soil borne and severity of disease is markedly related to environmental conditions such as higher pH and water deficiency in the soil during the period of tuber initiation. Thus scab-resistant cultivars are the goal of several breeding programs.

Cultivars Highly resistant scab cultivars were bred several decades ago and they have been often used as sources of resistance in breeding programs. Among cultivars resistant to scab mentioned by Ross (1986) there were three old German cultivars—Jubel (introduced in 1908), Hindenburg (introduced in 1916), and Ackersegen (introduced in 1929), or cultivars bred in Great Britain or the USA—King Edward and Pentland Crown (bred in 1902 and 1958 respectively), or Carmen and Cayuga (introduced in the USA in 1946). Jansky (2000) listed several scab-resistant cultivars bred in the USA and Canada in 1990s, e.g. AC Chaleur, AC Novachip, Andover, Genesee, Kranz, Ontario, Pike, and Russet Nugget.

Among 137 cultivars characterized by Swiezynski et al. (1998) of major importance in Europe and North America from two points of view—economical (larger seed area or cultivated in five countries) or genetic (parents of at least five of the presently grown cultivars), five were scored grade 8 for resistance to scab: Hydra, Jubel, Loshitsky, Pentland Crown, and Saturna. Cultivars expressing ER to common scab were not found (Haynes et al. 1997). Kranz and Eide (1941) assumed simple inheritance for resistance to scab based on progenies of Accession 123 X cv. Lookout Mountain; however, later studies by Pfeffer and Effmert (1985) indicated it as a polygenically inherited trait.

Wild and Primitive Cultivated Species Resistance to common scab was identified in *Solanum chacoense* (Dionne and Lawrence 1961), *S. commersonii,* and *S. yungasense* (Hawkes 1990). Resistance to scab was also identified in some accessions of cultivated species *S. phureja* (De Maine et al. 1993). In a three-year evaluation for scab resistance, about 3,000 clones representing 100 accessions of 18 species were assessed, of which Hosaka et al. (2000) identified 300 clones as resistant. The highest selection rate was found for accessions of *S. multidissectum, S. canasense, S. bukasovii,* and *S. kurtzianum.* In the CIP collection, 520 accessions of cultivated germplasm representing Andean cultivars were tested for scab resistance. None of the accessions was found with resistance on a high or moderate level (Huaman and Schmiediche 1999).

STRATEGIES OF BREEDING FOR RESISTANCE

All strategies of breeding for resistance include identification of diseases and their causal agents, establishing their respective screening and detection methods, search for resistance sources both in cultivated and wild germplasm pool, acquiring knowledge on a genetic basis of identified resistance, and application of appropriate methods for efficient transfer of disease resistance into advanced breeding lines (Helgeson 1989; Wastie et al. 1993; Whalen 1991). Application of a particular strategy in breeding programs depends on the effective resistance genes available vis-à-vis prevailing and arising pathogen races (Koppel 2000).

Conventional potato breeding methods are based on sexual gene transfer between crossed plants at both tetraploid and diploid level. The transfer of a desired trait from a particular genetic resource to cultivar level is time consuming, taking up to 25-30 years. Progress in breeding for resistance to bacterial diseases at the tetraploid level using conventional methods is slow because traits are mostly quantitatively inherited, and the tetrasomic genetic nature of cultivated potato is complicated, involving significant genotype × environment interactions of the targeted traits. Recent technological advances enable speeding up the conventional breeding process based on sexual gene transfer through application of marker-assisted selection and molecular-based proper selection of parental forms or multicategorical sensitive diagnostics (Watanabe and Watanabe 2000). Unconventional breeding methods based on asexual gene transfer focus on both somatic hybridization using protoplast fusion of incompatible species and genetic engineering of potato by insertion and expression of foreign genes. Genetic engineering technology may have a significant impact on breeding for resistance against bacterial pathogens in addition to its contribution to widening our understanding of molecular base of resistance.

Resistance Screening Methods

Screening for Resistance to Tuber Soft Rot and Blackleg

As resistance to tuber soft rot and blackleg is not closely related, it is necessary to test for both diseases in breeding for resistance to *Erwinia* spp. Several methods have been used but the ranking of genotypes is dependent on both screening methods and prevailing environmental conditions (Lapwood et al. 1984; Watie and Mackay 1985; Lojkowska and Kelman 1994; Kankila et al. 1994).

Resistance to tuber soft rot is usually assessed by laboratory-based tests. These involve inoculation of whole tubers (Austin et al. 1988), tuber slices (Henniger 1965), or parts of tubers (Taylor et al. 1993) with a suspension containing varying numbers of the bacteria (10^3-10^8 cfu ml^{-1}). Lojkowska and Kelman (1994) reported the tuber slice-inoculation procedure to be variable, but Allefs (1995) and Haynes et al. (1997) found variability was reduced by testing under standardized oxygen concentration and using a large enough sample size. Different inoculation methods can be used, namely tuber inoculation by jet injection (De Boer and Kelman 1978), inoculation with pipette tips when limited numbers of tubers are available (Maher and Kelman 1983), immersion of tubers in bacterial suspension (Tzeng et al. 1990), and deposition of inoculum in holes drilled in the tuber cortex (Bourne et al. 1981). The immersion inoculation method can be improved by vacuum infiltration (Wastie et al. 1994).

Most tests are done under aerobic conditions, or in mist chambers with high relative humidity (RH) ca. 100% (Reeves et al. 1999), which results in low O_2 within the tubers. More extreme test conditions are obtained under anaerobic conditions, e.g. in nitrogen (Bourne et al. 1981; Allefs 1995) or when tubers are dipped in rapeseed oil (Lebecka and Zimnoch-Guzowska, pers. comm.). Under these conditions, only O_2-independent resistance mechanisms are involved, e.g. calcium concentration of cell walls, the level of cell wall pectin esterification (Elphinstone 1994), glycoalkaloid content, and starch content (Subrtova et al. 1993; Zimnoch-Guzowska pers. comm.). The optimum incubation period is closely related to temperature, usually in the range of 20°C to 27°C, and inoculum concentration.

The pipette tip method was proposed as the standard method when evaluating tuber resistance to soft rot, using cultivars Bintje and Desirée as susceptible and intermediate resistant controls respectively (Priou et al. 1992). The method is currently in use at several breeding centers, including INRA (France), SCRI (UK), IHAR (Poland) and CIP (Peru). However, although resistant and susceptible clones are readily identified, ranking of clones falling in between is more variable.

Blackleg resistance is determined on inoculated potted plants in the greenhouse (Munzert 1975; Dobias 1977; Hidalgo and Echandi 1982; Wastie 1984; Pietrak et al. 1988; Zimnoch-Guzowska et al. 1999b) or under field conditions using inoculated seed tubers, when the effect of disease on yield can also be determined (Hossain and Logan 1983; Lapwood and Legg 1983; Lapwood and Read 1986a). Together with resistance of the mother tuber to rotting, stem-based resistance is an important component of overall resistance to blackleg. However, field assessment of resistance is difficult and tends to give inconsistent results (Lapwood and Read,

1986b). The method used to assess blackleg resistance in the greenhouse was described by Munzert (1975) and modified by Pietrak et al. (1988). The base of a stem of a potted four-week-old potato plant grown from a seed piece was inoculated with a toothpick smeared with a culture of *E. carotovora* subsp. *atroseptica* on a solid growth medium. The percentage of plants with visual symptoms of stem rot was recorded three to four weeks after inoculation. Covering plants with polyethylene during the incubation period increases humidity, thus creating more favorable conditions for stem infection (Zimnoch-Guzowska et al. 1999b). The greenhouse test for blackleg resistance was found to give more reliable results than field testing (Dobias 1977; Hossain and Logan 1983; Lojkowska and Kelman 1989), but temperature should not exceed 25°C as infection by *Erwinia carotovora* subsp. *atroseptica* is inhibited.

Resistance of detached stems or leaves to *Erwinia* spp. as a measure of blackleg resistance has been examined. Bisht et al. (1993) and Bains et al. (1999) inoculated fresh cut petioles with *Erwinia carotovora* subsp. *atroseptica* by inserting them into 50 g sterile sand drenched with 17 ml bacterial suspension containing 10^5 cfu ml^{-1} in Magenta jars. Zimnoch-Guzowska et al. (1999b) dipped detached leaves collected from six-week-old plants in a suspension of *E. carotovora* subsp. *atroseptica* containing 10^8 cfu ml^{-1} for 48 or 72 h in a mist chamber at 95% RH and 26°C. Assessment of the rotted area was done visually on a 1-5 scale, where 1 = 85-100% and 5 = 0-5% of leaf area rotted. Although detached leaf assay correlated weakly with the stem base test (r = 0.395*), it was not high enough to be a substitute for stem tests (Zimnoch-Guzowska et al. 1999b).

Screening for Resistance to Ring Rot

The search for resistance to bacterial ring rot has received little attention in recent years because high resistance to this disease has not been found and disease control based on zero tolerance level in certified seed programs in the USA, Canada, and most European countries is satisfactory. Development of more sensitive detection methods (IFAS, PCR), which allow easier differentiation of tolerant genotypes from noninfected planting material, would justify the search for high resistance (Kurowski and Manzer 1992).

Foliar and tuber symptoms of ring rot do not usually correlate positively (Kawchuk et al. 1998). Inoculum dose and environmental factors in addition to genetic resistance are important in determining the extent of disease symptom expression on tubers and foliage (Nelson and Kozub 1987; Westra et al. 1994). Resistance can be assessed by inoculating the mother tubers before planting (Langerfeld and Batz 1992) or testing plants grown in the field from seed pieces injected with *C. michiganensis* subsp. *sepedonicus* (Nelson et al. 1992). Inoculation of seed tubers by cutting with a knife contaminated with a high concentration of ring rot bacteria

(2×10^9 cfu ml^{-1}) simulated rather well the spread of the disease in the field (Nelson et al. 1992; Langerfeld and Batz 1992). Resistance assessment on potato cultivars grown in the field from seed pieces inoculated with *C. michiganensis* subsp. *sepedonicus* by vacuum infiltration was described by De Boer and McCann (1990). Symptoms are assessed on both foliage and progeny tubers. In addition, the second-generation crop can also be checked for disease development (Nelson et al. 1992). Disease incidences in different cultivars varied from year to year and barely correlated between the years ($r = 0.38$). Also, severity of foliage and tuber symptoms were not always similar in each cultivar (Langerfeld and Batz 1992). Detection of latent infection using IFAS, ELISA or PCR assay, is recommended for identifying high resistant genotypes expressing a lower level of latently infected tubers. The zero-tolerance for bacterial ring rot on seed potatoes eligible for certification is an added incentive to breeders to eliminate tolerant genotypes from breeding programs (Langerfeld and Batz 1992; De Boer and Hall 2000a).

Screening for Resistance to Bacterial Wilt

Screening methods used for assessing bacterial wilt resistance rely on measurements of incidence or severity of wilting symptoms. Cultivars were assessed on plants grown in fields naturally contaminated with Race 1, biovar, I or Race 3, biovar 2 and interplanted with susceptible and more resistant cultivars as control (Lopes et al. 1993). Resistant clones or cultivars have been selected by absence of disease symptoms when planted under these conditions; however, assessment of latent infection of stem and tubers of selected stocks has rarely been done in breeding programs (French et al. 1998).

In a three-year field study, Priou et al. (2001) preselected clones resistant and moderately resistant to bacterial wilt based on assessment of wilt symptoms and percentage of rotted tubers. Wilt incidence was assessed for two months at 15-day intervals on a 1-5 grade scale, where 1 = 0% and 5 = 76-100% wilted plants. Clones scored at 2.2 (below 30% wilted plants) were preselected for further evaluation. Weight of rotted tubers was recorded from total tuber yield and additionally tuber latent infection assessed by postenrichment enzyme-linked immunosorbent assay on nitrocellulose membrane (NCM-ELISA) of 30 tubers per clone randomly selected, as described by Priou et al. (1999). Currently available sources of resistance were selected in the CIP early 1990s breeding program based mainly on wilting symptoms; however, symptomless tuber infection was occasionally present.

No correlation between foliage wilting symptoms and latent tuber infection could be estabilished (Ciampi and Sequeira 1980; Priou et al. 2001). Therefore, selection for both traits has to be considered in breeding programs. Latent infection by *Ralstonia solanacearum* was previously detected by methods such as incubation of tubers for 3 weeks at 28°C to stimulate bacterial oozing, and tissue extract inoculated on diagnostic growth medium (Ciampi and Sequeira 1980). Recently developed methods include NCM-ELISA, which has a high sensitivity level ranging from 10^3 cells ml^{-1} (Elphinstone et al. 1996; Weller et al. 2000) to 2-10 cells ml^{-1} (Priou et al. 1999), and postenrichment TaqMan PCR (Elphinstone 1996).

A sampling strategy to estimate the frequency of latently infected tubers was proposed by Priou et al. (2001). Tuber size had no effect on latent infection. Sample size varied, depending on the level of resistance of tested material and testing conditions. In general, samples of ca 30 tubers taken from 10 randomly selected plants were tested for latent infection. The estimated proportion of infected tubers varied between 0.095 and 0.15 for various infection levels (Priou et al. 2001).

Potato material grown from true or botanical seed was tested for resistance by inoculating seedlings grown in trays by dipping in a bacterial suspension (10^8 cfu ml^{-1}) for 10 min in a greenhouse. The incidence of wilted seedlings was monitored for the next two months. The seedling method was recommended by Lima et al. (1996) for early screening in breeding programs.

Screening for Resistance to Common Scab

Both field and greenhouse-based tests have been used to assess common scab resistance. Field screening is usually done in fields with a long history of scab infection (McKee 1963). However, results are often of little value due to variation in rainfall during the critical period of scab development, especially while screening cultivars at different timings for tuber initiation, since infection correlated with soil moisture (Wiersema 1974). Covering test plots with plastic sheets may improve reliability (Booth 1970; Jellis 1974). Schöber (1974) found that field screening could be improved by filling ditches with inoculated sand in fields planted with potato. The uniformity of infection pressure in screening fields was evaluated by increasing the number of replicates, especially of susceptible cultivar controls. Moreover, testing was often done at more than one location, which was shown to have an affect on results (O'Keefe 1965).

Comparing field and greenhouse methods of disease assessment, Miller et al. (1965) showed that the latter is more reliable, in particular when using tubers infected with common scab as a source of inoculum. These tubers were blended and mixed into a bed of vermiculite in which test clones were then planted. In a modification of the method, Wiersema (1970) and Bjor and Roer (1980) recommended selection of highly resistant clones by growing potatoes in naturally contaminated dry sand or soil in 31 pots placed on top of a control irrigated subsoil. In assessing infection level, two parameters were used by Bjor and Roer (1980): (a) lesion severity on an increasing 1-3 scale and (b) relative surface cover in a 1-9 grade scale in which 1 is < 5% surface covered by scab and 9 = 100% diseased surface. Based on these two parameters, a scab index (SI) with a maximum value of 100 could be computed: SI = Lesion Severity × Relative Surface Cover × 100/27. SIs for 17 accessions tested by aforesaid method in pot trials and in contaminated fields correlated highly ($r = 0.82$) (Marais and Vorster 1988). In field studies, Goth et al. (1993) scored lesion severity on a 0-5 scale, where 0 = no scab and 5 = pitted scab of all diameters. Tuber surface area with scab was also estimated on a 0-5 scale, in which 0 = none and 5 = 76-100% surface area affected. A lesion index (LI) and surface area infected (SAI) were calculated as the sum of lesion rating + diseased surface area rating for each tuber taken from each plot respectively, divided by number of tubers and multiplied by the number of plants per plot. The unweighted pair-group method based on arithmetic averages was used to cluster the tested clones and control cultivars to work out clonal mean LI and SAI. Cluster analysis on LI and SAI effectively classified relative resistance of tested clones in relation to known standards, even when conditions for scab development were poor or severe, and data among tested clones were similar (Goth et al. 1995).

Introduction of Resistance by Sexual Crosses in Traditional Breeding Programs

Up-to-date progress in resistance to bacterial pathogens in presently grown cultivars was achieved with conventional techniques in sexual crosses introducing the resistance character into the cultivated genepool. Resistance to bacterial diseases is often listed among the important traits in breeding programs (Mendoza 1987; Tarn et al. 1992; Bradshaw et al. 2000); however, in regions of temperate climate its importance is relatively less than features related to yield quality and resistance to late blight and viruses. Thus in breeding programs elimination of the most susceptible forms, naturally infected, is more often practiced. With global climate warming the importance of genetic control for crop protection against bacterial diseases, mainly those on quarantine lists, will grow rapidly. Bacterial wilt resistance is already a major constraint for the potato crop in hot climates.

Breeding for resistance to bacterial diseases is effected by a few major strategies. The first is introgression of resistant genes identified in potato relatives into the cultivated breeding pool. This is usually accomplished by backcrossing the phenotype selected for resistance against undesired traits to the wild species. However, this is a time-consuming process and 5-6 generations are needed to realize the new cultivar with introgressed trait (Bradshaw et al. 2000). The backcross method is applied mainly for quick transfer of a specific trait from the donor germplasm to a cultivar and has been applied mostly for transfer of qualitatively inherited resistances (Tarn et al. 1992).

Traditional breeding relies almost entirely on phenotypic recurrent selection performed on progenies of parental clones or cultivars with a complementary set of characters. Breeders identify desired recombinants in progenies which combine features of both parents. Expectedly use of MAS should improve the efficiency of these breeding processes.

The strategy of recurrent mass selection includes progeny testing of best crosses from which candidates for new cultivars can be selected. Bacterial disease resistance is not tested in early generations of the breeding cycle. For this purpose appropriate methods of resistance screening have to be elaborated. However, this is handicapped by the need for production of numerous tubers for testing and often, due to significant interactions between genotype and environment (soft rot resistance, common scab resistance) or genotype and pathogen (bacterial wilt resistance), testing has to be repeated for several seasons and/or at various locations.

Resistances to major bacterial diseases in potato are quantitatively inherited. Resistance to common scab is only a simple model of inheritance assumed with high estimates of heritabilities; despite a high significance of genotype × environment interactions (Haynes et al. 1997), which appeared encouraging insofar as breeding for resistance is concerned. Thus representation of scab-resistant cultivars is quite high. This trait is closely related to tuber quality of table cultivars and selection for it continues to be an important factor for breeding programs in Europe and the United States.

Estimates on heritabilities of resistance to soft rot were moderate in various programs (Wolters and Collins 1991) Lebecka and Zimnoch-Guzowska 2001) and significant for both GCA (Lellbach 1978; Lebecka et al. 1998) and SCA (Lebecka et al. 1998). Although ranking of cultivars according to resistance to erwinias correlates strongly with the method of testing for resistance, there is lack of unity among the standards applied. Nonetheless, the European database (*www.europotato.org*) shows a few cultivars released in the last two-three years, classified as highly resistant to blackleg. These cultivars should be approached with caution in future breeding strategies, however.

The program for population breeding developed by Mendoza (1987) at CIP is unique in scale and philosophy. It focused on germplasm improvement for better adaptability to hot and cool environments along with desired combination of pest and disease resistances. For hot climates, heat tolerance was selected together with resistance to bacterial wilt and root-knot nematode. The initial genepool included commercial cultivars, breeding lines, landraces, and a set of wild species with resistance or tolerance to biotic and abiotic factors. Bacterial wilt resistance was one of the most important goals in this program directed toward breeding potatoes for developing countries (Schmiediche 1986). A long-term program initiated in 1974 resulted in selection of clones, progenitors, and populations carrying various traits, such as heat tolerance and resistance to major diseases, which have been introduced into many tropical and subtropical countries where potato is grown under hot and warm seasons. From these materials, new varieties, were released and many clones tested in countries such as Brazil, Kenya, and Uganda. None of the cultivars tested in Kenya after natural and artificial infection with *Ralstonia solanacearum* appeared resistant; however significant differences in bacterial wilt incidence and severity among the tested cultivars were noted. Contrarily, cultivars Kenya Dhamana, Mauritius, and Cruza (CIP-720118) had low bacterial wilt severity as well as incidence and were therefore rated tolerant (Ateka et al. 2001). With the development of more sensitive detection methods, applied to screen-selected resistant breeding lines and cultivars, it was demonstrated that highly resistant cultivars selected during the breeding process lacked symptoms of wilting and, under favorable disease development conditions, might serve as sources of inoculum since latently infected tubers are, in fact, tolerant.

In North Carolina, mass and pedigree selection were conducted in populations of cultivated diploid potatoes *Solanum phureja* and *S. stenotomum* which, after adaptation to long-day conditions, proved valuable sources of starch content and resistance to tuber soft rot (Wolters and Collins 1994).

A similar approach based on mass selection and open pollination was applied by SCRI breeders to create a long-day-adapted population of *S. phureja* resulting in high-yielding progeny from interploid crosses to *S. tuberosum*. Selected diploid clones *S. phureja* expressed high resistance to blackleg in tests under field and lab conditions (Lees et al. 2000).

Utilization of 2n Gametes for Transfer of Resistance from 2x to 4x Level

Diploid potato (2n = 2x = 24) is important in potato breeding for two major reasons: firstly, about 70% wild and cultivated primitive germplasm of *Solanum*, the reservoir of traits valuable for breeders, is represented by diploid species, and secondly, prebreeding at the diploid level is the best possible way for introgressing desired important traits into the breeding genepool. Several *Solanum* species identified as possible sources of resistance to bacterial diseases are diploid: *S. chacoense, S. phureja, S. tarijense, S. yungasense,* etc. Several negative characters are frequently present in the wild species, and hence these sources are mainly used in prebreeding programs for intercrossing to cultivated *S. tuberosum*. There are other cogent reasons for realization of a prebreeding program at the diploid rather than tetraploid level, such as: operating rules of disomic inheritance, relatively easy haploidization of cultivars and tetraploid clones, relatively good crossability between *Solanum* species and dihaploids of *S. tuberosum*, possible utilization of 2n gametes for transfer of genetic potential of bred diploids on the tetraploid level with limited recombination and lastly, the often encountered effect of selected traits in tetraploid progenies of interploid crosses: 2x-4x, 4x-2x or 2x-2x, in which both diploid parents produce 2n gametes (Chase 1963; Mok and Peloquin 1975; Iwanaga 1982; Zimnoch-Guzowska and Dziewonska 1989).

For over 30 years diploids have been bred at Agriculture and Agri-Tool Canada at Fredericton. This program has sought improvement of parents for cultivar breeding using *Solanum* species for better adaptation, processing quality, resistance to diseases and 2n gamete formation (De Jong and Tai 1977). In the early 1990s high resistance to common scab (*S. scabies*) was transmitted from 2x clones to 4x clones using 4x-2x crosses (Murphy et al. 1995).

Introgression of resistance to soft rot and blackleg from *S. stenotomum* and *S. phureja* into the tetraploid breeding pool was reported by Iwanaga (1985), when diploids selected at North Carolina State University, USA having 2n female gametes were used in 4x-2x crosses as pollinators.

Since the early 1970s a prebreeding program at the diploid level aimed at introgression into cultivars for breeding valuable genes from the *Solanum* genepool has been underway at the Plant Breeding and Acclimatization Institute, Mlochow Research Center, Poland. Diploid breeding has focused on improvement of agronomic characteristics in diploid hybrid clones bearing desired resistance against the main potato pathogens (Zimnoch-Guzowska and Dziewonska 1989). Groups of diploid clones with a high level of resistance to tuber soft rot and blackleg have been identified. Their pedigree comprised *Solanum chacoense, S. yungasense,* and *S. phureja,* known sources of resistance to *Erwinia* spp. (Zimnoch-Guzowska and Lojkowska 1993; Zimnoch-Guzowska et al. 1999a). Diploid hybrids able to produce 2n male gametes and possessing high resistance to tuber soft rot and blackleg were used in 4x-2x crosses as pollen parents. After a two-year evaluation of 24 tetraploid progenies from these crosses, over 130 tetraploid clones could be selected whose tuber resistance to soft rot had increased by 1 or 2 grades (in a 1 to 9 scale) compared to cultivars presently grown in Poland from the most resistant groups, possessing agronomic traits at a level acceptable for cultivar breeding (Lebecka and Zimnoch-Guzowska, pers. comm.).

Carputo et al. (1995, 1997) and Barone et al. (2001) used a breeding scheme which involved ploidy level and EBN manipulations in order to overcome the interspecific barriers existing between cultivated *Solanum tuberosum* and *S. commersonii (cmm),* a wild 1 EBN species identified as a source of resistance to *Erwinia* spp. In their five-step work on in vitro chromosome doubling, *S. commersonii* was crossed to 2x hybrid of *S. phureja*, producing 3x F_1 hybrids. F_1 hybrids able to form 2n eggs were backcrossed to *S. tuberosum* (3x-4x), resulting in 5xBC_1 progeny. Second (5x-4x) and third (4x-4x) backcrosses produced 4x progeny in which the presence of the wild genome was confirmed by *cmm*-specific RFLP and AFLP markers. Among those tested for soft rot resistance, BC_2 and BC_3 clones in the five most resistant forms showed the presence of *cmm*-specific markers in the range of 23-30%. This is an example of successful introgression of resistance genes through backcrossing and use of 2n gametes from sexually incompatible species into cultivated *S. tuberosum*.

Capo et al. (2002) used 2x-4x crosses to transfer good chipping ability and resistance to soft rot to the tetraploid level from diploid hybrid CP6 originating from the cross dH Chippewa (W 18131) × *S. chacoense* (PI 230582) × *S. phureja* (1.22). The clone CP6 produced 2n eggs by SDR mechanisms. The tetraploid parent *S. tuberosum* clone V2 was susceptible to tuber soft rot. After a two-year primary evaluation using the tuber point inoculation method, of the 15 tetraploid clones that resulted from the 2x-4x cross, only one clone, CPG 9, was classified as resistant and three others evaluated, similar to cv. Spunta, as partially resistant.

Qu et al. (1996) compared the efficiency between diploid and tetraploid clones in 4x-2x, 2x-2x, and 4x-4x crosses with respect to transmission of resistance to *Ralstonia solanacearum*. The highest frequency of resistant forms was found in progenies from the 4x-2x crosses.

At the CIP in a scheme designed for a long-term breeding program directed toward improvement of resistance to bacterial wilt, resistance sources from *Solanum phureja*, *S. sparsipilum*, *S. raphanifolium*, and *S. chacoense* were intercrossed to enhance the agronomic value of populations of diploid cultivated species *S. stenotomum*, *S. phureja*, and *S. goniocalyx*. Subsequently, their hybrids were used as pollen parents (with 2n gametes) and advanced tetraploid clones/cultivars as seed parents in 4x-2x crosses. After laboratory and field screening, the selected clones expressed high and moderate levels of resistance to bacterial wilt, notably superior to levels achieved by *S. phureja*-derived cultivars Molina and Cruza 148, varieties long considered less susceptible to this disease (Priou et al. 2001).

Resistance to bacterial wilt found in tuber-bearing *Solanum* species was transferred by Watanabe et al. (1992) via a diploid breeding program into the tetraploid level by 4x-2x crosses using resistant diploid parents which produced FDR 2n pollen. Some 4x-2x families from resistant diploid genotypes demonstrated a high survival rate after inoculation with *Ralstonia solanacearum*. The influence of female × male interaction effect on the evaluated progeny was assessed and left no doubt that proper selection of 4x and 2x parents is of paramount importance in obtaining a high frequency of transmission of resistance to bacterial wilt in the progeny.

Protoplast Fusion

Wild *Solanum* species are a valuable source of desirable agronomic characteristics for the development of potato cultivars resistant to plant pathogens. However, a large number of species have not been wholly utilized due to sexual incompatibilities between gametic cells of *S. tuberosum* and wild *Solanum* species. A potential alternative is interspecific somatic hybridization between sexually incompatible species. Two factors are essential for incorporation of somatic hybrids into breeding programs: full expression of the traits and fertility in somatic hybrids enabling subsequent sexual transfer of the desirable traits.

For the last twenty years somatic hybrids between sexually incompatible *Solanum* species have been obtained by several research groups (Butenko and Kuchko 1980; Melchers et al. 1978; Binding et al. 1982; Barsby et al. 1984; Austin et al. 1985; Puite et al. 1986). In 1981, Evans et al. described the potential to obtain disease resistant lines of *Nicotiana* through somatic hybridization of sexually imcompatible lines. Research on the introduction of resistance to plant pathogenic bacteria in potato cultivars through somatic hybridization was limited (Austin et al. 1988; Rokka et al. 1994; Carputo et al. 1997; Laferriere et al. 1999; Fock et al. 2000, 2001; Gang et al. 2000; Ahn et al. 2001). This is associated with the fact that the genetic and molecular bases of resistance to the most important bacterial pathogens of potato has not been fully investigated. Moreover, the mechanisms of transmission of desirable traits of somatic hybrids are not clearly understood. Also, well-documented data on wild *Solanum* species that might serve as a source of resistance, tolerance or immunity to plant pathogenic bacteria are limited.

Tubers in hexaploid somatic hybrids between diploid wild species *S. brevidens* and a tetraploid *S. tuberosum*, expressed a high level of resistance to soft rot caused by *Erwinia carotovora* subsp. *atroseptica*, *E. carotovora* subsp. *carotovora*, and *E. chrysanthemi* (Austin et al. 1988). Furthermore, some of the hybrids were fertile and pentaploid progeny obtained from sexual crossing between these hexaploid hybrids and the cultivar Katahdin (tetraploid) also had a high level of resistance to soft rot. This was the first evidence showing that resistance to bacterial disease can be successfully transferred following somatic hybridization. Despite several years of research on the mechanisms determining the somatic hybrid resistance to soft rot, the phenomenon has not yet been clarified. It was indicated that resistance originating from *S. brevidens* results from a structural modification of cell wall pectin (McMillan et al. 1993; Dorel et al. 1996) or from a very fast response to the infection and quick production of wound periderm (Austin et al. 1988; McGrath et al. 2002). Application of RFLP and RAPD analysis indicated that somatic hybrids contained at least one copy of each chromosome from each parent. Futhermore, loss of some *S. brevidens* DNA in the followed sexual progeny was explained by intergenomic pairing and recombination rather than loss of whole chromosome during meiosis (McGrath et al. 2002). This permits the conclusion that somatic hybridization may be a useful way for overcoming problems of sexual incompatibility and introducing stable resistance genes from wild species into new breeding lines or potato cultivars.

During the last few years other groups have also used *S. brevidens* as a potential source of resistance to soft rot and blackleg. In most of the studies, this wild, nontuber-forming species was confirmed as a stable source of resistance to pectinolytic *Erwinia* (Kankila et al. 1995; Rokka et al. 1994, 2000).

Somatic hybrids of *S. tuberosum* haploids and *S. commersonii* also present a high level of resistance to soft rot and blackleg caused by *E. carotovora* subsp. *atroseptica* and *E. carotovora* subsp. *carotovora* (Carputo et al. 1997, 2000). However, in the progeny derived from sexual crosses of somatic hybrids and *S. tuberosum*, average resistance was much reduced compared with the somatic parental hybrids and segregation for this trait was observed. The authors drew the hypothesis that normal meiosis in somatic hybrids caused disruption of chromosomes and reassortment of alleles (Carputo et al. 2000).

Somatic hybridization was also an efficient tool in the transfer of resistance to bacterial wilt caused by *Ralstonia solanacearum* to potato breeding lines. Plants from three sets of somatic hybrids, obtained in different laboratories, might be useful as sources of bacterial wilt resistance in potatoes. The somatic hybrid lines of *S. commersonii* and *S. tuberosum*, expressed disease resistance level after inoculation with Race 3, similar to that of the resistant *S. commersonii* parental line (Laferriere et al. 1999). The somatic hybrids were male and female fertile and could be crossed with *S. tuberosum* to produce viable seeds. The results obtained indicated that somatic hybridization rapidly provided material which could be crossed directly with the cultivated potato, without the use of bridge species.

Two other examples of the application of somatic hybridization for introduction of resistance to bacterial wilt in potatoes were presented by Fock et al. (2000, 2001). Most of the tetraploid somatic bybrids of *Solanum tuberosum* and *S. phureja* were tolerant to Race 1 and susceptible to Race 3 strain of *Ralstonia solanacearum* (Fock et al. 2000). However, the amphiploid hybrid clone showed a good tolerance to both races. Even more promising are somatic hybrids of *S. stenotomum* and *S. tuberosum*. All tested clones showed a resistance level as high as the wild parental line (Fock et al.

2001). Studies published during the last few years are encouraging and suggest that stable wilt resistance could be introduced to somatic hybrids and later to the commercial potato cultivars.

Transgenesis

Over 20 years old now, genetic engineering in potato is relatively quick these days compared to the first protocols of plant transformation. The early commercialized transgenic cultivars were mainly concerned with improved insect resistance, herbicide tolerance, and/or virus resistance. Conventional breeding for bacterial resistance has not achieved appreciable progress because several difficulties are encountered due to the complex genetic background of host resistance and strong interactions between the host, pathogen, and the environment. The main bacterial diseases of potato (soft rot, blackleg) are expressed by complex quantitative factors with low frequency of heritability. Novel bacterial resistances governed by alien genes will expectedly be monogenically inherited in a simple Mendelian manner when transgenics are used as parental lines for sexual gene transfer in future breeding processes. Thus expectations for application of novel sources of resistance with quick introduction into cultivated breeding material appear rather high. Whalen's statement (1991) is still valid, namely that genetic engineering technologies are likely to have a significant impact on breeding for resistance against bacterial pathogens, in addition to their important contribution to understanding the underlying molecular basis of the resistance mechanism. However, it seems obvious that in the last decade transgenics have contributed to better understanding of the functioning of resistance at the molecular level compared to conventional breeding for resistant germplasm.

The first strategy for creation of transgenic bacterial resistant potato applied in the last decade was based on introducing genes of antimicrobial proteins of various origin.

Two major insect genes were transferred to plants to enhance their resistance to bacterial or fungal diseases. Cecropin B and attacin E-antibacterial peptides isolated from the giant silk moth (*Hyalophora cecropia*) were widely used (Casteels et al. 1989; Montanelli and Nascari 1989, 1991; Destefano-Beltran et al. 1990, Allefs 1995). Strong antimicrobial activity of cecropins *in vitro* against pathogenic bacteria, including *Erwinia* spp. has been reported frequently (Destefano-Beltran et al. 1990; Nordeen et al. 1992; Mill and Hammerschlag 1993). No cecropin B peptide could be detected in transgenic plants. In field experiments, transgenic potatoes did not reveal efficient protection from bacterial diseases caused by *Ralstonia solanacearum* and *Erwinia* spp., however, preliminary results in laboratory testing were conflicting. Transgenic plants of tobacco (with Shiva-1 gene-synthesized peptide having properties similar to natural cecropin B) expressed delayed symptoms of bacterial wilt and lower mortality after infection with *R. solanacearum* compared to transgenic control plants (Jaynes et al. 1993). Allefs (1995) did not report enhanced tuber soft rot resistance to *E. carotovora* subsp. *atroseptica* and *E. chrysanthemi* on testing 48 transgenic clones with gene-encoding cecropin B derived from transformation of five cultivars. Contradictory results were obtained by Arce et al. (1999), when ca. 20% transgenic clones, having genes encoding attacin protein or synthetic cecropin SB-37 peptide, expressed increased resistance to *E. carotovora* subsp. *atroseptica* with reduced severity of blackleg or soft rot symptoms in the laboratory. The lack of resistance in field experiments was explained as due to little difference in lethal concentraiton of cecropin for bacteria in plant cells (Nordeen et al. 1992), or rapid degradation of cecropin B by proteolytic enzymes in plant cells (Florack et al. 1995; Allefs 1995).

Allefs (1995) tested transgenic potatoes carrying sequences of other antimicrobial peptides. Three cultivars—Bintje, Karnico and Kondor—were transformed with four constructs encoding different precursors of tachyplesin I protein (TPNI) under control of CaMV 35S promotor. TPNI protein was purified from Southeast Asian horseshoe crabs (*Tachypleus tridentatus*) which expressed strong antimicrobial activity *in vitro*. Some transgenic clones expressing TPNI peptide (having additional sequence of hordothionin signal peptide) showed less maceration of tuber tissue by *E. carotovora* subsp. *atroseptica* than control under aerobic conditions. Under anaerobic conditions results were less consistent even in clones positively identified in aerobic conditions. Potato transformed with a alpha-hordothionin precursor from barley showed a high level of expression in transgenic clones for the protein not associated with enhanced resistance to tuber soft rot upon testing both with *E. carotovora* subsp. *atroseptica* and *E. chrysanthemi* under aerobic as well as anaerobic conditions (Allefs 1995).

Lysozyme proteins are known for their antibacterial activity because of their lytic activity against peptidoglycan bacterial cell wall. The gene coding for bacteriophage T_4 lysozyme protein was used for potato transformation and partial resistance to *E. carotovora* expressed as evident from reduction in tuber tissue maceration by bacteria (Düring et al. 1993; De Vries et al. 1999). Serrano et al. (2000) reported enhanced resistance to *E. carotovora* subsp. *atroseptica* in over 20% of transgenic clones of cv. Desirée. The clones were transformed with chicken lysozyme chimeric gene and assessed in greenhouse assays for soft rot and blackleg symptoms. In a set of 69 transgenic clones tested, the level of transgene expression correlated with the level of resistance.

Exploring plant defense mechanisms, Wu et al. (1995) found significantly enhanced tuber resistance to *E. carotovora* subsp. *carotovora* under aerobic and anaerobic conditions in transgenic potato clones with a fungal gene encoding H_2O_2-generating glucose oxidase.

Pectate lyase (PL) enzymes degrade plant cell wall pectin into unsaturated oligogalacturonates which are known elicitors of plant defense responses. Therefore, a gene of isoenzyme PL3 of *E. carotovora* subsp. *atroseptica* was used by Wegener (2002) for transformation of cultivar Desirée. Four-year field testing of four PL-transgenic potato lines showed that the heterologous PL enzyme mediated an enhanced resistance to soft rot in field-grown tubers compared to the nontransgenic control. In addition, the threshold density of bacteria causing progressive soft rot was up to 19-fold higher on tuber tissue containing the PL enzyme. An induction of plant defense responses in PL-transgenic potatoes might be indicated by an increased resistance of tuber tissue cell walls to *Erwinia*-derived enzymes, an increased PPO and PAL-activity in tuber tissue, as well as by formation of necrosis on the wound surface of tubers after infection with bacteria.

A promising approach for the development of novel resistance to bacterial diseases in potato is the appliction of a programmed cell death system (Ghislain and Golmirzaie 1998). It was proposed to apply an *avr-*R gene cassette, which could be engineered by using the existing system *avr9-cf9* (Hammond-Kosack and Jones 1996) or the *avr Pto-Pto* (Thilmony et al. 1995), the latter functioning in heterologous plant species such as tobacco. The expression of *avrPto* could be modulated by promoter genes induced during the hypersensitive reaction of *R. solanacearum* resulting from incompatible plant/pathogen interactions (Pontier et al. 1994).

Recent achievements in molecular biotechnology have made possible obtaining and modifying genes useful for generating disease-resistant crops. One possible strategy is an overexpression of antimicrobial peptides in transgenic plant. Antimicrobial peptides—thionins, defensins, hevin- and knottin-like peptides, and snakins—have long been thought to play a key role in plant defense, both as part of constitutive, developmentally regulated defense barriers and as components of defense responses induced upon infection (Garcia-Olmedo et al. 1998). The important role of this group of peptides was proved: they have an antimicrobial activity *in vitro*, the higher concentration of them observed in resistant plant-microbe interaction. Recent papers by several research groups have presented molecular evidence of the important role of antimicrobial peptides in plant defense. Overexpression of some peptides enhances plant tolerance to pathogens (Molina and Garcia-Olmedo 1997; Titarenko et al. 1997). Construction of peptide-sensitive mutants showed that their virulence toward plant tissue in which these peptides are present, is very low (Titarenko et al. 1997).

A new strategy for obtaining pathogen-resistant plants is an induction of key plant defense pathways through signaling molecules (e.g. salicylic acid, jasmonic acid or ethylene). Recently, Asai et al. (2002) described an additional signaling pathway, the flagellin mitogen-activated protein kinase (MAPK) cascade, that has considerable potential for engineering crops with broad-spectrum resistance against bacterial and fungal pathogens. MAPK are intracellular mediators of information because they shuttle between cytoplasm and nucleus, and their targets are transcription factors. Asai et al. (2002) further demonstrated that the targets of the MAPK pathway are two plant-specific transcription factors of the WRKY family. One could conclude that direct engineering of crop plants with variants of active MAPK or WRKY transcription factors will make possible obtaining a modified signaling pathway that confers resistance to various pathogens. This strategy enables the creation of crop plants resistant to a wide range of pathogens.

Another promising strategy for obtaining pathogen-resistant plants is an expression of antibodies in the transgenic plant. It was shown that polyclonal and monoclonal antibodies can neutralize virus, bacteria, and selected fungi. This approach has been improved by the development of recombinant antibodies (Schillberg et al. 2001). Potato resistance to bacterial pathogens can be engineered by the expression of a pathogen-specific antibody or recombinant antibody. However, successful use of antibodies to generate plant pathogen resistance relies on appropriate target selection, antibody design, efficient antibody expression, stability, and targeting appropriate cellular components.

PROSPECTUS FOR DEVELOPING RESISTANCE TO BACTERIAL DISEASES IN POTATO

Further progress in enhanced resistance to bacterial pathogens in new potato cultivars is expected in the near future, given the new valuable knowledge and tools made available to potato breeders in the last two decades, such as: (i) advances in pooling information on available resources of resistance among potato cultivars and their wild relatives; (ii) elaboration of several new sensitive and sophisticated detection methods for distinguishing highly resistant forms from tolerant ones; (iii) continual research on molecular and biochemical aspects of host-pathogen interaction; (iv) obtaining primary data on localization of resistance genes and their mode of operation in the potato genome using DNA markers in marker-assisted selection (MAS); and (iv) besides conventional sexual crosses, applying new methods of genetic improvement of potato using somatic hybridization and transgenic approaches.

Modern potato cultivars are selected for more than 50 traits as required by various markets: processors, supermarkets, sustainable farmers, and seed growers. All these traits are expected to be of the highest grade possible. To create a new cultivar with durable high resistance to bacterial diseases or to quarantine bacterial diseases will become possible only when all the genepools created by conventional breeding programs have been supplemented with new genepools created by molecular breeding.

REFERENCES

Ahn, Y.K., J.C. Kang, H.Y. Kim, S. D. Lee, and H.G. Park. 2001. Resistance to blackleg and tuber soft rot in interspecific somatic hybrids between *S. brevidens* and *S. tuberosum* ("*Superior*". "Dejima" and dihaploid of "Superior"). J Korean Soc Hortic Sci 42: 430–434.

Akeley, R.V., F.J. Stevenson, P.T. Blood, E.S. Schultz, R. Bonde, and K.F. Nielsen. 1955. Merrimack. A new variety of potato resistant to late blight and ring rot and adapted to New Hampshire. Amer Potato J 32: 93.

Alam, Z. 1972. Inheritance of scab resistance in 24-chromosome potatoes. PhD thesis, Univ, Wisconsin, Madison, WI, USA.

Aldon, D., B. Brito, Ch. Boucher, and S. Genin. 2000. A bacterial sensor of plant cell contact controls the transcriptional induction of *Ralstonia solanacearum* pathogenicity genes. Amer J Potato Res 19: 2304–2314.

Alfano, J.R. and A. Collmer. 1996. Bacterial pathogens in plants: Life up against the wall. Plant Cell 8: 1683–1698.

Allefs, S. 1995. Resistance to *Erwinia* spp. in potato (*Solanum tuberosum* L.). PhD thesis, Landbouwuniversiteit Wageningen, Netherlands.

Araud-Razou, I., J. Vasse, H. Montrozier, Ch. Etchebar, and A. Trigalet. 1998. Detection and visualization of the major acidic exopolysaccharide of *Ralstonia solanacearum* and its role in tomato root infection and vascular colonization. Europ J Plant Path 104: 795–809.

Arce, P., M. Moreno, M. Gutierrez, M. Gebauer, P. Dell'Orto, H. Torres, I. Acuna, P. Oliger, A. Vanegas, X. Jordana, J. Kalazich, and L. Holuique. 1999. Enhanced resistance to bacterial infection by *Erwinia carotovora* subsp. *atroseptica* in transgenic potato plants expressing the attacin or the cecropin SB-37 genes. Amer J Potato Res 76: 169–177.

Asai, T., G. Tena, J. Plotnikova, M.R. Willmann, W.L. Chiu, L. Gomez-Gomez, T. Boller, F.M. Ausubel, and J. Sheen. 2002. MAP kinase signaling cascade in *Arabidopsis* innate immunity. Nature 415: 977–983.

Ateka, E.M., A.W. Mwang'ombe, and J.W. Kimenju. 2001. Reaction of potato cultivars to *Ralstonia solanacearum* in Kenya. African Crop Sci J 9, 1: 251–256.

Austin, S., M. Baer, H. Ehlenfeldt, P.J. Kazimierczak, and J.P. Helgeson. 1985. Interspecific fusion in *Solanum tuberosum*. Theor Appl Genet 71: 172–176.

Austin, S., E. Lojkowska, M.K. Ehlenfeldt, A. Kelman, and J.P. Helgeson. 1988. Fertile interspecific somatic hybrids of *Solanum*: a novel source of resistance to *Erwinia* soft rot. Phytopath 78: 1216–1220.

Baer, D. and N.C. Gudmestad. 1993. Serological detection of nonmucoid strains of *Clavibacter michiganensis* subsp. *sepedonicus* in potato. Phytopath 83: 157–163.

Bains, P.S., V.S. Bisht, D.R. Lynch, L.M. Kawchuk, and J.P. Helgeson. 1999. Identification of stem soft rot (*Erwinia carotovora* subsp. *atroseptica*) resistance in potato. Amer J Potato Res 76: 137–141.

Bain, R.A., M.C.M. Pérombelon, L. Tsror, and A. Nachmias. 1990. Blackleg development and tuber yield in reaction to numbers of *Erwinia carotovora* subsp. *atroseptica* on seed tubers. Plant Path 39: 125–133.

Barone, A., A. Sebastiano, D. Carputo, F. Della Rocca, and L. Frusciante. 2001. Molecular marker-assisted introgression of the wild *Solanum commersonii* genome into the cultivated *S. tuberosum* gene pool. Theor Appl Genet 102: 900–907.

Barras, F., F. Gijsegem, and A.K. Chatterjee. 1994. Extracellular enzymes and pathogenesis of soft-rot *Erwinia*. Ann Rev Phytopath 32: 201–234.

Barsby, T., J. Shepard, R. Kemble, and R. Wong. 1984. Somatic hybridization in the genus *Solanum*: *S. tuberosum* and *S. brevidens*. Plant Cell Rep 3: 1165–1167.

Basham, H.G. and D.F. Bateman. 1975. Relationship of cell death in plant tissue treated with a homogeneous endopectate lyase to cell wall degradation. Physiol Plant Path 5: 249–262.

Bauer, D.W., A.J. Bogdanove, S.V. Beer, and A. Collmer. 1994. *Erwinia chrysanthemi hrp* genes and their involvement in soft rot pathogenesis and elicitation of the hypersensitive response. Molec Plant Micr Interact 7: 573–581.

Beaulieu, C., M. Boccara, and F. van Gijsegem. 1993. Pathogenic behaviour of pectinase-defective *Erwinia chrysanthemi* mutants on different plants. Molec Plant Micr Interact 6: 197–202.

Bentsink, L., G.O. Leone, J.R. van Beckhoven, H.B. van Schijndel, B. van Gemen, and J.M. van der Wolf. 2002. Amplification of RNA by NASBA allows direct detection of viable cells of *Ralstonia solanacearum* in potato. J Appl Microbiol 93: 647–655.

Bicamumpaka, M. and A. Devaux. 1984. Programme de selection au Rwanda pour l'otention des varieties resistantes au mildious (*P. infestans*) et a la bacteriose (*P. solanacearum*). *In:* Abstract 9th Triennial Conf. EAPR, pp. 330–331.

Binding, H., S. Jain, J. Finger, G. Mordhorst, R. Nehls, and J. Gressel. 1982. Somatic hybridization of an atrazine resistant biotype of *Solanum nigrum* and *Solanum tuberosum*. Theor Appl Genet 63: 273–277.

Bisht, V.S., P.S. Bain, and J.R. Letal. 1993. A simple and efficient method to assess susceptibility of potato to stem rot by *Erwinia carotovora* subspecies. Amer Potato J 70: 611–616.

Bjor, T. and L. Roer. 1980. Testing the resistance of potato varieties to common scab. Potato Res 23: 33–47.

Boccara, M., A. Diolez, M. Rouve, and A. Kotoujansky. 1988. The role of individual pectate-lyase of *Erwinia chrysanthemi* strain 3937 in pathogenicity of *Saintpavlia* plants. Physiol Molec Plant Path 33: 95–104.

Bonde, R., F.J. Stevenson, and R.V. Akeley. 1947. Breeding potatoes for resistance to ring rot. Phytopath 37: 539–555.

Booth, R.H. 1970. Testing varietal reaction of potatoes to common scab (*Streptomyces scabies*) under control condition. J Natl Inst Bot 12: 119–123.

Boudazin, G., A.C. le Roux, K. Josi, P. Labarre, and B. Jouan. 1999. Design of division specific primers of *Ralstonia solanacearum* and application to the identification of European isolates. Europ J Plant Path 105: 373–380.

Bourne, W.F., D.C. McCalmont, and R.L. Wastie. 1981. Assessing potato tubers for susceptibility to bacterial soft rot (*Erwinia carotovora* subsp. *atroseptica*). Potato Res 24: 409–415.

Bradshaw, J.E., A.K. Lees, and H.E. Stewart. 2000. How to breed potatoes for resistance to fungal and bacterial diseases. Plant Breed Seed Sci 44, 2: 3–20.

Bramwell, P.A., P. Wiener, A.D.L. Akkermans, and E.M.H. Wellington. 1998. Phenotypic, genotypic and pathogenic variation among *Streptomyces* implicated in common scab disease. Lett. Appl Microbiol 27: 255–260.

Broekaert, W.F., B.P.A. Cammue, M.F.C. de Bolle, K.Thevissen, G.W. de Samblanx, and R.W. Osborne. 1997. Antimicrobial peptides from plants. Crit Rev Plant Sci 16: 297–323.

Brown, S.E., A.A. Reilley, D.L. Knudson, and C.A. Ishimaru. 2002. Genomic fingerprinting of virulent and avirulent strains of *Clavibacter michiganensis* subspecies *sepedonicus*. Curr Microbiol 44: 112–119.

Bushkova, L.N. 1983. Resistance o f potato varieties to the blackleg pathogen. Potato Abstracts 1986, 11: 70.

Bushkova, L.N. and N.M. Vlasov. 1983. Resistance of potato to *Erwinia atroseptica* [abstr.]. Kartofel' i Ovosci 11: 17.

Butenko, R. and A. Kuchko. 1980. Somatic hybridization of *Solanum tuberosum* L. and *S. chacoense* Bitt. by protoplast fusion. *In*: L. Ferenczy and GL Farkas (eds), Advances in Protoplast Research. Akad. Kiado, Budapest, pp. 146–154.

Capo, A., M. Cammareri, D. Rocca, A. Errico, A. Zoina, and C. Conicella. 2002. Evaluation for chipping and tuber soft rot (*Erwinia carotovora*) resistance in potato clones from unilateral sexual polyploidization ($2x \times 4x$). Amer J Potato Res 79: 139–145.

Carputo, D., T. Cardi, L. Frusciante, and S.J. Peloquin. 1995. Male fertility and cytology of triploid hybrids between tetraploid *Solanum commersonii* (2n = 4x = 48) and Phureja-Tuberosum haploid hybrids (2*n* = 2*x* = 24, 2EBN). Euphytica 83: 123–129.

Carputo, D., B. Basile, T. Cardi, and L. Frusciante. 2000. *Erwinia* resistance in backcross progenies of *Solanum tuberosum* × *S. tarijense* and *S. tuberosum* (+) *S. commersonii* hybrids. Potato Res 43: 135–142.

Carputo, D., T. Cardi, M. Speggiorin, A. Zoina, and L. Frusciante. 1997. Resistance to blackleg and tuber soft rot in sexual and somatic interspecific hybrids with different background. Amer Potato J 74: 161–172.

Caruso, P., M.T. Gorris, M. Cambra, J.L. Palomo, L. Collar, and M.M. Lopez. 2002. Enrichment double-antibody sandwich indirect enzyme-linked immunosorbent assay that uses a specific monoclonal antibody for sensitive detection of *Ralstonia solanacearum* in asymptomatic potato tubers. Appl Environ Microbiol 68: 3634–3638.

Casteels P., C. Ampe, F. Jacobs, M. Vaeck, and P. Tempst. 1989. Apidaecins: antibacterial peptides from honeybees. EMBO J 8: 2387–2391.

Chakrabarti, S.K., A.V. Gadewar, J. Gopal, and G.S. Shekhawat. 1994. Performance of tetraploid×diploid (TD) crosses of potato for bacterial wilt resistance in India. Bacterial Wilt Newsl 10: 7.

Chase, S.S. 1963. Analytic breeding in *Solanum tuberosum* L.—A scheme utilizing parthenotes and other diploid stocks. Can J Genet Cytol 5, 4: 359–363.

Ciampi, L. and L. Sequeira. 1980. Multiplication of *Pseudomonas solanacearum* in resistant potato plants and the establishment of latent infections. Amer Potato J 57: 319–329.

Ciampi-Panno, L. and A. Andrade-Soto. 1984. Preliminary evaluation of bacterial soft rot resistance in native Chilean potato clones. Amer Potato J 61: 109–112.

Cipar, M.S. and C.H. Lawrence. 1972. Scab resistance of haploids from two *Solanum tuberosum* cultivars. Amer Potato J 49: 117–120.

Clark, C.F., F.J. Stevenson, and L.A. Schaal. 1938. The inheritance of scab resistance in certain crosses and selfed lines of potato. Phytopath 28: 878–890.

Collmer, A. and N.T. Keen. 1986. The role of pectic enzymes in plant pathogenesis. Ann Rev Phytopath 24: 383–409.

Conn, K.L., E. Leci, G. Kritzman, and G. Lazarovits. 1998. A quantitative method for determining soil populations of *Streptomyces* and differentiating potential potato scab-inducing strains. Plant Dis 82: 631–638.

Cook, D. and L. Sequeira. 1991. Genetics and biochemical characterization of a *Pseudomonas solanacearum* gene cluster required for extracellular polysaccharide production and for virulence. J Bacteriol 173: 1654–1662.

Cook, D., E. Barlow, and L. Sequeira. 1989. Genetic diversity of *Pseudomonas solanacearum*: detection of restriction fragment length polymorphism with DNA probes that specify virulence and hypersensitive response. Molec Plant Micr Interact 2: 113–121.

Cook, D., E. Barlow, and L. Sequeira. 1991. DNA probes as tools of the study of host-pathogen evolution: the example of *Pseudomonas solanacearum*. *In*: H. Henneke and D.P.S. Verma (eds.), Advances in Molecular Genetics of Plant-Microbe Interaction. Kluwer Acad. Publ., Dordrecht, Netherlands, pp. 103–108.

Cullen, D.W., A.K. Lees, I.K. Toth, K.S. Bell, and J.M. Duncan. 2000. Detection and quantification of fungal and bacterial potato pathogens in plants and soil. EPPO/OEPP Bull 30: 485–488.

Dangl, J.L., R.A. Dietrich, and M.H. Richberg. 1996. Death don't have no mercy: cell death programs in plant-microbe interactions. Plant Cell 8: 1793–1807.

Darrasse, A., S. Priou, A. Kotoujansky, and Y. Bertheau. 1994. PCR and Restriction Fragment Length Polymorphism of a *pel* gene as a tool to identify *Erwinia carotovora* in relation to potato diseases. Appl Environ Microbiol 60: 1437–1443.

Darvill, A., C. Bergmann, F. Cervone, G. de Lorenzo, K.S. Ham, M.D. Spiro, W.S. York, and P. Albersheim. 1994. Oligosaccharins involved in plant growth and host-pathogen interactions. Biochem Soc Symp 60: 89–94.

De Boer, S.H. 1990. Control of bacterial ring rot. *In*: G. Boiteau, R.P. Singh, and R.H. Parry (eds.), Potato Pest Management in Canada. Fredericton, New Brunswick, Canada, pp. 242–253.

De Boer, S.H. and A. Kelman. 1978. Influence of oxygen concentration and storage factors on the susceptibility of potato tubers to soft rot. Potato Res. 21: 65–80.

De Boer, S.H. and S.A. Slack. 1984. Current status and prospects for detecting and controlling bacterial ring rot of potatoes in North America. Plant Dis 68: 841–844.

De Boer, S.H. and A. Wieczorek. 1984. Production of monoclonal antibodies of *Corynebacterium sepedonicum*. Phytopath 74: 1431–1433.

De Boer, S.H. and M.E. Mc Naughton. 1986. Evaluation of immunofluorescence with monoclonal antibodies for detecting latent bacterial ring rot infections. Amer Potato J 63: 533–542.

De Boer, S.H. and M.E. Mc Naughton. 1987. Monoclonal antibodies to the lipopolysaccharide of *Erwinia carotovora* subsp. *atroseptica* serogroup I. Phytopath 77: 828–832.

De Boer, S.H. and M. McCann. 1990. Detection of *Corynebacterium sepedonicum* in potato cultivars with different propensities to express ring rot symptoms. Amer Potato J 67, 10: 685–694.

De Boer, S.H. and L.J. Ward. 1995. PCR detection of *Erwinia carotovora* subsp. *atroseptica* associated with potato tissue. Phytopath 85: 854–858.

De Boer, S.H. and J.W. Hall. 2000a. Proficiency testing in a laboratory accreditation program for the bacterial ring rot pathogen of potato. Plant Dis 84: 649–653.

De Boer, S.H. and J.W. Hall. 2000b. Reproducibility of enzyme-linked immunosorbent assay and immunofluoroscence for detecting *Clavibacter michiganensis* subsp. *sepedonicus* in multiple laboratories. EPPO/OEPP Bull 30: 397–401.

De Jong, H. and G.C.C. Tai. 1977. Analysis of tetraploid-diploid hybrids in cultivated potatoes. Potato Res 20: 111–121.

De Lorenzo, G. and S. Ferrari. 2002. Polygalacturonase-inhibiting proteins in defense against phytopathogenic fungi. Curr Opin Plant Biol 5: 295–299.

De Maine, M.J., C.P. Caroll, H.E. Stewart, R.M. Solomon, and R.L. Wastie. 1993. Disease resistance in *Solanum phureja* and diploid and tetraploid *S. tuberosum* × *S. phureja* hybrids. Potato Res 36: 21–28.

De Maine, M.J., A.K. Lees, D.D. Muir, J.E. Bradshaw, and G.R. Mackay. 2000. Long-day-adapted Phureja as a resource for potato breeding and genetic research. *In*: S.M.P. Khurana, G.S. Shekhawat, B.P. Singh, and S.K. Pandey (eds.), Potato Global Research and Development. Proc. Global Conf. Potato, IARI, New Delhi, India, 6–11 December, 1999. pp. 134–137.

De Vries, J., K. Harms, L. Broer, G. Kriete, A. Mahn, K. During, and W. Wakernagel. 1999. The bacteriolytic activity in transgenic potatoes expressing a chimeric T4 lysozyme gene and the effect of T4 lysozyme on soil-and phytopathogenic bacteria. System Appl Microbiol 22 (2): 280–286.

Delanay, T.P., S. Uknes, B. Vernooij, L. Friedrich, K. Weymann, D. Negrotto, T. Geffeney, M. Gut-Cella, H. Kessmann, E. Ward, and J. Ryals. 1994. A central role of salicylic acid in plant disease resistance. Science 266: 1247–1250.

Destefano-Beltran, L., P.G. Nagpala, M.S. Cetiner, J.H. Dodds, and J.M. Jaynes. 1990. Enhancing bacterial and fungal disease resistance in plants: application to potato. *In*: M.E. Vayda and W.D. Park (eds.), The Molecular and Cellular Biology of the Potato. CAB Intl., Wallingford, UK, pp. 205–221.

Dionne, L.A. and C.H. Lawrence. 1961. Early scab resistant derivatives of *Solanum chacoense* × *Solanum phureja*. Amer Potato J 38: 6–8.

Dobias, K. 1976. Methoden zur Prüfung der Resistenz von Kartoffeln gegen den Erreger der Knollenasfaule. Tagungsbericht, Akademie der Landwirtschaftswissenschaften der Deutschen Demokratischen Republik 140: 221–230.

Dobias, K. 1977. Provocation tests for resistance to bacterial soft rot. Ochrana Rostlin 13: 101–106.

Dobias, K. 1978. Resistance of certain species of potatoes to blackleg. Vedecke prace Vyzkumneho a slechtitelskeho ustavu bramborarskeho v Havlickove Brode 7: 53–61.

Doering-Saad, C., P. Kampfer, S. Manulis, G. Kritzman, J. Schneider, J. Zakrzewska-Czerwińska, H. Schrempf, and I. Barash. 1992. Diversity among *Streptomyces* strains causing potato scab. Appl Environ Microbiol 58: 3932–3940.

Dorel, C., N. Hugouvieux-Cotte-Pattat, J. Robert-Baudouy, and E. Lojkowska. 1996. Production of *Erwinia chrysanthemi* pectinases in potato tubers showing high or low level of resistance to soft rot. Europ J Plant Path 102: 511–517.

Düring, K., P. Porsch, M. Fladung, and H. Lörz. 1993. Transgenic potato plants resistant to the phytopathogenic bacterium *Erwinia carotovora*. Plant J 3: 587–598.

Elphinstone, J.G. 1987. Soft rot and blackleg of potato: *Erwinia* ssp. Tech Inform Bull 21. International Potato Center, Lima, Peru, 18 pp.

Elphinstone, J.G. 1994. Inheritance of resistance to bacterial diseases. *In*: J.E. Bradshaw and G.R. Mackay (eds.), Potato Genetics. CAB Intl., Wallingford, UK, pp. 429–446.

Elphinston, J.G. 1996. Survival and possibilities for extinction of *Pseudomonas solanacearum* (Smith) in cool climates. Potato Res 39: 403–410.

Elphinstone, J.G., H. Stanford, and D.E. Stead. 1998. Survival and transmission of *Ralstonia solanacearum* in aquatic plants of *Solanum dulcamara* and associated surface water in England. EPPO/OEPP Bull 28: 93–94.

Elphinstone, J.G., J. Hennessy, J.K. Wilson, and D.E. Stead. 1996. Sensitivity of different methods for the detection of *Pseudomonas solanacearum* (Smith) in potato tuber extracts. EPPO/OEPP Bull 26: 663–678.

Elphinstone, J.G., D.E. Stead, D. Caffier, J.D. Janse, M.M. López, U. Mazzucchi, P. Müller, P. Persson, E. Rauschher, E. Schiessendoppler, M. Sousa Santos, E. Stefani, and J. van Vaerenbergh. 2000. Standardization of methods for detection of *Ralstonia solanacearum* in potato. EPPO/OEPP Bull 30: 391–395.

Evans, D., C. Flick, and R. Jensenet. 1981. Disease resistance: incorporation into sexually incompatible somatic hybrids of the genus *Nicotiana*. Science 23: 907–909.

Fessehaie, A., S.H. de Boer, and C.A. Levesque. 2002. Molecular characterization of DNA encoding 16S-23S rRNA intergenic spacer regions and 16S rRNA of pectolytic *Erwinia* species. Can J Microbiol 48: 387–398.

Firrao, G.R. and R. Lozzi. 1994. Identification of *Clavibacter michiganensis* subsp. *sepedonicus* using polymerase chain reaction. Can J Microbiol 40: 148–151.

Florack, D., S. Allefs, R. Bollen, D. Bosch, B. Visser, and W. Siekema. 1995. Expression of giant silkmoth cecropin B encoding genes in transgenic tobacco. Transgenic Res 4: 132–141.

Fock, I., C. Collonnier, A. Purwito, J. Luisetti, V. Souvannavong, F. Vedel, A. Servaes, A. Ambroise, H. Kodja, G. Ducreux, and D. Sihachakr. 2000. Resistance to bacterial wilt in somatic hybrids between *Solanum tuberosum* and *Solanum phureja*. Plant Sci 160: 165–176.

Fock, I., C. Collonnier, J. Luisetti, A. Purwito, V. Souvannavong, F. Vedel, A. Servaes, A. Ambroise, H. Kodja, G. Ducreux, and D. Sihachakr. 2001. Use *Solanum stenotomum* for introduction of resistance to bacterial wilt in somatic hybrids of potato. Plant Physiol Biochem 39: 899–908.

Fréchon, D., P. Exbrayat, V. Helias, L.J. Hyman, B. Jouan, P. Llop, M.M. Lopez, N. Payet, M.C. M. Pérombelon, I.K. Toth, J.R.C.M. van Beckhoven, J.M. van der Wolf, and Y. Bertheau. 1998. Evaluation of a PCR kit for the detection of *Erwinia carotovora* subsp. *atroseptica* on potato tubers. Potato Res 41: 163–173.

French, E.R., R. Anguiz, and P. Aley. 1998. The usefulness of potato resistance to *Ralstonia solanacearum* for the integrated control of bacterial wilt. *In:* P. Prior, C. Allen, and J.G. Elphinstone (eds.), Bacterial Wilt Disease: Molecular and Ecological Aspects. Springer Verlag, Berlin, Germany, pp. 381–385 (INRA ed.).

Gang, G., D.-Y. Qu, L. Yong, J. Liping, and F. Lanxiang. 2000. Identification molecular markers linked with resistance to bacterial wilt (*Ralstonia solanacearum*) in diploid potato. Acta Hortic. Sinica 27 (1): 37–41.

Garcia-Olmedo, F., A. Molina, A. Segura, and M. Moreno. 1995. The defensive role of nonspecific lipid-transfer proteins in plants. Trends Microbiol 3: 72–74.

Garcia-Olmedo, F., A. Molina, J.M. Alamillo, and P. Rodriguez-Palenzuela. 1998. Plant defense peptides. Biopolymers (Peptide Sci) 47: 479–491.

Genin, S. and C. Boucher. 2002. *Ralstonia solanacearum*: secrets of a major pathogen unveiled by analysis of its genome. Molec Plant Path 3: 111–118.

Ghislain, M. and A. Golmirzaie. 1998. Genetic engineering for potato improvement. *In:* S.M.P. Khurana, R. Chandra, and M.D. Upadhya (eds.), Comprehensive Potato Biotechnology. Malhotra Publ. House, New Delhi, India, pp. 115–162.

Gorris, M.T., B. Alarcon, M.M. Lopez, and M. Cambra. 1994. Characterization of monoclonal antibodies specific for *Erwinia carotovora* subsp. *atroseptica* and comparison of serological methods for its sensitive detection on potato tubers. Appl Environ Microbiol 60: 2076–2085.

Goth, R.W., K.G. Haynes, and D.R. Wilson. 1993. Evaluation and characterization of advanced potato breeding clones for resistance to scab by cluster analysis. Plant Dis 77: 911–914.

Goth, R.W., K.G. Haynes, R.J. Young, D.R. Wilson, and F.I. Lauer. 1995. Relative resistance of the potato cultivar Kranz to common scab caused by *Streptomyces scabies* as determined by cluster analysis. Amer Potato J 72: 505–511.

Gueneron, M., A.C. Timmers, C. Boucher, and M. Arlat. 2000. Two novel proteins, PopB, which has functional nuclear localisation signals, and PopC, which has a large leucine-rich repeat domain, are secreted though the Hrp-secretion apparatus of *Ralstonia solanacearum.* Molec Microbiol 36: 261–273.

Hammond-Kosack, K.E. and J.D.G. Jones. 1996. Disease resistance gene-dependent plant defense mechanisms. Plant Cell 8: 1773–91.

Hartman, G.L. and J.G. Elphinstone. 1994. Advances in the control of *Pseudomonas solanacearum* race 1 in major food crops. *In:* A.C. Hayward and G.L. Hartman (eds.), Bacterial Wilt. The Disease and Its Causative Agent, *Pseudomonas solanacearum.* CAB Intl., Wallingford, UK, pp. 157–177.

Hauben, L., E.R.B. Moore, L. Vauterin, M. Steenackers, L. Verdonck, and J. Swings. 1998. Phylogenetic position of phytopathogens within the Enterobacteriacae. System Appl Microbiol 21: 384–397.

Hawkes, J.G. 1990. The Potato, Evolution, Biodiversity and Genetic Resources. Belhaven Press, London, UK.

Haynes, K.G., R.W. Goth, and R.J. Young. 1997. Genotype × environment interactions for resistance to common scab in tetraploid potato. Crop Sci 37: 1163–1167.

Hayward, A.C. 1991. Biology and epidemiology of bacterial wilt caused by *Pseudomonas solanacearum.* Ann Rev Phytopath 29: 65–87.

Hayward, A.C. 1994. The hosts of *Pseudomonas solanacearum. In*: A.C. Hayward and G.L. Hartman (eds.), Bacterial Wilt: the Disease and Its Causative Agent, *Pseudomonas solanacearum.* CAB Intl Wallingford, UK, pp. 9–24.

Helgeson, J.P. 1989. Postharvest resistance through breeding and biotechnology. Phytopath 79: 1375–1377.

Helias, V., A. C. Roux, Y. Bertheau, D. Andrivon, J.P. Gauthier, and B. Jouan, 1998. Characterization of *Erwinia carotovora* subspecies and detection of *Erwinia carotovora* subsp. *atroseptica* in potato plants, soil and water extracts with PCR-based methods. Europ J Plant Path 104: 685–699.

Henniger, H. 1965. Untersuchungen über Knollen und Lagerfaulen der Kartoffel. I. Zur Methodik der Resistenzprüfung mit dem Errer der bakteriellen Knollenäsfaule (*Pectobacterium carotovorum*) var. *atrosepticum* (van Hall) Dowson. Der Züchter 35: 174–180.

Hidalgo, O.A. and E. Echandi. 1982. Evaluation of potato clones for resistance to tuber and stem rot induced by *Erwinia chrysanthemi*. Amer Potato J 59: 585–592.

Hidalgo, A.O. and E. Echandi. 1983. Influence of temperature and length of storage on resistance of potato to tuber rot induced by *Erwinia chrysanthemi*. Amer Potato J 60: 1–15.

Hide, G.A., S.J. Welham, P.J. Read, and A.E. Ainsley. 1996. The yield of potato plants as affected by stem cancer (*Rhizoctonia solani*), blackleg (*Erwinia carotovora* subsp. *atroseptica*) and by neighboring plants. J Agri Sci 126: 429–440.

Hosaka, K., H. Matsunaga, and K. Senda. 2000. Evaluation of several wild tuber-bearing *Solanum* species for scab resistance, Amer J Potato Res 77: 41–45.

Hossain, M. and C. Logan. 1983. A comparison of inoculation methods for determining potato cultivar reaction to blackleg. Ann Appl Biol 103: 63–70.

Howard, H.W. 1978. The production of new varieties. *In*: P.M. Harris (ed.). The Potato Crop. Chapman & Hall, London, UK, pp. 607–646.

Huaman, Z. and P. Schmiediche. 1999. The potato genetic resources held in trust by the International Potato Center (CIP) in Peru. Potato Res 42: 413–426.

Huaman, Z., L. de Lindo, and J.G. Elphinstone. 1988. Resistance to blackleg and soft rot and its potential use in breeding. *In*: E.R. French (ed.). Report of the Planning Conference on Bacterial Diseases of the Potato 1987. International Potato Center, Lima, Peru, pp. 215–228.

Huaman, Z., B. Tivoli, and L. de Lindo. 1989. Screening for resistance to *Fusarium* dry rot in progenies of cultivars of *S. tuberosum* ssp. *andigena* with resistance to *Erwinia chrysanthemi*. Amer Potato J 66: 357–364.

Huang, Q. and C. Allen. 2000. Polygalacturonases are required for rapid colonization and full virulence of *Ralstonia solanacearum* on tomato plants. Physiol Molec Plant Path 57: 77–83.

Husain, A. and A. Kelman. 1958. Relationship of slime production to mechanism of wilting and pathogenicity of *Pseudomonas solanacearum*. Phytopath 48: 155–164.

Hyman, L.J., A. Wallace, M.M. Lopez, M. Cambra, M.T. Gorris, and M.C.M. Pérombelon. 1995. Characterization of monoclonal antibodies against *Erwinia carotovora* subsp. *atroseptica* serogroup I: specificity and epitope analysis. J Appl Bacteriol 78: 437–444.

Ishimaru, C.A., N.L. Lapitan, A. VanBuren, A. Fenwick, and K. Pedas. 1994. Identification of parents suitable for molecular mapping of immunity and resistance genes in *Solanum* species. Amer Potato J 71: 517–533.

Iwanaga, M. 1982. Breeding at *2x* level for combined pest and disease resistance using wild species and extracted haploids from selected tetraploid clones. *In*: Report of Planning Conference: Utilization of Genetic Resources of the Potato, III, 1980. CIP, Lima , Peru, pp. 110–124.

Iwanaga, M. 1985. Ploidy level manipulation approach development of diploid populations with specific resistance and FDR *2n* pollen production. *In*: International Potato Center (CIP). Present and Future Strategies for Potato Breeding and Improvement: Report of the 26th Planning Conference, Lima (Peru), 12–14 Dec. 1983. CIP, Lima, Peru, pp. 57–70.

Jackson, M.T., L.C. Gonzalez, and J.A. Aquilar. 1979. Avances en el combate de la marchitez bacteriana de la papa en Costa Rica. Fitopatologia 14: 46–53.

Janse, J.D. 1996. Potato brown rot in Western Europe—history, present occurrence and some remarks on possible origin, epidemiology and control strategies. EPPO/OEPP Bull 26: 679–695.

Jansky, S. 2000. Breeding for disease resistance in potato. *In*: J. Janick (ed.), Plant Breeding Reviews. John Wiley and Sons, Inc., New York, NY, USA, vol. 19, pp. 69–155.

Javier, E.Q. 1994. Foreword. *In*: A.C. Hayward and G.L. Hartman (eds.), Bacterial Wilt: The Diseases and Its causative Agent, *Pseudomonas solanacearum*. CAB Intl., Wallingford, UK, p. xi.

Jaynes, J.M., P. Nagpala, L. Destefano-Beltran, J.H. Huang, J. Kim, T. Denny, and S. Cetiner. 1993. Expression of a cecropin B lytic peptide analog in transgenic tobacco confers enhanced resistance to bacterial wilt caused by *Pseudomonas solanacearum*. Plant Sci 89: 43–53.

Jellis, G.J. 1974. Improving the reliability of screening in the field for resistance to common scab. Potato Res 17: 356.

Jones, S., B. Yu, J.N. Bainton, M. Birdsall, B.W. Baycroft, S.R. Chabra, A.J.R. Cox, P. Golby, J.P. Reevs, S. Stephens, M.K. Winson, G.P.C. Salmond, G.S.A.B. Stewart, and P. Williams. 1993. The *lux* autoinducer regulates the production of exoenzyme virulence determinants in *Erwinia carotovora* and *Pseudomonas aeruginosa*. EMBO J 12: 2477–2482.

Kankila, J., V.-M. Rokka, and A. Kuusela. 1995. Disease resistance testing in a potato breeding programme utilizing somatic fusion of dihaploids. Nordisk Jordbruksforskning 77: 126.

Kankila, J., H.G. Kirk, T. Bjor, and K. Tolstrup 1994. Testing the Resistance of Potato Varieties to Soft Rot and Blackleg. Nordiska Genbanken, Alnarp, Denmark.

Kawchuk, L.M., D.R. Lynch, G.C. Kozub, G.A. Nelson, F. Kulcsar, and D.K. Fujimoto. 1998. Multi-year evaluation of *Clavibacter michiganensis* subsp. *sepedonicus* disease symptoms in cultivated potato genotypes. Amer J Potato Res 75 (6): 235–243.

Kelman, A. 1953. The bacterial wilt caused by *Pseudomonas solanacearum*. A literature review and bibliography. North Carolina Agric Exp Stn. Tech Bull 99: 1–194.

Kelman, A. 1954. The relationship of pathogenicity in *Pseudomonas solanacearum* to colony appearance on a tetrazolium medium. Phytopath 44: 693–695.

Kinkel, L.L., J.H. Bowers, K. Shimizu, E.C. Neeno-Eckwall, and J.L. Schottel. 1998. Quantitative relationships among composition in *Streptomyces*. Can J Microbiol 44: 768–776.

Koppel, M. 2000. Problems and achievements in breeding potatoes for disease resistance. Trans Estonian Agric Univ, Agron 209: 70–72.

Kotoujansky, A. 1987. Molecular genetics of pathogenesis by soft-rot *Erwinia*. Ann Rev Phytopath 25: 405–430.

Kranz, F.A. and C.J. Eide. 1941. Inheritance of reaction to common scab in the potato. J Agric Res 63: 219–231.

Krause, B., T. Koczy, J. Komorowska-Jedrys, and E. Ratuszniak. 1982. Laboratory assessment of tuber resistance of world potato cultivar collection to main causes of storage rots. Biuletyn Instytutu Ziemniaka 27: 111–134.

Kriel, C.J., S.H. Jansky, N.C. Gudmestad, and D.H. Ronis. 1995. Immunity to *Clavibacter michiganensis* subsp. *sepedonicus*: screening of exotic *Solanum* species. Euphytica 82 (2): 125–132.

Kurowski, J. and F.E. Manzer. 1992. Reevaluation of *Solanum* species accessions showing resistance to bacterial ring rot. Amer Potato J 69 (5): 289–297.

Laferriere, L.T., J.P. Helgeson, and C. Allen. 1999. Fertile *Solanum tuberosum* + *S. commersonii* somatic hybrids as sources of resistance to bacterial wilt caused by *Ralstonia solanacearum*. Theor Appl Genet 98: 1272–1278.

Lambert, D.H. and R. Loria, 1989. *Streptomyces acidiscabies* sp. nov. Intl J System Bacteriol 39: 393–396.

Langerfeld, E. and W. Batz. 1992. Reaction of potato cultivars to the causal agent of bacterial ring rot of potatoes (*Clavibacter michiganensis* subsp. *sepedonicus* (Spieck. et Kotth.) Davis et al.). Nachrichtenblatt des Deutschen Pflanzenschutzdienstes 44 (7): 157–161.

Lansade, M. 1950. Recherches sur le fletrissment bacterien de la pomme de terre en France *Corynebacterium sepedonicum* (S. et K.). Skapt. et Burkh. Annales des Epiphyties 2: 69–156.

Lapwood, D.H. and P.R. Legg. 1983. The effect of *Erwinia carotovora* subsp. *atroseptica* (blackleg) on potato plants. I. Growth and yield of different cultivars. Ann. Appl Biol 103: 71–78.

Lapwood, D.H. and P.J. Read. 1986a. A comparison of methods of seed tuber inoculation for assessing the susceptibility of potato cultivars to blackleg (*Erwinia carotovora* subs. *atroseptica*) in the field. Ann Appl Biol 109: 287–297.

Lapwood, D.H. and P.J. Read, 1986b. The susceptibility of stems of different potato cultivars to blackleg caused by *Erwinia carotovora* subsp. *atroseptica*. Ann Appl Biol 109: 555–560.

Lapwood, D.H., P.J. Read, and J. Spokes, 1984. Methods for assessing the susceptibility of potato tubers of different cultivars to rotting by *Erwinia carotovora* subsp. *atroseptica* and *carotovora*. Plant Path 33: 13–20.

Lauer, F.I. and C.J. Eide. 1963. Evaluation of parent clones of potato for resistance to common scab by the "highest scab" method. Europ Potato J 6: 35–44.

Lebecka, R. and E. Zimnoch-Guzowska. 2001. Resistance to tuber soft rot (*Erwinia carotovora* subsp. *atroseptica*) in potato unselected progenies of 4x-2x crosses. *In*: Abstracts, EAPR Pathology Section Meeting, Poznań, 10-15 July, 2001, pp. 35–36.

Lebecka, R., L. Domanski, and E. Zimnoch-Guzowska. 1998. Studies on the inheritance of resistance to soft rot in diploid potato families. *In*: 7th Intl Cong Plant Pathology: Papers, Abstracts, Edinburgh, Scotland, 9–16 August 1998, vol. 3, p. 3.4.67.

Lees, A.K., M.J. de Maine, M.J. Nicolson, and J.E. Bradshaw. 2000. Long-day-adapted *Solanum phureja* as a source of resistance to blackleg caused by *Erwinia carotovora* subsp. *atroseptica*. Potato Res 43: 279–286.

Lellbach, H. 1978. Schätzung genetischer Parameter aus diallelen kreuzungen bei der Nasfäuleanfälligkeit der Kartoffel. Archiv für Züchtungsforschung 3 (8): 193–199.

Lewosz, W. 1984. Blackleg of potatoes in the light of the literature and personal research. Biuletyn Instytutu Ziemniaka 31: 103–123.

Lima, M.F., C.A. Lopes, and P.E. de Melo. 1996. Selection of potato genotypes at the seedling stage for resistance to bacterial wilt. Pesquisa Agropecuária Brasileira 31 (4): 249–257.

Lojkowska, E. and A. Kelman. 1989. Screening of seedlings of wild *Solanum* species for resistance to bacterial stem rot caused by soft rot *Erwinias*. Amer Potato J 66: 379–390.

Lojkowska, E. and A. Kelman. 1994. Comparison of the effectiveness of different methods of screening for soft rot resistance of potato tubers. Amer Potato J 71: 99–113.

Lojkowska, E., C. Dorel, P. Reignault, N. Hugouvieux-Cotte-Pattat, and J. Robert-Baudouy. 1993. Use of GUS fusion to study the expression of *Erwinia chrysanthemi* pectinase genes during infection of potato tuber. Molec Plant Micr Interact 6: 488–494.

Lojkowska, E., C. Masclaux, M. Boccara, J. Robert-Baudouy, and N. Hugouvieux-Cotte-Pattat. 1995. Characterization of the *pelL* gene encoding a secondary pectate lyase of *Erwinia chrysanthemi* 3937. Molec Microbiol 16: 1183–1195.

Lopes, C.A., J.A. Buso, and P. Accatino. 1993. Screening CIP potato germplasm for resistance to bacterial wilt in Brazil: methods and preliminary results. Bacterial Wilt Newsl 9: 3–5.

Lopez, M.M., M.J. Lopez-Lopez, R. Marti, J. Zamora, J. Lopez-Sanchez, and R. Beltra. 2001. Effect of acetylsalicylic acid on soft rot produced by *Erwinia carotovora* subsp. *carotovora* in potato tubers under greenhouse conditions. Potato Res 44: 197–206.

López-Solanilla, E., F. Garcia-Olmedo, and P. Rodriguez-Palenzuela. 1998. Inactivation of the *sapA* to *sapF* locus of *Erwinia chrysanthemi* reveal common features in plant and animal bacterial pathogenesis. Plant Cell 10: 917–924.

Loria, R., R.A. Bukhalid, B.A. Fry, and R.R. King. 1997. Plant pathogenicity in the genus *Streptomyces*. Plant Dis 81: 836–846.

Loria, R., C.A. Clark, R.A. Bukhalid, and B.A. Fry. 2001. *Streptomyces. In*: N.W. Schaad, J.B. Jones and W. Chun (eds.), Laboratory Guide for Identification of Plant Pathogenic Bacteria. Amer Phytopath Soc, St. Paul, MN, USA pp. 236–249.

Loria, R., R.A. Bukhalid, R.A. Creath, R.H. Leiner, M. Olivier, and J.C. Steffens. 1995. Differential production of thaxtomins by pathogenic *Streptomyces* species in vitro. Phytopath 85: 537–541.

Louws, F.J., J. Bell, C.M. Medina-Mora, C.D. Smart, D. Opgenorth, C.A. Ishimaru, M.K. Hausbeck, F.J. de Bruijn, and D.W. Fulbright. 1998. Rep-PCR-mediated genomic fingerprinting: a rapid and effective method to identify *Clavibacter michiganensis.* Phytopath 88: 863–868.

Lyon, G.D. 1989. The biochemical basis of resistance of potatoes to soft rot *Erwinia* ssp.—a review. Plant Path 38: 313–339.

Lyons, N., L. Cruz, and M. Sousa Santos. 2001. Rapid field detection of *Ralstonia solanacearum* in infected tomato and potato plants using the *Staphylococcus aureus* slide agglutination test. EPPO/OEPP Bull 31: 91–93.

Maher, E.A. and A. Kelman, 1983. Oxygen status of potato tuber tissue in relation to maceration by pectic enzymes of *Erwinia carotovora.* Phytopath 73: 536–539.

Mäki-Valkama, T. and R. Karjalainen. 1994. Differentiation of *Erwinia carotovora* subsp. *atroseptica* and *carotovora* by RAPD-PCR. Ann Appl Biol 125: 301–309.

Marais, L. and R. Vorster. 1988. Evaluation in pot field trials of resistance of potato cultivars and breeding lines to common scab caused by *Streptomyces scabies.* Potato Res 31: 401–404.

McGrath, J.M., Ch. E. Williams, G.T. Haberlach, S.M. Wielgus, T.F. Uchytil, and J.P. Helgeson. 2002. Introgression and stabilization of *Erwinia* tuber soft rot resistance into potato after somatic hybridization of *Solanum tuberosum* and *S. brevidens.* Amer J Potato Res 79: 19–24.

Mc Kee, R.K. 1963. Scab resistance of potato varieties. Plant Path 12: 106–109.

McMillan, G.P., D. Hedley, L. Fyffe, and M.C.M. Pérombelon. 1993. Potato resistance to soft rot erwinias is related to cell wall esterification. Physiol Molec Plant Path 42: 279–289.

Melchers, A., M.D. Sacristan, and S.A. Holder. 1978. Somatic hybrid plants of potato and tomato regenerated from fused protoplasts. Carlsberg Res Comm 43: 203–218.

Mendoza, H.A. 1987. Advances in population breeding and its potential impact on the efficiency of breeding potatoes for developing countries. *In*: G.J. Jellis and D.E. Richardson (eds.), The Production of New Varieties: Technological Advances. Cambridge Univ. Press, Cambridge, UK, pp. 235–245.

Metzler, M.C., M.J. Laine, and S.H. de Boer. 1997. The status of molecular biological research on the plant pathogenic genus *Clavibacter.* FEMS Microbiol Lett 150: 1–8.

Mill, D. and F.A. Hammerschlag. 1993. Effect of cecropin B on peach pathogens, protoplasts, and cells. Plant Sci 93: 143–150.

Miller, J.C., J.F. Fontenot, and F.J. Dainello. 1965. Techniques used in screening potato clones for resistance to *Streptomyces scabies* (abstr.). Amer Potato J 42: 294.

Mills, D., B.W. Russell, and J.W. Hanus. 1997. Specific detection of *Clavibacter michiganensis* subsp. *sepedonicus* by amplification of three unique DNA sequences isolated by subtraction hybridization. Phytopath 87: 853–861.

Mok, D.W.S. and S.J. Peloquin. 1975. Breeding value of 2*n*-pollen (diplandroids) in tetraploid × diploid crosses in potatoes. Theor Appl Genet314.

Molina, A. and F. Garcia-Olmedo. 1997. Enhanced tolerance to bacterial pathogens caused by the transgenic expression of barley lipid transfer protein LTP2. Plant J 12: 669–675.

Montanelli, C. and G. Nascari. 1989. Transformation of potato for bacterial disease resistance using antibacterial genes from insects. Rivista di Agricoltura Subtropicale e Tropicale 83: 375–385.

Montanelli, C. and G. Nascari. 1991. Introduction of an antibacterial gene in potato (*Solanum tuberosum* L.) using a binary vector in *Agrobacterium rhizogenes*. J Genet Breed 45: 307–316.

Mukherjee, A., Y. Cui, Y. Liu, and A.K. Chatterjee. 1997. Molecular characterization and expression of the *Erwinia carotovora hrp*N deECC gene, which encodes an elicitor of the hypersensitive reaction. Molec Plant Micr Interact 10: 462–471.

Munzert, M. 1975. Eine Methode zur Prüfung der Resistenz der Kartoffelpflanze gegenüber dem Erreger der Schwarzbeinigkeit (*Erwinia carotovora* var. *atroseptica*/van Hall/Dye). Potato Res 18: 308–313.

Munzert, M. 1984. The breeding possibilities for improving resistance to blackleg (*Erwinia atroseptica*). *In*: Abstr Conf Papers 9th Triennial Conf. EAPR, Interlaken, Switzerland, p. 385.

Murphy, A.M., H. de Jong, and G.C.C. Tai. 1995. Transmission of resistance to common scab from the diploid to the tetraploid level via 4x-2x crosses in potatoes. Euphytica 82: 227–233.

Nassar, A., A. Darrasse, M. Lemattre, A. Kotoujansky, C. Dervin, R. Vedel, and Y. Bertheau. 1996. Characterisation of *Erwinia chrysanthemi* by pectinolytic isozyme polymorphism and Restriction Fragment Length Polymorphism analysis of PCR-amplified fragments of *pel* genes. Appl Environ Microbiol 62: 2228–2235.

Nelson, G.A. and G.C. Kozub. 1987. Effect of temperature and latent viruses on atypical ring rot symptoms of Russet Burbank potatoes. Amer Potato J 64: 589–597.

Nelson, G.A., D.R. Lynch, and G.C. Kozub. 1992. Ring rot symptom development on potato cultivars and lines in southern Alberta. Potato Res 35 (2): 133–142.

Nielsen, L.W. and F.L. Haynes, Jr. 1960. Resistance in *Solanum tuberosum* to *Pseudomonas solanacearum*. Amer Potato J 37: 260–267.

Nissinen, R., S. Kassuwi, R. Peltola, and M.C. Metzler, 2001. In planta—complementation of *Clavibacter michiganensis* subsp. *sepedonicus* strains deficient in cellulase production or HR induction restores virulens. Europ J Plant Path 107: 175–182.

Nordeen, R.O., S.L. Sinden, J.M. Jaynes, and L.D. Owens. 1992. Activity of cecropin SB37 against protoplast from several plant species and their bacterial pathogens. Plant Sci 82: 101–107.

O'Keefe, R.B. 1965. A field method for genetic studies and selection of high-yielding, scab-resistant potato clones (abstr.). Amer Potato J 42: 295.

Osiru, M.O., P.R. Rubaihayo, and A.F. Opio. 2001. Inheritance of resistance of tomato bacterial wilt and its implication for potato improvement in Uganda. African Crop Sci J 9(1): 9–16.

Palva, T.K., M. Hurting, P. Saindrenan, and E.T. Palva. 1994. Salicylic acid induced resistance to *Erwinia carotovora* subsp. *carotovora* in tobacco. Molec Plant Micr Interact 7: 356–363.

Pastrik, K.-H. 2000. Detection of *Clavibacter michiganensis* subsp. *sepedonicus* in potato tubers by multiplex PCR with coamplification of host DNA. Europ J Plant Path 106: 155–165.

Pastrik, K.-H. and E. Maiss. 2000. Detection of *Ralstonia solanacearum* in potato tubers by polymerase chain reaction. J Phytopath 148: 619–626.

Pawlak, A., J.J. Pavek, and D.L. Corsini. 1987. Resistance to storage diseases in breeding stocks. *In*: G.J. Jellis and D.E. Richardson (eds.). The Production of New Potato Varieties. Cambridge Univ. Press, Cambridge, UK, pp. 96–98.

Pérombelon, M.C.M. 1982. The impaired host and soft rot bacteria. *In*: G.N. Lacy and M.S. Mount (eds.). Phytopathogenic *Procaryotes*. Acad. Press, New York, NY USA, vol. II, pp. 56–59.

Pérombelon, M.C.M. 2000. Blackleg risk potential of seed potatoes determined by quantification of tuber contamination by the causal agent and *Erwinia carotovora* subsp. *atroseptica*: a critical review. EPPO/OEPP Bull 30: 413–420.

Pérombelon, M.C.M. 2002. Potato diseases caused by soft rot erwinias: an overview of pathogenesis. Plant Path 51: 1–12.

Pérombelon, M.C.M. and A. Kelman. 1980. Ecology of the soft rot erwinias. Ann Rev Phytopath 18: 361–387.

Pérombelon, M.C.M. and A. Kelman. 1987. Blackleg and other potato diseases caused by soft rot erwinias: Proposal for revision of terminology. Plant Dis 71: 283–285.

Pérombelon, M.C.M., V.M. Lumb, and D. Zutra. 1987. Pathogenicity of soft rot erwinias to potato plants in Scotland and Israel. J Appl Bacteriol 63: 73–84.

Pérombelon, M.C.M., Y. Bertheau, M. Cambra, D. Fréchon, M.M. Lopez, F. Niepold, P. Persson, A Sletten, I.K. Toth, J.W.L. van Vuurde, and J.M. van der Wolf. 1998. Microbiological, immunological and molecular methods suitable for commercial detection and quantification of the blackleg pathogen, *Erwinia carotovora* subsp. *atroseptica* on seed potato tubers: a review. EPPO/OEPP Bull 28: 141–155.

Pfeffer, C. and M. Effmert. 1985. Die Züchtung homozygoter Eltern für Resistenz gegen Kartoffelschorf, verursacht durch *Streptomyces scabies* (Thaxt.) Waksman & Henrici. Archiv für Zuchtungsforschung 15: 325–333.

Pietkiewicz, J.B. 1980. Variation in the reaction of potato tubers to diseases. Potato Res 23: 473.

Pietrak, J., L. Jakuczun, H. Zarzycka, and J. Komorowska-Jedrys. 1988. Testing potato clones for resistance to blackleg (*Erwinia carotovora* subsp. *atroseptica*) after artificial inoculation in the greenhouse and in the field. *In*: Genetic Principles of Potato Breeding. Potato Research Institute, Bonin, pp. 158–167.

Pirhonen, M., D. Flego, R. Heikinheimo, and E.T. Palva. 1993. A small diffusible signal molecule is responsible for the global control of virulence and exoenzyme production in the plant pathogen *Erwinia carotovora*. EMBO J 12: 2467–2476.

Pontier, D., L. Godiard, Y. Marco, and D. Roby. 1994. HSR 203J, a tobacco gene whose activation is rapid, highly localized and specific for incompatible plant/pathogen interactions. Plant J 5: 507–521.

Popkova, K., Ibarra Romero, I.L. Kukushkina, and N. Bashilova. 2000. Investigation of different strains of the ring rot pathogen (*Corynebacterium sepedonicum*) and blackleg pathogen (*Pectobacterium phytophthorum*) isolated from potato tubers. Mezhdunarodnyi Sel'skokhozyaistvennyi Zhurnal 4: 54–55.

Poussier, S. and J. Luisetti. 2000. Specific detection of biovars of *Ralstonia solanacearum* in Plant Tissues by Nested-PCR-RFLP. Europ J Plant Path 106: 255–265.

Poussier, S., D. Trigalet-Damery, P. Vandewalle, B. Goffinet, J. Luisetti, and A. Trigalet. 2000. Genetic diversity of *Ralstonia solanacearum* as asessed by PVR-RFLP of the *hrp* gene region, AFLP and 16S rRNA sequence analysis, and identification of an African subdivision. Microbiol 146: 1679–1692.

Priou, S., K. Al-Ani, and B. Jouan. 1992. Comparison of the effectiveness of two methods of screening potato to soft rot induced by *Erwinia carotovora* ssp. *atroseptica* (van Hall, 1902). *In*: F. Rousselle-Bourgeois, P. Rousselle (eds.). Proc. Joint Conf. EAPR Breeding & Varietal Assessment Section and the EUCARPIA Potato Section, Landerneau, France, 12-17 January 1992, Station de Pathologie vegetale. INRA, Le Rheu, France, pp. 139–140.

Priou, S., L. Gutarra, and P. Aley, 1999. Highly sensitive detection of *Ralstonia solanacearum* in latently infected potato tubers and soil by post-enrichment ELISA on nitrocellulose membrane. EPPO/OEPP Bull 29: 117–125.

Priou, S., C. Salas, F. de Mendiburu, P. Aley, and L. Gutarra. 2001. Assessment of latent infection frequency in progeny tubers of advanced potato clones resistant to bacterial wilt: a new selection criterion. CIP Program Report 1999-2000, pp. 105–116.

Puite, K.J., S. Roest, and L.P. Pijnacker, 1986. Somatic hybrid potato plants after electrofusion of diploid *Solanum tuberosum* and *S. phureja*. Plant Cell Rept 5: 262–264.

Qu, D.-Y., D.-W. Zhu, M.S. Rammana, and E. Jacobsen. 1996. A comparison of progeny from diallel crosses of diploid potato with regard to the frequencies of 2*n* pollen. Euphytica 92: 313–320.

Rasmussesn, J.B., R. Hammerschmidt, and M.N. Zook. 1991. Systemic induction of salicylic acid accumulation in cucumber after inoculation with *Pseudomonas syringae* pv. *syringae*. Plant Physiol 97: 1342–1347.

Rayls, J.A., U.H. Neuenschwander, M.G. Willits, A. Molina, H.Y. Steiner, and M.D. Humt. 1996. Systemic acquired resistance. Plant Cell 8: 1809–1819.

Reeves, A.F., O.M. Olanya, J.H. Hunter, and J.M. Wells. 1999. Evaluation of potato varieties and selections for resistance to bacterial soft rot. Amer J Potato Res 76: 183–189.

Riedl, W.A., F.J. Stevenson, and R. Bonde. 1946. The Teton potato, a new variety resistant to ring rot. Amer Potato J 23: 379–390.

Rokka, V.-M., Y.S. Xu, J. Kankila, A. Kuusela, S. Polli, and E. Pehu. 1994. Identification of somatic hybrids of dihaploid *Solanum tuberosum* lines and *S. brevidens* by species specific RAPD patterns and assessment of disease resistance of the hybrids. Euphytica 80: 207–217.

Rokka, V.-M., J.P.T. Valkonen, A. Tauriainen, L. Pietila, R. Lebecka, E. Zimnoch-Guzowska, and E. Pehu. 2000. Production and characterization of "second generation" somatic hybrids derived from protoplast fusion between interspecific somatohaploid and dihaploid *Solanum tuberosum* L. Amer J Potato 77: 149–159.

Romanenko, A.S., E.V. Rymareva, T.N. Shafikova, T.A. Konenkina, and A.M. Sobenin. 1997. Affinity of glycolipoprotein fraction from potato suspension cells for the toxin produced by the pathogen causing ring rot. Russian J Plant Physiol 44: 766–770.

Ross, A.F. 1961. Localised aquired resistance to plant virus infections in hypersensitive hosts. Virology 14: 329–339.

Ross, H. 1986. Potato Breeding—Problem and Perspectives. Adv. Plant Breed (suppl. 13). Verlag Paul Parey, Berlin, 132 pp.

Rousselle-Bourgeois, F. and S. Priou. 1995. Screening tuber-bearing *Solanum* ssp. for resistance to soft rot caused by *Erwinia carotovora* ssp. *atroseptica* (van Hall) Dye. Potato Res 38: 111–118.

Rowe, P.R. and L. Sequeira. 1970. Inheritance of resistance to *Pseudomonas solanacearum* in *Solanum phureja*. Phytopath 60: 1499–1501.

Salanoubat, M., S. Genin, F. Artiguenave, J. Gouzy, S. Mangenot, M. Arlat, A. Billault, P. Brottier, J.C. Camus, L. Cattolico, M. Chandler, N. Choisne, C. Claudel-Renard, S. Cunnac, N. Demange, C. Gaspin, M. Lavie, A. Moisan, C. Robert, W. Saurin, T. Schiex, P. Siguier, P. Thebault, M. Whalen, P. Wincker, M. Levy, J. Weissenbach, and C.A. Boucher. 2002. Genome sequence of the plant pathogen *Ralstonia solanacearum*. Nature 415: 497–502.

Salmond, G.P.C., B.W. Bycroft, G.S.A.B. Stewart, and P. Wiliams P. 1995. The bacterial 'enigma': cracking the code of cell-cell communication. Molec Microbiol 16: 615–624.

Samson, R., N. Ngwira, and N. Rivera. 1989. Biochemical and serological diversity of *Erwinia chrysanthemi*. *In*: Z. Klement (ed.). Proc. 7th Intl Conf. on Plant Pathogenic Bacteria. Akademie Kiadó és Nyomoda Vállalat, Budapest, Hungary, pp. 895–901.

Schaad, N.W., Y. Berthier-Schaad, A. Sechler, and D. Knorr. 1999. Detection of *Clavibacter michiganensis* subsp. *sepedonicus* in potato tubers by BIO-PCR and an automated real-time fluorescence detection system. Plant Dis 83: 1095–1100.

Schick, R. and A. Hopfe. 1962. Die Züchtung der Kartoffel. *In*: R. Schick and M. Klinkowski (eds.). Die Kartoffel. VEB Deutscher Landwirtschaftsverlag, Berlin, Germany pp. 1462–1563.

Schillberg, S., S. Zimmermann, M.-Y. Zhang, and R. Fischer. 2001. Antibody–based resistance to plant pathogens. Transgenic Res 10 (1): 1–12.

Schmiediche, P. 1983. Breeding bacterial wilt, *Pseudomonas solanacearum*, resistant germplasm. *In*: Present and Future Strategies for Potato Breeding and Improvement. Report Planning Conference, CIP, Lima, Peru, pp. 45–55.

Schmiediche, P. 1986. Breeding potatoes for resistance to bacterial wilt caused by *Pseudomonas solanacearum*. *In:* G.J. Persley (ed.). Bacterial Wilt Disease in Asia and the South Pacific. Proc. Intl Workshop PCARRD, Los Banos, Philippines, 8-10 October 1985. ACIAR Proceedings 13, pp. 105–111.

Schineder, B.J., J.L. Zhao, and C.D. Orser. 1993. Detection of *Clavibacter michiganensis* subsp. *sepedonicus* by DNA amplification. FEMS Microbiol Lett 109: 207–212.

Schöber, B. 1974. Methoden zur Testung von Kartoffelknollen auf Resistenz gegen Braunfäule (*Phytophthora infestans*) and Schorf (*Streptomyces scabies*). Potato Res 17: 354–355.

Seal, S. 1995. PCR-based detection and characterization of *Pseudomonas solanacearum* for use in less developed countries. EPPO/OEPP Bull 25: 227–231.

Seo, S.T., N. Furuya, C.K. Lim, Y. Takanami, and K. Tsuchiy. 2002. Phenotypic and genetic diversity of *Erwinia carotovora* subsp. *carotovora* strains from Asia. J Phytopath 150: 120–127.

Serrano, C., P. Arce-Johnson, H. Torres, M. Gebauer, M. Gutierrez, M. Moreno, X. Jordana, A. Venegas, J. Kalazich, and L. Holuique. 2000. Expression of the chicken lysozyme gene in potato enhances resistance to infection by *Erwinia carotovora* subsp. *atroseptica*. Amer J Potato Res 77: 191–199.

Skaptason, J.B. and W.H. Burkholder. 1942. Classification and nomenclature of the pathogene causing "ring rot" of potatoes. Phytopath 32: 439–441.

Śeledż, W., S. Jafra, M. Waleron, and E. Lojkowska. 2000. Genetic diversity of *Erwinia carotovora* strains isolated from infected plants grown in Poland, EPPO/OEPP Bull 30: 403–407.

Śeledż, W., 2002. Differentiation of Polish isolates of *Erwinia carotovora* subsp. *atroseptica* by molecular methods. PhD. thesis, Univ. G dousle, Poland.

Smid, E.J., A.H.J. Jansen, and L.G.M. Gorris. 1995. Detection of *Erwinia carotovora* subsp. *atroseptica* and *Erwinia chrysanthemi* in potato tubers using polymerase chain reaction. Plant Path 44: 1058–1069.

Smith, J.J., L.C. Offord, M. Holderness, and G.S. Saddler. 1995. Pulsed-field gel electrophoresis analysis of *Pseudomonas solanacearum*. EPPO/OEPP Bull 25: 163–167.

Smith, J.J., J.M. van der Wolf, R. Feuillade, A. Trigalet, L.C. Offord, and G.S. Saddler. 1998. Genetic diversity amongst *Ralstonia solanacearum* isolates of potato in Europe. EPPO/OEPP Bull 28: 83–84.

Smith, N.C., J. Hennessy, and D.E. Stead. 2001. Repetitive sequence-derived PCR profiling using the BOX-A1R primer for rapid identification of the plant pathogen *Clavibacter michiganensis* subspecies *sepedonicus*. Europ J Plant Path 107: 739–748.

Staskawicz, B.J., M.B. Mudgett, J.L. Dangl, and J.E. Galan. 2001. Common and contrasting themes of plant and animal diseases. Science 292: 2285–2289.

Staskawicz, B.J., F.M. Ausubel, B.J. Baker, J.G. Ellis, and J.D. Jones. 1995. Molecular genetics of plant disease resistance. Science 268: 661–617.

Sticher, L., B. Mauch-Mani, and P. Matraux. 1997. Systemic acquired resistance. Ann Rev Phytopath 35: 235–270.

Strobel, G.A. 1970. A phytotoxic glycopeptide from potato plants infected with *Corynebacterium sepedonicum*. J. Biol Chem 245: 32–38.

Subrtova, D., A. Hejtmankova, Z. Vankova, and J. Hubacek. 1993. Changes in the content of volatile substances, solanine and starch caused by soft rot infection. Potravin Vedy 11: 3–41.

Swiezynski, K.M., M.T. Sieczka, I. Stypa, and E. Zimnoch-Guzowska. 1998. Characteristics of major potato varieties from Europe and North America. Plant Breed Seed Sci 42, suppl 2.

Tarn, T.R., G.C.C. Tai, H. de Jong, A.M. Murphy, and J.E.A. Seabrook. 1992. Breeding potatoes for long-day temperate climates. Plant Breed Rev 9: 217–332.

Taylor, R.J., G.A. Secor, C.L. Ruby, and P.H. Orr. 1993. Tuber yield, soft rot resistance, bruising resistance and processing quality in a population of potato (cv. Crystal) somaclones. Amer Potato J 70: 117–130.

Thilmony, R.L., Z. Chen, R.A. Bressan, and G.B. Martin. 1995. Expression of the tomato *Pto* gene in tobacco enhances resistance to *Pseudomonas syringae* pv *tabaci* expressing *avrPto*. Plant Cell 7: 1529–1536.

Thurston, H.D. and T.L. Lozano. 1968. Resistance to bacterial wilt of potato in Colombian clones of *Solanum phureja*. Amer Potato J 45: 51–65.

Timms-Wilson, T.M., K. Bryant, and M.J. Bailey, 2001. Strain characterization and 16S-23S probe development for differentiating geographically dispersed isolates of the phytopathogen *Ralstonia solanacearum*. Environ Microbiol 3: 785–797.

Titarenko, E, E. Lopez-Solanilla, F. Garcia-Olmedo, and P. Rodriguez-Palenzuela. 1997. Mutants of *Ralstonia (Pseudomonas) solanacearum* sensitive to antimicrobial peptides are altered in their lipopolysaccharide structure and are avirulent in tobacco. J Bacteriol 179: 6699–6704.

Toth, I.K., A.O. Avrova, and L.J. Hyman. 2001. Rapid identification and differentiation of the soft rot erwinias by 16S-23S intergenic transcribed spacer-PCR and restriction fragment length polymorphism analyses. Appl Environ Microbiol 67: 4070–4076.

Toth, I.K., Y. Bertheau, L.J. Hyman, L. Laplaze, M.M. Lopez, J. McNicol, F. Niepold, P. Persson, G.P.C. Salmond, A. Sletten, J.M. van der Wolf, and M.C.M. Pérombelon. 1999a. Evaluation of phenotypic and molecular typing techniques for determining diversity in *Erwinia carotovora* subsp. *atroseptica*. J Appl Microbiol 87: 770–781.

Toth, I.K., L.J. Hyman, and J.R. Wood. 1999b. A one step PCR-based method for the detection of economically important soft rot *Erwinia* species on micropropagated potato plants. J Appl Microbiol 87: 158–166.

Tung, P.X., J.G.T. Hermsen, P. van der Zaag, and P.E. Schmiediche. 1993. Inheritance of resistance to *Pseudomonas solanacearum* E.F. Smith in tetraploid potato. Plant Breed 111, 1: 23–30.

Tzeng, K., R.G. McGuire, and A. Kelman, 1990. Resistance of tubers from different potato cultivars to soft rot caused by *Erwinia carotovora* subsp. *atroseptica*. Amer Potato J 67: 287–305.

Van der Wolf, J.M., W. Sledz, J.D. van Elsas, L. van Overbeeck, and J.H.W. Bergervoet. 2002. Flow cytometry to detect *Ralstonia solanacearum* and to assess viability. Proc. 3rd Intl . Bacterial Wilt Symp., 4-8 February 2002, South Africa, pp. 37.

Van der Wolf, J.M., P.J. van Bekkijm, J.D. van Elsas, E.H. Nijhuis, S.G.C. Vriend, and M.A. Ruissen. 1998. Immunofluorescence colony staining and selective enrichment in liquid medium for studying the population of *Ralstonia solanacearum* (race 3) in soil. EPPO/ OEPP Bull 28: 71–79.

Van Soest, L.J.M. 1983. Evaluation and distribution of important properties in the German-Netherlands potato collection. Potato Res 26: 109–121.

Vayda, M.E. and H.J. Schaffer. 1988. Hypoxic stress inhibits the appearance of wound-response proteins in potato tubers. Plant Physiol 88: 805–809.

Vlasov, N.M. 1983. Segregation of F_1 potato hybrids for resistance to blackleg (*Pectobacterium phytophthorum*). Nauchnotekhnicheskii Byulleten Vsesoyuznovo Ordena Lenina i Ordena Druzhby Narodov Nauchnoissledovatelskogo Instituta Rastenievodstva Imeni N.I. Vavilova 134: 57–58.

Vlasov, N.M., I. N. Buskova, and D.S. Pereverzev. 1987. Genetic sources for breeding potato varieties resistant to blackleg. Sbornik Nauchnykh Trudov po Prikladnoi Botanike, Genetike i Selektsii 115: 54–59.

Waleron, M., K. Waleron, A.J. Podhajska, and E. Lojkowska. 2002. Genotyping of bacteria belonging to the former *Erwinia* genus by PCR-RFLP analysis of *rec*A gene fragment. Microbiol 148: 583–595.

Wastie, R.L. 1984. Inoculating plant material by jet injection. Plant Path 33: 61–63.

Wastie, R.L. and G.R. Mackay. 1985. Breeding for resistance to blackleg—the present and future. *In*: D.C. Graham and M.D. Harrison (eds.). Report Intl Conf. Potato Blackleg Disease. Potato Marketing Board, Oxford, UK, pp. 75–76.

Wastie, R.L., P. Rousselle, and R. Waugh. 1993. A review of techniques for acquiring pest and disease resistance. *In:* Proc. 12th Triennial Conf. EAPR, Paris, 18-23 July, 1993. EAPR, pp. 57–74.

Wastie, R.L., G.R. Mackay, and A. Nachmias. 1994. Effect of *Erwinia carotovora* subsp. *atroseptica* alone and with *Altenaria solani* or *Verticillium dahliae* on disease development and yield of potatoes in Israel. Potato Res 37: 113–120.

Watanabe, J.A. and K.N. Watanabe. 2000. Pest resistance controlled by quantitative loci and molecular breeding strategies in tuber-bearing *Solanum.* Plant Biotech 17 (1): 1–16.

Watanabe, K., H.M. El-Nashaar, and M. Iwanaga. 1992. Transmission of bacterial wilt resistance by first division restitution (FDR) $2n$ pollen *via* $4x \times 2x$ crosses in potatoes. Eupytica 60 (1): 21–26.

Weber, J. 1990. *Erwinia*—a review of recent research. *In:* Proc. 11th Triennial Conf. EAPR, Edinburgh, UK, 8-13 July, 1990, pp. 112–121.

Wegener, C.D. 2002. Induction of defence responses against *Erwinia* soft rot by an endogenous pectate lyase in potatoes. Physiol Mol Plant Path 60 (2): 91–100.

Weller, S.A., J.G. Elphinstone, N.C. Smith, and D.E. Stead. 2000. Detection of *Ralstonia solanacearum* from potato tissue by post-enrichment TaqMan PCR. Bull EPPO/OEPP 30: 381–383.

Westra, A.A.G. and S.A. Slack. 1992. Isolation and characterization of extracellular polysaccharide of *Clavibacter michiganensis* subsp. *sepedonicus.* Phytopath 82: 1193–1199.

Westra, A.A.G., C.P. Arneson, and S.A. Slack. 1994. Effect of interaction of inoculum dose, cultivar and geographic location on the development of foliar symptoms of bacterial ring rot of potato. Phytopath 84 (4): 410–415.

Whalen, M.C. 1991. Advances in breeding for resistance to bacterial pathogens. *In*: R. Murray (ed.), Advanced Methods in Plant Breeding and Biotechnology. Biotechnology in Agriculture, No. 4. CAB Intl., Wallingford, UK, pp. 299–318.

Wiersema, H.T. 1974. Testing for resistance to common scab. Potato Res 17: 356–357.

Wolters, P.J. and W.W. Collins. 1991. Inheritance of *Erwinia* resistance and the correlation between resistance and specific gravity in diploid potatoes (abstr.). Amer Potato J 68 (9): 641–642.

Wolters, P.J. and W.W. Collins. 1994. Evaluation of diploid potato clones for resistance to tuber soft rot induced by strains of *Erwinia carotovora* subsp. *atroseptica, E. carotovora* subsp. *carotovora* and *E. chrysanthemi.* Potato Res 37: 143–149.

Wu, G., B.J. Shortt, E.B. Lawrence, E.B. Levine, K.C. Fitzsimmons, and D.M. Shah, 1995. Disease resistance conferred by expression of a gene encoding H_2O_2–generating glucose oxidase in transgenic potato plants. Plant Cell 7: 1357–1368.

Wullings, B.A., A.R. van Beuningen, J.D. Janse, and A.D.L. Akkermans. 1998. Detection of *Ralstonia solanacearum,* which causes brown rot of potato, by fluorescent *in situ* hybridiziation with 23S rRNA-targeted probes. Appl Environ Microbiol 64: 4546–4554.

Zadina, J. 1958. A genetical evaluation of some potato varieties resistant to scab (*Actinomyces scabies* (Thaxter) Gussow). Vedecke Prace Vyzkumneho Ustavu Bramborarskeho Csazv v Havlickove Brode-1958: 59–79.

Zadina, J. and K. Dobias. 1976. Moglichkeiten der Resistenzuchtung gegen die Knollenasfaule bei Kartoffeln. Tagungsbericht, Akademie der Landwirtschaftswissenschaften der Deutschen Demokratischen Republik 140: 207–219.

Zimnoch-Guzowska, E. and M.A. Dziewonska. 1989. Breeding of potatoes at the diploid level. *In*: Parental Line Breeding and Selection in Potato Breeding. EAPR Breeding and EUCARPIA Potato Section Conference, 11–16 December 1988, IAC Wageningen. Pudoc, Wageningen, Netherlands, pp. 163–171.

Zimnoch-Guzowska, E. and E. Lojkowska. 1993. Resistance to *Erwinia* in diploid potatoes with high starch content. Potato Res 36: 177–182.

Zimnoch-Guzowska, E., R. Lebecka, and J. Pietrak. 1999a. Soft rot and blackleg reactions in diploid potato hybrids inoculated with *Erwinia* spp. Amer J Potato Res 76: 199–207.

Zimnoch-Guzowska, E., R. Lebecka, and S. Sobkowiak, 1999b. An attempt to evaluate potato resistance to *Erwinia carotovora* ssp. *atroseptica* by inoculation of detached leaves. Plant Breed Seed Sci 43 (1): 101–112.

Zimnoch-Guzowska, E., W. Marczewski, R. Lebecka, B. Flis, R. Schafer-Pregl, F. Salamini, and C. Gebhardt. 2000. QTL analysis of new sources of resistance to *Erwinia carotovora* ssp. *atroseptica* in potato done by AFLP, RFLP and resistance-gene-like markers. Crop Sci 40: 1156–1167.

15

Resistance to Late Blight and Other Fungi

Richard W. Jones[1] and Ivan Simko[1,2]

[1]*USDA-ARS, Vegetable Laboratory, Plant Sciences Institute, Henry Wallace Beltsville Agricultural Research Center, 10300 Baltimore Avenue, Building 010A, Beltsville, MD 20705 USA*

[2]*Dept. Natural Resource Sciences and Landscape Architecture, University of Maryland, College Park, MD 20742 USA*

INTRODUCTION

Potato is afflicted with a number of fungal pathogens; however, major losses are generated by a limited number of these pathogens. Of greatest importance has been *Phytophthora infestans* (Mont.) de Bary, causal agent of late blight. The ability of *P. infestans* to develop new races has been countered by introgression of numerous resistance genes from wild germplasm. Sexual recombination and asexual genome plasticity allow *P. infestans* to overcome current varietal resistance and to develop greater virulence. Among potato pathogens, *P. infestans* is unique in its global prevalence and high level of crop loss.

Another significant disease is *Verticillium* wilt, or early dying disease. The pathogen lacks the level of diversity found in *P. infestans*; however, its soil-borne nature makes it a persistent problem. Resistance is not readily overcome, so new cultivars likely harbor the same resistance gene(s). Molecular analysis of resistance has enabled the cloning of two *Verticillium* resistance genes, followed by heterologous expression of each gene with concomitant resistance. Identification of the genes can also allow for their use as gene-specific molecular markers beneficial to breeding programs.

A third fungal pathogen of importance is *Alternaria solani* (Soraner), causal agent of early blight. Efforts to select for field resistance have been variable and might reflect poor understanding of the physiological and molecular aspects of the host-pathogen interaction. Early blight resistance is complicated by limited heritability and multiple gene effects.

Corresponding author: Richard W. Jones, Tel.: 01-301-504-7380; Fax: 01-301-504-5555, *e-mail: jonesr@ba.ars.usda.gov*

Current and future progress in understanding and manipulating fungal resistance will be facilitated by cloning of resistance genes and through use of *Solanum* comparative genomics information.

RESISTANCE TO LATE BLIGHT

Late blight, caused by *Phytophthora infestans* (Mont.) de Bary, is the most devastating disease of potato worldwide. *P. infestans* is a specialized pathogen, primarily causing disease on solanaceous species (Erwin and Ribeiro 1996), the most economically important of which are potato and tomato. The pathogen is a limiting factor in potato cultivation in many developing countries. When uncontrolled, late blight epidemics can completely kill the foliage of a crop within a few weeks. In 1845 and 1846, *P. infestans* totally destroyed Ireland's potato crop, which resulted in poverty and mass starvation—the infamous Irish potato famine.

Symptoms of late blight appear on leaves as pale green, water-soaked spots, often beginning at the tips or edges. A pale yellowish-green border that merges with the healthy tissue often surrounds the circular or irregularly shaped lesions. The lesions enlarge rapidly and turn brown or purplish-black. During periods of high relative humidity (RH) and leaf wetness, the underside of the leaf is covered with masses of sporangia. In dry weather, infected leaf tissues turn brown and quickly dry out. Infected stems and petioles turn brown to black, and entire vines may be killed and blackened in a short time when damp weather persists (Rowe 1993). Tubers become infected later in the season, from inoculum generated on foliar and stem tissues. Infection of tubers results in a shallow, coppery brown dry rot that spreads irregularly from the surface through the outer tissues. Late blight lesions on tubers frequently become infected by various bacteria and fungi, which often cause secondary breakdown of the tissues (Rowe 1993). Use of infected tubers for seed pieces can lead to pathogen transport.

P. infestans is a heterothallic *Oomycete*. The pathogen evolved in Toluca Valley, in the central highlands of Mexico (Niederhauser 1956). Biochemical analyses and comparison of sequences of ribosomal and mitochondrial genes suggest that *Oomycetes* share little taxonomic affinity to fungi (Fig. 15.1), but are more closely related to heterokont algae (van de Peer and de Wachter 1997; van de Peer et al. 2000). Before the 1980s, occurrence of two known mating types (A1 and A2) were restricted to Mexico; whereas in the rest of the world, only the A1 mating type was present (Goodwin et al. 1994). Since both mating types are needed for sexual reproduction, the presence of only the A1 mating type confined its reproduction mode to the asexual form (Fig. 15.2). After 1980, this situation changed, first in Europe and later in the USA and Canada. New

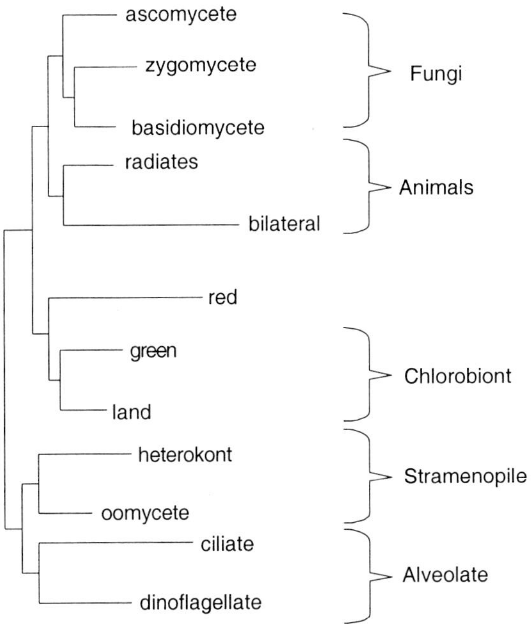

Fig. 15.1 Simplified phylogenetic tree of (crown) Eukaryotes (adapted from van de Peer and de Wachter 1997, van de Peer et al. 2000).

populations of *P. infestans* detected were both the A1 and A2 mating types (Fry et al. 1992, 1993; Deahl et al. 1991). The mixed presence of the two mating types led to sexual reproduction and consequently to dramatic increases in genetic variation.

The dramatic increase of genetic variation has consequently led to major changes in resistance to fungicides. Prior to 1990, strains of the late blight fungus resistant to the systemic fungicide metalaxyl were rarely encountered. Metalaxyl was first introduced in 1977 and before long extensively used to control late blight disease (Schwinn and Margot 1991). Matters suddenly changed with the advent of new populations. By 1992, 45 % *P. infestans* isolates from the USA and Canada were resistant to metalaxyl fungicide, and two years later this figure jumped to 87% (Deahl and Jones 1999). Resistant strains of the pathogen had been identified in most potato producing areas in 1998 and potato farmers were forced to use older, contact fungicides to protect the crop against the disease (Deahl and Jones 1999). It appears that metalaxyl resistance in *P. infestans* is a monogenic dominant trait while variation in sensitivity involves minor (probably recessive) genes (Lee et al. 1999).

Late blight on potato is usually controlled by frequent application of contact fungicides. To be effective, protectant materials must be applied

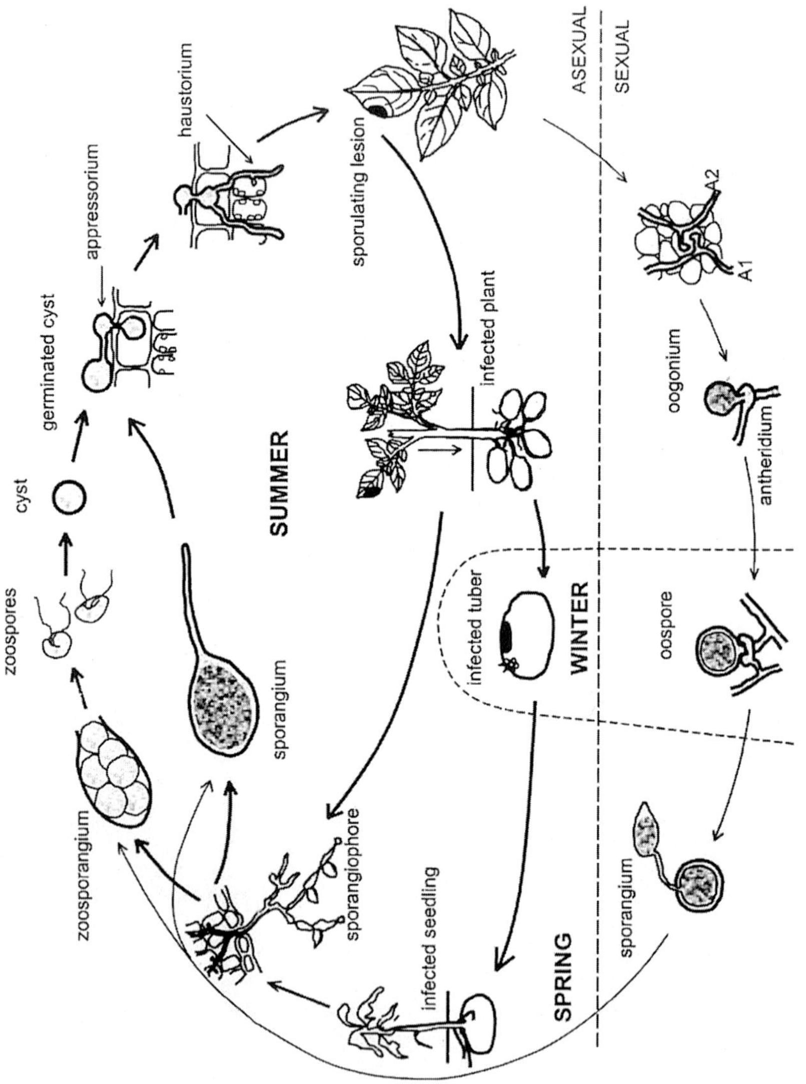

Fig. 15.2 Life-cycle of *Phytophthora infestans* (reprinted with permission from van West 2000).

before infection and the application repeated regularly as the plants grow. Under conditions favorable to late blight development it is not uncommon to make 10 applications on late-maturing cultivars (Rowe 1993). Reductions in the cost of crop protection and decreased environmental hazard can be achieved by cultivating resistant varieties. Three types of natural resistance to late blight can be distinguished in *Solanum* species:

— Monogenic resistance (vertical resistance)
— Polygenic resistance (horizontal resistance)
— Nonhost resistance.

Monogenic Resistance

Monogenic (vertical, race-specific) resistance is characterized by interactions between the products of dominant resistance R-gene alleles in the host and corresponding avirulence (*Avr*) gene alleles in the pathogen—the gene-for-gene hypothesis (Flor 1971). The result of the interaction is the hypersensitive response (HR) that prevents further spread of the pathogen. HR generally occurs as a rapid, localized cell death, and could be considered a form of programmed cell death in plants (Heath 1998).

Hypersensitivity to late blight in cultivated potato varieties is controlled by eleven R-genes, all of which were introgressed from Mexican species *Solanum demissum* Lindl. (Wastie 1991). To date, five of the eleven R-genes for foliage resistance to late blight have been mapped on a molecular map. The *R1* gene is located on chromosome 5 (El-Kharbotly et al. 1994; Leonards-Shippers et al. 1992), *R2* on chromosome 4 (Li et al. 1998), and *R3*, *R6*, and *R7* clustered on chromosome 11 (El-Kharbotly et al. 1994, 1996a). Three other R-genes originating from *Solanum pinnatisectum* Dunal (R_{pi1}), *Solanum bulbocastanum* Dunal (R_{blc}), and *Solanum berthaultii* Hawkes (R_{ber}) and conferring resistance to complex races were mapped to chromosomes 7 (Kuhl et al. 2001), 8 (Naess et al. 2000), and 10 (Ewing et al. 2000) respectively. More R-genes are present in wild species (Trognitz 1998, Micheletto et al. 1999) but have not yet been mapped.

Recently, the first gene for resistance to late blight (*R1*) was cloned by using a combined positional cloning and candidate gene approach method (Ballvora et al. 2002). The structure of the *R1* gene is similar to other plant genes conferring resistance to different types of pathogens in having a leucine zipper motif, a putative nucleotide binding domain, and a leucine-rich repeat domain. The deduced amino acid sequence of the *R1* gene is most similar (36% identity) to the *Prf* gene for resistance to *Pseudomonas syringae* (van Hall) of tomato (Salmeron et al. 1996). *R1* (GenBank: AF447489) is present as an extra copy in a DNA insertion in the *R1*-bearing chromosome compared to the *r1* (susceptible) chromosome. The resistance gene has likely been introgressed into the *Solanum tuberosum* L. genome from the wild species *S. demissum* through heterogenetic chromosomal crossing-over (Ballvora et al. 2002). Work on cloning the second resistance gene originating from *S. demissum* (*R7*) is well advanced (Wirtz et al. 2002). *RB* (GenBank: AY303171)—a late blight resistance gene originating from *S. bulbocastanum*—was cloned using a map-based approach in combination with a long-range PCR strategy (Song et al. 2003). The *RB* is more closely related to the *I2* protein of tomato (30% identity) than to any other known R protein, including the *R1* resistance protein (22% identity).

Resistance based on a single R-gene, or combinations of few R-genes has not proven durable. New pathotypes of *P. infestans* quickly evolve that are virulent against the formerly resistant host (Wastie 1991). Molecular studies on *P. infestans* revealed that single *Avr* genes for most of the interactions condition specificity for R-genes in potato. Six dominant *Avr* loci have been placed on the molecular map of *P. infestans* (van der Lee et al. 2001). Further studies at the molecular level are needed to unravel more aspects of the interaction between *P. infestans* and the host plant.

To avoid the problem of short durability of R-gene conferred resistance, many breeders advocate elimination of R-genes from breeding populations (Wastie 1991). The opposite strategy of incorporating many different R-genes into a single genotype (gene pyramiding) has also been advanced (Mohan et al. 1997). The presence of multiple resistance genes in a single genotype increases the probability of achieving the durable resistance. However, it is also possible that there will be selection of pathogens that can tolerate the loss of multiple compatible alleles. Moreover, there are data suggesting that the presence of multiple R-genes could lead to reduced plant fitness and loss of yield in the absence of the pathogen (Jones 2001).

It was suggested that using South America's *Solanum* species in a breeding program might provide a better alternative approach to achieve durable resistance. Colon et al. (1995a) hypothesized that those species may harbor a race nonspecific resistance because they have not evolved in the presence of *P. infestans*. This idea was probably influenced by the presence of race-specific R-genes in Mexican *S. demissum*. However, it appears that Mexican races of *P. infestans* from the Toluca Valley can even overcome at least some R-gene resistance derived from South American potato species. In laboratory tests on detached leaves, the isolate MX990005 (isolated in Mexico in 1999) was compatible with the R-gene originating from *S. berthaultii* (W.E. Fry and H. Mayton, unpubl. data; Ewing et al. 2000; Simko 2002).

Another problem confronting breeders when using R-genes as a source of resistance, are suppressors of the R-gene function. Ordonez et al. (1997) observed the presence of specific suppressors elicited by specific isolates of the fungus. These dominant suppressors segregated in the host independently from the R-gene (assuming Mendelian inheritance). The presence of a dominant suppressor was independently described by El-Kharbotly et al. (1996b). Resistance segregation in the F_1 population clearly indicated that the R-gene was present but not expressed in parental plants.

Polygenic Resistance

Polygenic (horizontal) resistance is controlled by minor genes with additive effects and considered to be race nonspecific (Simmonds and Wastie 1987). It was long known that polygenic resistance contributes to disease suppression but until the advent of molecular tools, it was nearly intractable to analysis. The availability of molecular markers, however, enables genetic analysis of complex traits, such as resistance to late blight. Identification of polygenes that contribute significantly to the resistance in a particular progeny population is therefore possible (Meyer et al. 1998).

The first molecular linkage map of potato was constructed with RFLP markers (Bonierbale et al. 1988). Since then, AFLP (van Eck et al. 1995), SSR (Milbourne et al. 1998), RAPD (Hosaka 1999), and transposons (Jacobs et al. 1995) have been used to construct linkage maps based on segregation in diploid populations. These molecular linkage maps provide the framework for the location of qualitative and quantitative resistance genes in the potato genome. Factors significantly affecting quantitative resistance were detected over all 12 chromosomes (Gebhardt and Valkonen 2001, Simko 2002). Two-thirds of the detected QTLs showed a relatively small effect on resistance (10%-20%). A large effect on trait variation (30%-45%) was observed in about one-tenth of the mapped QTLs (Simko 2002). Comparative analysis across many different mapping populations revealed three highly active genomic regions on the distal parts of chromosomes 3, 4, and 5 (Simko 2002). The region that most consistently conferred foliage resistance was located on chromosome 5, near marker locus GP21 (Fig. 15.3). Polygenic resistance at this locus was detected in populations originating from numerous *Solanum* species, using various *P. infestans* races and different resistance testing methods. The susceptible allele appears to be dominant over those conferring foliage resistance (Leonards-Schippers et al. 1994).

Polygenic resistance in potato seems to be durable since the general level of resistance is stable over time (Colon et al. 1995b), genotypes react similarly to different isolates of *P. infestans* (Inglis et al. 1996), and resistance has good phenotypic stability in different environments (Haynes et al. 1998, Haynes and Christ 1999). Broad-sense and narrow-sense heritability of resistance in diploid populations of *Solanum phureja* Juz. and Bukasov × *Solanum stenotomum* Juz. and Bukasov were high (0.79 and 0.78 respectively), suggesting additive genetic variance (Haynes and Christ

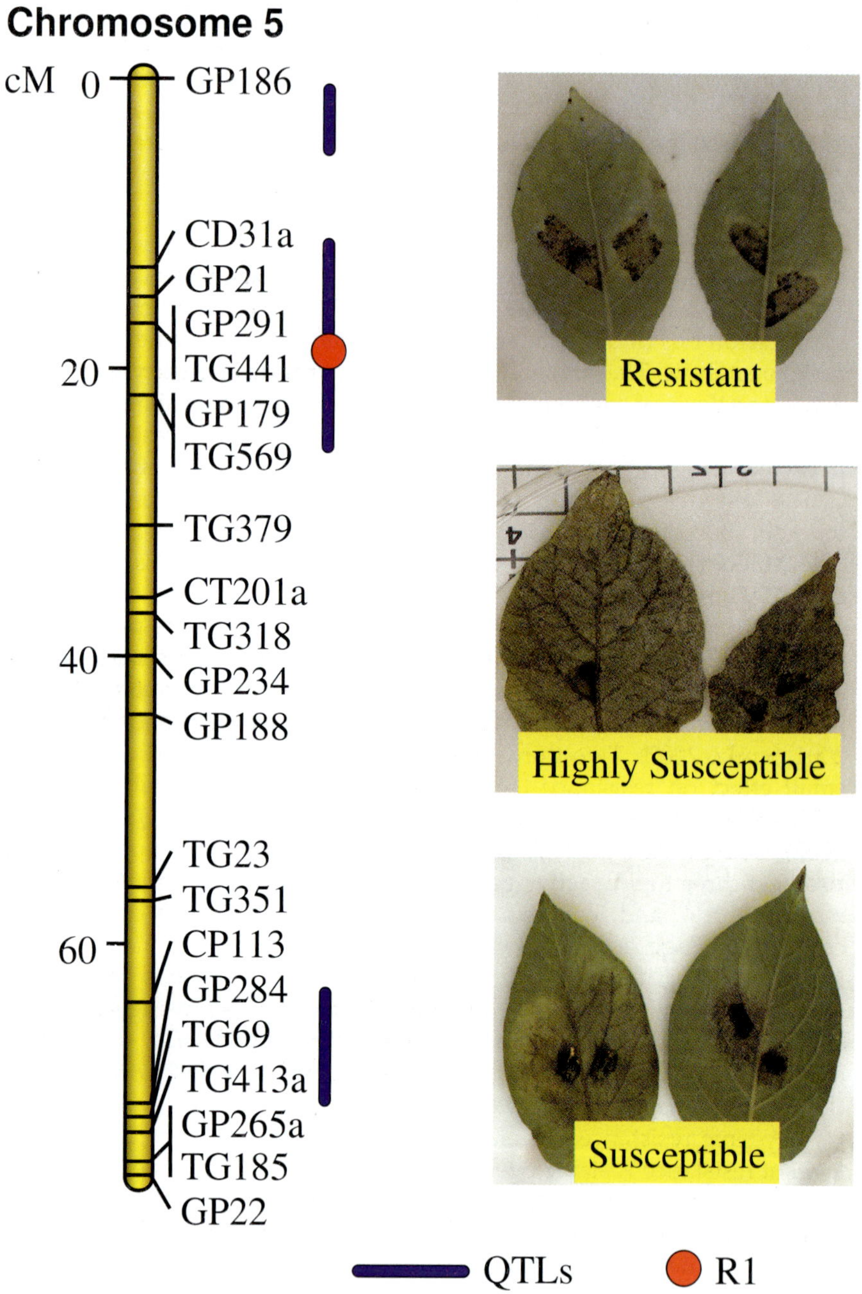

Fig. 15.3 Location of QTLs and *R1* mediated resistance to late blight on potato chromosome 5 (reprinted with permission from American Journal of Potato Research, 2002, 79, no. 2, cover page).

1999). Since additive effects can readily be sexually transmitted via 4x-2x crosses (Haynes 1990), there is a high expectation of recovering late blight resistant tetraploids.

Unfortunately, quantitative resistance to late blight tends to be associated with late maturity (Toxopeus 1958). Analysis of European and North American potato varieties indicated that almost 60% of variation in resistance can be explained by indirect effect of maturity (Fig. 15.4) (Simko unpubl. data). There are several possible explanations for this association. It may be that genes for late maturity incidentally accompanied genes for late blight resistance when selections were made from *S. demissum* (Wastie 1991), or that there was a stronger selection for resistance in late maturing cultivars (Swiezynski 1990). Another possibility is that one or more of the polygenes contributing to late maturity have a direct pleiotropic effect on late blight resistance (Ewing et al. 2000). There is evidence that the potato plant becomes more susceptible to infection as the foliage matures. When the same cultivar was planted on successive dates and inoculated with *P. infestans* at the same time, the earliest plantings were most susceptible. The age effect on epidemic development was most noticeable for late-season cultivars and least noticeable for early-season cultivars (Fry and Apple 1986). According to observations on potato leaves, the growth rate of late blight lesions generally increased linearly from the apex to the base of plants, except that such differences were usually absent on very young plants (Carnegie and Colhoun 1982). It appears that factors that tend to hasten maturity, such as short photoperiod (Colon 1994) and low nitrogen level (Lowings and Acha 1959), make plants more susceptible to late blight.

Results from molecular mapping confirmed linkage between resistance to late blight and late maturity. Of the five chromosomes on which Ewing et al. (2000) detected QTLs for late blight resistance, four also carried QTLs governing maturity, coincident with or in close position to the QTLs for late blight resistance. Also, plants having a resistance allele at marker locus GP21 on chromosome 5, which increased foliage resistance, always exhibited late maturity (Collins et al. 1999; Ewing et al. 2000; Oberhagemann et al. 1999; Simko et al. 1999; van den Berg et al. 1996; van Eck and Jacobsen 1996). In the present map resolution, whether these QTLs for resistance to late blight are pleiotropic with late maturity or are closely linked cannot be distinguished. Gebhardt et al. (2001) genotyped approximately 600 European and North American potato varieties with PCR-based DNA markers from a 1 cM interval in the resistance region on chromosome 5. The markers were tested for association with maturity type and foliage and tuber resistance to late blight. A highly significant association with all three traits was detected with two markers tightly linked to the *R1* resistance gene.

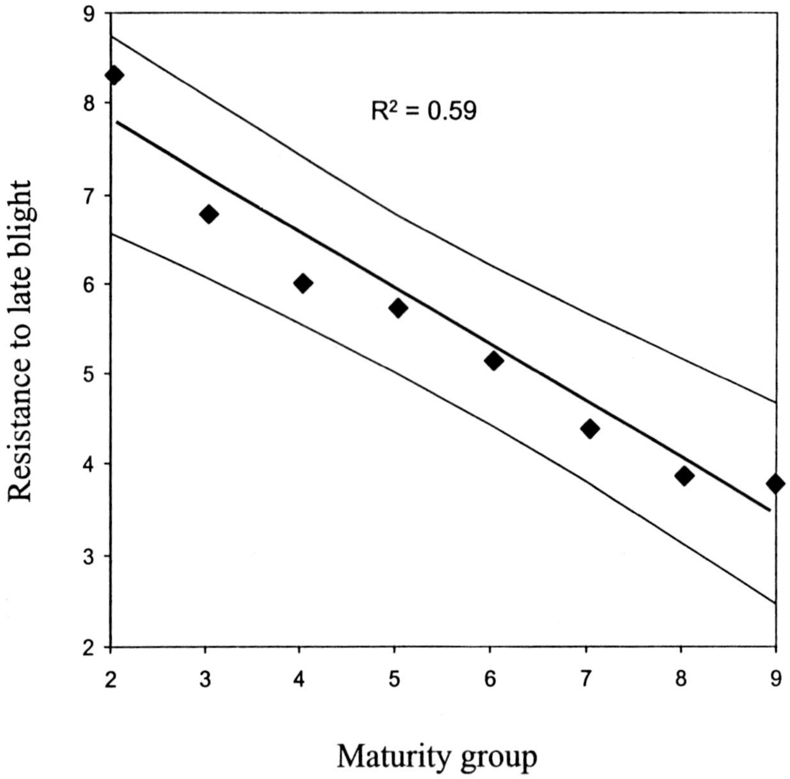

Fig. 15.4 Relationship between maturity and resistance to late blight in European and North American varieties. A higher value indicates earlier maturity and higher resistance to late blight respectively. Diamond indicates the mean resistance value for each maturity group and 95% confidence interval is plotted for the trend line. One hundred and seventy varieties have been used to calculate linear correlation (data were obtained from Central Potato Association of Czech Republic 1998, NIVAA CPRO-DLO 1991).

A combination of high polygenic resistance with earliness in a potato cultivar is difficult, and has not yet been achieved though breeding for resistance to late blight begun nearly 100 years ago (Wastie 1991). Intercrossing the most resistant cultivars might result in better resistance, although it may also result in further deterioration of earliness (Colon et al. 1995b), since most resistant cultivars are late maturing. Further work at the molecular level may provide an answer to the question of the relationship between maturity and resistance to late blight. If the two traits are closely linked, rather than pleiotropic, molecular markers may be developed and used for marker-assisted selection. Alternatively, only resistance QTLs not linked with late maturity might be used in breeding programs. It was hypothesized that not all QTLs for late blight resistance and maturity are congruent because substantial differences in resistance exist within each potato maturity class (Ewing et al. 2000).

The distinction made between two types of resistance—qualitative (vertical) and quantitative (horizontal), controlled by major R-genes or minor polygenes, is often not clear-cut at the phenotypic level (Gebhardt and Valkonen 2001). It has been suggested that there is no real difference between quantitative and qualitative resistance to late blight and that the differences observed at the phenotypic level may be the result of extreme allelic variation (Gebhardt 1994). By this hypothesis it would be expected that genomic regions containing R-genes should show high QTL activity as well. Analysis of QTL mapping results showed mixed support for the hypothesis that quantitative and qualitative resistance are related. Two mapped R-genes (*R1* and *R2*) are located in genomic regions with high QTL activity, but six (*R3*, *R6*, *R7*, R_{pi1}, R_{blc}, and R_{ber}) are found on chromosome parts exhibiting low polygenic resistance to late blight (Simko 2002). It should also be noted that the mapped R-genes and polygenes do not originate from the same *Solanum* species, which could complicate interpretation of the results with respect to the hypothesis. On the other hand, cytological observations indicated that the difference between compatibility and incompatibility is quantitative rather than qualitative (Vleeshouwers et al. 2000). In highly resistant clones, all cells responded with a rapid HR upon *P. infestans* hyphae penetration, whereas in fully susceptible clones only a low percentage of cells displayed HR. The timing of HR induction differed remarkably between clones and quantification of the number of HR responding cells suggested a correlation between resistance level and HR effectiveness (Vleeshouwers et al. 2000). Another possibility is that R-genes and factors controlling quantitative resistance to late blight may be members of a clustered gene family (Leonards-Schippers et al. 1994; Oberhagemann et al. 1999, Ballvora et al. 2002). If a close relationship between quantitative and qualitative resistance is confirmed, the conserved sequence motifs from R-genes could be used to map functional polygenes. The first interesting results on potato were achieved with primers designed to such conserved motifs, when at least 20 resistance gene-like (RGL) markers were identified (Leister et al. 1996). Nevertheless, more molecular information about functional polygenes is needed before this potentially useful method can be fully realized.

Nonhost Resistance

P. infestans exhibits specialization and can only infect a limited number of plant species. This implies that the majority of nonhost plants possess a series of either preformed or inducible mechanisms to successfully preclude infection by *P. infestans* (Birch and Whisson 2001). Niks (1987) defined a nonhost as a plant species in which most genotypes cannot be infected by most genotypes of the pathogen. Platt (1999) observed the response of solanaceous plants to inoculation with three races of *P. infestans* (US-1, US-8, and US-11). He concluded that black nightshade (*Solanum nigrum* L.), eggplant (*Solanum melongena* L.), green pepper (*Capsicum annum* L.), and tobacco (*Nicotiana tabacum* L.) are nonhosts for *P. infestans* since no disease symptoms or hypersensitive responses were observed in stem piece tissues of these plant species.

Transfer of resistance in nonhost to cultivated potato has received relatively little attention in efforts to produce resistant potato varieties. Part of the problem being the difficulty in producing sexual hybridization between nonhost species and *S. tuberosum*. However, *in vitro* culture techniques, in particular embryo rescue, have opened a way to study transfer of nonhost resistance (Eijlander and Stiekema 1994). Colon et al. (1993) analyzed crosses of *S. tuberosum* and *S. demissum* with nonhost *Solanum* species *S. nigrum* Linn. (black nightshade) and *Solanum villosum* Mill. (bittersweet). Resistance was transferred to hybrids, in which it behaves as a dominant character. The authors recommend use of nonhost resistance in breeding resistant potato germplasm provided that recombination events occur between the two genomes.

MOLECULAR MAP SYNTENY

High levels of synteny among molecular maps of potato, tomato, and pepper allow comparison of resistance gene position across genera (Grube et al. 2000; Pan et al. 2000). In pepper, two QTLs for resistance to *Phytophthora capsici* Leonian (Lefebvre and Palloix 1996) are located in the genomic regions corresponding to the potato R-genes on chromosomes 4 (*R2*) (Li et al. 1998; Grube et al. 2000) and 11 (*R3, R6, R7*) (El-Kharbotly et al. 1994, 1996a) respectively. In tomato, two of the three known R-genes also map to genomic positions corresponding to potato monogenic resistance. The *Ph-1* gene from tomato on chromosome 7 (Pierce 1971) appears to be positioned similar to the R_{pi1} gene from *S. pinnatisectum* (Kuhl et al. 2001), while the *Ph-2* gene (Moreau et al. 1998) is located on chromosome 10 in the same genomic region as R_{ber} from *S. berthaultii* (Ewing et al. 2000). However, the *Ph-2* and R_{ber} genes do not appear to be orthologous since they showed different race specificity when tested with US-7 race of *P. infestans* (Ewing et al. 2000). Nevertheless, similarities in location of genes for resistance to *Phytophthora* species in potato, tomato, and pepper suggest possible evolutionary conservation of R-genes. Grube et al. (2000) hypothesized that the general function of R alleles (e.g. initiation of resistance response) may be conserved at homologous loci in related plant genera, though the taxonomic specificity of host R-genes may be evolving rapidly. In the case of potato, tomato, and pepper, it is possible that resistance gene specificity across genera remains relatively conserved and that homologous genes are still conferring resistance to *Phytophthora* species. Cloning and sequencing of resistance genes should provide valuable tools for testing the hypothesis.

VERTICILLIUM WILT OR EARLY DYING DISEASE

Early dying disease is associated with infection by the deuteromycete fungi *Verticillium dahliae* Kleb. and *Verticillium albo-atrum* Reinke & Berth. In cooler climates or growing seasons *V. albo-atrum* may predominate, while warmer conditions favor infection by *V. dahliae*. Colonization of host xylem occurs after infection through the roots, resulting in reduced rates of water and nutrient flow through the water-conducting tissues. Symptoms appear as interveinal chlorosis, a mild degree of wilt, and premature senescence beginning with the older leaves.

Verticillium is a soil-borne fungal pathogen that survives multiple years in the soil as melanized microsclerotia in the absence of a host. Hyaline conidia are produced in the infected host and provide a mechanism for movement of the fungus through the xylem vessels. *Verticillium* is known to infect a wide range of dicot hosts, making crop rotation a less useful tool in disease management.

Probabilities of infection by the soil-borne microsclerotia are based upon the number of propagules in the soil and the growth form of the root system. The root must grow in close proximity to the microsclerotia to stimulate germination and penetration. This can influence apparent resistance scores during field evaluations, yet soil microsclerotia quantification is not often included in field studies. Greenhouse studies often use root-dipping inoculations, or introduction of quantified levels of spores into the soil, that should provide more uniform screening values. Unfortunately, greenhouse assays sometimes do not take into account soil variables such as nematode populations that might well influence plant susceptibility (Bowers et al. 1996).

Resistance has been rated either through symptom scoring or quantification of spore loads in the xylem. It is generally assessed as a reduction in symptoms or spore load, but not a complete lack of colonization; thus ratings might be viewed as indications of tolerance that reflect the form in which the resistance gene functions. Recent cloning of the tomato *Ve* gene revealed a protein with endocytotic domains, suggesting the transfer of an elicitor from the xylem fluid. There would have to be a minimal initial inoculum load to provide sufficient levels of elicitor. Interestingly, another solanaceous resistance gene, *Cf9*, shares the endocytotic feature and is believed to transport the small elicitor peptide *Avr9* produced in the apoplast by *Cladosporium fulvum* Cooke, causal agent of tomato leaf mold disease.

The resistance gene *Ve*, mapped to linkage group 9 in tomato (Diwan et al. 1999), was cloned after development of SCAR markers located within 290 kB of *Ve* (Kawchuk et al. 1994, 2001). SCAR markers were used to probe a lambda library and positive genomic fragments were tested by transformation into potato. The two inverted resistance gene copies, *Ve1* and *Ve2*, were each capable of conferring tolerance in the highly susceptible Desiree potato. This test involved race 1 of *V. albo-atrum*; results for *V. dahliae* would expectedly be similar but were not reported.

Identification of the *Ve* gene from tomato provides two methods for developing R-gene mediated *Verticillium* resistant potato cultivars. The first method is production of transgenic potato through introduction of the tomato-encoded resistance gene. Development of resistant potato using a tomato gene provides additional support for the idea that heterologous gene expression among solanaceous crops is quite feasible. Similar results were previously shown in transfer of a resistance gene from pepper to tomato (Tai et al. 1999).

The second method for development of *Verticillium* resistant potato would be use of gene-specific marker-based selection. Development of new cultivars with resistance gene(s) against the pathogen can be assisted with molecular marker technology that allows identification and tracking of resistance genes. We employed primers that amplify the leucine-rich repeat (LRR) domain from tomato *Ve1* and *Ve2* genes. *Verticillium* resistance gene homologues have been detected in resistant potato cv. Reddale when using these primers and genomic DNA as a template. Deduced amino acid sequence shared high similarity with *Ve1* (83%-90%) and *Ve2* (88%-91%) genes from tomato. The *StVe1*, a potato homologue to the LRR region of the *Ve1* gene, appears to be located on chromosome 9, corresponding to the position of the tomato *Ve* gene. Microsatellite markers linked to the *StVe1* have been used to screen tetraploid genotypes of various pedigrees. One of the tested markers (STM1051) showed good association with potato resistance to *Verticillium* wilt. We observed a strong correlation between absence of the marker specific amplicon and high susceptibility of potato genotypes ($p < 0.001$). The STM1051 marker has potential for use in detection of genotypes highly susceptible to *Verticillium*. Currently we are developing a new tetraploid mapping population to test linkage between the marker locus and *Verticillium* disease resistance (Simko et al. 2003, 2004).

EARLY BLIGHT

The disease early blight, contrary to its name, may occur anytime from mid to late season. Symptoms of infection by the deuteromycete fungus

Alternaria solani Sorauer, can be found principally on the foliage, with additional infections on the petioles and stem. Infections are initiated after germination of multicellular melanized conidia. Slowly expanding, often concentric lesions are dry and dark brown, reflecting the necrotrophic nature of the infection. As lesions expand they may be preceded by a chlorotic halo, possibly representing *Alternaria* toxin production.

Infection and secondary spread is favored by cycles of wet and dry conditions in which infection is initiated during periods of free moisture and spore dispersal favored by dry periods. Initial infections appear on lower leaves and result in the largest lesions. Young foliage may have only tiny lesions which may rapidly coalesce due to the large number of infection sites. Older, lower leaves can often be lost due to infection, resulting in lower tuber bulking. Infected lower leaves provide an inoculum source for infection of tubers at harvest, which then develop dry-darkened, slightly depressed lesions.

One factor influencing the severity of early blight is the level of nitrogen available to the plant. Low nitrogen levels increase disease severity evidenced in the higher susceptibility of older leaves in which nitrogen levels are the lowest. This is a critical factor when evaluating cultivars or germplasm, as they may translocate and utilize nitrogen differently. The physiological basis for nitrogen-mediated differences in susceptibility has not been suitably investigated. It is not known if the nitrogen levels regulate the host resistance or the pathogen virulence. In either case, genes involved in nitrogen utilization could prove an interesting target for marker development.

A second factor influencing apparent susceptibility is maturity, with late maturing potatoes having the highest apparent resistance (Johanson and Thurston 1990; Christ 1991). How maturity factors relate with nitrogen levels remains to be determined.

Unlike the late blight disease, there are no cultivar differentials that indicate specific R-genes exist for resistance to the early blight pathogen. The lack of specific R-genes may be expected due to the necrotrophic nature of *Alternaria* infections. No published efforts to distinguish possible isolates of *A. solani* are known. Various attempts have been made to assess the form of resistance present (Ortiz et al. 1993). Studies on diploid populations indicate narrow-sense heritability values of 0.61 to 0.78, supporting the premise that additive effects prevail in diploid populations (Gopal 1998, Christ and Haynes 2001). The lack of any clear R-gene mediated resistance, coupled with physiological variables of the host, make breeding efforts difficult. A more feasible method may be through engineering of potato with inhibitory gene products.

NOVEL FORMS OF FUNGAL DISEASE RESISTANCE

Resistance to fungal diseases has progressed through continued efforts to introgress and select for genes of plant origin. With current progress in genetic transformation and gene expression technologies, the ability to develop new resistant varieties can be expanded and accelerated. As previously mentioned, cloning of specific resistance genes from solanaceous germplasm and both homologous and heterologous expression have clearly demonstrated the usefulness of these technologies.

In some cases resistance can be generated by incorporation of genes that confer a fungistatic or fungicidal property to the plant cells. One example of this is found in the efforts to control *Verticillum* through expression of an antimicrobial peptide originating from alfalfa seed (Gao et al. 2000). Antimicrobial peptides are found in a wide array of organisms and act principally by permeabilizing the membranes of target organisms (van der Biezen 2001). Peptides with inhibitory activity toward *P. infestans* and *A. solani* have been reported (Cavallarin et al. 1998; Ali and Reddy 2000). The mode of action suggests that they would be race nonspecific, ideally suited for controlling the race-prolific *P. infestans*. Antimicrobial peptides found in potato, such as the snakins (Segura et al. 1999) are not very effective in their native form, although they may contribute a modest degree of protection. It would be interesting to determine whether snakin expression levels correlate with cultivar resistance to pathogens. Overexpression of potato-specific snakin or other antimicrobial peptides should prove to be a valuable addition to current efforts in breeding for resistance.

ACKNOWLEDGEMENTS

The authors would like to thank Dr. E.E. Ewing and Dr. K. Lewers for critically reading the manuscript.

REFERENCES

Ali, G.S. and A.S.N. Reddy. 2000. Inhibition of fungal and bacterial plant pathogens by synthetic peptides: In vitro growth inhibition, interaction between peptides and inhibition of disease progression. Molec Plant Micr Interact 13: 847–859.

Ballvora, A., M.R. Ercolano, J. Weiss, K. Meksem, C.A. Bormann, P. Oberhagemann, F. Salamini, and C. Gebhardt. 2002. The *R1* gene for potato resistance to late blight (*Phytophthora infestans*) belongs to the leucine zipper/NBS/LRR class of plant resistance genes. Plant J 30: 361–371.

Birch, P.R.J. and S.C. Whisson. 2001. *Phytophthora infestans* enters the genomic era. Molec Plant Path 2: 257–263.

Bonierbale M., R.L. Plaisted, and S.D. Tanksley. 1988. RFLP maps based on a common set of clones reveal modes of chromosomal evolution in potato and tomato. Genetics 120: 1095–1103.

Bowers, J.H., S.T. Nameth, R.M. Riedel and R.C. Rowe. 1996. Infection and colonization of potato roots by *Verticillium dahliae* as affected by *Pratylenchus penetrans* and *P. crenatus*. Phytopath 86: 614–621.

Carnegie, S.F. and J. Colhoun. 1982. Susceptibility of potato leaves to *Phytophthora infestans* in relation to plant age and leaf position. Phytopath Z 104: 157–167.

Cavallarin, L., D. Andreu, and B.S. Segundo. 1998. Cercropin A-derived peptides are potent inhibitors of fungal plant pathogens. Molec Plant Micr Interact 11: 218–227.

Central Potato Assoc. Czech Republic. 1998. The Catalogue of Potato Varieties of the Czech Republic. Central controlling and testing institute for agriculture—variety testing department in Brno and Potato research institute in Havlíčkův Brod, Czech Republic.

Christ, B.J. 1991. Effect of disease assessment method on ranking potato cultivars for resistance to early blight. Plant Dis 75: 353–356.

Christ, B.J. and K.G. Haynes. 2001. Inheritance of resistance to early blight disease in a diploid potato population. Plant Breed 120: 169–172.

Collins, A., D. Milbourne, L. Ramsay, R. Meyer, C. Chatot-Balandras, P. Oberhagemann, W. De Jong, C. Gebhardt, E. Bonnel, and R. Waugh. 1999. QTLs for field resistance to late blight in potato are strongly correlated with maturity and vigour. Mol. Breed. 5: 387–398.

Colon, L. 1994. Resistance to *Phytophthora infestans* in *Solanum tuberosum* and wild *Solanum* species. PhD thesis, Wageningen Agric Univ., Wageningen, The Netherlands.

Colon, L.T., R.C. Jansen, and D.J. Budding. 1995a. Partial resistance to late blight (*Phytophthora infestans*) in hybrid progenies of four South American *Solanum* species crossed with diploid *S. tuberosum*. Theor Appl Genet 90: 691–698.

Colon, L.T., L.J. Turkensteen, W. Prummel, D.J. Budding, and J. Hoogendoorn. 1995b. Durable resistance to late blight (*Phytophthora infestans*) in old potato cultivars. Eur J Plant Path 101: 387–397.

Colon, L.T., R. Eijlander, D.J. Budding, M.T. Van Ijzendoorn, M.M.J. Pieters, and J. Hoogendoorn. 1993. Resistance to potato late blight (*Phytophthora infestans* (Mont.) de Bary) in *Solanum nigrum*, *S. villosum* and their sexual hybrids with *S. tuberosum* and *S. demissum*. Euphytica 66: 55–64.

Deahl, K.L. and R. Jones. 1999. The occurrence of late blight in North America. *In:* Late Blight: A Threat to Global Food Security, Vol. 1. International Potato Center, Lima, Peru, pp. 15-18.

Deahl, K.L., R.W. Goth, R. Young, S.L. Sinden, and M.E. Gallegly. 1991. Occurrence of the A2 mating type of *Phytophthora infestans* in potato fields in the United States and Canada. Amer Potato J 68: 717–725.

Diwan, N., R. Fluhr, Y. Eshed, D. Zamir and S.D. Tanksley. 1999. Mapping of *Ve* in tomato: a gene conferring resistance to the broad-spectrum pathogen, *Verticillium dahliae* race 1. Theor Appl Genet 98: 315–319.

Eijlander, R. and W.J. Stiekema. 1994. Biological containment of potato (*Solanum tuberosum*)—outcrossing to the related wild species Black nightshade (*Solanum nigrum*) and Bittersweet (*Solanum dulcamara*). Sex Plant Reprod 7: 29–40.

El-Kharbotly, A., C. Palomino-Sanchez, F. Salamini, E. Jacobsen, and C. Gebhardt. 1996a. R6 and R7 alleles of potato conferring race-specific resistance to *Phytophthora infestans* (Mont.) de Bary identified genetic loci clustering with the *R3* locus on chromosome XI. Theor Appl Genet 92: 880–884.

El-Kharbotly, A., A. Pereira, W.J. Stiekema, and E. Jacobsen. 1996b. Race specific resistance against *Phytophthora infestans* in potato is controlled by more genetic factors than only R-genes. Euphytica 90: 331–336.

El-Kharbotly, A., C. Leonards-Schippers, D. Huigen, E. Jacobsen, A. Pereira, W. Stiekema, F. Salamini, and C. Gebhardt. 1994. Segregation analysis and RFLP mapping of the *R1* and *R3* alleles conferring race-specific resistance to *Phytophthora infestans* in progeny of dihaploid potato parents. Molec Gen Genet 242: 749–754.

Erwin, D.C. and Ribeiro, O.K. 1996. Phytophthora Diseases Worldwide. American Phytopath Soc, St. Paul, MN, USA.

Ewing, E.E., I. Simko, C.D. Smart, M.W. Bonierbale, E.S.G. Mizubuti, G.D. May, and W.E. Fry. 2000. Genetic mapping from field tests of quantitative and qualitative resistance to *Phytophthora infestans* in a population derived from *Solanum tuberosum* and *Solanum berthaultii*. Molec Breed 6: 25–36.

Flor, H.H. 1971. Current status of the gene-for-gene concept. Annu Rev Phytopath 9: 275–296.

Fry, W.E. and A.E. Apple. 1986. Disease management implications of age-related changes in susceptibility of potato foliage to *Phytophthora infestans*. Amer Potato J 63: 47–56.

Fry, W.E., S.B. Goodwin, J.M. Matuszak, L.J. Spielman, M.G. Milgroom, and A. Drenth. 1992. Population genetics and intercontinental migrations of *Phytophthora infestans*. Annu Rev Phytopath 30: 107–130.

Fry, W.E., S.B. Goodwin, A.T. Dyer, J.M. Matuszak, A. Drenth, P.W. Tooley, L.S. Sujkowski, Y.J. Koh, B.A. Cohen, L.J. Spielman, K.L. Deahl, D.A. Inglis, and K.P. Sandlan. 1993. Historical and recent migrations of *Phytophthora infestans*: chronology, pathways and implications. Plant Dis 77: 653–661.

Gao, A-G., S.M. Hakimi, C.A. Mittanck, Y. Wu, B.M. Woerner, D.M. Stark, D.M. Shah, J. Liang and C.M.T. Rommens. 2000. Fungal pathogen protection in potato by expression of a plant defensin peptide. Nature Biotech 18: 1307–1310.

Gebhardt, C. 1994. RFLP mapping in potato of qualitative and quantitative genetic loci conferring resistance to potato pathogens. Amer Potato J 71: 339–345.

Gebhardt, C. and J.P.T. Valkonen. 2001. Organization of genes controlling disease resistance in the potato genome. Annu Rev Phytopath 39: 79–102.

Gebhardt, C., K. Schueler, B. Walkemeier, and A. Ballvora. 2001. Association mapping in potato of a QTL for late blight resistance. *In:* Proc Plant and Animal Genomes IX Conf, January 13-17, 2001, PAGC, San Diego, CA, USA, Vol. 9, p. 39.

Goodwin, S.B., B.A. Cohen, and W.E. Fry. 1994. Panglobal distribution of a single clonal lineage of the Irish potato famine fungus. Proc Natl Acad Sci USA 91: 11591–11595.

Gopal, J. 1998. Heterosis and combining ability analysis for resistance to early blight (*Alternaria solani*) in potato. Potato Res 41: 311–317.

Grube, R.C, E.R. Radwanski, and M. Jahn. 2000. Comparative genetics of disease resistance within the Solanaceae. Genetics 155: 873–887.

Haynes, K.G. 1990. Covariances between diploid parent and tetraploid offspring in tetraploid x diploid crosses of *Solanum tuberosum* L. J Hered 81: 208–210.

Haynes, K.G. and B.J. Christ. 1999. Heritability of resistance to foliar late blight in a diploid hybrid potato population of *Solanum phureja* x *Solanum stenotomum*. Plant Breed 118: 431–434.

Haynes, K.G., D.H. Lambert, B.J. Christ, D.P. Weingartner, D.S. Douches, J.E. Backlund, G. Secor, W.E. Fry, and W. Stevenson. 1998. Phenotypic stability of resistance to late blight in potato clones evaluated at eight sites in the United States. Amer J Potato Res 75: 211–217.

Heath, M.C. 1998. Apoptosis, programmed cell death and the hypersensitive response. Eur J Plant Path 104: 117–124.

Hosaka K. 1999. A genetic map of *Solanum phureja* clone 1.22 constructed using RFLP and RAPD markers. Amer J Potato Res 75: 97–102.

Inglis, D.A., D.A. Johnson, D.E. Legard, W.E. Fry, and P.B. Hamm. 1996. Relative resistance of potato clones in response to new and old populations of *Phytophthora infestans*. Plant Dis 80: 575–578.

Jacobs, J.M.E., H.J. van Eck, P. Arens, B. Verkerk Bakker, B. Te Lintel Hekkert, H.J.M. Bastiaanssen, A. El-Kharbotly, A. Pereira, E. Jacobsen, and W.J. Stiekema. 1995. A genetic map of potato (*Solanum tuberosum*) intergrating molecular markers, including transposons and classical markers. Theor Appl Genet 91: 289–300.

Johanson, A. and H.D. Thurston. 1990. The effect of cultivar maturity on the resistance of potatoes to early blight caused by *Alternaria solani*. Amer Potato J 67: 615–623.

Jones, J.D.G. 2001. Putting knowledge of plant disease resistance genes to work. Curr Opin Plant Biol 4: 281–287.

Kawchuk, L.M., D.R. Lynch, J. Hachey, P.S. Bains, and F. Kulcsar. 1994. Identification of a co-dominant amplified polymorphic DNA marker linked to the *Verticillium* wilt resistance gene in tomato. Theor Appl Genet 89: 661–664.

Kawchuk, L.M., J. Hachey, D.R. Lynch, F. Kulcsar, G. van Rooijen, D.R. Waterer, A. Robertson, E. Kokko, R. Byers, R.J. Howard, R. Fischer, and D. Prufer. 2001. Tomato Ve disease resistance genes encode cell surface-like receptors. Proc Natl Acad Sci USA 98: 6511–6515.

Kuhl, J.C., R.E. Hanneman Jr., and M.J. Harvey. 2001. Characterization and mapping of *Rpi1*, a late-blight resistance locus from diploid (*1EBN*) Mexican *Solanum pinnatisectum*. Molec Genet Genomics 265: 977–987.

Lee, T.Y., E. Mizubuti, and W.E. Fry. 1999. Genetics of Metalaxyl resistance in *Phytophthora infestans*. Fungal Genet Biol 26, 118–130.

Lefebvre, V. and A. Palloix. 1996. Both epistatic and additive effects of QTLs are involved in polygenic induced resistance to disease: a case study, the interaction pepper*Phytophthora capsici* Leonian. Theor Appl Genet 96: 503–511.

Leister, D., A. Ballvora, F. Salamini, and C. Gebhardt. 1996. A PCR based approach for isolating pathogen resistance genes from potato with potential for wide application in plants. Nat Genet 14: 421–429.

Leonards-Schippers, C., W. Gieffers, F. Salamini, and C. Gebhardt. 1992. The *R1* gene conferring race-specific resistance to *Phytophthora infestans* in potato is located on chromosome V. Molec Gen Genet 223: 278–283.

Leonards-Schippers, C., W. Gieffers, R. Schäfer-Pregl, E. Ritter, S.J. Knapp, F. Salamini, and C. Gebhardt. 1994. Quantitative resistance to *Phytophthora infestans* in potato: a case study for QTL mapping in an allogamous plant species. Genetics 137: 67–77.

Li, X., H.J. van Eck, J.N.A.P. Rouppe van der Voort, D.J. Huigen, P. Stam, and E. Jacobsen. 1998. Autotetraploids and genetic mapping using common AFLP markers: the *R2* allele conferring resistance to *Phytophthora infestans* mapped on potato chromosome 4. Theor Appl Genet 96: 1121–1128.

Lowings, P.H. and I.G. Acha. 1959. Some factors affecting growth of *Phytophthora infestans* (Mont.) de Bary. 1. *P. infestans* on living potato leaves. Trans Br Mycol Soc 42: 491–501.

Meyer, R.C., D. Milbourne, C.A. Hackett, J.E. Bradshaw, J.W. McNichol, and R. Waugh. 1998. Linkage analysis in tetraploid potato and association of markers with quantitative resistance to late blight (*Phytophthora infestans*). Molec Gen Genet 259: 150–160.

Micheletto, S., M Andreoni, and M.A. Huarte. 1999. Vertical resistance to late blight in wild potato species from Argentina. Euphytica 110: 133–138.

Milbourne D., R.C. Meyer, A.J. Collins, L.D. Ramsey, C. Gebhardt, and R. Waugh. 1998. Isolation, characterization and mapping of simple sequence repeat loci in potato. Molec Gen Genet 259: 233–245.

Mohan, M., S. Nair, A. Bhagwat, T.G. Krishna, M. Yano, C.R. Bhatia, and T. Sasaki. 1997 Genome mapping, molecular markers and marker-assisted selection in crop plants. Molec Breed 3: 87–103.

Moreau, P., P. Thoquet, J. Olivier, H. Laterrot, and N. Grimsley. 1998. Genetic mapping of *Ph-2*, a single locus controlling partial resistance to *Phytophthora infestans* in tomato. Molec Plant Micr Interact 11: 259–269.

Naess, S.K., J.M. Bradeen, S.M. Wielgus, G.T. Haberlach, J.M. McGrath, and J.P. Helgeson. 2000. Resistance to late blight in *Solanum bulbocastanum* is mapped to chromosome 8. Theor Appl Genet 101: 697–704.

Niederhauser, J.S. 1956. The blight, the blighter, and the blighted. T. NY Acad Sci 19: 55–63.

Niks, R.E. 1987. Nonhost plant species as donors for resistance to pathogens with narrow host range. I. Determination of nonhost status. Euphytica 36: 841–852.

NIVAA, CPRO-DLO. 1991. Netherlands Catalogue of Potato Varieties. NIVAA, CPRO-DLO, Wageningen, The Netherlands.

Oberhagemann, P., C. Chatot-Balandras, R. Schafer-Pregl, D. Wegener, C. Palomino, F. Salamini, E. Bonnel, and C. Gebhardt. 1999. A genetic analysis of quantitative resistance to late blight in potato: towards marker-assisted selection. Mol Breed 5: 399–415.

Ordonez, M.E., G.A. Forbes, and B.R. Trognitz. 1997. Resistance to late blight in potato. A putative gene that suppresses R genes and is elicited by specific isolates. Euphytica 95: 167–172.

Ortiz, R., C. Martin, M. Iwanaga, and H. Torres. 1993. Inheritance of early blight resistance in diploid potatoes. Euphytica 71: 15–19.

Pan, Q., Y-S. Liu, O. Budai-Hadrian, M. Sela, L. Carmel-Goren, D. Zamir, and R. Fluhr. 2000. Comparative genetics of nucleotide binding site-leucine rich repeat resistance gene homologues in the genomes of two dicotyledons: tomato and arabidopsis. Genetics 155: 309–322.

Pierce, L.C. 1971. Linkage test with *Ph* conditioning resistance to race 0, *Phytophthora infestans*. Rep Tomato Genet Coop 21: 30.

Platt, H.W. 1999. Response of solanaceous cultivated plants and weed species to inoculation with A1 or A2 mating type strains of *Phytophthora infestans*. Can J Plant Path 21: 301–307.

Rowe, R.C. (ed.). 1993. Potato Health Management. APS Press, Amer. Phytopath. Soc., St. Paul, MN, USA.

Salmeron, J.M., G.E. Oldroyd, C.M. Rommens, S.R. Scofield, H.S. Kim, D.T. Lavelle, D. Dahlbeck, and B.J. Staskawicz. 1996. Tomato *Prf* is a member of the leucine-rich repeat class of plant disease resistance genes and lies embedded within the *Pto* kinase gene cluster. Cell 86: 123–133.

Schwinn, F.J. and P. Margot. 1991. Control with chemicals. *In:* D.S. Ingram and P.H. Williams (eds.). Advances in Plant Pathology, vol. 7: *Phytophthora infestans*, the Cause of Late Blight of Potato. Acad Press, London, UK, pp. 225–265.

Segura, A., M. Moreno, F. Madueno, A. Molina, and F. Garcia-Olmedo. 1999. Snakin-1, a peptide from potato that is active against plant pathogens. Molec Plant Micr Interact 12: 16–23.

Simko, I. 2002. Comparative analysis of quantitative trait loci for foliage resistance to *Phytophthora infestans* in tuber-bearing *Solanum* species. Amer J Potato Res 79: 125–132.

Simko, I., D. Vreugdenhil, C.S. Jung, and G.D. May. 1999. Similarity of QTL detected for *in vitro* and greenhouse development of potato plants. Molec Breed 5: 417–428.

Simko, I., S. Costanzo, K.G. Haynes, B.J. Christ, and R.W. Jones (2003). Identification of molecular markers linked to the Verticillium wilt resistance gene homologue in potato (*Solanum tuberosum* L.). Acta Hort. 619: 127–133.

Simko, I., S. Costanzo, K.G. Haynes, B.J. Christ, and R.W. Jones (2004). Linkage disequilibrium mapping of a *Verticillium dahliae* resistance quantative trait locus in tetraploid potato (*Solanum tuberosum*) through a candidate gene approach. Theor Appl Genet 108: 217–224.

Simmonds, N.W. and R.L. Wastie. 1987. Assessment of horizontal resistance to late blight of potatoes. Ann Appl Biol 111: 213–221.

Song, J., J.M. Bradeen, S.K. Naess, J.A. Raasch, S.M. Wielgus, G.T. Haberlach, J. Liu, H. Kuang, S. Austin-Phillips, C.R. Buell, J.P. Helgeson, and J. Jiang. 2003. Gene *RB* cloned from *Solanum bulbocastanum* confers broad spectrum resistance to potato late blight. Proc Natl Acad Sci USA 100: 9128–9133.

Swiezynski, K.M. 1990. Resistance to *Phytophthora infestans* in potato cultivars and its relation to maturity. Genet Pol 31: 99–106.

Tai, T.H., D. Dahlbeck, E.T. Clark, P. Gajiwala, R. Pasion, M.C. Whalen, R.E. Stall, and B.J. Staskawicz. 1999. Expression of the Bs2 pepper gene confers resistance to bacterial spot disease in tomato. Proc Natl Acad Sci USA 96: 14153–14158.

Toxopeus, H.J. 1958. Some notes on the relationship between field resistance to *Phytophthora infestans* in leaves and tubers and ripening time in *Solanum tuberosum* subsp. *tuberosum*. Euphytica 7: 123–130.

Trognitz, B.R. 1998. Inheritance of resistance in potato to lesion expansion and sporulation by *Phytophthora infestans*. Plant Path 47: 712–722.

van de Peer, Y. and R. De Wachter. 1997. Evolutionary relationship among the Eukaryotic crown taxa taking into account site-to-site rate variation in 18S rRNA. J Molec Evol 45: 619–630.

van de Peer, Y., S.L. Baldauf, W.F. Doolittle, and A. Meyer. 2000. An updated and comprehensive rRNA phylogeny of (crown) Eukaryotes based on rate-calibrated evolutionary distances. J Molec Evol 51: 565–576.

van den Berg, J.H., E.E. Ewing, R.L. Plaisted, S. McMurry, and M.W. Bonierbale. 1996. QTL analysis of potato tuberization. Theor Appl Genet 93: 307–316.

van der Biezen, E. A. 2001. Quest for antimicrobial genes to engineer disease-resistant crops. Trends Plant Sci 6: 89–91.

van der Lee, T., A. Robold, A. Testa, J.W. van't Klooster, and F. Govers. 2001. Mapping of avirulence genes in *Phytophthora infestans* with AFLP markers selected by bulked segregant analysis. Genetics 157: 949–956.

van Eck, H.J. and E. Jacobsen. 1996. Application of molecular markers in the genetic analysis of quantitative traits. *In:* P.C. Struik, J. Hoogendoorn, J.K. Kouwenhoven, L.J. Mastenbroek, L.J. Turkensteen, A. Veerman A, and J. Vos (eds.), Abstracts Papers, Posters and Demonstrations of 13th Triennial Conference EAPR. European Association for Potato Research, Wageningen, The Netherlands.

van Eck, H.J., J. Rouppe van der Voort, J. Draaistra, P. van Zandvoort, E. van Enckevort, B. Segers, J. Peleman, E. Jacobsen, J. Helder, and J. Bakker. 1995. The inheritance and chromosomal localization of AFLP markers in a non-inbred offspring. Molec Breed 1: 397–410.

van West, P. 2000. Molecular tools to unravel the role of genes from *Phytophthora infestans*. PhD thesis, Wageningen Agric. Univ., Wageningen, The Netherlands.

Vleeshouwers, V.G.A.A., W. van Dooijeweert, F. Govers, S. Kamoun, and L.T. Colon. 2000. The hypersensitive response is associated with host and non-host resistance to *Phytophthora infestans*. Planta 210: 853–864.

Wastie, R.L. 1991. Breeding for resistance. *In:* D.S. Ingram and P.H. Williams (eds.). Advances in Plant Pathology, vol. 7: *Phytophthora infestans*, the Cause of Late Blight of Potato. Acad Press, London, UK, pp. 193–223.

Wirtz, U., F. Wei, C.E. Tornqvist, C. Ronning, R. Buell, P. Zhang, M. Hernandez, C. Smart, W. Fry, and B. Baker. 2002. Genome structure of *R1* and *R7* potato late blight disease resistance loci of wild potato species *Solanum demissum*. *In:* Proc Plant, Animal and Microbe Genomes X Conference, January 12-16, 2002, PAGC San Diego, CA, USA, Vol. 10, p. 17.

Author Index

Subject Index